# MYSTERIOUS CREATURES

# MYSTERIOUS CREATURES
## A Guide to Cryptozoology

George M. Eberhart

A B C ● C L I O

Santa Barbara, California    Denver, Colorado    Oxford, England

Library of Congress Cataloging-in-Publication Data

Eberhart, George M.
    Mysterious creatures : a guide to cryptozoology / George M. Eberhart.
        p.   cm.
    Includes index.
    ISBN 1-57607-283-5 (set)
    1. Parapsychology. 2. Occultism. 3. Animals, Mythical. I. Title.
BF1031 .E312 2002
001.944—dc21                                    2002013785

06  05  04  03  02    10  9  8  7  6  5  4  3  2  1

This book is also available on the World Wide Web as an e-book. Visit http://www.abc-clio.com for details.

ABC-CLIO, Inc.
130 Cremona Drive, P.O. Box 1911
Santa Barbara, California 93116-1911

This book is printed on acid-free paper ∞.
Manufactured in the United States of America

To
Bernard Heuvelmans (1916–2001)

# Contents

## Cryptids A–Z

# Preface

"Why are so few students interested in science?" This commonly posed question reflects a continuing worry among educators and cultural pundits. Here is a slight paraphrase:

"Why do so few youngsters want to become biologists, when so many are interested in cryptozoology?"

The purpose of my paraphrase is to suggest that such matters as Loch Ness monsters—or unidentified flying objects (UFOs) or psychic phenomena—offer a way of getting students interested in science. These topics are *mysteries,* and human beings are naturally curious about mysteries. In trying to get to the bottom of them, we find ourselves learning about science along the way. Moreover, we learn about it in a way that shows science not to be a boring, cut-and-dried subject as it is sometimes portrayed in popular culture.

One doesn't need to be a formal student, of course. I was already a teacher when I became curious about whether Loch Ness monsters could be real, and my curiosity led me to learn about—among other things—biology and geology and the history, philosophy, and sociology of science. The latter interests eventually led to fruitful changes of career and intellectual activity, for which I have long been grateful. So one value of cryptozoology lies in its ability to stimulate curiosity and the good things that tend to follow on that.

Cryptozoology also has value for science itself. Though most cryptozoological claims may never be validated, the few that are vindicated are likely to be of exceptional interest, as is the case with the now accepted Giant squid (*Architeuthis*) that was long regarded as a purely mythical creature (the KRAKEN) or the almost certainly existing GIGANTIC OCTOPUS for which we have only historical evidence and a few bits of preserved tissue.

Further, cryptozoological investigations sometimes have beneficial side effects. At Loch Ness, it was side-scan sonar looking for NESSIES that discovered (in 1976) a World War II–era Wellington bomber worth recovering for preservation in a museum. Earlier sonar quests for NESSIES had, in 1960, revealed previously unsuspected shoals of Arctic char (*Salvelinus alpinus*) in Loch Ness. The realization, spurred by NESSIE hunting, that very little was known about the ecology of Loch Ness has led to a variety of useful discoveries and continuing research there.

Scientists can benefit, like everyone else, from needing to rethink long accepted facts. When there are persistent reports that people have seen creatures supposedly extinct, ignored issues must be faced, whether those creatures are Pumas (*Puma concolor*) that may be roaming the eastern United States or the plesiosaur-like FRESHWATER MONSTERS reported from many lakes besides Loch Ness:

• How sure can we really be that no pumas are alive east of the Mississippi? Can a lack of captured or killed specimens be decisive, even as very few people have gone looking?

• If we can still, in 1976, discover something such as a Megamouth shark (*Megachasma pelagios*), is it inconceivable that there are real SEA MONSTERS, some of which became landlocked in lakes such as Loch Ness and Loch Morar, whose depths reach below 700 feet?

And so on. It cannot be a bad thing, every now and again, to reassess long held conclusions. Never did I teach freshman chemistry classes without gaining better clarity or a new insight, through needing to find answers to the naive questions posed by neophyte students.

Furthermore, cryptozoology is useful to social science as well as to natural science. The authority that science wields in contemporary society has made it an object of study by historians, philosophers, political scientists, and sociologists. For nigh on a century, philosophy of science has grappled with the "demarcation" issue: How do we distinguish *real* science from *pseudo*science? In seeking to answer that question through examining specific claims that have been sometimes pronounced pseudoscientific, one inevitably learns more precisely what real science actually is. To paraphrase Rudyard Kipling, "What should they know of science, who only science know?" Wondering why science has ignored NESSIES led me, when I was already a practicing professional chemist, to better insight into what science actually is and does.

Cryptozoology affords practice in the most difficult sort of thinking. In established disciplines, peer review and accepted approaches and paradigms assist in solving puzzles and problems. By contrast, seeking to solve mysteries outside the mainstream disciplines means trying to think critically with the minimum of formulaic guidelines, for the eventual solution may be unlike anything previously encountered. (In the realm of detective mysteries, an analogy may be G. K. Chesterton's Father Brown, who could find perfectly rational explanations for events that seem at first to be utterly inexplicable.)

Cryptozoology, then, is valuable on a number of counts, and many besides myself will rejoice that this encyclopedia has become available. Available *at last,* I might add, for it would

have been very useful to me over the last couple of decades. Of course, the field has seen several compendiums, even the recent *Cryptozoology A to Z* by Loren Coleman and Jerome Clark, which gives useful summaries about the most common topics. *Mysterious Creatures,* however, is without precedent in being comprehensive and a genuinely scholarly reference work. Everyone interested in cryptozoology—and many others as well—will want to have this readily at hand. Nowhere else can one look up a cryptid (a merely claimed or mythical or supposedly extinct creature) and find reliable information about the etymology of the name and variant names, physical description, behavior, tracks, habitat, distribution, significant sightings, and, far from least, sources and possible explanations.

Over the years, I have appreciated the several bibliographies about unorthodox subjects that George Eberhart has prepared. This encyclopedia is an even more valuable contribution.

**Henry H. Bauer** is emeritus professor of chemistry and science studies at the Virginia Polytechnic Institute and State University, Blacksburg. He is the author of *The Enigma of Loch Ness: Making Sense of a Mystery* (Chicago: University of Illinois, 1986) and edits the *Journal of Scientific Exploration,* for which he wrote "The Case for the Loch Ness 'Monster': The Scientific Evidence" in the summer 2002 issue.

# How to Use This Book

My first brush with cryptozoology was in 1960 when I read *On the Track of Unknown Animals* by Belgian zoologist Bernard Heuvelmans. It was a life-changing experience. Heuvelmans's masterful scientific, historical, and literary sleuthing in quest of elusive fauna was both exciting and scholarly: he seemed a combination of Sherlock Holmes and *The Lost World*'s Professor Challenger. His volume prodded my ten-year-old brain to take a keen interest in not only science and history but also different languages and cultures, the evaluation of evidence, and the rich discoveries that await the fearless explorer of large library collections. In that sense, I have been writing this book ever since, and I hope that, in turn, it may stimulate others to seek out new species or identify the animals that are lurking just behind the myths.

## What Constitutes a Cryptid?

Cryptids are the alleged animals that a cryptozoologist studies. Obviously, someone—either an ethnic group familiar with a specific habitat, a traveler to a remote region, or a surprised homeowner who sees an ALIEN BIG CAT or SKUNK APE in the backyard—first has to allege that such animals exist. (Words set in SMALL CAPITALS refer to entries in the text.) The examination and evaluation of ethnographic, testimonial, and physical evidence to determine the identity of a cryptid is what cryptozoology is all about.

Some would say that only those animals with a reasonable chance of one day becoming recognized as new species should be included in this volume and that bizarre, red-eyed ENTITIES such as MOTHMAN or mythical creatures such as DRAGONS and UNICORNS are beyond its scope. This is a practical approach for the zoologist whose aim is to add to knowledge of the world's biodiversity, and it is one of cryptozoology's primary goals as well. However, I have taken a broader view in this encyclopedia, for it can be equally important to show how known animals can pose as cryptids or how people's belief systems and expectations can color their observations of the natural world. Do Golden eagles (*Aquila chrysaetos*) occasionally get reported as BIG BIRDS or THUNDERBIRDS? Are witnesses of HAIRY BIPEDS or EASTERN PUMAS in certain parts of Maryland influenced by the tales of GOATMAN and SNALLYGASTER in those areas?

Solving historical puzzles also seems relevant to cryptozoology. Just what animals were responsible for medieval BASILISK lore? Could the Columbian mammoth (*Mammuthus columbi*) or the Giant short-faced bear (*Arctodus simus*) have survived somewhat later in time than is currently supposed and thus be responsible for Native American legends of the STIFF-LEGGED BEAR?

Most of the mystery animals in this book fall into one of the following ten categories:

1. *Distribution anomalies,* or well-known animals found in locales where they have not previously been found or are thought extinct, such as the EASTERN PUMA.

2. *Undescribed, unusual, or outsize variations of known species,* such as the BLUE TIGER, HORNED HARE, or GIANT ANACONDA.

3. *Survivals of recently extinct species,* such as the IVORY-BILLED WOODPECKER in the southern United States, thought extinct since the 1960s.

4. *Survivals of species known only from the fossil record into modern times,* such as the ROA-ROA of New Zealand, which might be a surviving moa.

5. *Survivals of species known only from the*

*fossil record into historical times* but found to have existed later than currently thought, such as the Muskox of Noyon Uul.

6. *Animals not known from the fossil record but related to known species,* such as the Andaman Wood Owl or Beebe's Manta.

7. *Animals not known from the fossil record or bearing a clear relationship to known species,* such as Bigfoot and some Sea Monsters.

8. *Mythical animals with a zoological basis,* such as the Golden Ram.

9. *Seemingly paranormal or supernatural entities with some animal-like characteristics,* such as Black Dogs or Cannibal Giants.

10. *Known hoaxes or probable misidentifications* that sometimes crop up in the literature, such as the Coleman Frog and *Bothrodon Pridii.*

**What Do Cryptozoologists Do?**

Ultimately, the job of the cryptozoologist is to strip away the myth, the misidentification, and the mystery from reports of animals undescribed by science. When confronted with a new sighting, the investigator's first task is to see what local fauna might account for it. The accuracy and validity of eyewitness testimony must be ascertained. (For more on this, see Jack Rabbit's "Native and Western Eyewitness Testimony in Cryptozoology," on pp. xxxv–xliii.) Then, the potential for a hoax must be evaluated. If a coherent body of evidence accumulates to indicate that a real animal not native to the area is involved, the next step is to determine whether any living animals fitting the description were introduced or have lived there all along unnoticed by compilers of field guides. Failing that, an examination of relevant animals in the fossil record is warranted, with an emphasis on groups that are known in the region.

Even if nothing in the fossil record matches, a case could be made for an evolved version of a known fossil. What plesiosaurs looked like 65 million years ago can only serve as a basic guide to what they might have evolved into had they survived the cometary impact at the end of the Cretaceous period. Our own physical characteristics have changed greatly even in the past 10 million years. In any event, the fossil record is incomplete and in most cases can tell us little about what the outward appearance of an ex-

tinct animal might have been. This leaves much room for speculation.

Cryptozoologists are sometimes accused of never wanting to solve a mystery, perhaps because of the glamour and romance of the unknown. However, mystery mongering is much more frequently found in treatments by the media. Most of us would rather have one less Yeti or Mokele-Mbembe to worry about, whether it winds up in a museum or in a long list of animals that never were.

Mysteries are both a bother and a challenge to cryptozoologists. They are a bother because we wonder what some journalist or observer "got wrong" about an animal that really exists; after all, we can tolerate only so many "head like a goat, body like a lion" stories. And mysteries are a challenge because we feel compelled to use deductive reasoning and a vast amount of specialized and interdisciplinary knowledge to find out what animal—known, unknown, or supposedly extinct—could be the stimulus for a sighting. The triumph of a solution outweighs the uncertainty of an incomplete puzzle.

Fieldwork is a crucial though often thankless part of that solution. Cryptid hunting is expensive, often dangerous, always time-consuming, usually frustrating, and potentially hazardous to one's scientific credibility. But if it weren't for the dragon-hunting exploits of W. Douglas Burden in the 1920s, the Komodo dragon (*Varanus komodoensis*) might still be a rumor. Investigators such as Loren Coleman and Roy Mackal have taken much criticism along the way, but they are the ones who are searching for the hairs and tracks, asking the right questions, making the plaster casts, and waiting for just the right Kodak moment when a cryptid's head and neck rise above the water. We who write reference books salute them!

**How Is This Book Organized?**

The first part of this book contains descriptions of 1,085 unknown animals, arranged alphabetically in a field-guide format. Each of these falls into one of forty major cryptid categories (shown in capitals), most of them based on existing classes and orders of known animals. Each major category offers a brief description of the animal group associated with it, as well as a list of the cryptids included. The major categories are a good place to begin a general search for specific mystery beasts:

AMPHIBIANS (Unknown)
BATS (Unknown)
BEARS (Unknown)
BIRDS (Unknown)
CATS (Unknown)
CEPHALOPODS (Unknown)
CETACEANS (Unknown)
CIVETS AND MONGOOSES (Unknown)
CROCODILIANS (Unknown)
DINOSAURS (Living)
DOGS (Unknown)
ELEPHANTS (Unknown)
ENTITIES
FISHES (Unknown)
FLYING REPTILES
FRESHWATER MONSTERS
GIANT HOMINIDS
HOOFED MAMMALS (Unknown)
HYENAS (Unknown)
HYRAXES (Unknown)
INSECTIVORES (Unknown)
INVERTEBRATES (Unknown)
LIZARDS (Unknown)
MARSUPIALS (Unknown)
MERBEINGS
OTTERS (Unknown)
PANGOLINS (Unknown)
PRIMATES (Unknown)
RABBITS (Unknown)
RODENTS (Unknown)
SEA MONSTERS
SEALS (Unknown)
SEMIMYTHICAL BEASTS
SIRENIANS (Unknown)
SLOTHS (Unknown)
SMALL HOMINIDS
SNAKES (Unknown)
TURTLES (Unknown)
WEASELS (Unknown)
WILDMEN

A few mystery animals of uncertain taxonomy are included in more than one category, such as the NANDI BEAR, a cryptid with varying characteristics that turns up under BEARS, HYENAS, and PRIMATES.

Most of the entries are structured in a similar fashion, with a brief identification of the cryptid followed by information arranged under as many as twelve of the following sections:

*Etymology.* The derivation or meaning of the cryptid's name. In a few cases, the date of the name's first appearance is provided, as well as information on the person who coined the term. If the cryptid's name is not an English word, the language is given. The language family is shown in parentheses for non-Western languages—for example, Lingala (Bantu), "water monster." Current names for ethnic groups, their languages, and language families were identified or verified in Ethnologue: Languages of the World, at http://www.ethnologue.com, or Andrew Dalby's *Dictionary of Languages* (1998). Numerous phrase books and dictionaries were also consulted.

*Scientific name.* In some cases, a cryptid has been assigned a Latin or Greek scientific name by a researcher who has investigated it, such as *Nessiteras rhombopteryx* for NESSIE, given by Peter Scott and Robert Rines in 1975. When a cryptid is welcomed into the ranks of known animals, such a name could become the genus and species designation used for the formal scientific description, unless an existing genus is more appropriate. The binomial method of naming living creatures was proposed in the eighteenth century by Carl von Linné. The first name is generic, as in *Homo;* the second is specific, as in *sapiens.* Animals with the same generic name are said to belong to the same genus, while the specific name identifies the species.

*Variant names.* Other names by which a cryptid is known are found in this section. These may include alternate spellings and geographic variants. Other languages, language families (following the slash mark), and meanings (in quotation marks) are given in parentheses—for example, Meshe-adam (Azerbaijani/Turkic, "forest man").

*Physical description.* This section provides a summary of the appearance of the cryptid. Information is listed in the following order: general appearance, length, height, diameter, weight, color or coat, head, face, eyes, ears, nostrils, cheeks, mouth and teeth, chin, neck and shoulders, chest and torso, arms, hands, wings, back, legs, feet, and tail. Since the description is often derived from multiple sources (sighting reports and other testimony), there is a possibility that some erroneous data are included.

*Behavior.* A summary of the habits and inter-actions of the cryptid is offered in this section. In-formation is listed in the following order: period of activity (such as nocturnal or diurnal), pre-ferred area of operations (such as aquatic or arbo-real), stance and locomotion (such as bipedal movement or vertical undulations), vocalizations, sensory capabilities, odor, food, sleep and nesting habits, reproductive strategies, social interactions, interspecies interactions, human interactions, and technology. Uncertain or doubtful behaviors are often introduced by the phrase *is said to.*

*Tracks.* Dimensions and characteristics of footprints or other impressions left by a cryptid on the ground or in snow are described in this section.

*Habitat.* Here, details are provided about the specific environment in which a cryptid lives, whether in the sea (abyssal, coastal, etc.) or on land (forests, desert, scrubland, etc.).

*Distribution.* This section describes the geo-graphic range where sightings of the cryptid are said to occur. Specific landscape features (such as mountains and lakes) are provided when known; otherwise, country names and their sub-divisions (such as states, provinces, and depart-ments) are given. Place-names are those in use as of 2001; all previous political designations (for instance, Rhodesia, Yakutia, or Jaffa) have been updated to their modern equivalents (Zim-babwe, Sakha Republic, or Yafo). The Mi-crosoft Encarta Atlas 2000 was used to verify present status in most cases; current political di-visions were identified in Gwillim Law's *Admin-istrative Subdivisions of Countries* (1999).

*Significant sightings.* Capsule summaries of either important or typical observations of a cryptid are arranged in chronological order in this section. The examples are by no means comprehensive. Most of these observations are anecdotal in nature, although in some instances, the "sighting" involves an artifact, petroglyph, or sonar contact. The observed characteristics of the animal are not repeated, unless they are atypical or more detailed than those given in the preceding physical description section. All older place-names have been modernized.

*Present status.* This section contains notes on whether a cryptid is likely to be extinct, as well as other comments and data that do not fit elsewhere.

*Possible explanations.* This section lists more or less reasonable hypotheses as to what the cryptid might be, either as a misidentification of a known species, as an unknown species, or as a survival of an extinct species. An explanation's position in the list does not reflect the likeli-hood of its validity. Sometimes, there is more than one probable hypothesis for sightings of a given cryptid. In most cases, I have avoided making personal judgments, preferring instead to wait until a definitive answer has emerged; however, I have pointed out the ways in which certain arguments are weak. Both common and scientific names are given for known animals. The lack of this section for a given cryptid may mean either that there is too little information for anyone to make an informed guess or that a discussion of the possibilities is found elsewhere; for example, CHEMISIT explanations are found under NANDI BEAR, and KSY-GYIK candidates are discussed generally under WILDMEN.

*Sources.* This section offers a selected list of references for further consultation, with an em-phasis on firsthand, scientific sources, as well as the most informative books and journal articles. The sources are arranged by the date of original appearance, which puts ancient and medieval sources at the beginning of the list despite later imprint dates.

A geologic timescale appears on p. xlv to aid in visualizing the periods of the earth's history and the development of life.

The second part of this book, "Animals Dis-covered since 1900" (pp. 623–654), is an anno-tated list of 431 species or other taxonomic groups described or rediscovered since the turn of the twentieth century, arranged by type of animal. One of the criticisms leveled at crypto-zoology is that large, noticeable animals are not likely to have remained unknown to science for centuries. However, the wide variety of organ-isms that have turned up only in recent years—which were previously unnoticed by scientists—is extraordinary when viewed en masse. Among those animals are forty-seven new primates, twenty-nine new hoofed mammals, and fifteen new cetaceans.

The third part, "Lake and River Monsters" (pp. 655–690), is a list of 884 bodies of water worldwide said to contain FRESHWATER MON-STERS or other large aquatic animals. Some are named and appear in the A-Z part of the book (such as NESSIE or CHAMP), while others are unnamed, vaguely defined, semimythical, or little more than rumor. Brief descriptions are given when known. To my knowledge, this is the most comprehensive and accurate list compiled to date.

## What Further Resources Are Available?

Several sources keep cryptozoologists up-to-date on sightings, discoveries, and theories. The monthly British periodical *Fortean Times* (distributed in the United States by Eastern News Distributors, 2020 Superior St., Sandusky, OH 44870) regularly contains news and features on cryptids. Its Web site (http://www.forteantimes. com) offers breaking news on mystery animals.

The approximately annual *Anomalist* (P.O. Box 12434, San Antonio, TX 78212) often features cryptozoological topics. It also has a newsline (http://www.anomalist.com).

Several relevant discussion groups are available on the Yahoo! Groups site (http://groups. yahoo.com), both public ones and those for members only. The members-only cryptozoology group (cz) is one of the best. There are also several BIGFOOT and NESSIE groups.

The monthly *Fate* magazine (P.O. Box 460, Lakeville, MN 55044) has been publishing cryptozoological news and articles since 1948, though its focus is primarily on psychic phenomena. Some features are available on line (http://www.fatemag.com).

The Centre for Fortean Zoology in England publishes the quarterly *Animals and Men* (15 Holne Court, Exeter, U.K. EX4 2NA) and a yearbook with longer features. Back volumes are available (http://www.eclipse.co.uk/cfz/).

The Eastern Puma Research Network (P.O. Box 3562, Baltimore, MD 21214) has a quarterly newsletter that provides information on sightings and statistics.

The British Columbia Scientific Cryptozoology Club (Suite 2305, 8805 Hudson St., Vancouver, BC, Canada V6P 4M9) has a quarterly newsletter and an on-line site (http://www. ultranet.ca/bcscc/).

Mark A. Hall's *Wonders* (407 Racine Dr., Box E, Wilmington, NC 28403) is published four times a year. Back issues are available, and their contents are listed on his Web site (http:// home. att.net/~mark.hall.wonders/).

The Web site of the Institut Virtuel de Cryptozoologie in France (http://www.cryptozoo. org) has excellent news reports and analysis. A portion of the site offers English translations. Other Web sites of interest include:

- The British Big Cat Society, http://www.britishbigcats.org.
- Dick Raynor's Loch Ness site, http://www.lochnessinvestigation.org.
- Australian Yowie Research, http://www.yowiehunters.com.
- Jan-Ove Sundberg's Swedish cryptozoology site, http://www.cryptozoology.st.
- The Bigfoot Field Researchers Organization, http://www.bfro.net, has news and a comprehensive sightings database.
- Chad Arment's cryptozoology site (http://www.strangeark.com) is a good jumping-off point for the on-line *North American BioFortean Review* and Craig Heinselman's *Crypto* newsletter.
- Pib Burns maintains an excellent assortment of links at http://www.pibburns. com/cryptozo.htm.

Unfortunately, some excellent journals are no longer published, and back issues are difficult to find. The International Society of Cryptozoology is gone, along with its *ISC Newsletter* and refereed journal *Cryptozoology*. *Pursuit*, *INFO Journal*, *Exotic Zoology*, and *Cryptozoology Review* have ceased publication as well, and the future of *Strange Magazine* is uncertain. It is almost always a good idea to obtain whatever is currently available before it becomes unfindable.

The same can be said of many cryptozoology books, especially those that are self-published or put out by small or alternative presses. Academic and public libraries do not collect this material. Once it's out of print, you are out of luck, unless you are willing to put up with inflated prices on eBay. One excellent mail-order source for current and out-of-print books and journals is Arcturus Books (1443 S.E. Port St. Lucie Blvd., Port St. Lucie, FL 34952). Though primarily devoted to UFO books, its catalog regularly contains crypto titles.

Many people have been fooled into thinking that everything is available on the Web and that it is a vast, free library accessible at the click of a search engine. This just isn't true, even if you add in the resources on what has been called the Invisible Web, which contains data that are not directly findable by search engines. The Web is the biggest encyclopedia in the world, and it is constantly updated, but there are huge gaps in its coverage that make it only a supplement to printed books and journals and not a replacement.

If I had relied exclusively on the Internet in preparing this book, it would have been only about 15 percent of its current length and probably would have included much misinformation. And if I had relied solely on print resources, it would have been only 85 percent as long, would have taken four years to complete instead of two, and would not have been as up-to-date at the point of publication.

When setting out to research a cryptozoological topic, begin by examining the sources given for cryptids in this book. Focus on specific animals or topics. Figure out how much you want to know about the subject, and narrow or widen your searches accordingly. Be forewarned that one source may lead you to many others, sometimes only to answer new questions that have been raised. Go where the information is, whether it's on the Web or in the library. The answer you are looking for may be in a 1995 issue of *Fortean Times,* a 1903 issue of the *Chicago Tribune,* a field guide to Indonesian birds, a Tibetan-English dictionary, a 1966 article in an Australian herpetological journal, or the on-line FishBase resource.

Always evaluate and question the information you find. Double-check specific facts, if possible. When you find a new source of information, ask these questions:

- Is the source scholarly, popular, governmental, or commercial?
- What are the author's credentials?
- When was the information originally published?
- In what country did it originate?
- What is the reputation of the publisher, distributor, or Web site?
- Does the source show any specific biases?
- Does it offer a bibliography or adequate

documentation for the information it provides?
- Are there a large number of misspelled words and names? Authors who are sloppy about spelling are often sloppy with facts.
- For what audience is the material intended?
- Is it suitable for your level of understanding of the subject?
- Does it have the features you need: illustrations, graphs, charts, tables, definitions, maps?
- Is it current?
- Are various points of view represented?
- Are the conclusions justified by the facts presented?

When you run across a new account of a cryptid sighting, ask the six journalistic questions:

- *Who* reported the sighting? Are they trained observers or knowledgeable about the local fauna?
- *What* actually was seen? Are there enough details for you to be certain that it could not have been a known animal?
- *Where* was the sighting reported? Can you find the location on a map?
- *When* did it occur? Is the information specific (for example, citing day and time) or vague (making reference, perhaps, to an event several summers ago)?
- *How* did the event unfold? Are the behaviors of the observers and the cryptid accounted for and credible?
- *Why* did the sighting get reported? Did the witness contact a newspaper, local authorities, a scientific organization, or a cryptozoologist?

Finally, determine whether the information you have found is consistent with what you have located in other sources. If it's not, don't automatically assume that the new material is wrong; it may well be that the older sources were in error.

**Acknowledgments**
I wish to thank everyone who has provided information and encouragement for this project

over the past two years, especially Loren Coleman, Henry Bauer, Jack Rabbit, Frank J. Reid, Chad Arment, Craig Heinselman, Ben Roesch, Janet Bord, Bill Rebsamen, Michael Swords, Karl Shuker, and Russell Maylone (curator of special collections at Northwestern University Libraries).

*George M. Eberhart*
*Chicago, Illinois*
February 2002

# Introduction
## If We Don't Search, We Shall Never Discover

Passion and cryptozoology go hand in hand. Enthusiasm and zeal fill my mind and body when I think about getting into the field in pursuit of real, flesh-and-blood animals waiting to be discovered.

Did excitement dwell within me, you might wonder, when a game warden and I trekked for hours in the mud on a hot midwestern afternoon in 1963, looking for signs of a black panther? Was it fun during the nights of cold in that tent in the Trinity-Shasta area of California, as I tracked an elusive BIGFOOT through those forests in 1974? Was it enjoyable to experience the biting rain on myself and my lads, Malcolm and Caleb, during the daylong soaking we received in an open boat on Loch Ness in 1999?

Needless to say, the answer for a cryptozoologist is "Yes!" With a fervor that flourished in another time, groups of women and men spend their todays searching for cryptids that may tomorrow be new species, pursuing creatures that may not even exist, looking for animals that the thinnest of evidence says are real, and listening to rumors and tales of others just over the horizon. The late Bernard Heuvelmans wrote in 1988: "Cryptozoological research should be actuated by two major forces: patience and passion." While he may have never caught a single cryptid in his life, he knew all too well about the search.

Cryptozoologists are reliving a time two centuries ago when all of zoology was in an age of discovery. This field preserves the spirit of those days. But by the beginning of the twentieth century, zoology seemed to have slipped into a period in which new species were fully revealed only as a circumstance of taxonomy and cladistic debates were far in the future. Animal discoveries were incidental, certainly not the true mission.

That would all change, first with a quiet tradition of examining the curiosities of natural history beginning at the end of the nineteenth century, as seen in the writings of, for example, Philip Henry Gosse and Francis T. Buckland. With the advent in the twentieth century of a modern generation of zoology authors, such as Willy Ley and Henry Wendt, the time was ripe for a renewed interest in fauna whispered about but not acknowledged. It was then that two gentlemen came along whom I knew personally and who would inspire a fresh cohort of searchers.

Ivan T. Sanderson, a Scottish zoologist living in the United States, wrote an article for the January 3, 1948, *Saturday Evening Post* titled "There Could Be Dinosaurs." In France, Belgian zoologist Bernard Heuvelmans read this essay on the possible survival of extinct animals in Africa, and it changed his life forever. Sanderson had trekked through tropical jungles (we call them rain forests now) in South America, Africa, and Asia. Heuvelmans had spent years reviewing the scientific literature and gleaning the zoological treasures hidden there. In the 1995 revision of his *On the Track of Unknown Animals* (pp. XXIII–XXIV), Heuvelmans expressed the stirrings he found inside himself that would call for a release in cryptozoology:

In the 1950s, I was an angry young zoologist, indignant at the ostracism imposed by official science—we would say today the scientific Establishment—on those animals known only through the reports of isolated travelers, or through often fantastic native legends, or from simple but mysterious footprints, or the recital of sometimes bloody depredations, or through traditional images, or even a few ambiguous photos.

Instances of this sort were, in fact, quite numerous. These were attested to by files, often quite thick, which in general gathered dust at the bottom of drawers or, at the most, were considered as "amusing curiosities." It would have been much better to term them "the secret archives of zoology," or even, since they were in some way shameful in the eyes of correct thinkers, "the Hades of zoological literature." It had in fact been decreed on high and, moreover, in a totally arbitrary fashion, that only those species for which there existed a representative specimen, duly registered in some institution, or at the least an identifiable fragment of a specimen, could be admitted into zoological catalogues.

Lacking this, they were banned from the Animal Kingdom, and zoologists were morally constrained to speak of them only with an exasperated shrug of the shoulder or a mocking smile.

To propose devoting a profound study to sea-serpents, the Abominable Snowman, the Loch Ness monster, or to all such-like, amounted to straightforward provocation. Furthermore, no scientific publication would have accepted it for printing unless, of course, it ended with the conclusion that the being in question was the result of popular imagination, founded on some misapprehension, or the product of a hoax. As for myself, however, in spite of my status as professional zoologist and my university degrees, I dreamed of delivering all of those condemned beasts from the ghetto in which they had so unjustly been confined, and to bring them to be received into the fold of zoology.

Independently, these men—one of the field and one of the library—would invent the same word and go on to become the mutual godfathers of cryptozoology, the study of unknown or hidden animals. The new science would formalize and rescue "romantic zoology" from the days of discovery during the Victorian era, bringing it into the twentieth and twenty-first centuries. Heuvelmans's books, *On the Track of Unknown Animals* (1958) and *In the Wake of the Sea Ser-*

*pents* (1965), and Sanderson's *Abominable Snowmen: Legend Come to Life* (1961) became the canvases on which many of us first saw the beauty of the pursuit.

While I was growing up, I had a strange notion that I wanted to become a naturalist, in the oldest meaning of the word. Instead, I did one better: I became a cryptozoologist.

## Cryptozoologists All

There are many different kinds of cryptozoologists. Some do fieldwork, some do archival research, and others are chroniclers. The field is open to a wide variety of disciplines because the essence of cryptozoology is multidisciplinary.

You can become a cryptozoologist in many ways, but applying for the job through a newspaper advertisement is not likely to be one of them. Nobody hires a cryptozoologist to investigate whether a cryptid in a nearby lake is really there. Instead, cryptozoologists tend to seek positions that give them some freedom to pursue their research, whether in university careers (Grover Krantz, Roy Mackal, Jeff Meldrum, myself), in wildlife management roles (Bruce S. Wright), in government service (Mark A. Hall), or as editors and writers (Ivan T. Sanderson, Bernard Heuvelmans, John Green). I teach at a university, consult, research, and write. I believe my accountant puts "professor/author" on my tax forms. Various educational backgrounds (anthropology, linguistics, zoology, biology, and especially other life sciences) are helpful, and other talents and training go into making one a cryptozoologist.

In 2002, while discussing a case of newly discovered tracks in Pennsylvania, Mark A. Hall and I were identified by the media as "cryptozoologists" and "scientists." It is usually someone else who labels cryptozoologists as scientists, and this incident led to an exchange between Hall and myself on this matter. As Hall noted:

Science is done by people who are paid to perform science and they are the scientists. However, we are amateur scientists in the old sense of the phrase. The sciences grew out of people who were amateurs who established something new. "Amateur" in the modern sense is not so complimentary. When looking backward, scientists can be understanding about the value of amateurs in their fields, such as the expert in rattlesnakes (now dead) who wrote what are considered the best books on the subject. Or E. A. Hooton who was a top name in primates even though his degree was something like English literature. . . . Our culture is going to determine whether we are scientists, not us. At present someone sees us as scientists. It is not my inclination to say that they are wrong.

The example of J. L. B. Smith is also worth noting. Smith was a chemistry professor in South Africa when, one could say, cryptozoology discovered him in 1938 by mere chance when a young museum curator asked him a question about a peculiar fish, which turned out to be the first living coelacanth discovered. Smith was an amateur ichthyologist who became an amateur cryptozoologist, and in 1952, he caught the second coelacanth through cryptozoological methods—by talking to locals, looking for the animal where they said it might be found, spreading the word that he was interested, and applying all his passion and patience to a fruitful end.

Cryptozoology entails a vast amount of important but tedious work, such as searching through newspaper microfilm, library archives, or researchers' old files: not all of the work is spent in hot pursuit of animals in the field. There is also the labor of tracking down witnesses and double-checking their credibility. But the ultimate goal is thrilling. To be seriously involved in chasing mystery animals and investigating extraordinary incidents that have happened to ordinary people is, indeed, exciting.

Modern cryptozoology is also international in scope, thanks to Vietnamese, French, Russian, Spanish, and other non-English-speaking researchers, and it is seen today as the study of the *evidence* for hidden animals. This definition emphasizes the forensics that have become so important to cryptozoology—for example, casting footprints, gathering hair and fecal samples, and collecting relevant cultural artifacts.

## Can You Study to Become a Cryptozoologist?

I am sorry to say that very few cryptozoology classes are given. I taught a full-credit one in 1990 and since that time have used large doses of cryptozoology in my 100-student university course for juniors and seniors on documentary film. Today, no formal cryptozoology degree programs are available anywhere. So my advice would be to pick whatever subject you are most interested in (primates? felids? native tales? giant squids? fossil humans?) and then match it up with the field of study that is linked to that subject (anthropology, zoology, linguistics, marine science, paleoanthropology). Pursue that subject, pick the college or university that is highly regarded in the field, and you just might develop a niche in cryptozoology. I studied anthropology and zoology, then moved on to a more psychologically based graduate degree to understand the human factor. I also took doctoral courses in anthropology.

Existing zoology and anthropology departments cover many subjects, and there is no reason why cryptozoological topics cannot occasionally be addressed in such venues. In addition, more and more professors are opening their minds to cryptozoology and hominology. Some young people who grew up with SASQUATCH are, believe it or not, becoming professors, and a few are involved in cryptozoology. This is a good sign, and it makes it easier to pass on environmental concerns (habitat loss can thwart animal discoveries) to a new generation of students.

Several departmental courses around the country have already included some cryptozoological topics, and guest speakers have occasionally been invited to lecture. Such choices at universities are still rare, but cryptozoology in the twenty-first century appears ready for a growth spurt.

With passion and patience, more animals will be discovered and more cryptozoologists will be born. You could be tomorrow's Ruth Harkness, the discoverer of the Giant panda (*Ailuropoda melanoleuca*), or Hans Schomburgk, the discoverer of the Pygmy hippopotamus (*Hexaprotodon liberiensis*).

**Loren Coleman,** a cryptozoologist for over forty years, is adjunct associate professor at the University of Southern Maine in Portland. He is the author or coauthor of seventeen books, the most recent being *Tom Slick: True Life Encounters in Cryptozoology* (2002) and *The Field Guide to Lake Monsters and Sea Serpents* (2003). His Web site is http://www.lorencoleman.com.

# Native and Western Eyewitness Testimony in Cryptozoology

On an Internet-based cryptozoology forum, this question was recently posed: *How should we evaluate the validity of eyewitness accounts from native peoples?* I shall attempt to answer that question. The issue of native eyewitness testimony is of considerable importance to cryptozoologists, as such accounts are a major component of the body of evidence for many purportedly undiscovered animal species. Native testimony typically receives either one of two opposite and inappropriate treatments in cryptozoological literature, depending on the author's agenda: wholesale dismissal or wholesale acceptance.

In this article, I'll present examples that illustrate why neither wholesale dismissal nor wholesale acceptance of native testimony is reasonable. Then I'll discuss the factors that are known to affect the validity of eyewitness testimony in general (gleaned from the substantial body of published research on the topic), with comments on how those factors may bear on native eyewitnesses. Finally, I'll offer my own thoughts on a few factors that may apply mainly or exclusively to native eyewitnesses, born in cultures and environments different from our own.

## The Invalidity of Wholesale Dismissal of Native Testimony

Historically, Westerners have viewed native peoples as inferior—at best, like naïve children; at worst, like base animals. Regrettably, scientists have often reinforced this popular misconception (Durant and Durant, 1968; Gould, 1981). Native "folk tales" were regarded as prattle, without scientific significance of any sort— as products of "the overheated imagination of natives, which is sometimes influenced by alcohol or the love of rousing sensation" (Kittenberger, 1929). Consequently, the considerable wisdom (including, but by no means limited to,

knowledge about local animals and plants) accumulated by various non-Western societies was largely ignored.

In recent years, Western researchers have come to realize the error of their earlier thinking with regard to native peoples. Scientific studies have shown certain outlandish-sounding claims by native observers to be true, or at least to have a basis in fact. Two interesting examples: the Matsés Indians' tales of a frog that produces a "magic potion" that can be used to enhance hunting prowess; and the assertions by New Guinean tribesmen that certain local birds are poisonous.

The Matsés (Panoan) Indians of Peru claim that *sapo,* a sticky substance excreted from the skin of the Giant monkey tree frog (*Phyllomedusa bicolor*), lends a hunter superhuman endurance and renders him invisible to game animals. Western biochemists have assayed the frog's skin secretions, and found that they contain chemicals that suppress pain, thirst, and hunger. A hunter under the influence of *sapo* may be able to withstand physical hardships that would otherwise distract him from his game-tracking. *Sapo* also contains powerful emetics, diuretics, and laxatives. Researchers speculate that these agents flush the hunter's body of odorous compounds, thereby making him "invisible" (in an olfactory, not optical, manner) to his quarry (Erspamer et al., 1993).

The New Guineans' claim that the Hooded pitohui or "garbage bird" (*Pitohui dichrous*) is poisonous seemed highly unlikely when it was first recorded in *Birds of My Kalam Country,* a compilation of the New Guinea highlanders' folklore (Majnep and Bulmer, 1977). Western scientists had been acquainted with these common birds for over a century, and had not discovered any evidence of chemical defense. Further, of the approximately 9,000 known species

*Giant monkey tree frog* (Phyllomedusa bicolor*). (Painting [acrylic on canvas] by Jack Rabbit, © 1999)*

of birds, not one was known to produce a poison or venom of any sort (Diamond, 1992). However, in 1990, Western ornithologists independently and accidentally discovered that handling the live Hooded pitohui caused "numbness, burning, and sneezing" (Dumbacher et al., 1992). Subsequent analysis of pitohui tissues revealed the presence of homobatrachotoxin, the same poisonous compound secreted by a genus of Poison-dart frogs (*Phyllobates*) from Central and South America. In the concluding paragraph of his pitohui commentary in *Nature* (1992), Jared M. Diamond asks: "What other treasures of biological knowledge are becoming lost with the rapid acculturation of the world's few remaining Stone Age hunters?"

## The Invalidity of Wholesale Acceptance of Native Testimony

In the light of these findings and many others like them, we can see that our previous arrogant dismissal of native wisdom was unwarranted. In our recent reassessment of indigenous cultures, however, we now tend to go too far the other way. A substantial body of recent popular liter-

ature portrays all natives as sages—infinitely wise about their environments, infallible in matters regarding local flora and fauna. See for example such magazines as *Pangaia* and *Green Egg*, and Marlo Morgan's controversial novel *Mutant Message Down Under* (Morgan, 1994). This new attitude, while perhaps less offensive than the old one, is equally absurd. The following examples—the Apris of Somalia and the Biscobra of India—illustrate that while natives may exhibit considerable knowledge about the animals among which they live, that knowledge is sometimes faulty.

Spawls (1979) tells us that the Somalis fear the Apris, a snake so venomous that its mere touch causes death within seconds. Spawls himself has positively identified a specimen of the snake—before an audience of terrified Somali witnesses—as *Gongylophis colubrinus,* a nonvenomous and inoffensive sand boa.

Minton and Minton (1969) report that natives of northern India tell chilling stories of the Bis-cobra, whose name indicates it has the killing power of 20 cobras. The culprit turns out to be the harmless gecko *Eublepharis hard-*

*Head of the harmless gecko* Eublepharis hardwickii, *mistaken for the venomous* BIS-COBRA *in northern India. (Drawing [pencil on paper] by Jack Rabbit, © 1995)*

*wickii.* In Pakistan, natives call the related gecko *Eublepharis macularius* the Hun-khun, and hold it to be "the deadliest creature . . . more dangerous than the Cobra." Similar superstitions surround likewise innocuous geckos in various regions all over the world, including Egypt, Java, Mexico, and Argentina (Minton and Minton, 1969; Goodman and Hobbs, 1994; personal observation).

Both errors in evaluation of native accounts—wholesale dismissal and wholesale acceptance—stem from the same flaw in thinking: the belief that native peoples are fundamentally different from Westerners. Although they live in different environments and have different beliefs, they are not beings wholly unlike us. Like us, they have the capacity for wisdom and logic; like us, they have the capacity for folly and superstition. In spite of cultural differences, New Yorkers and New Guineans share the same sensory apparatus and the same sort of brain with which to process sensory input. It follows, therefore, that all of us share the same limitations in our ability to perceive, to interpret, and to recall objects and events. In considering how to evaluate native eyewitness accounts of cryptid animals, then, I propose that we should look at how experts evaluate Western eyewitness testimony, and note those factors in which a witness's cultural background may play a significant role.

Authors of cryptozoological literature often adopt an indignant and contemptuous tone when they discuss attempts to explain eyewitness testimony in terms of ordinary phenomena. In such an attempt, the cryptozoologists claim, is the implicit assertion that the witnesses are lying, or insane, or merely stupid. I hope that after you've read through the following findings on eyewitness reliability, you'll realize that a witness can represent a falsehood as truth *without lying;* that a witness can hallucinate *without being insane;* and that a witness can misinterpret what he has seen *without being stupid.* All humans—even the most honest, the most level-headed, the keenest-eyed, the smartest—have imperfect perception and imperfect memory. A variety of factors bear upon our ability to perceive an event correctly, and later to recall correctly what we have perceived.

## Factors Affecting Reliability of Eyewitness Testimony

Researchers into the validity of eyewitness testimony identify the following factors as affecting reliability:

*Slippage of memory.* Witnesses recall details more accurately immediately after an event than they do after a long period of time has elapsed (Loftus, 1979). This phenomenon is called slippage of memory, and its effect is progressive. A memory accurate an hour after an event will be less accurate after a week, less accurate still after a month, and even less accurate after a year.

*Period of observation.* Witnesses notice more details and recall them more accurately with increased observation time (Buckhout, 1974; Williams et al., 1992). Also noteworthy when considering period of observation is the fact that witnesses almost always overestimate the duration of a recalled event (Loftus, 1979).

*Observation conditions.* Witnesses are able to make more accurate observations at close range than at long range, and in bright light than in dim light (Buckhout, 1974). In cryptozoological eyewitness accounts, obstructions (like foliage) and weather conditions (like rain or fog) may further impede accurate observation.

*Fear and stress.* Witnesses are less accurate in recalling details of events during which they experienced fear or high stress (Buckhout, 1974; Dent and Stephenson, 1979; Williams et al., 1992). A witness who is confronted with a large, unknown animal is more likely to be concerned with escaping harm than with making accurate observations of the animal's anatomical features—features critical for making a positive identification later. Peter Byrne (1975), who catalogs scores of SASQUATCH sightings in his book *The Search for Bigfoot,* comments: "The reaction of most people who encounter a Bigfoot seems fairly standard. The usual pattern is one of shock, surprise, often followed by near-panic and rapid flight."

One can well imagine that a witness in a state of "near panic" and in the act of "rapid flight" might have difficulty recalling the finer details of his BIGFOOT encounter.

*Expectancy.* Witnesses tend to see what they expect to see (Buckhout, 1974; Williams et al., 1992). I consider this factor to be hugely significant in evaluating native eyewitness accounts. If the witness has been raised from infancy hearing folk stories about a terrible beast that lurks in the nearby forests, he's apt to make minimal observed data (a loud crash in the brush, a quick blurry glimpse of *something*) fit his expectations. This phenomenon occurs in Western cultures as well—although probably with less frequency, because belief in monsters is generally discouraged. Binns (1984) records the following eyewitness account from Loch Ness: "I saw a heavy wash or wake such as a motor-boat might produce, and I thought: 'Now when I get round that rocky promontory I'll possibly see the Monster.' But when I rounded the bend I saw a couple of swans. On the smooth water the waves appeared all out of proportion to their source."

On any other lake, the witness would probably have immediately thought of a mundane explanation for unusual surface turbulence; but since the incident occurred on notorious Loch Ness, he thought first of the elusive NESSIE.

*Want or need on the part of the witness.* Witnesses tend to see what they want to see (Buckhout, 1974). This factor is perhaps more significant for enthusiastic cryptozoologists than for natives. In an article for *Fortean Times,* self-described "armchair cryptozoologist" Ronald Rosenblatt (1996) describes his encounter with the "nearly extinct" Rhinoceros iguana (*Cyclura cornuta*) in the parking lot of the Miami Seaquarium: "Although I am no expert, [the lizard] looked to me like a giant iguanid. In fact, it looked most like the now nearly extinct Rhinoceros Iguana."

The photograph that accompanies the article does indeed depict an adult *Cyclura* (probably *C. cornuta*). However, *C. cornuta* is only "nearly extinct" on its native Hispaniola. In the United States, it is a popular cage pet, and escapees are not uncommon in Miami. I myself have collected two specimens during my two-year stay in the area.

Rosenblatt continues: "When I looked into the matter, I discovered that . . . while some large lizards have turned up in southern Florida, they have been monitor lizards, not iguanids like the animal I saw. . . . This odd experience changed my attitude toward people who report strange animals. When one has had such an experience, it is no longer possible to accept the derision of the skeptics at face value. What could be more unlikely than seeing a giant lizard in the middle of a huge city? It would be easy to doubt the truth of my experience, yet I know it happened and the photographs back me up. I didn't imagine the lizard and I didn't exaggerate its size."

Rosenblatt did not exercise due diligence in his research. *The Audubon Society Field Guide to North American Reptiles and Amphibians* (Behler and King, 1979) lists two large introduced iguanids, the Common iguana (*Iguana iguana*) and the Spiny-tailed iguana (*Ctenosaura pectinata*), as occurring in Miami. *Iguana iguana* is so abundant that city officials have posted a prominent "Iguana Crossing" sign less than a kilometer from the location of Rosenblatt's sighting.

Seeing a giant lizard in Miami is like seeing a stray cat in any other big metropolitan area. Seeing a *Cyclura cornuta* in Miami is like seeing a stray purebred Siamese cat—unusual, to be sure, but not newsworthy and certainly not inexplicable.

*Fabrication of memories.* Witnesses sometimes remember events that, quite simply, never happened (Buckhout, 1974; Dent and Stephenson, 1979; Williams et al., 1994). They aren't lying; they just remember incorrectly. This phenomenon, called confabulation, is well known and extensively verified experimentally.

Witnesses seem particularly prone to confabulate their presence at "historically significant events" (Buckhout, 1974). In *Cryptozoology A to Z* (1999), Loren Coleman and Jerome Clark record the following account of a huge, dinosaur-like MOKELE-MBEMBE's attack on a small West African village: "Pascal Moteka, who lived near Lake Télé, said his people had once constructed a barrier of wooden spikes across a river to keep the giant beasts from interfering with their fishing. When Mokele-mbembe tried to break through the barrier, the assembled villagers managed to kill it with spears. Celebrating their triumph, the people butchered and cooked the carcass, but everyone who ate the dinosaur meat died shortly afterwards."

Moteka does not claim to have witnessed the incident, so this is not a confabulated tale. However, the described event provides fertile material for confabulation. The story is both "historically significant" and highly improbable. Upon hearing a witness claim his presence at such an incident, the interviewer is faced with two immediately intuitive possibilities: that the witness is giving an accurate account of an actual event, or that the witness is lying. Findings on fabricated memories, however, suggest a third possibility: that the witness is telling the truth—in all sincerity, and to the best of his recollection—about an event that never took place. The details of the fabricated memory may be pieced together from folk stories, from vivid childhood dreams, from an actual but dimly remembered conflict between villagers and a hippo or elephant, or even from information accidentally imparted by the interviewer himself.

*Completion of fragmentary pictures.* Witnesses, over time, may "fill in the gaps" if their observation is incomplete (Buckhout, 1974). "I saw a big black object, apparently moving, in the water" can become "I saw a big black *animal* in the water" in an observer's memory after a while. With the passage of time, the "black animal in the water" may develop eyes, fins, and other features and attributes that the witness didn't claim to see immediately following the event.

*Conformity.* Witnesses sometimes alter their observations to fit those of other witnesses (Buckhout, 1974; Luus and Wells, 1994). Witnesses feel a greater degree of certainty about their observations if they hear that other witnesses have made similar, substantiating claims. This factor is noteworthy because it undermines the notion that an incident involving multiple witnesses is necessarily more credible than an event involving only one witness. If three witnesses thought they saw BIGFOOT, and a fourth is pretty sure that what he saw was just a bear, the loner is likely to lose confidence in his perception and to change it to agree with that of his companions.

*Avoidance of saying "I don't know."* Witnesses are reluctant to admit ignorance or inability to recall, and will sometimes invent details in order to avoid saying "I don't know" (Buckhout, 1974). This factor is important to consider in devising a proper interview of a witness. Reports in which the witness is prompted with questions tend to be more detailed but less accurate, because much of the detail is unconsciously invented.

*Significance of the detail or event.* Witnesses usually remember "important" things and forget "trivial" things (Buckhout, 1974; Williams et al., 1992). If an armed robber orders a bank teller to surrender the contents of the cash drawer, the teller's attention may be so fixed upon the gun that he does not at the time notice, nor does he later recall, the color of the bandit's eyes. The detail simply isn't important to the witness in the context of the event (although it may become very important later, in identifying the criminal). Likewise, a witness confronted by a big, unknown, possibly fierce animal is very likely to overlook subtle field marks.

*Age.* Witnesses may be more or less reliable depending on their age. For various physiological and psychological reasons, the elderly and children are generally less reliable than young and middle-aged adults (Buckhout, 1974; Dent and Stephenson, 1979).

The elderly are subject to various impairments to sensory perception (cataracts, glaucoma, hearing loss, etc.), to memory loss, and to senile dementia, any of which can detract from the accuracy of their observations and recollections (Dent and Stephenson, 1979).

Children are more vulnerable to suggestion than are adults (Dent and Stephenson, 1979; Williams et al., 1992), and are more likely to fill in missing details from imagination (Loftus, 1979). Additionally, children exhibit a near-universal and possibly innate fear of the dark and of "monsters" that might prowl in the dark (Sagan, 1977). In Western cultures, this fear is discouraged as shameful and irrational. In cultures wherein children wandering unsupervised at night might fall victim to predatory mammals, venomous snakes, and other sorts of natural hazards, the "irrational" fear may be actively encouraged, and reinforced by nightly repetitions of scary folktales.

*Sex.* Witnesses may be more or less reliable depending on their sex. Older studies show that men are more reliable in all instances; more recent studies show that women are more reliable except when they are afraid or under stress (Dent and Stephenson, 1979). Again, fear is an important factor to consider in many cryptozoological reports.

*Physical condition.* Witnesses may suffer from physical ailments (near-sightedness, cataracts, colorblindness, etc.) that affect their ability to describe accurately what they have seen (Buckhout, 1974; Dent and Stephenson, 1979). Physical impairments are probably particularly important in native witnesses, many of whom may have undiagnosed problems with their vision, and few of whom have access to first-rate corrective treatment.

Even witnesses who are free from permanent disabilities are vulnerable to temporary physical stresses that can affect their reliability. Long-term lack of food or sleep, for example, can impede a witness's ability to interpret perceived objects or events; in extreme instances, hunger and exhaustion can cause hallucinations (Sagan, 1995).

Roy Mackal (1976) recounts a NESSIE sighting by H. L. Cockrell, who had spent three consecutive nights in a kayak trying to photograph the monster: "Two unsuccessful night hunts led to a third which was also unsuccessful until dawn. At first light, a breeze had dropped and the loch was very calm. Cockrell noticed something to his left about fifty yards away. The object appeared to be swimming very steadily and converging on him. . . . Cockrell said it looked like a very large flat head that was wide and four or five feet long. . . . He took two pictures, but then a slight squall came up. After it was over, he closed in on the object and found a four-foot stick, one inch thick. . . . I am quite content to accept Cockrell's assessment that he photographed a stick or small log and assume that a combination of fatigue from three nights' activity on Loch Ness and a tremendous psychological bias of belief and expectation produced the recorded experience."

*Training.* Witnesses with training in fields that require accurate observation often recall descriptive details better than untrained witnesses; witnesses with such training may also be less prone to suggestion (Williams et al., 1992). The reported findings deal with policemen observing humans and their activities. I submit that a similar situation may exist with trained zoologists, experienced hunters, or even avid birdwatchers, observing animals and their activities. Natives who rely on their local animals and plants for sustenance obviously have more relevant training than the average Western suburbanite, and this factor must be considered in any evaluation of native testimony.

*Biased interviewing.* Witnesses are extremely subject to influence by interviewers (Buckhout, 1974; Dent and Stephenson, 1979; Williams et al., 1992). Leading questions and presentation of photographs for comparison ("Did it look like *this?*") can warp an observer's recollection. Witnesses are also sensitive to nonverbal cues that indicate the interviewer's satisfaction or dissatisfaction with certain answers, and the witness may unconsciously tailor his story in order to appear competent and helpful to the questioner (Buckhout, 1974).

## Factors Affecting Reliability of Native Eyewitness Testimony

In addition to the aforementioned factors that apply to analysis of any eyewitness testimony, I suggest a few others that apply primarily to the testimony of natives:

*Language barrier.* The description a native gives is only as good as his command of English, your command of his language, or your interpreter's command of both languages. In any translation, errors can occur.

*Alternative taxonomies.* Native peoples have their own classification schemes for animals and plants. Their methods of categorization are sometimes very different from our own (Durkheim and Mauss, 1963; Lévi-Strauss, 1966). Ours is based on common descent—which, until the very recent introduction of DNA analysis, has been evaluated primarily by physical similarity. Other cultures' taxonomies are based on the ways in which animals are used (deer and alligators might be grouped together, because they both furnish leather); on the time of day when animals are active (bats and owls might be grouped together, because they are both active at night); or on where the animal lives (parrots and monkeys might be grouped together, because they both dwell in trees). When a native says, "The animal is in the family of the crocodile and the monitor lizard," he may not be indicating the fact that the animal is large and reptilian, but rather some native taxonomic similarity—the fact that, like a crocodile or monitor lizard, it lives near the water; or the fact that, like a crocodile or monitor lizard, it is eaten by the locals. Language barriers can amplify misunderstandings of this sort.

*Overconfidence on the part of the witness in his own expertise.* I've personally encountered this problem in talking with hunters and outdoorsmen in the United States. I believe it may be common to hunters and outdoorsmen in all cultures—and, of course, it would be more prevalent in cultures wherein a greater proportion of the population are outdoorsmen. Witnesses with extensive experience in the woods convince themselves that when they encounter an animal they've not seen before, the animal must be something extraordinary and alien—because, after X number of years in the woods, *surely* the witness knows every animal out there. In the mind of the witness, unknown *to him* means unknown *period*. This assumption is likely to be false especially among native people for a number of reasons.

1. In any region, there are bound to be known animals so rare or secretive that even an experienced hunter could go an entire lifetime without seeing them once.
2. Native peoples usually have limited access to electricity, flashlights, batteries, etc., and their nighttime foraging activities are therefore restricted; many nocturnal animals could escape notice for generations.
3. Native peoples frequently have no written language, and have limited access to television, books, the Internet, and other information resources; so they have no way of learning about wildlife except by direct experience or by word of mouth. While a native hunter may have fantastically thorough knowledge of the wildlife within a few days' walk from his village, he may at the same time be largely ignorant of animals found only 100 miles away. What happens when, due to some unusual circumstance, a lone specimen of some strange-looking animal wanders from its accustomed range? Someone with access to the Discovery Channel would say, "Oh. That's a rhino. I've seen those on TV. What's it doing *here?*" A native hunter might well believe he's seen a monster, and might have great difficulty describing a creature so completely foreign to his experience.

*Incomplete separation of science, history, and myth.* In our society, science, history, religious allegory, and fictional entertainment are fairly distinct. Individuals who fail to recognize the distinctions are in the minority, and are generally held in scorn. In native cultures, however, the lines between these different sorts of information are blurred—when, indeed, there are lines at all (Lévi-Strauss, 1978). Lack of a written language almost assures distortion of information as it gets passed orally from generation to generation.

Skepticism is encouraged in Western scientific tradition—even the most fundamental principles of science are periodically questioned and subjected to testing (Hawking, 1988). Skepticism is discouraged in native societies, wherein unquestioning acceptance of inherited tradition and wisdom is a virtue (Lévi-Strauss, 1978). Belief in monsters, never actually seen but frequently talked about, therefore seems likelier in native cultures than in Western cultures—and, as previously discussed, belief profoundly affects eyewitnesses by creating expectancy.

*Different attitudes toward sense data.* In Western society, we are encouraged through formal education to recognize the fallibility of our senses. Observation is the *beginning* of the process that leads to proof; observation alone does not constitute proof. This mode of thought, however, is unnatural, counterintuitive, and only recently developed. For peoples who rely heavily on their keen eyesight or acute hearing to secure food and to avoid dangers, *seeing is believing*. Observation *is* proof (Lévi-Strauss, 1978).

## The Bottom Line

In evaluating eyewitness testimony from anyone, from any culture, always consider the limitations of human perception and memory. Always consider how your questions may affect the witness's recollection. And always ask yourself: *Which is more likely—that the incident occurred exactly as described? Or that the witness has misinterpreted or misrepresented the data?*

Let's close with an illustrative anecdote. A frightened neighbor once called upon me to rescue her from a cryptozoological menace in her back yard. The creature, which she described as a "furry lobster, about two feet long," had been on her patio when she first encountered it. In their mutual fear, both furry lobster and neighbor fled the scene. The furry lobster took shelter under a shrub in the garden; my neighbor hurried indoors, to telephone first the police (who weren't interested), and then me. I found the mystery animal right away. It was a juvenile Spiny-tailed iguana (*Ctenosaura* sp.).

The spiny-tailed iguana is not native to South Florida, but introduced specimens have established breeding populations throughout the region and the lizard is by no means uncommon here. I was not surprised to find the animal—but I *was* surprised at the woeful inaccuracy of my neighbor's description. In no way did this lizard resemble a lobster; in no way was it furry; and its total length was about one foot, half the size reported. What further distortions might have been introduced if I'd heard the report a year after the incident? If the report had been imperfectly translated from another language? If I'd shown the witness pictures of animals approximately matching the "furry lobster" description? If I'd asked her to draw for me what she'd seen? And how might my perception of the report have been different if the event had taken place not in suburban Miami, but in uncharted Amazonia?

Sorry to disappoint anyone who's been on the trail of the Florida Furry Lobster. To the rest of you, happy hunting.

## Acknowledgments

Chris Orrick posed the question that led to the composition and publication of this article, and generously shared research materials with me. Peter Hynes provided valuable editorial comments on an early draft of the article. Chad Arment moderates the online cryptozoology forum, and also contributed research materials necessary for this article's completion. Matt Bille provided relevant documents that my local libraries could not. Ben Roesch and John Moore edited the later drafts of the article, and prevented many errors of print, omission, and fact from appearing in the final published version. I thank you all.

## References

Behler, John L., and F. Wayne King. *The Audubon Society Field Guide to North Ameri-*

can Reptiles and Amphibians. New York: Alfred A. Knopf, 1979.

Binns, Ronald. *The Loch Ness Mystery Solved.* Buffalo, N.Y.: Prometheus, 1984.

Buckhout, Robert. "Eyewitness Testimony," *Scientific American* 231 (December 1974): 23–31.

Byrne, Peter. *The Search for Bigfoot.* Washington: Acropolis, 1975.

Coleman, Loren, and Jerome Clark. *Cryptozoology A to Z.* New York: Fireside, 1999.

Dent, H. R., and G. M. Stephenson. "Identification Evidence: Experimental Investigations of Factors Affecting the Reliability of Juvenile and Adult Witnesses." In *Psychology, Law, and Legal Processes,* edited by David P. Farrington, Keith Hawkins, and Sally M. Lloyd-Bostock. Atlantic Highlands, N.J.: Humanities Press, 1979.

Diamond, Jared M. "Rubbish Birds are Poisonous," *Nature* 360 (1992): 19–20.

Dumbacher, John P., et al. "Homobatrachotoxin in the Genus *Pitohui:* Chemical Defense in Birds?" *Science* 258 (1992): 799–801.

Durant, Will, and Ariel Durant. *The Lessons of History.* New York: Simon and Schuster, 1968.

Durkheim, Emile, and Marcel Mauss. *Primitive Classification.* Chicago: University of Chicago Press, 1963.

Erspamer, Vittorio, et al. "Pharmacological Studies of 'Sapo' from the Frog *Phyllomedusa bicolor* Skin: A Drug Used by the Peruvian Matses Indians in Shamanic Hunting Practices," *Toxicon* 31 (1993): 1099–1111.

Goodman, Steven M., and Joseph Hobbs. "The Distribution and Ethnozoology of Reptiles in the Northern Portion of the Egyptian Eastern Desert," *Journal of Ethnobiology* 14 (1994): 75–100.

Gould, Stephen Jay. *The Mismeasure of Man.* New York: Norton, 1981.

Hawking, Stephen W. *A Brief History of Time.* New York: Bantam, 1988.

Kittenberger, Kálmán. *Big Game Hunting and Collecting in East Africa, 1903–1926.* New York: Longmans, Green, 1929.

Lévi-Strauss, Claude. *The Savage Mind.* Chicago: University of Chicago Press, 1966.

———. *Myth and Meaning.* Toronto: University of Toronto Press, 1978.

Loftus, Elizabeth F. *Eyewitness Testimony.* Cambridge, Mass.: Harvard University Press, 1979.

Luus, C. A. Elizabeth, and Garry L. Wells. "The Malleability of Eyewitness Confidence: Co-Witness and Perseverance Effects," *Journal of Applied Psychology* 79 (1994): 714–723.

Mackal, Roy P. *The Monsters of Loch Ness.* Chicago: Swallow Press, 1976.

Majnep, Ian Saem, and Ralph Bulmer. *Birds of My Kalam Country.* Auckland, N.Z.: Auckland University Press, 1977.

Minton, Sherman A., and Madge Rutherford Minton. *Venomous Reptiles.* New York: Charles Scribner's Sons, 1969.

Morgan, Marlo. *Mutant Message Down Under.* Thorndike, Me.: Thorndike Press, 1994.

Rosenblatt, Ronald. "Car Park Lizard," *Fortean Times,* no. 92 (November 1996): 50.

Sagan, Carl. *The Dragons of Eden.* New York: Random House, 1977.

———. *The Demon-Haunted World.* New York: Random House, 1995.

Spawls, Stephen. *Sun, Sand, and Snakes.* London: Collins, 1979.

Williams, K. D., Elizabeth F. Loftus, and Kenneth Deffenbacher. "Eyewitness Evidence and Testimony." In *Handbook of Psychology and Law,* edited by Dorothy Kagehiro and William S. Laufer. New York: Springer-Verlag, 1992.

Reprinted with permission from *Cryptozoology Review* 4, no. 1 (Summer 2000): 11–18.

**Jack Rabbit** is an independent researcher in Virginia. His interests are zoology, wildlife painting, writing short fiction, and playing the fretted dulcimer.

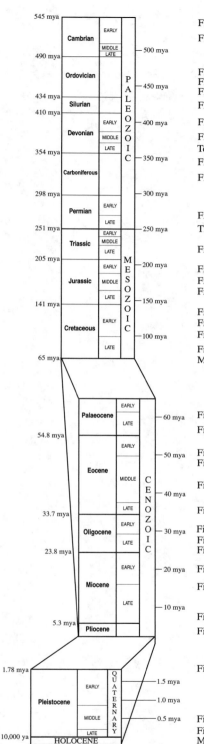

First appearance of invertebrates with shells and other hard parts

First appearance of jawless fishes

First appearance of cephalopods
First appearance of jawed fishes
First appearance of land plants (similar to liverworts and mosses)

First appearance of land animals (arthropods)

First appearance of lobe-finned fishes

First appearance of plants with leaves and roots
Tetrapods evolve from lobe-finned fishes

First appearance of winged insects

First appearance of diapsid and synapsid reptiles

First appearance of parareptiles

The biggest mass extinction of all time, with 90% of all species disappearing

First appearance of crocodiles, pterosaurs, dinosaurs, turtles, and frogs

First appearance of salamanders and caecilians
First appearance of scincomorph and anguimorph lizards
First appearance of mammals and birds

First appearance of flowering plants
First appearance of gekkotan lizards and snakes
First appearance of marsupials and monotremes

First appearance of ungulates and iguanian lizards
Mass extinction of the dinosaurs, plesiosaurs, and many marine invertebrates

First appearance of primates and rodents

First appearance of proboscideans

First appearance of bats and horses
First appearance of whales and sirenians

First appearance of grasses

First appearance of dogs

First appearance of cats, civets, and seals
First appearance of apes
First appearance of hyenas

First appearance of bears and otters

First appearance of hominids

First appearance of *Gigantopithecus*

First appearance of *Australopithecus*

First appearance of *Homo erectus*

First appearance of archaic *Homo sapiens*
First appearance of Neanderthals
Megafaunal extinctions

## GEOLOGIC TIMESCALE

This chart divides the last 545 million years of earth history into named intervals of geological time, many of which are referred to in this book. The three major eras (Paleozoic, Mesozoic, and Cenozoic) are subdivided into the periods shown on the left (Cambrian, Ordovician, and so on), from the oldest at the top to the present time at the bottom. Note that this consists of only the last 12% of earth's 4.6-billion-year history.

Time is given in millions of years ago (mya), with the boundaries of each period on the left and a graduated scale on the right. There are three different scales: 50-million-year markers for the Paleozoic and Mesozoic, 10-million-year markers for the Cenozoic, and .5-million-year markers for the Quaternary.

Notable milestones in the history of life are listed on the right.

# Defining the Field

The word *cryptozoology* (in French, "la crypto-zoologie") was coined by Bernard Heuvelmans in the late 1950s. It comes from the Greek *kryptos* ("hidden") + *zoon* ("animal") + *logos* ("discourse"), which when combined yield "the science of hidden animals." More accurately, cryptozoology is the study of the evidence for animals that are undescribed by science. The word first appeared in print in 1959 when Chief Game Inspector of the French Overseas Territories Lucien Blancou dedicated his book *Géographie cynégétique du monde* to Heuvelmans: "Bernard Heuvelmans, maître de la cryptozoologie" (Bernard Heuvelmans, master of cryptozoology). Heuvelmans has also credited Ivan T. Sanderson with inventing the word independently when Sanderson was a student, which would have been at Eton College in the 1920s.

The use of the word *cryptid* for unknown animals was proposed by John E. Wall of Altona, Manitoba, in a summer 1983 letter to *ISC Newsletter.*

*Dracontology,* now in use for the study of both FRESHWATER MONSTERS and SEA MONSTERS, was coined by French Canadian Jacques Boisvert in the early 1980s. It was accepted by l'Office de la Langue Française du Québec (as *dracontologie*) in 1984 and by the *American Heritage Dictionary* in 1985.

The term *hominology* was invented by Russian researcher Dmitri Bayanov in the early 1970s to describe the study of existing HOMINIDS that do not belong to our own species, *Homo sapiens.* In a letter to primatologist John Napier in 1973, Bayanov said the term was intended to "bridge the gap between zoology and anthropology."

# MYSTERIOUS CREATURES

## VOLUME 1: A–M

# Abnauayu

WILDMAN of West Asia.

*Etymology:* Abkhaz (Northwest Caucasian), "forest man" or "shy boy."

*Variant names:* Bnahua (Abaza/Northwest Caucasian), Ochokochi (Mingrelian/Caucasian).

*Physical description:* Covered with reddish-black hair. Dark skin. Thick head-hair that hangs down the back like a mane. Low forehead. Eyes with a reddish tinge. Flat nose. High cheekbones. Enormous teeth. Muscular arms and legs. Females have large breasts and buttocks. Fingers long and thick. Splayed feet.

*Behavior:* Skilled runner and swimmer. No speech but makes muttering noises. Sharp sense of hearing. Food includes grapes, hominy, and meat. Sleeps in a hole in the ground. Can apparently breed successfully with humans. Washes newborn infants in cold-water springs. Uses improvised weapons of sticks and stones. Habitually plays with stones, grinding and smashing them.

*Distribution:* Caucasus Mountains, Abkhazia Autonomous Republic, Georgia.

*Significant sighting:* A female Abnauayu, nicknamed "Zana," was captured in the mid-nineteenth century, possibly in Ajaria, Georgia. The nobleman Edgi Genaba took her to his farm near Tkhina in Abkhazia, where she lived until her death in the 1880s or 1890s. At first, she was kept shackled in a strong enclosure; later, as she became tame, Zana was let loose to wander about. She was trained to do simple tasks such as grinding grain and fetching firewood. Zana was survived by two sons and two daughters fathered by local human males; these offspring grew up and became relatively normal citizens. Two of Zana's grandchildren were interviewed by Boris Porshnev in 1964. (One of them, Sha-

likula, was said to have been able to pick up a chair, along with a man sitting on it, with his teeth.) Zana's grave has not been found, but the skeleton of her son Khwit has been exhumed; the skull combines "modern and ancient features," according to a 1987 Russian study. Grover Krantz had an opportunity to examine Khwit's skull, and he says it is a modern *Homo sapiens,* though with slightly stronger jaws and flaring cheekbones.

*Possible explanations:*

(1) Neanderthal (*Homo neanderthalensis*) sites are known at Sakhazia and Dzhruchula in Georgia. The large teeth and low forehead are characteristic of these West Asian and European hominids.

(2) Zana's ability to breed successfully with modern humans is intriguing, to say the least, and makes it more likely that she represents an anatomically modern human with some archaic retentions, particularly with regard to lifestyle and material culture.

*Sources:* Boris F. Porshnev, "Bor'ba za Trogloditov," *Prostor* (Alma-Ata), 1968, no. 6, pp. 113–116; Bernard Heuvelmans and Boris F. Porshnev, *L'homme de Néanderthal est toujours vivant* (Paris: Plon, 1974), pp. 171–177; John Colarusso, "Ethnographic Information on a Wild Man of the Caucasus," in Marjorie Halpin and Michael M. Ames, eds., *Manlike Monsters on Trial* (Vancouver, Canada: University of British Columbia Press, 1980), pp. 255–264; Dmitri Bayanov, *In the Footsteps of the Russian Snowman* (Moscow: Crypto-Logos, 1996), pp. 46–52; Grover S. Krantz, *Bigfoot Sasquatch Evidence* (Blaine, Wash.: Hancock House, 1999), p. 210.

## Abominable Snowman

English name for the YETI of Central Asia.

*Etymology:* Coined by *Calcutta Statesman* columnist Henry Newman in 1921 as a translation of the Sherpa (Sino-Tibetan) METOH-KANGMI, which a telegraphist miscoded as "Metch kangmi." Newman claimed it meant "abominable snowman." The phrase became a popular term with journalists from the 1920s through the 1960s. The name does not come from the creature's supposed horrible odor, as some have alleged. The term also serves as a generic name for unknown Asian hominids.

*Variant names:* ABSM, Snowman.

*Physical description:* See YETI.

*Distribution:* Himalaya Mountains of Nepal and Tibet.

*Sources:* Charles K. Howard-Bury, *Mount Everest: The Reconnaissance, 1921* (London: Edward Arnold, 1922), p. 241; Henry Newman, *Indian Peepshow* (London: G. Bell and Sons, 1937), pp. 156–160; Ralph Izzard, *The Abominable Snowman* (Garden City, N.Y.: Doubleday, 1955), pp. 28–29.

## Abonesi

SMALL HOMINID of West Africa.

*Distribution:* Northern Togo.

*Source:* Bernard Heuvelmans, *Les bêtes humaines d'Afrique* (Paris: Plon, 1980), p. 496.

## ABSM

*See* ABOMINABLE SNOWMAN.

*Etymology:* Ivan T. Sanderson's initialism for ABOMINABLE SNOWMAN, which he used as a generic name for any unknown PRIMATE or HOMINID.

*Source:* Ivan T. Sanderson, *Abominable Snowmen: Legend Come to Life* (Philadelphia: Chilton, 1961).

## Abyssal Rainbow Gar

One of BEEBE'S ABYSSAL FISHES of the North Atlantic Ocean.

*Physical description:* Length, 4 inches. Scarlet head. Long beak. Blue body. Yellow tail.

*Behavior:* Abyssal. Swims with a stiff, upright posture.

*Distribution:* North Atlantic Ocean.

*Significant sighting:* Observed only once at 2,500 feet by William Beebe in a bathysphere off Bermuda in the early 1930s.

*Source:* William Beebe, *Half Mile Down* (New York: Harcourt, Brace, 1934).

## Acorn Worm (Giant)

Undiscovered marine INVERTEBRATE.

*Physical description:* Adult Acorn worms (Class Enteropneusta) of the type *Planctosphaera pelagica* have never been observed. The larvae (tornariae) are larger than those of other hemichordates, and if the size ratio is the same as in other species, the adults could grow to 9 feet long. The larvae are large, transparent spheres with arborescently branched, ciliated feeding bands and a U-shaped alimentary tract.

*Habitat:* Oceanic mud at depths of 250–1,660 feet.

*Distribution:* Eastern North Atlantic and North Pacific Oceans.

*Possible explanation:* The larvae may be the abnormally enlarged tornariae of another family of Acorn worms (Ptychoderidae) that fail to metamorphose into adults.

*Sources:* C. J. van der Horst, "Planctosphaera and Tornaria," *Quarterly Journal of Microscopial Science* 78 (1936): 605–613; M. G. Hadfield and R. E. Young, "Planctosphaera (Hemichordata: Enteropneusta) in the Pacific Ocean," *Marine Biology* 73 (1983): 151–153; M. W. Hart, R. L. Miller, and L. P. Madin, "Form and Feeding Mechanism of a Living *Planctosphaera pelagica* (Phylum Hemichordata)," *Marine Biology* 120 (1994): 521–533.

## Adam-Ayu

WILDMAN of Central Asia.

*Etymology:* Kazakh (Turkic), "bear man."

*Distribution:* Tian Shan Mountains, northern Xinjiang Uygur Autonomous Region, China.

*Source:* Odette Tchernine, *The Yeti* (London: Neville Spearman, 1970), p. 178.

## Adam-Dzhapais

WILDMAN of Central Asia.

*Etymology:* Kyrgyz (Turkic), "wild man."

*Variant names:* Adam-japayis, Adam-yapay-isy, Adam-yavei, Japayi-kishi, Zhabayi-adam (Kazakh/Turkic).

*Distribution:* Eastern Pamir Mountains, Tajikistan; Kunlun Mountains, southern Xinjiang Uygur Autonomous Region, China.

*Sources:* Odette Tchernine, *The Yeti* (London: Neville Spearman, 1970), pp. 178, 181; Bernard Heuvelmans and Boris F. Porshnev, *L'homme de Néanderthal est toujours vivant* (Paris: Plon, 1974), pp. 86, 108.


## Adjulé

DOG-like animal of North Africa.

*Etymology:* Tamahaq (Berber) name.

*Variant names:* Kelb-el-khela ("bushdog," in Mauritania), Tarhsît (for the female).

*Physical description:* Like a wolf.

*Distribution:* Sahara Desert.

*Possible explanation:* An African wild dog (*Lycaon pictus*) outside its usual range. These dogs stand up to 2 feet 7 inches at the shoulder and are normally found in protected sub-Saharan savanna areas. Earlier in the twentieth century, there may have been enough gazelles in sub-desert areas for scattered packs to subsist.

*Source:* Théodore Monod, "Sur la présence du Sahara du *Lycaon pictus* (Temm.) (Résultats scientifiques de la Mission Saharienne Augiéras-Draper)," *Bulletin de la Société Zoologique de France* 53 (1928): 262–264.


## Adlekhe-Titin

WILDMAN of West Asia.

*Etymology:* Ubykh (Circassian), "mountain man."

*Variant names:* Lakhatet, Lakshir.

*Physical description:* Covered with hair.

*Distribution:* Northwestern Caucasus Mountains, Russia.

*Sources:* Georges Dumézil and Aytek Namitok, "Récits oubykh," *Journal Asiatique* 243 (1955): 1–47; John Colarusso, "Ethnographic Information on a Wild Man of the Caucasus," in Marjorie Halpin and Michael M. Ames, eds., *Manlike Monsters on Trial* (Vancouver, Canada: University of British Columbia Press, 1980), pp. 255–264.


## Afa

Unknown LIZARD of the Middle East.

*Etymology:* Madan (Marsh Arab) word.

*Physical description:* Large lizard.

*Distribution:* Marshes at the mouth of the Tigris River, Iraq.

*Possible explanation:* An undescribed species of Monitor lizard (Family Varanidae), large carnivorous reptiles that live in tropical areas.

*Source:* Wilfred Thesiger, *The Marsh Arabs* (New York: Dutton, 1964), p. 115.


## Afanc

FRESHWATER MONSTER of Wales.

*Etymology:* Welsh, "beaver." The cognate Irish word *abhac* ("dwarf") derives from *abha* ("river"), which may signify a water spirit.

*Variant name:* Addanc.

*Physical description:* Variously described as a giant beaver or crocodile.

*Behavior:* Causes flooding. Drags people into the water.

*Distribution:* Llyn yr Afanc (Beaver Pool), Betws-y-coed, Conwy, Wales; Llyn Barfog and Llyn-y-cae in Gwynedd, Wales; Llyn Glaslyn, Powys, Wales.

*Significant sightings:* King Arthur is said to have slain an Afanc in Llyn Barfog.

Oliver Vaughan saw the pale head of an animal in Llyn Glaslyn from the slope of Snowdon in the 1930s.

*Sources:* John Rhys, *Celtic Folklore, Welsh and Manx* (Oxford: Clarendon, 1901), p. 130; F. W. Holiday, *The Great Orm of Loch Ness* (New York: W. W. Norton, 1969), pp. 131–132; F. W. Holiday, *The Dragon and the Disc* (New York: W. W. Norton, 1973), p. 85; Susan Cooper, *Silver on the Tree* (London: Chatto and Windus, 1977); James MacKillop, *Oxford Dictionary of Celtic Mythology* (New York: Oxford University Press, 1998), p. 5.

## Afonya

GIANT HOMINID of Northern Europe.

*Etymology:* Six Russian teenagers gave a large, hairy hominid this nickname, which is a diminutive of the Russian name Afanasii; however, the term may be derived from the name of a popular 1975 Russian film about a drunken Soviet plumber.

*Physical description:* Height, 7–8 feet. Body-hair mostly light gray with lighter and darker patches. Dark skin. Round head. Wide forehead. Face wrinkled. Reddish eyes, set wide apart. Arms hang to the knees. Light-colored buttocks.

*Behavior:* Primarily nocturnal. Stooped-over stance. Sometimes knuckle-walks. Runs very quickly and smoothly. Climbs trees with some agility. Call is a mooing sound. May live in cabins when they are deserted. May steal dogs for hunting or companionship. Throws rocks and sticks as weapons.

*Tracks:* Length, 15 inches. Stride measures over 4 feet.

*Distribution:* Kola Peninsula, European Russia.

*Significant sightings:* A group of teenagers on a fishing expedition to Lake Lovozero in the Murmansk Region of Russia were pestered in their cabin and chased for several days in August and September 1988 by an aggressive creature they nicknamed "Afonya." It was also seen by a local game warden. Maya Bykova and a team of researchers visited the area shortly afterward and succeeded in catching a glimpse of Afonya. They returned the following summer and uncovered tracks, hair, feces, and additional testimony. Bykova developed a specific call that Afonya responded to and answered, and she was able to entice it to the cabin where the teenagers had stayed. Her assistant, Nikolai Damilin, used a different call equally successfully. The team carried out experiments using tape recordings of animal sounds that included the calls of primates. One of the creatures went to the cabin in response to the sounds and left footprints. Strange whistling was recorded several times and analyzed by Leonid Yershov.

*Sources:* Dmitri Bayanov, *In the Footsteps of the Russian Snowman* (Moscow: Crypto-Logos, 1996), pp. 190–206; Anatoli Schmidt, Karl C. Beyer, and Andreas Braun, "The Books by Dmitri Bayanov and Their Secrets," 2000, http://www.stgr-primates.de/site10.html.

## Agatch-Kishi

WILDMAN of West Asia.

*Etymology:* Karachay-Balkar (Turkic), "wild man."

*Distribution:* Caucasus Mountains, Russia.

*Source:* Dmitri Bayanov, *In the Footsteps of the Russian Snowman* (Moscow: Crypto-Logos, 1996), p. 24.

## Agogwe

SMALL HOMINID of East Africa.

*Etymology:* Kuria or Chagga (Bantu) word.

*Physical description:* Height, about 4 feet. Brown or russet-colored hair.

*Behavior:* Upright gait. Said to barter for goods with local tribes.

*Habitat:* Dense forests.

*Distribution:* North-central Tanzania.

*Significant sighting:* William Hichens briefly observed two hairy men in north-central Tanzania in the 1920s. They walked upright across a clearing in the forest.

*Possible explanation:* Surviving gracile australopith, suggested by Bernard Heuvelmans. (*See* KAKUNDAKARI for a more detailed explanation.) The Laetoli fossil beds that contain perfectly preserved *Australopithecus afarensis* footprints are in north-central Tanzania, as is Olduvai Gorge where the robust fossil *Paranthropus boisei* was discovered by Louis Leakey in 1959. Since East Africa is the probable birthplace of early hominid species, its traditions of small hairy men are tantalizing.

*Sources:* William Hichens, "African Mystery Beasts," *Discovery* 18 (1937): 369–373; Bernard Heuvelmans, *Les bêtes humaines d'Afrique* (Paris: Plon, 1980), pp. 515–516, 530–535.

## Agrios Anthropos

WILDMAN of North Africa.

*Etymology:* Greek, "wild man."

*Physical description:* Covered with hair.

*Distribution:* Western Libya.

*Possible explanation:* Distorted accounts of Gorillas (*Gorilla gorilla*) or Chimpanzees (*Pan troglodytes*) living in the forest much farther to the south.

*Source:* Herodotus, *The Histories,* trans. Aubrey de Sélincourt (London: Penguin, 1996), p. 276 (IV. 191).

## Ah-Een-Meelow

SEA MONSTER of the South Pacific Ocean.

*Etymology:* Barok (Austronesian), "fish eel."

*Physical description:* Length, 50 feet. Head like a python's. Neck, 10–15 feet long, 2 feet thick. Four gray-green body loops, 10 feet apart. Frill on the back. Vertical, segmented tail, 2 feet long.

*Behavior:* Moves with vertical undulations.

*Habitat:* Seen close to the coast.

*Distribution:* Ramat Bay, New Ireland, Papua New Guinea.

*Source:* Paul Cropper and Malcolm Smith, "Some Unpublicized Australasian 'Sea Serpent' Reports," *Cryptozoology* 11 (1992): 51–69.

## Ahool

Giant BAT-like creature of Southeast Asia.

*Etymology:* From its call.

*Variant name:* Aul.

*Physical description:* In flight, looks like a flying-fox bat but larger. Dark-gray fur. Monkey-like face. Large, black eyes. Flat forearms topped by claws. Batlike wings. Wingspan, 11–12 feet. Feet said to point backward.

*Behavior:* Nocturnal. Squats on the forest floor with its wings pressed against its flanks. Flies low over rivers in search of fishes. Call is "AH-OOooool," repeated three times. Said to kill people with its claws.

*Distribution:* Mountains in the western part of Java.

*Significant sighting:* Ernst Bartels was sleeping near Cijengkol, Java, Indonesia, in 1927 when he was awakened by the sound of flapping wings and the call of an animal that sounded like "a-hool."

*Possible explanation:* An unknown large bat with an enormous wingspan, possibly a micro-bat, suggested by Karl Shuker.

*Sources:* Ernst Bartels and Ivan T. Sanderson, "The One True Batman," *Fate* 19 (July 1966): 83–92; Karl Shuker, "A Belfry of Crypto-Bats," *Fortean Studies* 1 (1994): 235–245.

## Ahuítzotl

Legendary OTTER-like animal of Mexico.

*Etymology:* Nahuatl (Uto-Aztecan), "water dog."

*Physical description:* Looks like a small dog. Smooth, black coat. Small, pointed ears. Paws like a raccoon's. A bony spur projects underneath its tail. Tip of the tail looks like a human hand.

*Behavior:* Amphibious. Makes a sound like a baby crying. Said to drag humans into the water with its tail.

*Habitat:* Rivers or lakes.

*Distribution:* Mexico.

*Present status:* Known to the Aztecs but probably extinct now.

*Possible explanations:*

(1) The Coyote (*Canis latrans*), suggested by Ferdinand Anders. However, coyotes do not like water.

(2) The Mexican hairy porcupine (*Coendu mexicanus*), proposed by Eduard Seler, though this is an arboreal animal, not an aquatic one.

(3) The Marine otter (*Lontra felina*), although it is only found on the Pacific coast from Peru to Tierra del Fuego.

(4) The Sea otter (*Enhydra lutris*), but it does not range farther south than California.

(5) The Neotropical otter (*Lontra longicaudis*), which is found in rivers throughout much of Mexico. Except for the odd tail and aggressive manner, this would be an excellent candidate. Andrew Gable writes that the Aztecs knew this otter as the Aitzcuintli; however, the Nahuatl name for the Domestic dog (*Canis familiaris*) was Itzcuintli, so there may be differing interpretations in Aztec texts.

(6) An unknown species of prehensile-tailed otter, proposed by Andrew Gable.

*Sources:* Bernardino de Sahagún, *Florentine Codex: General History of the Things of New Spain* [1577?], trans. Arthur J. O. Anderson and Charles E. Dibble (Salt Lake City: University of Utah Press, 1950–1982), vol. 11, pp. 68–69; Eduard Seler, "Die Tierbilder der mexikanischen und der Maya-Handschriften," *Zeitschrift für Ethnologie* 41 (1909): 390–393; Ferdinand Anders, "Der altmexikanische Federmosaikschild in Wien," *Archiv für Völkerkunde* 32 (1978): 67, 79–80; Andrew D. Gable, "Two Possible Cryptids from Precolumbian Mesoamerica," *Cryptozoology Review* 2, no. 1 (Summer 1997): 17–25.

## Aidakhar

FRESHWATER MONSTER of Central Asia.

*Etymology:* Possibly Kazakh (Turkic) word, said to mean "huge snake."

*Physical description:* Length, 45–50 feet. Head, 6 feet long and 3 feet wide. Long neck. One hump.

*Behavior:* Said to have a trumpeting call.

*Distribution:* Lake Kök-köl, Zhambyl Region, Kazakhstan.

*Significant sighting:* Anatolii and Volodya Pecherskii saw the animal in 1975 from about 25 feet away.

*Possible explanation:* Lake water being sucked into underground caverns is said to create noisy, monsterlike whirlpools.

*Sources: Denver Post,* January 31, 1977; "Muddying the Waters," *ISC Newsletter* 5, no. 4 (Winter 1986): 10.

## Aképhalos

WILDMAN of North Africa.

*Etymology:* Greek, "headless man."

*Variant names:* Blemyes, Blemmyes.

*Physical description:* No head. Eyes are located in the chest.

*Distribution:* Western Libya.

*Possible explanation:* Confused account about members of a nomadic tribe who looked either headless due to their distinctive headdresses or directionless because they did not have a fixed homeland.

*Sources:* Herodotus, *The Histories,* trans. Aubrey de Sélincourt (London: Penguin, 1996), p. 276 (IV. 191); Pliny the Elder, *Natural History: A Selection,* ed. John F. Healy (New York: Penguin, 1991), p. 57 (V. 8, 46); Etienne Quatremère, *Mémoires géographiques et historiques sur l'Egypte* (Paris: F. Schoell, 1811), vol. 2, pp. 127–161; Bernard Heuvelmans, *Les bêtes humaines d'Afrique* (Paris: Plon, 1980), pp. 148, 161.

## Äläkwis

CANNIBAL GIANT of western Canada.

*Etymology:* Bella Coola (Salishan) word.

*Physical description:* Covered with hair.

*Distribution:* Bella Coola Inlet, British Columbia.

*Possible explanation:* Said to be a Native American living alone in the woods.

*Source:* Thomas F. McIlwraith, "Certain Beliefs of the Bella Coola Indians Concerning Animals," *Archaeological Reports of the Ontario Department of Education* 35 (1924–1925): 17–27.

## Alan

Mythical FLYING HUMANOID of Southeast Asia.

*Etymology:* Itneg or Kalinga (Austronesian) word.

*Variant names:* Balbal (Tagbanwa/Austronesian), Manananggal.

*Physical description:* Woman's face and body. Long tongue. Feathered neck. Wings. Curved nails. Scaly arms with talons. Toes and fingers are said to be reversed.

*Behavior:* Friendly but mischievous. Hangs batlike from a tree. Lives in a golden house. Said to raise foster human children.

*Habitat:* Forests.

*Distribution:* Northern Luzon and Palawan Islands, Philippines.

*Sources:* Dean C. Worcester, *The Philippine Islands and Their People* (New York: Macmillan, 1899), p. 109; Fay-Cooper Cole, "Traditions of the Tinguian: A Study in Philippine Folklore," *Fieldiana: Anthropology* 14, no. 1 (1915); Maria Leach, ed., *Funk and Wagnalls Standard Dictionary of Folklore, Mythology, and Legend*

(New York: Funk and Wagnalls, 1949–1950), vol. 1, p. 33; Joe Nigg, *A Guide to the Imaginary Birds of the World* (Cambridge, Mass.: Apple-Wood, 1984), pp. 113–115; "The Manila Vampire," *Fortean Times,* no. 64 (August-September 1992): 11.

## Algerian Hairy Viper

Mystery SNAKE of North Africa.

*Physical description:* Hairy, like a caterpillar. Length, 22 inches. Brownish-red.

*Distribution:* Vicinity of Algiers, Algeria.

*Significant sighting:* Seen only once in January 1852, coiled around a tree near Draria, Algeria.

*Possible explanation:* A large caterpillar of some kind.

*Source:* Karl Shuker, "Hairy Reptiles and Furry Fish," *Strange Magazine,* no. 18 (Summer 1997): 26–27.

## Alien Big Cat

Large puma- or leopardlike CAT of Europe.

*Etymology: Alien* is used in the sense of "out-of-place."

*Variant names:* ABC, Babette, Beast of Cézallier, Beast of Estérel, Beast of Noth, Beast of Valescure, Black panther, BRITISH BIG CAT, Chapalu (in Wales), Elli (in Finland), Hannover puma, Monster of Pindray, Odenwald beast, Pornic panther.

*Physical description:* Many are described as jet-black cats, a melanistic morph common only in Asian leopards and American jaguars.

*Behavior:* Attacks livestock.

*Distribution:* Most common in Great Britain (*see* BRITISH BIG CAT). Scattered reports occur throughout Europe. Its existence in the British Isles especially seems unlikely from an ecological standpoint. A partial list of European places where Alien big cats have been reported follows:

*Czech Republic*—Jinačovice.

*Denmark*—Meldungen.

*Finland*—Imatra, Kekäleenmäki, Kristinestad, Ruokolahti, Vaasa.

*France*—Cézallier; Epinal, Vosges Department; Estérel; Forêt de Chize, near Niort; Noth, Creuse Department; Pindray near Poitiers; Pornic, Brittany Region; Valescure.

*Germany*—Bruchmühlbach-Miesau, Deggendorf, Erding, Ernsdorf, Fürth, Gelnhausen, Hannover, Heubach, Kalbach, Lindenfels, Odenwald, Rantrum, Saarland State, Schwalbach, Soest, Steinbach, Winterkasten.

*Italy*—Bari, Foggia.

*Switzerland*—Graubünden.

*Significant sightings:* Some 289 sheep and 3 cows were killed from February to November 1977 around Epinal, Vosges Department, France, by big cats or dogs with eyes like a lynx's and fur like a wolf's. In the summer of 1978, at least two animals that had survived the winter were seen by various witnesses, who described them as large and black with short legs and big paws. The animals disappeared from the region in 1979.

In July 1982, Uwe Sander of Rantrum, Schleswig-Holstein State, Germany, claimed to have been attacked by a puma rumored to be at large north of Hamburg. Hunters and police officers searched the area to no avail. Sander obtained some hair from the animal, but analysis showed it had come from a rabbit.

A lionlike cat the size of a calf terrorized the area around Noth, Creuse Department, central France, in November and December 1982, killing cattle and sheep.

Black panthers were sighted in the Odenwald, Hesse State, Germany, in August 1989 and near Heubach, Hesse State, in October of the same year. In the first two days of November, several people reported panther encounters in Fürth, Steinbach, Winterkasten, and Lindenfels. However, few tracks were found, no domestic animals were killed, and organized hunts yielded nothing.

On June 22, 1992, forestry official Martti Arvinen encountered a golden-brown lioness in the wilderness near Ruokolahti, Finland. The animal turned and ran. Numerous tracks were found, as well as the half-eaten carcass of a young moose (called an "elk" in Europe). So many other sightings in Finland took place over the next week that the newspapers nicknamed the animal "Elli."

*Possible explanations:*

(1) Leopards (*Panthera pardus*) were common from Africa to Indonesia before their

range began to shrink around 1800. They are still found in forested and rocky areas of Africa and East Asia. Melanism (black coloration caused by a recessive gene) is most common in India and Southeast Asia. Spots are still present but rendered less visible by the dark pigment. Males can measure 8 feet in total length and weigh up to 200 pounds. They are lone, nocturnal hunters, stalking their prey and killing swiftly with a bite to the throat.

(2) Lions (*Panthera leo*) lived in Southern and Eastern Europe from 700,000 years ago to around A.D. 100. Upper Paleolithic cave art, particularly that in Grotte Chauvet in France, features them in surprising detail, down to the black dots at the base of the whiskers. None are depicted as maned, leading some to speculate that European male lions were maneless; however, cave artists may have favored the dominant females of the pride.

(3) The much smaller American Jaguar (*Panthera onca*), found from Mexico to Argentina, is also prone to melanism.

(4) The Puma (*Puma concolor*) is only found in North and South America. Eradicated in the eastern United States by the early twentieth century, it is now making a comeback (*see* EASTERN PUMA). It ranges from light to dark brown in color and has no spots. Melanism is virtually unknown. The average length is 6–8 feet (including tail), and the animal is about 3 feet high at the shoulder. It is wary of humans and avoids contact. Its normal prey is deer, but it also eats fishes, rabbits, and game birds.

(5) The European wildcat (*Felis silvestris silvestris*) has been making a comeback in certain areas, particularly Switzerland, Belgium, the Czech Republic, Slovakia, France, and Germany.

(6) The Eurasian lynx (*Lynx lynx*) was reintroduced in eastern Switzerland, Austria, and Slovenia in the 1970s and has reoccupied about two-thirds of the Swiss Alps. Like the European wildcat, this smaller animal could be mistaken for a big cat from a distance.

(7) The Gray wolf (*Canis lupus*) is still found in the wilder parts of Europe, but it is so well known that misidentifications are unlikely.

(8) Paranormal ENTITIES without a zoological basis, perhaps having a psychic nature.

*Sources:* "British Report," *Doubt,* no. 18 (1947): 269; Jean-Louis Brodu and Michel Meurger, *Les félins-mystère: Sur les traces d'un mythe moderne* (Paris: Pogonip, 1984); Ulrich Magin, "Continental European Big Cats," *Pursuit,* no. 71 (1985): 114–115; Ulrich Magin, "The Odenwald Beast," *Fortean Times,* no. 55 (Autumn 1990): 30–31; Véronique Campion-Vincent, "Appearances of Beasts and Mystery-Cats in France," *Folklore* 103 (1992): 160–183; Sven Rosén, "Out of Africa: Are There Lions Roaming Finland?" *Fortean Times,* no. 65 (October-November 1992): 44–45; Ulrich Magin, "The Saarland Panther," *INFO Journal,* no. 68 (February 1993): 22–23; Ulrich Magin, *Trolle, Yetis, Tatzelwürmer* (Munich, Germany: C. H. Beck, 1993), pp. 51–59; Michel Meurger, "Leopards of the Great Turk: Exotic Felines in French Cultural History," *Fortean Studies* 1 (1994): 198–209.

## Alien Big Dog

The DOG equivalent of Europe's ALIEN BIG CAT. Some livestock-ravaging cryptids have a decidedly canid look, though in most respects they behave similarly to the mystery cats.

*Variant names:* BEAST OF GÉVAUDAN, Girt dog, Island monster (Isle of Wight), PHANTOM WOLF, Vectis monster (Isle of Wight).

*Physical description:* Like a large dog but with certain peculiar features. Dark color. Small ears. Long snout. Short legs. Long tail.

*Behavior:* Kills livestock but often only drinks the blood instead of eating the animal.

*Tracks:* Clawed.

*Distribution:* England; Ireland; Serbia; Russia.

*Significant sightings:* An unknown animal killed as many as seven or eight sheep each night by cutting their throats and drinking their blood near Ennerdale Water, Cumbria, England, from May to September 1810. Will Rotherby, who was knocked down by the beast, described it as lionlike, though most observers thought it a dog. A dog was killed on September 12, after which the killings stopped.

A mystery animal killed sheep, as many as thirty in one night, from January to April 1874, in County Cavan, Ireland, and later near Limerick. Throats were cut and blood sucked, but the sheep were not eaten.

From July to December 1893, a dog-sized animal with a long snout and a long tail attacked women and children near Trosna, Orël Region, Russia. At least one child and two women were said to have been killed. Repeated attempts by hunters to shoot or capture the animal failed, though it apparently ate some poisoned sheep set as bait and disappeared beyond the Vytebet' River. In fact, more than one beast may have been involved, possibly a big cat and a smaller dog.

In November 1905, a mystery animal killed sheep in the area around Great Badminton, South Gloucestershire, England, leaving the flesh almost untouched, but the blood had been lapped up.

A lion-headed, maned, hairless mystery animal on the Isle of Wight, England, was killed in 1940; it turned out to be a fox in an advanced state of mange.

A dog the size of a small pony was seen on Dartmoor, Devon, England, by policeman John Duckworth in 1969 and 1972.

In the mid-1990s, a pair of unusual animals was killed near Slatina, 9 miles southeast of Čačak in Serbia. Slightly bigger than pit bulls, they had short legs, long snouts, and no tails. They had been killing chickens and livestock and drinking their blood. A similar animal was killed near Malá Kopašnica, about 100 miles to the southeast.

Near Gornja Gorevnica, Serbia, in November 2000, many sheep were found killed by an animal that made a tiny incision in their necks and drank their blood. More than 150 hunters went to Jelica Mountain to hunt for the beast, but they found nothing. Some thought that North Atlantic Treaty Organization (NATO) forces had introduced predators to destroy Serbian livestock.

In the summer of 2001, a mystery animal killed as many as ten sheep a night in the region around Novi Kneževac, Serbia. Sheep weighing as much as 200 pounds were found slaughtered, and the guard dogs remained silent.

*Possible explanations:*

(1) The Gray wolf (*Canis lupus*) has been extinct in England since 1486, in Scotland since 1743, and in Ireland since about 1770. Russia has always been a stronghold for wolf populations, which have actually increased since World War II. Attacks on people by wolves are extremely rare, except for the occasional rabid specimen. In the twentieth century, the only evidence for such attacks involved some unconfirmed reports from Italy that wolves had attacked and killed unaccompanied young children. In the absence of natural wild prey, wolves will go after livestock, especially in the winter. Sheep, carrion, and domestic dogs were found to be their most frequent prey, according to one study in Spain.

(2) A feral Domestic dog (*Canis familiaris*), especially a hound or other large breed or crossbreed.

(3) Wolf × dog hybrids occur more frequently as wolf populations become more isolated. Hybrids have been reported throughout Southern Europe.

(4) Arctic foxes (*Alopex lagopus*) turned up in Yorkshire, England, in 1983 and North Wales in 1990.

(5) A few Coyote cubs (*Canis latrans*) are said to have been introduced around 1881 in Epping Forest, Essex, England.

(6) A giant variety of Pine marten (*Martes martes*), suggested by Andrew Gable.

*Sources:* "Wolves in Great Britain," *Land and Water* 17 (March 7, 1874): 190; "An Irish Wolf," *Land and Water* 17 (March 28, 1874): 245; R. G. Burton, "A Wild Beast in Russia," *The Field* 82 (December 9, 1893): 882; A. H. B., "A Wild Beast in Russia," *The Field* 82 (December 23, 1893): 973; "The Wild-Dog of Ennerdale," *Chambers's Journal,* ser. 6, 7 (1904): 470–472; "Sheep-Slaying Mystery," *Daily Mail* (London), November 1, 1905, p. 5; "The Badminton Jackal," *Gloucester Journal,* November 25, 1905, p. 8; "Badminton Jackal," *Daily Mail* (London), December 19,

1905, p. 5; R. G. Burton, "Wolf-Children and Were-Wolves," *Chambers's Journal,* ser. 7, 14 (1924): 306–310; Karl Shuker, *Mystery Cats of the World* (London: Robert Hale, 1989), pp. 93–95; Karl Shuker, *Extraordinary Animals Worldwide* (London: Robert Hale, 1991), pp. 177–179; Marcus Scibanicus, "Strange Creatures from Slavic Folklore," *North American BioFortean Review* 3, no. 2 (October 2001): 56–63, http://www.strangeark.com/nabr/NABR7.pdf.

## Almas

WILDMAN of Central Asia.

*Etymology:* Mongolian (Altaic), "wild man," though possibly derived from *ala* ("to kill") + *mal* ("animals"). The word is found in many southern Mongolian place-names.

*Variant names:* Albast, Albasty, Alboost, Almast (Kazakh/Turkic), Habisun mörtu ("edgewise going"), KHÜN GÖRÜESSÜ, Nühni almas ("burrow" almas), Zagin almas ("saxaul" almas), Zagitmegen ("old woman of the saxaul thickets").

*Physical description:* Adult height, 5 feet–6 feet 6 inches. Covered with 6-inch-long, curly, reddish-brown hair except for hands and face. Dark skin. Prominent browridges. Small, flat nose. Pronounced cheekbones. Jutting jaw. No chin. Short neck. Females have pendulous breasts. Long arms. Long fingers. Short thumb. Fingernails and toenails present. Bare, callused knees. Short legs. Broad feet. Big toe shorter than others but massive and projecting inward.

*Behavior:* Walks with knees bent and legs spread apart (at least in snow). Females throw breasts over their shoulders when running. Said to be able to outrun camels. No known language but can produce some bloodcurdling shrieks. Eats grass, wild plants, and perhaps small mammals. Lives in caves. Possibly engages in primitive barter with humans (will leave skins at prearranged places and pick up items left by the nomads) and may interbreed with them (a lama at the Lamaiin Gegeenii Hüryee Monastery in Mongolia was said to be a half-breed Almas). Said to occasionally suckle human infants. Can use only simple tools. Apparently has no knowledge of fire.

The ALMAS, wildman of Central Asia. (Richard Svensson/Fortean Picture Library)

*Tracks:* Rarely seen but slightly longer than a human's and much wider. No arch present.

*Distribution:* Altai Mountains, Mongolia; Gobi Desert of Mongolia and Nei Mongol Autonomous Region, China; Tian Shan Mountains of Xinjiang Uygur Autonomous Region, China; Qilian Shan Mountains of Gansu and Qinghai Provinces, China; Sayanskiy Range, Tuva Republic, Siberian Russia.

*Significant sightings:* Bavarian soldier Johannes Schiltberger was captured by Turks at the Battle of Nikopol, Bulgaria, in 1396; after the Turks lost to Timur at the Battle of Ankara in 1402, Schiltberger became a slave to various Mongol warlords, migrating all the way from Armenia to Mongolia itself and finally returning to Europe in 1427. While in the Tian Shan Mountains in the retinue of the Mongol prince Egidi, he became the first Westerner to see an Almas, two of which had been caught in the mountains. They were covered with hair except on their hands and faces.

Sometime in the late nineteenth century, a caravan was resting in the southern part of the Mongolian province of Övörhangay on the way to Hohhot, Nei Mongol Autonomous Region, China, when one of the men in the party went to collect the camels that had been set loose to graze. When he did not return, the others went off into the saxaul thickets to look for him. At the entrance to a cave, they found evidence of a struggle and figured an Almas had abducted him. One of the elders suggested they pick him up on the way back from Hohhot, which they did, waiting until the creature emerged from the cave at sundown and shooting it. The rescued man seemed to be insane and died two months afterward.

In April 1906, Soviet scholar Badzar Baradiin reportedly had a brief encounter with an Almas while he was traveling in the Gobi Desert near Badain Jaran, Nei Mongol Autonomous Region, China. However, Michael Heaney considers this story a fiction, based on the fact that there is no mention of the incident in Baradiin's meticulous diary of the trip; moreover, the actual route was 150 miles east of where the event supposedly took place.

A seven-year-old Almas female was accidentally killed in the Gobi when she set off a crossbow attached to an animal snare. Many people in the sparsely populated area are said to have seen the body, but the locals begged investigators not to talk about it, since crossbow snares were illegal.

In 1927, travelers left a caravan unattended while they went to look for a camel that had dropped back. Upon their return at daybreak, they found several Almas warming themselves by the dying campfire. The creatures had eaten some dried dates and sweets but had left the jars of wine untouched.

A monk named Dambayorin was traveling across the Gobi in 1930 when he saw a naked child in the distance. When he got closer, he saw it was covered with red hair, realized it was an Almas, and fled in terror.

An entire skin of an Almas is said to have hung in the temple of the monastery at Baruun Hural, Mongolia, in 1937. It had humanlike legs and arms and long hair hanging from its head. The Almas had been killed in the Gobi by the hunter Mangal Durekchi and given to the lamas.

A Mongolian pharmacist named Nagmit was in the mountains with two Kazakhs when they came upon an Almas. They shouted at it, offering it food and clothing, but it kept its distance. When they shot at it, intentionally missing, the creature merely seemed curious, then departed.

Russian pediatrician Ivan Ivlov was traveling in the Altai Mountains of western Mongolia in 1963 when he saw a male, a female, and a young Almas on a mountain slope. He observed them through binoculars at a distance of about a mile until they moved out of sight. Afterward, he queried a number of his child patients about the Almas and obtained some detailed stories.

*Present status:* Vanished or severely reduced over much of its range.

*Possible explanations:*

(1) Surviving *Homo erectus,* suggested by Mark Hall and Loren Coleman. The nearest known fossils are the Zhoukoudian Peking man remains found north of Beijing in the 1920s. The browridge, flat nose, absent chin, and robust jaw match Almas descriptions. *H. erectus* used a primitive (Acheulean) toolkit of hand axes and other bifacial stone tools. The youngest level of *erectus* remains at Zhoukoudian date from about 300,000 years ago.

(2) Surviving Neanderthals (*Homo neanderthalensis*), proposed by Myra Shackley. Neanderthal fossils are not known in Central Asia, though Shackley claims to have recovered, in Mongolia, Mousterian tools normally associated with them. Almas descriptions seem to indicate a more primitive morphology than known Neanderthal fossils, so Shackley has also theorized that they may represent a common ancestor to Neanderthals and modern humans.

*Sources:* Johannes Schiltberger, *The Bondage and Travels of Johann Schiltberger* (London: Hakluyt Society, 1879); Nikolai M. Przheval'skii, *Mongolia, the Tangut Country, and the Solitudes of Northern Tibet* (London: S. Low, Marston, Searle and Rivington, 1876), pp. 249–250; T. Douglas Forsyth, "On the Buried Cities in the Shifting Sands of the Great Desert of Gobi," *Royal Geographical Society Journal* 47 (1877): 1–17; Rinchen, "Almas: Mongol'skii rodich snezhnogo cheloveka," *Sovremennaya*

*Mongoliya* 5 (1958): 34–38; G. P. Dement'ev and D. Zevegmid, "Une note sur l'homme des neiges en Mongolie," *La Terre et la Vie* 4 (1960): 194–199; Ivan T. Sanderson, *Abominable Snowmen: Legend Come to Life* (Philadelphia: Chilton, 1961), pp. 318–320; Rinchen, "Almas Still Exists in Mongolia," *Genus* 20 (1964): 186–192; Boris A. Porshnev, "Bor'ba za Trogloditov," *Prostor* (Alma-Ata), 1968, no. 4, pp. 98–112, no. 5, pp. 76–101, no. 6, pp. 108–121, no. 7, pp. 109–127, translated into French as pt. 1 of Bernard Heuvelmans and Boris F. Porshnev, *L'homme de Néanderthal est toujours vivant* (Paris: Plon, 1974), see pp. 40–47, 141–142; Odette Tchernine, *The Yeti* (London: Neville Spearman, 1970), pp. 51–62, 177; Myra Shackley, "The Case for Neanderthal Survival: Fact, Fiction, or Faction?" *Antiquity* 56 (1982): 31–41; Michael Heaney, "The Mongolian Almas: A Historical Reevaluation of the Sighting by Baradin," *Cryptozoology* 2 (1983): 40–52; Myra Shackley, *Still Living? Yeti, Sasquatch and the Neanderthal Enigma* (New York: Thames and Hudson, 1983), pp. 91–108, 161–164; Chris Stringer, "Wanted: One Wildman, Dead or Alive," *New Scientist,* August 11, 1983, p. 422; Ra Rabjir, *Almas survalzhilsan temdeglel* (Ulaanbaatar, Mongolia: Ulsyn Khevleliin Gazar, 1990); Ivan Mackerle, *Mongolské Záhady* (Prague: Ivo Zelezny, 2001).

## Almasti

WILDMAN of West Asia.

*Etymology:* Kabardian (Circassian) word, said to mean "forest man." Seemingly derivative of the Mongolian ALMAS; possibly borrowed from the Mongolian-speaking Kalmyks to the north in Kalmykia.

*Variant names:* Almasty, Gubganana (for the female).

*Physical description:* Height, 5–6 feet. Weight, up to 500 pounds. Reddish, shaggy body-hair. Long, tangled head-hair. Slanted and reddish eyes. Flattened nose. Prominent cheekbones. Receding lower jaw. Females have breasts. Short, bowed legs. Splayed feet. Babies are allegedly born pink and hairless but are covered in short hair by the age of one.

*Behavior:* Active primarily at dusk and at night; sleeps in the daytime. Seen most frequently in July and August. Often mumbles. Call is a cry of tremendous power. Extremely bad smell. Omnivorous but primarily vegetarian, liking grasses, especially hemp and corn, and melons and cherries. Also known to eat frogs, lizards, rats, horse dung (possibly for the salt content), and the placenta of domestic animals. Rests in chance refuges in the winter (empty cabins, barns) and makes nests of weeds, rags, leaves, and grass in the summer. Has been observed braiding horse's manes. Sometimes wears tattered clothing around the waist, apparently acquired from local people.

*Habitat:* Remote mountains and woodlands.

*Distribution:* The Russian Caucasus Mountains, from Abkhazia in Georgia and the Kabardin-Balkar Republic in the west, south to Armenia, and east through the Dagestan Republic to Azerbaijan.

*Significant sightings:* Erjib Koshokoyev and other policemen nearly trapped a female in a hemp field in the Caucasus Mountains south of Nal'chik in October 1944.

In 1956, N. Ya. Serikova was staying at a collective farm in the Zolsk area of the Kabardin-Balkar Republic, Russia. She was listening to the sounds of a wedding party next door when an Almasti came into the room, screeched twice, and left the hut, slamming the door behind it. Apparently, it frequented a nearby house, where an old woman had befriended it.

Russian researcher Marie-Jeanne Kofman found a set of tracks in the Dolina Narzanov Valley in the north Caucasus in March 1978.

While he remained hidden in a barn in Kuruko ravine, Kabardin-Balkar Republic, Russia, on August 25, 1991, biologist Gregory Panchenko observed an Almasti enter through a window and plait a horse's mane. The horse did not offer any resistance. After a short time, during which it made high-pitched, twittering sounds, the Almasti departed through an open window above the barn door. Panchenko verified that the horse's mane had new and clumsily plaited braids that were not there the day before.

In the summer of 1992, French filmmaker Sylvain Pallix and Marie-Jeanne Kofman organized an expedition to the Kabardin-Balkar Republic under the auspices of the Russian Society of Cryptozoology to investigate recent Almasti reports. Although the organizers had a falling out, the French team managed to get some fieldwork done with the help of Kabardinian teacher Muaed Mysyrjan. Eyewitness Doucha Apsikova took the team to the place where she had seen an Almasti only a few days previously. Researcher Andrei Kozlov made plaster casts of the footprints found at the site.

*Possible explanations:* Though not a particularly rich region for fossil hominids, the area does have *Homo erectus* (Dmanisi in Georgia), archaic human (Azych in Azerbaijan), and Neanderthal (Sakhazia and Dzhruchula) sites.

*Sources:* Marie-Jeanne Kofman, "Sledy ostaiutsia," *Nauka i Religiia*, 1968, no. 4, pp. 105–124; Odette Tchernine, *The Yeti* (London: Neville Spearman, 1970), pp. 18–23, 159–165; Bernard Heuvelmans and Boris F. Porshnev, *L'homme de Néanderthal est toujours vivant* (Paris: Plon, 1974), pp. 178–190; John Colarusso, "Ethnographic Information on a Wild Man of the Caucasus," in Marjorie Halpin and Michael M. Ames, eds., *Manlike Monsters on Trial* (Vancouver, Canada: University of British Columbia Press, 1980), pp. 255–264; Myra Shackley, *Still Living? Yeti, Sasquatch and the Neanderthal Enigma* (New York: Thames and Hudson, 1983), pp. 109–116; Marie-Jeanne Kofman, "Brief Ecological Description of the Caucasus Relic Hominoid (Almasti) Based on Oral Reports by Local Inhabitants and on Field Investigations," in Vladimir Markotic and Grover Krantz, eds., *The Sasquatch and Other Unknown Hominoids* (Calgary, Alta., Canada: Western Publishers, 1984), pp. 76–86; "Interview: Does a Wildman Exist in the Caucasus? A Soviet Investigator Gives Her Views," *ISC Newsletter* 7, no. 2 (Summer 1988): 1–4; Marie-Jeanne Kofman, "L'Almasty, yeti du Caucase," *Archaeologia*, June 1991, pp. 24–43; Dmitri Bayanov, *In the Footsteps of the Russian Snowman* (Moscow: Crypto-Logos, 1996), pp. 24–31, 39–42, 53–62; Hans-M. Beyer, "With the President in the Caucasus," 1996, http://www.stgr-primates. de/caucasus1996.html; Anatoli Schmidt, Karl

C. Beyer, and Andreas Braun, "The Koffmann-Pallix-Expedition *Almasty 92* in 1992," 1999, http://www.stgr-primates.de/almasty92.html.

## Alovot
Mystery BIRD of Southeast Asia.

*Etymology:* Possibly Simeulue (Austronesian), from *ovot* ("old forest").

*Physical description:* Pheasant the size of a chicken. Dark-brown plumage with lighter spots. Small, comblike crest (perhaps only in one sex). Short legs.

*Behavior:* Nocturnal. Shy and cautious. Feeds on rice. Nests on stumps or logs. Egg is light brown and smaller than a hen's.

*Habitat:* Dense forest. Takes wing with a heavy, low flight when surprised.

*Distribution:* Simeulue Island, Sumatra, Indonesia.

*Possible explanations:*
(1) An unknown species of Peacock-pheasant (*Polyplectron* sp.), suggested by Karl Shuker. The description resembles the Mountain peacock-pheasant (*P. inopinatus*), which lives in undisturbed mountain forests of Malaysia.
(2) An unknown species of Gallopheasant (*Lophura* sp.). The description resembles a female Crested fireback (*L. ignita*), once common in Sumatra and Borneo. Females are brown with white-striped underparts.

*Sources:* Edward Jacobson, "The Alovot, a Bird Probably Living in the Island of Simalur (Sumatra)," *Temminickia* 2 (1937): 159–160; Karl Shuker, "Gallinaceous Mystery Birds," *World Pheasant Association News*, no. 32 (May 1991): 3–6.

## Altamaha-Ha
FRESHWATER MONSTER of Georgia.

*Etymology:* After the river.

*Physical description:* Length, 10–25 feet. Diameter, 10–12 inches. Smooth, gray-brown skin. Small head. Long neck. Two or three humps.

*Behavior:* Swims by undulations.

*Distribution:* Altamaha River, near Darien, Georgia.

*Significant sightings:* On January 16, 1983, Tim Sanders watched a 20- to 25-foot creature from the Champney River bridge.

On July 6, 1997, Jim and Mary Marshall were boating on the river when they saw an animal, 10–12 feet long, with three humps.

*Sources:* Ann Richardson Davis, *The Tale of the Altamaha "Monster"* (Waverly, Iowa.: G&R, 1996); Ann Richardson Davis, The Altamaha-Ha page, http://www.gabooks.com/ altahaha. shtml; Global Underwater Search Team (GUST), Operation River Search, http://www.bahnhof.se/~wizard/cryptoworld/ index30a.html.

## Alula Whale

Unknown CETACEAN of the Indian Ocean.

*Etymology:* After Alula, an alternate name for Caluula, a village on the Gulf of Aden in Somalia.

*Physical description:* Sepia-brown variety of killer whale. Length, 20 feet. High, rounded forehead. White, star-shaped scars. Dorsal fin is about 2 feet high.

*Distribution:* Gulf of Aden.

*Significant sighting:* W. F. J. Mörzer Bruyns watched as many as four of these whales at a time pass by his ship in the eastern Gulf of Aden.

*Possible explanation:* A subspecies of killer whale (*Orcinus orca*).

*Source:* W. F. J. Mörzer Bruyns, *Field Guide of Whales and Dolphins* (Amsterdam: Tor, 1971).

## Alux

LITTLE PEOPLE of Central America.

*Etymology:* Yucatec (Mayan) word. Plural, *Aluxob.*

*Variant names:* Ahlu't (Quiché/Mayan), Aluche (Spanish), A'lus, Barux, Kat.

*Physical description:* Height, 2 feet 6 inches–4 feet. Stout and squat. Disproportionately large head. Long black beard. Powerful muscles. Females have large breasts.

*Behavior:* Said to live in small votary shrines at Mayan temples. Whistles "chuii, chuii." Usually naked but sometimes wears a tunic. Some-times wears a wide-brimmed palm hat. Guards treasure. Has a tiny dog. Pushes people out of their hammocks. Said to inflict fevers. Local people sometimes leave food as a peace offering. Uses a machete. Throws pebbles or stonelike pellets. Carries a shotgun.

*Distribution:* Mayan archaeological sites in Yucatan State, Mexico; Guatemala.

*Sources:* Virginia Rodriquez Rivera, "Los duendes en Mexico (el alux)," *Folklore Americano* 10, no. 10 (1962): 68–85; Rolfe F. Schell, *Yank in Yucatan: Adventures and Guide through Eastern Mexico* (Fort Myers Beach, Fla.: Island Press, 1963); Bill Mack, "Mexico's Little People," *Fate* 37 (August 1984): 38–41; Loren Coleman, *Curious Encounters* (Boston: Faber and Faber, 1985), pp. 47–48, 50–51, 54–55, 57–58; John E. Roth, *American Elves* (Jefferson, N.C.: McFarland, 1997), pp. 96–105; Scott Corrales, "Aluxoob: Little People of the Maya," *Fate* 54 (June 2001): 30–34.

## Amali

Dinosaur-like animal of Central Africa, similar to the MOKELE-MBEMBE.

*Etymology:* Myene (Bantu), "fabulous animal." Linguistically similar to the N'YAMALA.

*Behavior:* Amphibious.

*Tracks:* Three claws. Prints are the size of frying pans.

*Habitat:* Lakes.

*Distribution:* Gabon.

*Significant sighting:* The adventurer Trader Horn allegedly discovered a cave painting of an Amali, chiseled it out, and sent it as a gift to President Ulysses S. Grant. If true, this incident most likely occurred in the early 1880s, after Grant left office.

*Sources:* Trader Horn, *Life and Works,* ed. Ethelreda Lewis (London: Jonathan Cape, 1927), vol. 1, pp. 272–273; Bernard Heuvelmans, *Les derniers dragons d'Afrique* (Paris: Plon, 1978), pp. 250–251, 268.

## Amaypathenya

LITTLE PEOPLE of the southwestern United States.

*Etymology:* Mohave (Hokan), "just like spirits."
*Variant name:* Amatpathenya.
*Physical description:* Height, 2 feet.
*Behavior:* Makes petroglyphs. Practices magic. Can shape-shift.
*Distribution:* Southwestern Arizona; southeastern California.
*Sources:* Kenneth M. Stewart, "The Amatpathenya—Mohave Leprechauns?" *Affword* 3, no. 1 (Spring 1973): 40–41; John E. Roth, *American Elves* (Jefferson, N.C.: McFarland, 1997), pp. 64–70.


## Amhúluk

FRESHWATER MONSTER of Oregon.
*Etymology:* Kalapuya (Penutian) word.
*Physical description:* Spotted. Long horns. Four legs.
*Distribution:* Lake near Forked Mountain, west of Forest Grove, Oregon.
*Sources:* Albert S. Gatschet, "Oregonian Folk-Lore," *Journal of American Folklore* 4 (1891): 139–143; Albert S. Gatschet, "Water-Monsters of American Aborigines," *Journal of American Folklore* 12 (1899): 255–260.


## Amikuk

SEA MONSTER of Alaska.
*Etymology:* Inuktitut (Eskimo-Aleut) word.
*Physical description:* Octopus-like.
*Distribution:* Bering Straits, Alaska.
*Source:* Edward William Nelson, "The Eskimo about Bering Strait," *Annual Report of the Bureau of American Ethnology* 18, pt. 1 (1896–1897): 442.


## AMPHIBIANS (Unknown)

The three types of animals popularly known as amphibians are newts and salamanders (Order Urodela), frogs and toads (Order Anura), and caecilians (Order Gymnophiona), all belonging to the Class Lissamphibia. They have moist skin, no claws, and a layer of fibrous tissue that separates the base and crown of their teeth. Each goes through a primarily aquatic larval stage before metamorphosing into an adult. The earliest modern amphibians arose more than 210 million years ago in the Triassic from an earlier group of amphibians known as lepospondyls.

The largest living amphibian is the Chinese giant salamander (*Andrias davidianus*); one specimen caught in Guizhou Province, China, in 1923 measured 5 feet 9 inches along the curve of its body. The largest frog is the rare Goliath frog (*Conraua goliath*) from Central Africa, which has been measured to an overall length, with legs extended, of 34.5 inches.

Of the six mystery amphibians listed here, three are frogs, one is a salamander, and two are possible caecilians (wormlike, legless animals that burrow into the soil in the tropics).

### Mystery Amphibians

BLUE-NOSED FROG; COLEMAN FROG; MINHOCÃO; MULILO; SAPO DE LOMA; TRINITY ALPS GIANT SALAMANDER


## Andaman Wood Owl

Mystery BIRD of Southeast Asia.
*Habitat:* Dense forest.
*Distribution:* Andaman and Nicobar Islands, Indian Ocean.
*Possible explanation:* An unknown species of Wood owl (*Strix* sp.). Other small owls, including the Andaman scops owl (*Otus balli*) and the Andaman hawk owl (*Ninox affinis*), live on the islands.
*Source:* Sálim Ali and S. Dillon Ripley, *Handbook of the Birds of India and Pakistan, Together with Those of Nepal, Sikkim, Bhutan and Ceylon* (New York: Oxford University Press, 1968–1974), vol. 3.


## Andean Wolf

Unrecognized mountain DOG of South America.
*Scientific name: Dasycyon hagenbecki,* given by Ingo Krumbiegel in 1949.
*Physical description:* Thick, blackish-brown fur. Back hair is 8 inches long. Small, round ears. Strong jaws. Short, solid legs. Powerful claws.
*Distribution:* Andes Mountains, Argentina.
*Significant sightings:* In 1927, Lorenz Hagen-

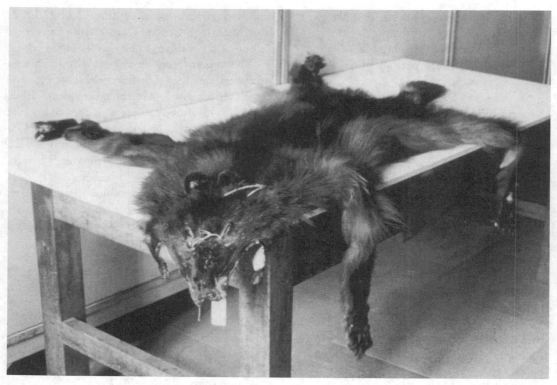

*ANDEAN WOLF pelt at the zoological museum in Munich. (Alan Pringle/Fortean Picture Library)*

beck discovered four pelts for sale in Buenos Aires, Argentina, similar to those of a maned wolf but thicker, darker in color, and with smaller ears.

Ingo Krumbiegel examined an odd skull in 1935, supposedly that of a maned wolf but larger and originating from the Andes Mountains outside the animal's range (lowlands farther to the east).

*Present status:* A 2000 attempt at DNA analysis of the remaining pelt at Munich's zoological museum proved unsatisfactory because it was contaminated with human, dog, wolf, and pig DNA. The pelt had also been chemically treated.

*Possible explanation:* The accidental pairing of a Maned wolf (*Chrysocyon brachyurus*) skull with a German shepherd (*Canis familiaris*) pelt.

*Sources:* Ingo Krumbiegel, "Der Andenwolf: Ein neuentdecktes Grosstier," *Umschau* 49 (1949): 590–591; Ingo Krumbiegel, "Der 'Andenwolf,' *Dasycyon hagenbecki* (Krumbiegel, 1949)," *Säugetierkundliche Mitteilungen* 1

(1953): 97–104; Fritz Dieterlen, "Über den Haarbau des Andenwolfes, *Dasycyon hagenbecki* (Krumbiegel, 1949)," *Säugetierkundliche Mitteilungen* 2 (1954): 26–31; Angel Cabrera, "Catalogo de los mamiferos de America del Sur, 1. (Metatheria—Unguiculata—Carnivora)," *Revista Museo Argentino de Ciencias Naturales Rivadavia (Zoologia)* 4 (1957): 1–307; Bernard Heuvelmans, *On the Track of Unknown Animals* (New York: Hill and Wang, 1958), pp. 68–69; Karl Shuker, "Pity about the Pelt," *Fortean Times,* no. 145 (May 2001): 23.

## Anfish

FRESHWATER MONSTER of the Middle East.

*Etymology:* Madan (Marsh Arab) word.

*Physical description:* Hairy skin.

*Distribution:* Marshes at the mouth of the Tigris River, Iraq.

*Source:* Wilfred Thesiger, *The Marsh Arabs* (New York: Dutton, 1964), p. 115.

## Angeoa

FRESHWATER MONSTER of northern Canada.

*Etymology:* Inuktitut (Eskimo-Aleut) word.

*Physical description:* Length, 50–60 feet. Black. Enormous fin.

*Distribution:* Dubawnt Lake, Nunavut.

*Significant sighting:* An Inuit man told Farley Mowat in the 1940s that his father had encountered the Angeoa at the end of the nineteenth century. It overturned his kayak, killing his companion.

*Source:* Farley Mowat, *People of the Deer* (Boston: Little, Brown, 1952), pp. 313–316.

## Angont

FRESHWATER MONSTER of Ontario and Michigan.

*Etymology:* Huron (Iroquoian) word.

*Physical description:* Serpentine. Horns. Flaming eyes.

*Behavior:* Venomous flesh, said to be used by the Indians as a poison.

*Distribution:* Southern Ontario around Georgian Bay; a pool near the Huron River, Michigan.

*Sources:* Paul Ragueneau, *Relation de ce qui s'est passé en la mission des peres de la Compaigne des Jesus aux Hurons, pays de la Nouvelle France, ès années 1647 et 1648,* in *Rélations des Jésuites* (Québec, Canada: A. Coté, 1858), vol. 2, pp. 45–82; Horatio Hale, "Huron Folklore," *Journal of American Folklore* 2 (1889): 249–254.

## Anka

Giant BIRD of West Asia.

*Etymology:* Turkish word, originally from the Arabic *'anká'.* (Incidentally, *anka* means "sparrow hawk" in the Quechuan language of the Andes.)

*Variant names:* Angka, Anka kuşu, Anka-mogrel, Zümrüt anka.

*Physical description:* Female has eight wings. Male is multicolored with a white ring around its long neck. Wingspan is the breadth of five elephants (roughly 20 feet).

*Behavior:* Terrifying call. Preys on large mammals, birds, and humans. Carries off children.

*Habitat:* High mountain peaks.

*Distribution:* Caucasus Mountains, Russia.

*Significant sightings:* The prophet Hanzala ibn Safwan saved his people by killing the Anka, to whom a youth was sacrificed every day.

An Anka was said to have been housed in the zoological gardens of the Fatimid dynasty in the eleventh century (possibly in the capital at al-Mahdiyah, Tunisia).

*Possible explanation:* Based originally on a species of heron in Egypt, possibly a surviving Giant heron (*Ardea bennuides*), which may have inspired the BENNU BIRD symbol.

*Present status:* After the introduction of Islam, the Anka became associated with the SIMURGH.

*Sources:* M. Th. Houtsma et al., eds., *The Encyclopaedia of Islam* (Leiden, the Netherlands: E. J. Brill, 1913), vol. 1, p. 356, and new ed., 1960, vol. 1, p. 509; Joe Nigg, *A Guide to the Imaginary Birds of the World* (Cambridge, Mass.: Apple-Wood, 1984), pp. 45–47.

## Anomalous Jaguar

Unknown CAT of South America.

*Physical description:* The size of a jaguar. Cinnamon-brown and white background color. Covered with solid black, irregular spots.

*Distribution:* Central Peru.

*Significant sighting:* Peter Hocking obtained the skull of a specimen shot in 1993 in the Yanachaga Mountains, Peru.

*Possible explanation:* Undescribed color morph of a Jaguar (*Panthera onca*).

*Source:* Peter J. Hocking, "Further Investigation into Unknown Peruvian Mammals," *Cryptozoology* 12 (1996): 50–57.

## Antarctic Killer Whale

Mystery CETACEAN of the Antarctic.

*Scientific name: Orcinus glacialis,* given by A. A. Berzin and V. L. Vladimirov in 1983.

*Distribution:* Antarctic waters.

*Significant sighting:* Recorded in 1980 in Prydz Bay in the Indian Ocean sector of Antarctica.

*Possible explanation:* Generally considered a subspecies of the Killer whale (*Orcinus orca*).

*Sources:* A. A. Berzin and V. L. Vladimirov,

"Nov'ĭ vid kosatki (Cetacea, Delphinidae) iz vod antarktiki," *Zoologicheskii Zhurnal* 62 (1983): 287–295; Michael A. Bigg et al., *Killer Whales: A Study of Their Identification, Genealogy, and Natural History in British Columbia and Washington State* (Nanaimo, B.C., Canada: Phantom, 1987).

## Antarctic Long-Finned Whale
Unknown CETACEAN of the Antarctic.

*Physical description:* Length, 20–30 feet. Black. Long, erect, slightly curved dorsal fin situated toward the tail.

*Distribution:* Antarctic waters.

*Significant sightings:* In 1841, commander of the *Erebus* James C. Ross and surgeon Robert McCormick reported seeing a high-finned whale in the Ross Sea off Ross Island, Antarctica.

Zoologist Edward A. Wilson observed groups of similar cetaceans on January 28 and February 8, 1902, during Robert Scott's *Discovery* expedition to the Antarctic. They were black, with some white around the mouth or chin. The dorsal fins were 3–4 feet long and sabre-shaped.

Cetologist Robert Clarke and colleagues logged eight sightings of a high-finned mystery whale about 20 feet long off the coast of Chile, November 24–27, 1964.

*Sources:* James Clark Ross, *A Voyage of Discovery and Research in the Southern and Antarctic Regions, during the Years 1839–43,* vol. 1 (London: John Murray, 1847); Edward Adrian Wilson, *Mammalia (Cetacea & Pinnipedia)* (London, 1907), pp. 4–5; Robert Clarke, Anelio Aguayo L., and Sergio Basulto del Campo, "Whale Observation and Whale Marking off the Coast of Chile in 1964," *Scientific Report of the Whales Research Institute,* no. 30 (1978): 117–177; Darren Naish, "Multitudinous Enigmatic Cetaceans, or 'Whales in Limbo,'" *Animals and Men,* no. 11 (December 1996): 28–34.

## Antipodes
WILDMAN of North Africa or India.

*Etymology:* Greek, "feet on the opposite side."

*Variant name:* Opisthodactyles.

*Physical description:* Feet said to point backward.

*Present status:* Interpreted both as men with feet pointing backward (toes to the rear) and as men on the opposite side of the world (whose feet would be pointing toward us through the earth). Significant for cryptozoology in that many WILDMEN are said to have their feet pointing backward.

*Sources:* Isidore of Seville, *Origines,* XI. 3, 24; Bernard Heuvelmans, *Les bêtes humaines d'Afrique* (Paris: Plon, 1980), pp. 149–150, 162.

## Apris
Venomous SNAKE of East Africa.

*Etymology:* Somali (Cushitic) word.

*Behavior:* So venomous that merely touching it causes death within seconds.

*Distribution:* Somalia.

*Possible explanation:* Positively identified as the East African sand boa (*Gongylophis colubrinus*), a nonvenomous, orange or yellow snake with chocolate-brown to black splotches.

*Source:* Stephen Spawls, *Sun, Sand, and Snakes* (London: Collins, 1979).

## Apsarās
MERBEING of Hindu mythology.

*Etymology:* Sanskrit (Indo-Aryan), "moving in water."

*Physical description:* Not fish-tailed. Depicted as a voluptuous, large-hipped young woman.

*Behavior:* Both aquatic and celestial. Has a sweet fragrance. Enjoys singing and dancing. Skilled lute and cymbal player. Often carries lotus flowers. Able to foretell the future. Promiscuous.

*Distribution:* India; Cambodia.

*Source:* Gwen Benwell and Arthur Waugh, *Sea Enchantress* (London: Hutchinson, 1961), pp. 31–32.

## Arabian Flying Snake
FLYING REPTILE of the Middle East.

*Physical description:* Looks like a water snake. Has wings like a bat.

*Behavior:* Flies from Arabia to Egypt in the spring. Attacked by ibises.

*Significant sighting:* In the fourth century B.C., Herodotus claimed to have seen their skeletons in large numbers at a place called Buto between Egypt and Arabia.

*Possible explanations:*

(1) The Desert locust (*Schistocerca gregaria*) is a large, flying grasshopper of Africa and Asia that periodically forms massive swarms hundreds of miles long, causing enormous crop damage. Though each locust is only 2.25 inches long, such a swarm would leave behind a large number of dead insects.

(2) The fossil bones of *Spinosaurus,* a large theropod dinosaur of the Late Cretaceous, 95 million years ago. The type specimen was first described in 1915 near Marsá Matruh, Egypt, on the Mediterranean coast. Its most striking feature is a set of dorsal spines that probably supported a sail-like membrane.

*Sources:* Herodotus, *The Histories,* ed. John Marincola (New York: Penguin, 1996), pp. 112–113 (II. 75–76); Adrienne Mayor, *The First Fossil Hunters: Paleontology in Greek and Roman Times* (Princeton, N.J.: Princeton University Press, 2000), pp. 135–136.

## Archie

FRESHWATER MONSTER of Scotland.

*Physical description:* Similar to an otter. Horselike head.

*Behavior:* Basks on the surface.

*Distribution:* Loch Arkaig, Highland, Scotland.

*Sources:* James Howard Harris, third earl of Malmesbury, *Memoirs of an Ex-Minister: An Autobiography* (London: Longmans, Green, 1885), pp. 406–407; Herbert Maxwell, "Highland Superstitions," *The Field,* February 3, 1934, p. 289.

## Arizona Jaguar

Subspecies of jaguar of uncertain status in Arizona.

*Scientific name: Panthera onca arizonensis.*

*Present status:* Hunted nearly to extinction by 1905, though breeding populations persisted at least through the 1940s and probably until the 1960s. Confirmed jaguar sightings took place in 1971 (east of Nogales), 1986 (Dos Cabezas Mountains), 1996 (Baboquivari Mountains), and 2001 (near the Mexican border), with scattered, unconfirmed reports occurring almost annually. Whether this constitutes a breeding population or occasional stragglers from Mexico is uncertain. The subspecies was declared endangered in the United States in 1997.

*Sources:* U.S. Fish and Wildlife Service, *Endangered and Threatened Wildlife and Plants: Final Rule to Extend Endangered Status for the Jaguar in the United States,* 50 CFR Part 17 (July 22, 1997), on line at http://endangered. fws.gov/r/fr97622.html; Mitch Tobin, "Wandering Jaguar Shakes Things Up," *Tucson Arizona Daily Star,* February 24, 2002.

## Artrellia

Dragonlike LIZARD of Australasia.

*Etymology:* Papuan (Austronesian), "tree dragon."

*Variant names:* AU ANGI-ANGI, Kaiaimunu, Rharhru.

*Physical description:* A huge lizard. Length, 12–40 feet.

*Behavior:* Arboreal. Said to eat humans.

*Distribution:* Papua New Guinea.

*Significant sightings:* In the 1930s, members of the American Museum of Natural History's Archbold Expedition to the interior of Papua New Guinea were told of a man-eating dragon.

Robert Grant and David George were exploring the Strachan Island District in 1961 when they encountered a gray lizard about 26 feet long. Its neck was more than 3 feet long.

An animal captured in swampland near the Gulf of Papua in 1980 by the Operation Drake Expedition proved to be a juvenile specimen of crocodile monitor. Its length has been variously reported as 6 feet 6 inches and 7 feet 3 inches long. One of the group's zoologists, Ian Redmond, later sighted a 12-foot animal.

*Possible explanation:* The Crocodile monitor (*Varanus salvadorii*) has curved teeth, elongated toes, strongly curved sharp claws, and rubbery pads on the bottom of the feet. Its long tail is up to two and a half times the length of the head and body. The color is black, with white speckles. The animal may regularly grow over 12 feet, making it the longest lizard in the world. One specimen was measured unofficially at 15 feet 7 inches, and there have been rumors of individuals 20 feet long.

*Sources:* David M. Davies, *Journey into the Stone Age* (London: Robert Hale, 1969); John Blashford-Snell, *Mysteries: Encounters with the Unexplained* (London: Bodley Head, 1983); John Blashford-Snell and Ann Tweedy, *Operation Raleigh: Adventure Challenge* (London: Collins, 1988); Mark K. Bayless, "The Artrellia, Dragon of the Trees: Meet New Guinea's Crocodile Monitor (*Varanus salvadorii*)," *Reptiles* 6 (June 1998): 32–47; Rex Gilroy, "Giant Lizards of the Australian Bush," *Australasian Ufologist* 4, no. 4 (2000): 17–20.

## Ashuaps

FRESHWATER MONSTER of Québec, Canada.

*Etymology:* Short for the Ashuapmouchouan River, which flows into Lac St.-Jean.

*Physical description:* Length, 30–100 feet.

*Distribution:* Lac St.-Jean, Québec.

*Significant sightings:* In 1978, Marcel Tardif and his wife saw a blackish animal, 50–60 feet long, off Scott Point. The same day, something overturned a canoe paddled by Michel Verreault, his wife, and daughter.

*Possible explanation:* California sea lions (*Zalophus californianus*) were released into the lake in 1975, though these only reach 8 feet in length.

*Source:* Michel Meurger and Claude Gagnon, *Lake Monster Traditions: A Cross-Cultural Analysis* (London: Fortean Tomes, 1988), pp. 98–103.

## Atahsaia

CANNIBAL GIANT of the southwestern United States.

*Etymology:* Zuni, "cannibal demon."

*Distribution:* Central New Mexico.

*Source:* Kyle Mizokami, Bigfoot-Like Figures in North American Folklore and Tradition, http://www.rain.org/campinternet/bigfoot/bigfoot-folklore.html.

## Atchen

CANNIBAL GIANT of eastern Canada.

*Etymology:* Montagnais and Atikamekw (Algonquian) word.

*Variant names:* Atcen, Atcheme, Atshen, Kokotshc, Mistabew.

*Behavior:* Bad-tempered, though the Mistabew is said to be benevolent.

*Distribution:* North-central Québec.

*Sources:* Joseph E. Guinard, "Witiko among the Tête-de-Boule," *Primitive Man* 3 (1930): 69–71; Michael Bradley, "Quebec Sasquatches, a Brief Note," *Pursuit*, no. 35 (Summer 1976): 66.

## Atlas Bear

Small BEAR of North Africa.

*Scientific name:* Ursus arctos crowtheri, proposed by Heinrich Schinz in 1844.

*Physical description:* Smaller than the American black bear (*Ursus americanus*) but more robust. Black or dark-brown hair, 4–5 inches long. Pointed muzzle. Short toes and claws.

*Behavior:* Feeds on roots and fruits.

*Distribution:* Atlas Mountains, Morocco; Djurdjura Mountains, Algeria.

*Significant sightings:* A specimen, described by Edward Blyth, was killed south of Tetouán, Morocco, in 1834 but was not preserved.

Jules René Bourguignat discovered fresh bear bones in caves in the Djurdjura Mountains, Algeria, in 1867 and collected stories about the living animal from local people.

*Present status:* Probably extinct.

*Possible explanations:*

(1) A surviving small subspecies of the Brown bear (*Ursus arctos faidherbi*) that lived in North Africa during the Pleistocene and later. The remains of one dating from A.D. 420–600 were found in a cave in the Djurdjura Mountains, Algeria, in 1989.

(2) Verbal accounts of the Striped hyena (*Hyaena hyaena*), which in Arabic is called

*dubbah,* might be misinterpreted as referencing a bear (Arabic, *debb*).

*Sources:* Edward Blyth (letter), *Proceedings of the Zoological Society of London* 9 (1841): 63–65; Heinrich Schinz, *Systematisches Verzeichniss aller bis jetzt bekannten Säugethiere* (Solothurn, Switzerland: Jent und Gassmann, 1844–45); Henri Aucapitaine, "Sur la question de l'existence d'ours dans les montagnes de l'Afrique Septentrionale," *Comptes Rendus de l'Académie des Sciences* 50 (1860): 655–656; Jules René Bourguignat, *Histoire du Djebel-Thaya et des ossements fossiles recueilles dans la grande Caverne de la Mosquée* (Paris: Challamel Aîné, 1870); Watik Hamdine, Michel Thevenot, and Jacques Michaux, "Histoire récente de l'ours brun au Maghreb," *Comptes Rendus de l'Académie des Sciences (Sciences de la Vie)* 321 (1998): 565–570.

## Atnan

LITTLE PEOPLE of western Canada.

*Etymology:* Carrier (Na-Dené), "little people."

*Variant names:* Etna-na-yaz, Kannawdzets.

*Physical description:* Height, 1–3 feet. Long hair.

*Behavior:* Fast runner. Wears a robe in the winter. Uses stones as weapons.

*Distribution:* François Lake and Nechako River area, central British Columbia.

*Present status:* Killed off by the Indians long ago.

*Sources:* Diamond Jenness, "Myths of the Carrier Indians of British Columbia," *Journal of American Folklore* 47 (1934): 97, 247–248; John E. Roth, *American Elves* (Jefferson, N.C.: McFarland, 1997), pp. 115–119.

## Atúnkai

FRESHWATER MONSTER of Oregon.

*Etymology:* Kalapuya (Penutian) word.

*Physical description:* Looks like a seal or sea otter.

*Distribution:* Northwestern Oregon.

*Source:* Albert S. Gatschet, "Water-Monsters of American Aborigines," *Journal of American Folklore* 12 (1899): 255–260.

## Au Angi-Angi

Large LIZARD of Australasia.

*Etymology:* Papuan (Austronesian) word.

*Physical description:* Length, 20 feet. Width, 6 feet. Brown and green, crocodile-like skin. Cowlike head. Large eyes. Long, sharp teeth. Long neck. Two small forelegs. Large, triangular scoops on the back. Thick hind legs. Long, slender tail.

*Behavior:* Amphibious. Bipedal. Said to act aggressively with fishing boats.

*Distribution:* Lake Murray, Papua New Guinea.

*Significant sighting:* A dinosaur-like reptile was spotted by two groups of people near Boboa, Lake Murray, Papua New Guinea, December 11–12, 1999.

*Present status:* Reports of smaller, arboreal lizards in the same area may refer to the Crocodile monitor (*Varanus salvadorii*). See ARTRELLIA.

*Possible explanation:* Exaggerations or hoaxes based on pictures of dinosaurs.

*Sources:* Bernard Heuvelmans, "An Annotated Checklist of Apparently Unknown Animals with Which Cryptozoology Is Concerned," *Cryptozoology* 5 (1986): 1–26; "Dinosaur-Like Reptile Sighted at Lake Murray," *Boroko Independent* (Papua New Guinea), December 30, 1999.

## Auli

Unknown SIRENIAN of East and Central Africa.

*Variant names:* Aila, Ia-bahr-tedcha ("water calf"), Mama fouta, Ourha-bieh.

*Physical description:* Size of a sheep.

*Behavior:* Drags its hind feet like a seal.

*Distribution:* Lake T'ana in Ethiopia; tributaries of the Mereb Wenz, Eritrea; Lake Chad, the Dagana Marshes, and the Ounianga Lakes in Chad.

*Significant sighting:* In 1900 or 1901, Dr. Auguste Morel was traveling in the Dagana marshes, Chad, with local fishermen in support of French colonial troops pacifying the region. A huge animal that left a tremendous disturbance in the water nearly upset their boats. Morel found a large area of crushed reeds but no tracks. The locals had no idea what the animal was.

*Possible explanations:*
(1) An extended range of the West African manatee (*Trichechus senegalensis*) is possible in Chad, though less likely in Ethiopia. Adults are generally 9–10 feet long. The animal is found in rivers, estuaries, swamps, and lagoons from the Senegal River in the north to the Cuanza River, Angola, in the south, and it occurs as far as 1,200 miles from the sea along the Niger River. Its presence in the Chari River and Lake Chad has been suspected but never confirmed.
(2) An unknown species of manatee or freshwater dugong that evolved from fossil forms found in Egypt.

*Sources:* Theodor von Heuglin, *Reise nach Abessinien, den Gala-Ländern, Ost-Sudán und Chartúm in den Jahren 1861 und 1862* (Jena, Germany: H. Costenoble, 1868), pp. 247, 289; Édouard-Louis Trouessart, "Existe-t-il dans les marais du lac Tchad un grand mammifère encore inconnu des naturalistes?" *La Nature* 76 (January 21, 1911): 120–121; *Neue Mannheimer Zeitung,* September 6, 1938, p. 7; Nigel Heseltine, *From Libyan Sands to Chad* (London: Museum Press, 1960), pp. 128–129; Bernard Heuvelmans, *Les derniers dragons d'Afrique* (Paris: Plon, 1978), pp. 142–144, 277–279, 358–363.

## Australian Big Cat

ALIEN BIG CAT of Australasia. Probably not a marsupial and similar but not identical to the leopard and puma, neither of which have been introduced on the continent.

*Variant names:* Briagolong tiger (Victoria), Brookton tiger (Western Australia), Cordering cougar (Western Australia), Emmaville panther (New South Wales), Jamberoo tiger (New South Wales), Kaiapoi tiger (New Zealand), Kangaroo Valley panther (New South Wales), Kingstown killer (New South Wales), Marulan tiger (New South Wales), Nightgrowler, Tallong tiger (New South Wales), Tanjil terror (Victoria), Tantanoola tiger (South Australia), Tantawanglo tiger (New South Wales), Warialda cougar (New South Wales), WARRIGAL.

*Physical description:* There seem to be two primary Australian varieties: about 60 percent of the incidents involve a jet-black, leopardlike cat, while the other 30 percent describe a light-brown felid resembling a North American puma. The leopardlike cat is a solid jet-black color, with powerful muscles; it is the size of a German shepherd dog or slightly larger. The pumalike cat is sandy-colored or fawn-gray; it is 3–4 feet long, with a tail of equal length and has a shoulder height of 2 feet 6 inches; white bands around the tail are occasionally seen. A maned variety has been reported in the Blue Mountains (WARRIGAL).

*Behavior:* Nocturnal. Most reports are of single animals, with only a few involving a female and cubs or an adult pair. Can run at great speed for long distances. Gives out terrifying howls and shrieks, especially at night. A solitary hunter, it kills sheep by biting the neck or choking them, unlike the messy kills of feral dogs or dingos. The sheep's internal organs and most of the bones (except the ribs) are often consumed, either by eating through a hole in the groin or peeling the entire skin back. Heavy carcasses are often moved elsewhere before being eaten. Dogs seem particularly terrified of them. Not afraid of humans or cars.

*Tracks:* Four-toed. Claw marks occasionally visible. Up to 5.5 inches in diameter. Some casts are said to closely match puma tracks. Leaves scratch marks in gum trees.

*Distribution:* Eastern New South Wales; western Victoria; southwestern Western Australia. Scattered reports elsewhere, including New Zealand.

*Significant sightings:* Tony Healy and Paul Cropper estimate they have collected more than 1,000 reports from 1885 to 1994 in every Australian state except Tasmania. Two photographs, both taken in Western Australia (by Barry Morris in 1978 and Alan Lawrence in 1982) only show silhouettes and are inconclusive.

A striped animal killed many sheep in the area around Tantanoola, South Australia, from 1893 to 1895, when an unusual-looking dog was killed. This stuffed and not very fearsome animal is still on display in the Tiger Hotel in Tantanoola. Descriptions of the beast were vague, and it's uncertain whether or not a THY-LACINE or big cat was involved.

A large, jet-black cat was seen prowling the hills around Jamberoo, New South Wales, in 1909.

A striped cat was held responsible for sheep killings in the area around Marulan and Tallong, New South Wales, between 1927 and 1930.

A big cat that could allegedly eat an adult sheep in one sitting was investigated by Fisheries and Game Officer Rod Estoppey near Briagolong, Victoria, from the mid-1930s to the mid-1950s.

A large, black, leopardlike animal was reported around Emmaville in the New England area of New South Wales, from 1956 to 1962, with comebacks in 1969, 1973, and 1995. Some incidents were also reported in the region before World War II. Known as the Emmaville panther, it was held responsible for many nocturnal sheep killings. During 1956 and 1957, some 340 sheep were killed on a single ranch owned by Clive Berry. The case was declared solved on at least two occasions after the killing of a large black boar and an old hairless dingo, but the depredations continued.

An odd carnivore was responsible for killing many sheep near Brookton, Western Australia, for two years in the 1960s. Hunter Harry Butler shot it, and it turned out to be a beat-up and scalped dingo that had lost its tail and left deformed tracks.

In 1969 at Byaduk, Victoria, Les Rentsch watched a pumalike cat with a glistening, silver-gray coat for five to six minutes. It had two large upper fangs.

In September 1972, George Moir of Kulja, Western Australia, found several of his piglets dead, with their hearts torn out and their throats ripped open. He also watched two black animals with long tails rounding up his sheep. Moir and a game warden chased them for 5 miles but could not catch them.

A black panther was seen by thirty-two witnesses around Cambewarra Mountain, New South Wales, in June 1975. Leopardlike tracks were examined by retired naval officer Raymond Noakes, and cows, dogs, chickens, goats, and sheep were reported missing or mutilated.

A woman reported a large cat around July 10, 1977, in the Kaiapoi area, South Island, New Zealand. Pawprints and droppings, but little else, were found on July 21 at Pines Beach.

A large, pumalike cat has been reported near Cordering, Western Australia, by many ranchers since 1977. It apparently could kill sheep with surgical precision. Many of them were not eaten, but those that were had their skins peeled back and the ribs stripped of all meat. Kangaroos were also found killed by puncture wounds to the head.

Peter Bruem observed a black, leopardlike cat and a brown, pumalike cat running together near Bendeela, New South Wales, in the summer of 1979. He waited in the shade to see whether they would return and they did, approaching within 100 yards.

A large, black, catlike animal was reported frequently in the Kangaroo Valley area, New South Wales, between 1968 and 1981. It gained particular notoriety when it killed a valuable pony near Budgong in June 1981. The case was declared "solved" twice, when a feral cat (in 1977) and melanistic wallaby (in 1981) were captured.

Norwegian zoologist Per Seglen encountered a dark, leopardlike animal near Badgingarra National Park, Western Australia, on August 21, 1982. It had a long, spotted or heavily striped tail.

Large brown or black, pumalike cats have been responsible for livestock depredations in the Grampians Mountain Range, Victoria, since the 1940s. Reports increased dramatically around 1969 and remained steady though the 1970s and 1980s. Rob Wallis saw a black, muscular cat near Moyston in August 1989 as it crossed the road in front of his vehicle. He estimated it was 8 feet long including the tail and weighed 250 pounds. He located its tracks the next morning and made a plaster cast of one clear track that resembled the print of a smallish puma, although claw marks were visible.

*Present status:* In 1987, the Victorian government added pumas to the list of predators that are known to attack livestock.

*Possible explanations:*

(1) Surviving Marsupial lion (*Thylacoleo*

*carnifex*), a leopard-sized, arboreal, carnivorous marsupial that lived as recently as 10,000 years ago. However, both its front and hind paws were fingerlike (with pseudo-opposable thumbs bearing a huge claw) and designed for climbing trees, and they would not have left anything resembling cat tracks behind. Its dentition was odd—it lacked canine teeth, compensating with huge incisors and two pairs of bladelike carnassial teeth that made it look more like a fierce badger than a panther. Its shoulder height was in the neighborhood of 2 feet–2 feet 6 inches.

(2) Imported black Leopards (*Panthera pardus*) that escaped from zoos or were brought as U.S. regimental mascots during World War II. Leopards are about 3 feet 6 inches–4 feet long, with a 2 feet 6 inch tail. They stand about 2 feet at the shoulder. Melanism in leopards is common in India and Southeast Asia. However, Australian big cat witnesses have never reported spots, which are visible in black specimens in bright daylight. Leopards are also not known for widespread slaughter of livestock—they generally kill only what they need to survive. Nor can they sustain a long-distance run. Also, black and tawny cats were being hunted in Victoria as long ago as the 1880s, long before World War II.

(3) Descendants of one or more Pumas (*Puma concolor*) supposedly brought as regimental mascots by U.S. forces during World War II or otherwise imported. The puma's coloration varies from sandy-brown to silver-gray, with a whitish belly. Melanistic pumas are virtually unknown. Its length is 3 feet 6 inches–4 feet 6 inches, with a 3 feet–3 feet 6 inch tail, tipped with dark brown. The average height at the shoulder is 2 feet 6 inches. Average weight is 80–200 pounds. Its eyes shine greenish-gold. These animals are excellent jumpers but cannot run long distances, and they are shy and elusive by nature. Some reports occurred in Australia prior to the 1940s.

(4) Dingos (*Canis familiaris* var. *dingo*), feral Australian dogs descended from early domesticated dogs brought to the continent by the Aborigines, are reddish-brown. They hunt larger animals in packs, not singly, and their kills are messy, with signs of a protracted struggle. Dingo tracks might be mistaken for cat tracks under poor conditions.

(5) A population of feral Domestic cats (*Felis silvestris catus*) that have grown large. Most ferals (except those in the Gibson and Simpson Deserts, which are said to be up to 3 feet long) are no bigger than house cats, however, and revert to a tabby pattern after a few generations in the wild. Adult pumas are seven times the weight of the heaviest recorded Australian feral.

(6) An unknown species of marsupial carnivore, suggested by Rex Gilroy.

(7) A mainland population of TASMANIAN DEVIL could account for some of the smaller black animals, especially in Victoria.

*Sources: Sydney Bulletin,* May 4, 1895, p. 24; Gilbert Whitley, "Mystery Animals of Australia," *Australian Museum Magazine* 7 (1940): 132–139; Neville Bonney, *The Tantanoola Tiger* (Blackwood, S. Australia: Lynton, 1976); Bruce L. Owens, "The Strange Saga of the Emmaville Panther," *Australian Outdoors and Fishing,* April 1977, pp. 17–19; Paul Cropper, "The Panthers of Southern Australia," *Fortean Times* 32 (Summer 1980): 18–21; David O'Reilly, *Savage Shadow: The Search for the Australian Cougar* (Perth, W. Australia: Creative Research, 1981); Karl Shuker, *Mystery Cats of the World* (London: Robert Hale, 1989), pp. 222–230; Tony Healy and Paul Cropper, *Out of the Shadows: Mystery Animals of Australia* (Chippendale, N.S.W., Australia: Ironbark, 1994), pp. 57–97; Rex Gilroy, *Mysterious Australia* (Mapleton, Queensl., Australia: Nexus, 1995); Malcolm Smith, *Bunyips and Bigfoots: In Search of Australia's Mystery Animals* (Alexandria, N.S.W., Australia: Millennium Books, 1996), pp. 116–142.

## Australian Camel

A native population of camel, a HOOFED MAMMAL, said to exist in Australia prior to 1840.

*Variant name:* Big one bullocky.

AUSTRALIAN GIANT MONITOR *seen in 1979 by herpetologist Frank Gordon in the Wattagan Mountains, New South Wales. (William M. Rebsamen/Fortean Picture Library)*

*Distribution:* Northern Territory, Australia.

*Significant sighting:* A solitary camel was occasionally seen by Aborigines in the north, perhaps as early as the 1830s. It was rumored to have been brought by an early white settler.

*Possible explanation:* It is possible that a few camels were brought to Australia prior to the beginning of official and typically strict importation records. The first camel officially imported came from the Canary Islands in 1840. The next major importation of Dromedaries (*Camelus dromedarius*) took place in 1860 for the ill-fated Bourke and Wills Expedition to Northern Australia. Between 1860 and 1907, an estimated 10,000–12,000 camels were imported into Australia for use as draft and riding animals in the dry interior, especially in the goldfields. By 1930, they had been replaced by motor vehicles, and most had escaped or been released into the wild. The establishment of small, naturalized camel herds over the years demonstrates that these animals could adapt readily to the terrain and the climate. The current population is variously estimated at 150,000–300,000, with approximately 50 percent in Western Australia, 25 percent in the Northern Territory, and 25 percent in western Queensland and northern South Australia.

*Sources:* E. Lloyd, *A Visit to the Antipodes, with Some Reminiscences of a Sojourn in Australia, by a Squatter* (London: Smith, Elder, 1846), pp. 140–141; Calamunnda Camel Farm: Information, History, and Facts on Camels, http://camelfarm.com/camel_information.html.

## Australian Giant Monitor

Unknown LIZARD of Australia.

*Variant names:* Burrunjor (in Northern Territory), Mungoon-galli, Murra murri (in the Blue Mountains), Whowie (in Riverina).

*Physical description:* Length, 20–30 feet or more.

*Behavior:* Attacks cattle.

*Distribution:* Northern New South Wales; Arnhem Land, Northern Territory; Cape York, Queensland.

*Significant sightings:* In 1975, a group of bushwalkers found large tracks and tail marks at the edge of the Wallangambe Wilderness in the Blue Mountains of New South Wales.

On December 27, 1975, a farmer near Cessnock, New South Wales, saw a bulky, 30-foot monitor lizard moving through scrub brush. It was mottled gray in color, with dark stripes along the back and tail, and stood 3 feet off the ground.

In early 1979, herpetologist Frank Gordon was driving his Land Rover in the Wattagan Mountains in New South Wales south of Canberra when he saw a reptile 27–30 feet long by the side of the road. It rose up and ran away on all four legs into the neighboring woods.

In July 1979, cryptozoologist Rex Gilroy was called to a freshly plowed field by a farmer. Across the field were thirty or so tracks that seemed to have been made by an enormous lizard. While most of the tracks had been ruined by rain, Gilroy was able to make a plaster cast of one that had been preserved.

*Possible explanations:*

(1) The Perentie (*Varanus giganteus*), Australia's largest lizard, grows to 8 feet long; some individuals might attain 10 feet. It is cream-colored, with dark-brown speckles, and it occurs from western Queensland to the coast of Western Australia.

(2) Surviving *Megalania prisca*, a 15- to 21-foot lizard that lived in central Australia in the Pliocene and Pleistocene (2 million–20,000 years ago). At 1,300 pounds, it weighed ten times as much as the Komodo dragon (*Varanus komodoensis*) and was probably an active predator and scavenger. Its teeth were nearly 1 inch long. At least some specimens had a sagittal crest.

*Sources:* Rex Gilroy, "Cessnock's Fantastic 30 Ft. Lizard Monsters," *Strange Phenomena and Psychic Australian,* March 1979, at http://www.internetezy.com.au/~mj129/strangephenomenonr.html; Rex Gilroy, "Australia's Lizard Monsters," *Fortean Times,* no. 37 (Spring 1982): 32–33; Rex Gilroy, "Giant Lizards of the Australian Bush," *Australasian Ufologist* 4, no. 4 (2000): 17–20.

## Aypa

WATER TIGER of South America.

*Physical description:* Covered in scales (or glossy fur). Head and neck like a tiger's. Extremely large teeth.

*Behavior:* Aquatic.

*Distribution:* Serra de Tumucumaque, Amapá State, Brazil.

*Source:* René Ricatte, *De l'Île du Diable aux Tumuc-Humac* (Paris: La Pensée Universelle, 1978).

## Aziza

LITTLE PEOPLE of West Africa, said to have given the knowledge of magic to humans.

*Distribution:* Benin.

*Source:* Bernard Heuvelmans, *Les bêtes humaines d'Afrique* (Paris: Plon, 1980), p. 496.

# B

## Badak Tanggiling

One-horned, rhinoceros-like HOOFED MAMMAL of Southeast Asia.

*Etymology:* Malay (Austronesian) word.

*Variant name:* Scaled rhinoceros.

*Physical description:* Length, 10 feet; slightly larger than the Sumatran rhino. Only one horn. The female is sometimes hornless.

*Distribution:* Sumatra, Indonesia.

*Significant sighting:* The hunter J. C. Hazewinkel shot eight of these animals in the 1920s.

*Present status:* The only currently known species of rhino in Sumatra is the two-horned Sumatran rhino (*Dicerorhinus sumatrensis*).

*Possible explanation:* The single-horned Javan rhinoceros (*Rhinoceros sondaicus*) may have persisted in Sumatra until the 1940s.

*Sources:* Joseph Delmont, *Catching Wild Beasts Alive* (London: Hutchinson, 1931); J. C. Hazewinkel, "A One-Horned Javanese Rhinoceros Shot in Sumatra, Where It Was Not Thought to Exist," *Illustrated London News* 93 (December 23, 1933): 1018–1019; Willy Ley, *The Lungfish and the Unicorn* (New York: Viking, 1941); Karl Shuker, *Extraordinary Animals Worldwide* (London: Robert Hale, 1991), pp. 162–163.

## Badigui

Dinosaur-like animal of Central Africa, similar to the MOKELE-MBEMBE.

*Etymology:* Banda-Bambari (Ubangi) word.

*Variant names:* Diba (Gbaya/Ubangi), Guaneru, Ngakula-ngu (Banda/Ubangi, "water devil"), Songo (Gbanziri/Ubangi).

*Physical description:* Serpentine. Snakelike markings, lighter underneath. Flat, snakelike head. Neck, 10–12 feet long.

*Behavior:* Aquatic. Browses on tree branches without leaving the water. Strangles hippopotamuses but does not eat them.

*Tracks:* Leaves a furrow 3–5 feet wide.

*Distribution:* The Brouchouchou (near Ippy) and Gounda Rivers, Central African Republic; possibly Equatorial Guinea.

*Significant sightings:* About 1890, a Banda-Mbrès tribesman named Moussa saw a Badigui eating the large leaves of a tree (genus *Mitragyna*) near a stream in the Bakala District of Central African Republic. Its head was a bit larger than a python's, and its neck was much longer than a giraffe's. The skin was as smooth as a snake's, with similar markings.

In 1928, a Badigui crushed a field of manioc belonging to the chief of Yetomane, Central African Republic, and left wide tracks. About the same time, it killed a hippopotamus in the River Brouchouchou.

Lucien Blancou's gun bearer Mitikata told him that, in about 1930 near Ndélé, Central African Republic, he had seen an Ngakula-ngu's tracks, which were as wide as a truck.

*Sources:* Bernard Heuvelmans, *On the Track of Unknown Animals* (New York: Hill and Wang, 1958), pp. 463, 466–467, 470, 475–477, 482; Bernard Heuvelmans, *Les derniers dragons d'Afrique* (Paris: Plon, 1978), pp. 262–266, 388, 395.

## Bagge's Black Bird

Unidentified BIRD of East Africa.

*Physical description:* Black. As large as a sheep.

*Behavior:* Alarm call is like the bellow of a bull.

*Distribution:* Lake Bujuku, south of Mount Speke in the Ruwenzori Range, Uganda.

*Significant sighting:* Only one observation. Stephen Bagge's guide saw a number of these birds in 1898 at an altitude of 9,000 feet.

*Source:* John Preston, *Touching the Moon* (London: Mandarin, 1990), p. 35.

## Bái-Xióng

White BEAR of East Asia.

*Etymology:* Mandarin Chinese (Sino-Tibetan), "white bear." The same term is sometimes used for the Giant panda (*Ailuropoda melanoleuca*).

*Variant names:* Bei-sheng, Bitchun.

*Physical description:* Creamy-white bear, smaller than the Polar bear (*Ursus maritimus*).

*Distribution:* Shennongjia Forest, Hubei Province, China; Mongolia.

*Significant sighting:* Four specimens, obtained since 1963, have been on display in the Wuhan and Beijing zoos.

*Present status:* Known from ancient Chinese writings. Possibly confused with the giant panda after it was discovered in 1868 or with the ALMAS in Mongolia.

*Possible explanations:*

(1) Albino specimens or a pale-color morph of the Brown bear (*Ursus arctos*). Shennongjia is known for a high incidence of albinism in other species, including deer and monkeys.

(2) Separate subspecies of the brown bear, rather than a color variation.

*Sources:* Emanuel Vlček, "Old Literary Evidence for the Existence of the 'Snow Man' in Tibet and Mongolia," *Man* 59 (1959): 133–134; Karl Shuker, *The Lost Ark* (London: HarperCollins, 1991), p. 45.

## Bakanga

SPOTTED LION of Central Africa.

*Physical description:* Looks like a cross between a lion and a leopard. Reddish-brown ground color. Dappled like a leopard.

*Behavior:* Aggressive. Barks like a dog.

*Distribution:* Central African Republic.

*Sources:* Émile Gromier, *Vie des animaux sauvages de l'Oubangi-Chari* (Paris: Payot, 1938); C. A. W. Guggisberg, *Simba: The Life of the Lion* (Cape Town, South Africa: H. Timmins, 1961).

## Balong Bidai

FRESHWATER MONSTER of Southeast Asia.

*Etymology:* Malay (Austronesian) word.

*Physical description:* Flat, like a mat.

*Behavior:* Engulfs people and drowns them.

*Habitat:* Deep pools in rivers.

*Distribution:* Peninsular Malaysia.

*Possible explanation:* Whirlpool or natural gas eruption.

*Source:* Ronald McKie, *The Company of Animals* (New York: Harcourt, Brace, World, 1966), pp. 191–195.

## Bangenza

Unknown PRIMATE of Central Africa.

*Etymology:* Lusengo (Bantu) word.

*Physical description:* Brown color. Larger than a chimpanzee.

*Distribution:* North of Lisala, Democratic Republic of the Congo.

*Possible explanation:* Large, solitary male Chimpanzee (*Pan troglodytes*).

*Source:* Bernard Heuvelmans, *Les bêtes humaines d'Afrique* (Paris: Plon, 1980), p. 591.

## Banib

Variant name for the BUNYIP in southeastern Australia.

*Etymology:* Wergaya dialect form of Wemba (Australian) word.

*Variant name:* Banib-ba-gunuwar.

*Physical description:* Large. Black. Long neck.

*Distribution:* Lake Albacuytya, Victoria.

*Source:* Luise A. Hercus, *The Languages of Victoria: A Late Survey* (Canberra: Australian Institute of Aboriginal Studies, 1969), vol. 2, p. 279.

## Ban-Jhankri

Name erroneously attributed to the YETI in Central Asia.

*Etymology:* Nepali (Indo-Aryan), "forest wizard."

*Variant names:* Bangjakri, Bhan-jakri.

*Distribution:* Nepal; India.

*Possible explanation:* Refers to hill-dwelling shamans or faith healers.

*Source:* Frank W. Lane, *Nature Parade* (London: Jarrolds, 1955), p. 235.


## Ban-Manush

WILDMAN of Central Asia.

*Etymology:* Hindi (Indo-Aryan), "forest man."

*Variant names:* Bang, Ban-manchhe (in Nepal), Bunmanus, Lidini (for the female), Van-manas (in India), Van-manusha.

*Physical description:* Height, 4–5 feet. Covered in grayish hair. Face, hands, and feet similar to a human's.

*Behavior:* Upright gait. Said to carry away both women and men.

*Tracks:* Found in fields near forests or in the snow.

*Distribution:* Garhwal Himalayas, India; Jammu and Kashmir State, India; Nepal; Bangladesh.

*Sources:* James Baillie Fraser, *Journal of a Tour through Part of the Snowy Range of the Himālā Mountains* (London: Rodwell and Martin, 1820), pp. 284, 334, 420; Clark B. Firestone, *The Coasts of Illusion: A Study of Travel Tales* (New York: Harper, 1924), p. 123; Odette Tchernine, *The Yeti* (London: Neville Spearman, 1970), pp. 14–15, 22, 40, 175; Kesar Lall, *Lore and Legend of the Yeti* (Kathmandu: Pilgrims Book House, 1988), pp. 33–37, 52–54.


## Barguest

BLACK DOG of northern England.

*Etymology:* Possibly from the German *Bargeist* ("spirit of the [funeral] bier"), the German *Berggeist* ("hill ghost"), the German *Bärgeist* ("bear ghost" or "bar [gate] ghost"), or the Old English *burh gast* ("town ghost"); alternatively, a derivative of the English *boggart*, a supernatural being. Another possibility is a derivative of "barrow ghost."

*Variant names:* Bargest, Barghaist, Barghest, Barn-ghaist.

*Physical description:* Black dog as large as a calf. Woolly. Large, luminous eyes. Big fangs.

*Behavior:* Howls and shrieks. Accompanied by the sound of chains. Can change its shape. Said to appear at the deaths of notable people. Domestic dogs will follow it, howling and baying.

*Habitat:* Dark lanes, churchyards.

*Distribution:* Yorkshire and Lancashire, England.

*Sources:* William Hone, *The Every-Day Book and Table Book* (London: T. Tegg, 1841), vol. 3, p. 655; William Henderson, *Notes on the Folk-Lore of the Northern Counties of England and the Borders* (London: Folk-Lore Society, 1879), pp. 274–275; John Harland and T. T. Wilkinson, *Lancashire Folk-Lore* (Manchester, England: John Heywood, 1882); Jessica Lofthouse, *North Country Folklore* (London: Robert Hale, 1976).


## Bar-Manu

WILDMAN of Central Asia.

*Etymology:* Probably Gujari (Indo-Aryan), "big, hairy one" or "strong, muscular man."

*Physical description:* Muscular. Sometimes squat and plump. Covered with dark-brown hair except for eyes, nose, cheeks, knees, palms, and soles. Huge head. Receding forehead. Protruding eyebrows. Large, flat nose. Prominent cheekbones. Wide mouth. Massive neck. Small chin. Wide feet.

*Behavior:* Agile. Short, gutteral, or high-pitched loud cries ("aha") but no articulate speech. Foul body odor. Omnivorous.

*Tracks:* Humanlike. Turned inward.

*Habitat:* Forested regions, especially at an altitude of 6,500–9,800 feet.

*Distribution:* Shishi Kuh Valley, near Chitral, North-West Frontier Province, Pakistan.

*Significant sightings:* In September 1977, goatherd Purdum Khan watched a Bar-manu for two hours at an altitude of 7,600 feet in the Hindu Kush Range in Pakistan's North-West Frontier Province. It was a young male, about 5 feet 8 inches tall, sitting and eating ant larvae. Its penis was large and erect.

Tracks were found in 1994 by a French expedition led by Jordi Magraner.

*Possible explanation:* Very few fossil hominid remains have been found in the area, with the exception of Selungur Cave, Kyrgyzstan, where possible *Homo erectus* remains dating to 125,000 years ago were found, and Darra-I-Kur Cave, Afghanistan, where archaic modern human bones and artifacts were recovered.

*Sources:* Jordi Magraner, *Notes sur les hominidés reliques d'Asie centrale, district de Chitral, NWFP, Pakistan* (Paris: Jordi Magraner, 1992); Éric Joly and Pierre Affre, *Les monstres sont vivants: Enquête sur des créatures "impossibles"* (Paris: Bernard Grasset, 1995), pp. 79–89; Jordi Magraner, "Oral Statements Concerning Living Unknown Hominids: Analysis, Criticism, and Implications for Language Origins," http://www.n2.net/prey/bigfoot/biology/jordi.htm; Michel Raynal, "Jordi Magraner's Field Research on the Bar-manu: Evidence for the Authenticity of Heuvelmans's *Homo pongoides*," *Crypto Hominology Special,* no. 1 (April 7, 2001), pp. 98–103, at http://www.strangeark.com/crypto/Cryptohominids.pdf.

## Basajaun

WILDMAN of Western Europe.

*Etymology:* Basque, "man of the woods"; from *baso* ("forest") + *jaun* ("man").

*Variant names:* Anxo, Basandere ("woods woman"), Bebrices, Iretges, Mono careto ("ugly ape"), Nonell de la neu (Catalan, "Nonell of the snows"), Peladits ("finger peeler"), Tártalo (cyclops), Torto, Yan del Gel.

*Physical description:* Height, 6–10 feet. Heavily built. Long head-hair down to the knees.

*Behavior:* Agile. Protects flocks. Forecasts storms. Wears animal skins. Generally benevolent, though other Basque entities are not. Carries a stick.

*Distribution:* Basque Country, Spain; the Maladeta Massif, Aragon, Spain; the Pyrénées Mountains of Spain and France.

*Significant sightings:* Two hairy men, brothers named Iretges, lived long ago in the woods near Bédeilhac-et-Aynat, Ariège Department, France,

wearing animal skins and abducting the occasional shepherdess. One day, the villagers lured them with a trap and killed them.

A 6-foot-tall apemanlike creature was seen in the Pyrénées Mountains of Huesca Province, Spain, in May 1979. Some workers saw it sitting in a tree and making animal noises. It came down and threw a tree trunk at them, whereupon they fled.

A mountain climber named Juan Ramó Ferrer saw a reddish apelike creature near Bielsa, Huesca Province, in the spring of 1994. It jumped from tree to tree and squealed.

*Sources:* Vicente de Arana, *Los últimos iberos: Leyendas de Euskaria* (Madrid: Librería de Fernando Fé, 1882); *ABC* (Spanish national newspaper), May 16, 1979; José María Satrústegui, *Mitos y creencias* (San Sebastián, Spain: Editorial Sendoa, 1983); Ulrich Magin, "The European Yeti," *Pursuit,* no. 74 (1986): 64–66; Sergio de la Rubia-Muñoz, "Wild Men in Spain," *INFO Journal,* no. 72 (Winter 1995): 22–25; Connaissances sur les Pyrénées, http://mageos.ifrance.com/EcoPyrene/; Le Pog des Iretges, http://perso.respublica.fr/ariegeois/iretges.html.

## Basilisk

Birdlike lizard or venomous snake of Europe and North Africa; *see* SEMIMYTHICAL BEASTS.

*Etymology:* From the Greek *basiliskos* ("little king").

*Variant names:* Cockatrice, from the Middle French *cocatris* and the Latin *calcatrix* ("tracker"); Regulus (Latin, "prince").

*Physical description:* Length, 12 inches. Ancient writers described a snakelike animal, with a bright white spot on the head. By the late Middle Ages, the animal had come to be called a Cockatrice and was described as a bird with a spotted rooster's comb and a snake's tail.

*Behavior:* Moves with its middle portion sticking up. Hisses. Said to be born from a cock's egg hatched by a toad or snake. Its stare can paralyze, and its breath (or odor) is fatal to snakes and humans. Its skin was used to deter spiders, snakes, and birds in Roman temples. It can be frightened by a crowing rooster and

*The* Basilisk, *a semimythical bird-like lizard. (© 2002 ArtToday.com, Inc., an IMSI Company)*

killed by a weasel. Seeing its own reflected image can also prove fatal.

*Distribution:* Cyrenaica Province, Libya; Europe.

*Significant sightings:* A Basilisk killed many people in Rome, Italy, in the mid-ninth century until Pope Leo IV destroyed it with prayer.

Another Basilisk was discovered in a well in Vienna, Austria, in June 1212.

In 1587, two children were killed by a Basilisk in Warsaw, Poland, while they were playing in an abandoned cellar. A servant who found them was also struck dead. Authorities finally sent in a condemned prisoner, outfitted with a leather suit and mirrors. The man emerged with a snake that officials judged to be a genuine Basilisk.

When the parish church of Renwick, Cumbria, England, was torn down in 1733, a huge, bat-winged creature angrily flapped at the workmen. One of them, John Tallantire, killed it with a tree branch, earning him and his descendants an exemption from fees to the manor.

*Possible explanations:*

(1) The Egyptian cobra (*Naja haje*) is yellowish-brown and becomes blue-black with age. Found in North Africa and Arabia, it can grow to 8 feet, though its more typical length is 5–6 feet. It is not a spitting cobra, but its venom can be deadly. This species was probably the famous asp that bit Cleopatra, and it is depicted on the crowns of the Egyptian pharaohs. The cobra's hood might conceivably be compared to a rooster's crest when erect. African cobras do not have hood markings.

(2) The King cobra (*Ophiophagus hannah*) is the largest of the venomous snakes. It is not a spitting snake, and its venom is less toxic than other cobras, but it injects much more venom per bite—6–7 milliliters, enough to kill an elephant or twenty people. The king cobra also has an unnerving ability to move forward while in a threatening, strike posture. It has a black head with four white crossbars. Body color varies from olive-green to black. Though it can attain a length of 19 feet, it is not aggressive and is often adopted as a village pet. It hunts other snakes in the daytime and is the only snake known to construct a nest. Its range is from India to the Philippines. The use of Mongooses (Family Herpestidae) in catching snakes in India may explain the reference to weasels as enemies of Basilisks.

(3) The Indian cobra (*Naja naja*) is a spitting cobra. The venom is spit out in a shower and directed toward the eyes of the victim, which can cause blindness or death. This trait may have given rise to the legend of the Basilisk's paralyzing stare. It can accurately hit a target as much as 10 feet away in a lunging spit. On its hood, it has two black-and-white spots connected by a curved line.

(4) The Horned viper (*Cerastes cerastes*) of North Africa and Arabia has a pair of horns over its eyes. The eleventh-century Arab physician Avicenna was one of the first to suggest this snake as a Basilisk candidate.

(5) David Heppell has suggested that beached Giant squids (*Architeuthis* sp.) may have contributed to Basilisk lore.

(6) The Tatzelwurm probably accounts for some Basilisk characteristics.

(7) The crest is similar to the Crowing Crested Cobra of East and Central Africa.

(8) The roosterlike Cockatrice might be de-

BATHYSPHAERA INTACTA, *a deep-sea fish seen only once off Bermuda in 1932 by William Beebe in his bathysphere.* (*William M. Rebsamen*)

rived from the lizardlike appearance of certain stages of a chicken embryo.

*Sources:* Bible, Old Testament (Pss. 91:13, Isa. 59:5); Pliny the Elder, *Natural History: A Selection,* trans. John F. Healy (New York: Penguin, 1991), pp. 117–118 (VIII. 78); Ælian, *De natura animalium* (II. 5–7, III. 31, V. 50, VIII. 28, XVI. 19); Alexander Neckam, *De naturis rerum libro duo* [ca. 1200] (London: Longman, Green, Longman, Roberts, and Green, 1863); Thomas Browne, *Pseudodoxia Epidemica* [1672] (Oxford: Clarendon, 1981), pp. 181–186, 808–814; Henry Phillips, *Basilisks and Cockatrices* (Philadelphia: E. Stern, 1882); Will-Erich Peuckert, *Schlesische Sagen* (Jena, Germany: E. Diederichs, 1924), pp. 242, 318; E. W. Gudger, "Jenny Hanivers, Dragons and Basilisks in the Old Natural History Books and in Modern Times," *Scientific Monthly* 38 (1934): 511–523; T. H. White, *The Bestiary: A Book of Beasts* (New York: G. P. Putnam's, 1960), pp. 168–169; Gerald Findler, *Ghosts of the Lake Counties* (Clapham, England: Dalesman, 1972); Marc Alexander, *Enchanted Britain* (London: Arthur Barker, 1981); Joe Nigg, *A Guide to the Imaginary Birds of the World* (Cambridge, Mass.: Apple-Wood, 1984), pp. 29–31; Karl Shuker, "From Flying Toads to Snakes with Wings," *Fate* 47 (September 1994): 31–36.

## Bathysphaera intacta

One of BEEBE'S ABYSSAL FISHES of the North Atlantic Ocean.

*Physical description:* Length, 6 feet. Has a row of strong, pale-blue lights along its side. Two ventrical tentacles, each tipped with a pair of red and blue lights.

*Distribution:* North Atlantic Ocean.

*Significant sighting:* Observed only once at 2,100 feet by William Beebe in a bathysphere off Bermuda on November 22, 1932.

*Possible explanation:* Beebe classified it with the Scaleless black dragonfishes (Subfamily Melanostomiidae), but he admitted this was a guess and that the largest known dragonfish was only 15 inches long.

*Sources:* William Beebe, *Half Mile Down* (New York: Harcourt, Brace, 1934); William Beebe and Jocelyn Crane, "Deep-Sea Fishes of the Bermuda Oceanographic Explorations: Family Melanostomiatidae," *Zoologica* 24 (1939): 65–238.

## BATS (Unknown)

Bats (Order Chiroptera) are the only group of mammals to have evolved powered flight. Many taxonomists now believe that bats evolved from a common ancestor with the PRIMATES. They are divided into two suborders, the Megabats (Megachiroptera), which are fruit eaters such as the giant Flying foxes (*Pteropus* spp.), and the Microbats (Microchiroptera), which catch insects in flight using a radarlike system of echolocation—emitting high-frequency sound pulses and listening to the echo as those pulses bounce back from solid objects. The earliest known fossil megabat is *Archaeopteropus transiens* from the Early Oligocene (30 million years ago) of Italy, while the earliest known fossil microbat is *Icaronycteris index* from the Early Eocene (52 million years ago) of Europe and North America.

Contenders for the largest living bat are the Great flying fox (*Pteropus neohibernicus*) of New Guinea and New Ireland, with a wingspan of 5 feet 6 inches–6 feet; the Large flying fox (*P. vampyrus*) of Malaysia and Indonesia, with a wingspan of 5 feet 6 inches; and the Indian flying fox (*P. giganteus*), with a wingspan of 4–5 feet. All are megabats and quite timid.

Although vampires are not strictly a subject for cryptozoology (since they are said to be either shape-shifting supernatural beings or reanimated dead members of our own species), their association with bats adds an eerie, symbolic element to the quest for Chiroptera that are unknown to science. Though feared or revered as creatures of the night in European folklore from ancient times, bats were first explicitly connected to the vampire mythos with Bram Stoker's novel *Dracula* in 1897. Since then, they have made a substantial contribution to the concept of the vampire as an exclusively nocturnal entity.

Only the microbats of the Subfamily Desmodontinae are blood feeders, and these are confined to tropical America. There are three known species: The Common vampire bat (*Desmodus rotundus*), the White-winged vampire bat (*Diaemus youngi*), and the Hairy-legged vampire bat (*Diphylla ecaudata*).

The largest carnivorous bat is the Australian false vampire bat (*Macroderma gigas*), with a wingspan of up to 32 inches. The largest carnivorous bat in Africa is the Yellow-winged bat (*Lavia frons*), with a wingspan of only 16 inches.

Winged cryptids are often difficult to classify unless they are observed up close. From a distance, the surprised observer might find it difficult to distinguish between an unknown giant bat, a BIG BIRD, a FLYING REPTILE, or even a FLYING HUMANOID. A pterodactyl might look very much like a large bat when swooping down an African river at dusk. Consequently, only the most mammalian of flying cryptids are included in this section. Of the eight listed here, five are giants and three are smaller, although they have peculiar habits. Five are African, two are South American, and one is Asian.

**Mystery Bats**

AHOOL; CAMAZOTZ; ELEPHANT-DUNG BAT; ETHIOPIAN VAMPIRE BAT; FANGALABOLO; GIANT VAMPIRE BAT; GUIAFAIRO; OLITIAU

## Batsquatch

FLYING HUMANOID of the northwestern United States.

*Etymology:* SASQUATCH with bat wings, coined by the witness.

*Physical description:* Height, 9 feet. Bright-blue fur. Yellowish eyes. Tufted ears. Long, wolflike muzzle. Sharp, straight teeth. Batlike wings. Clawed, birdlike feet.

*Behavior:* Apparently can affect car engines.

*Distribution:* Western Washington State.

*Significant sighting:* Brian Canfield's truck stalled as a winged creature dropped from the sky into the road in front of him, near Lake Kapowsin, Washington, on April 16, 1994. It stood still for several minutes, unfolded its wings, and started flapping them. Canfield could feel the turbulence in the air as this occurred. The creature slowly rose and flew off in the direction of Mount Rainier.

*Present status:* Only one known encounter.

*Sources: Tacoma (Wash.) News Tribune,* April 24, 1994; Phyllis Benjamin, "Batsquatch, Flap, Flap," *INFO Journal,* no. 73 (Summer 1995): 29–31.

## Batûtût

SMALL HOMINID of Southeast Asia.

*Etymology:* Unknown.

*Variant name:* Ujit.

*Physical description:* Height, 4 feet.

*Behavior:* Nocturnal. Bipedal. Feeds on river snails and breaks open their shells with a rock. Said to kill people and rip out their livers. Wary of fire.

*Distribution:* Sabah State, Malaysia, in the north of the island of Borneo.

*Tracks:* Length, 6 inches. Width, 4 inches. Toes and heel are humanlike, but sole is too short and broad for a man. Big toe is on the opposite side of the arch of the foot.

*Significant sighting:* British zoologist John MacKinnon found two dozen footprints in the Ulu Segama National Park, Sabah, in 1969.

*Possible explanation:* Niah Cave in Sarawak, northern Borneo, has yielded archaic human remains. The cranium of a young, adult female found in 1958 is known as "Deep Skull" and may be 40,000 years old, which may represent the earliest anatomically modern remains in Indonesia.

*Sources:* Frederick Boyle, *The Savage Life* (London: Chapman and Hall, 1876), p. 36; John MacKinnon, *In Search of the Red Ape* (New York: Ballantine, 1974), pp. 100–102.

## BEARS (Unknown)

Modern Bears (Family Ursidae) are the largest living members of the Order Carnivora. In general, they have heavy bodies but can stand on their hind legs and grab things with their front paws. Their feet are plantigrade and hairless (except for the polar bear), and they feature five toes with nonretractile claws. Worldwide, there are only eight species; along with the Giant panda (*Ailuropoda melanoleuca*), they descended from a common ancestor, the Dawn bear (*Ur-*

*savus elemensis*), about 20 million years ago, in the Early Miocene. The first true bears turned up in both the Old and New World about 5 million years ago, in the Pliocene. The Giant short-faced bear (*Arctodus simus*) of the Pleistocene must have been the most fearsome predator in North America, standing 11 feet tall on its long hind legs and capable of short bursts of speed in chasing its prey. The smaller Brown bear (*Ursus arctos*) replaced it at the end of the Ice Ages.

Brown bears vary widely in coloration and size. The Kodiak bear (*U. a. middendorffi*), which lives on several islands in the Gulf of Alaska, is the largest living terrestrial carnivore. Adult males average 8 feet from nose to tail, stand 4 feet 4 inches at the shoulder, and weigh 1,050–1,175 pounds. The greatest recorded weight for a wild specimen was 1,656 pounds for a male shot on Kodiak Island in 1894. The Polar bear (*Ursus maritimus*) is sometimes longer but less robustly built.

Of the sixteen bearlike cryptids listed here, three are North American, three are South American, seven are Asian, and three are African. Although the NANDI BEAR is included, it is here only because it has been incorrectly named and is more likely to turn out to be a HYENA or PRIMATE; it also appears in those lists. The STIFF-LEGGED BEAR, which might reasonably be considered a Native American folk memory of the giant short-faced bear, has also been colisted with the ELEPHANTS because some think it represents a legend of a living mammoth.

### Mystery Bears

ATLAS BEAR; BÁI-XIÓNG; BERGMAN'S BEAR; DRE-MO; DZU-TEH; IRKUIEM; MACARENA BEAR; MACFARLANE'S BEAR; MILNE; MURUNG RIVER BEAR; NANDI BEAR; NEPALESE TREE BEAR; PYGMY BROWN BEAR; QOQOGAQ; STIFF-LEGGED BEAR; TÔO

### Beast of Bardia

Mystery ELEPHANT of Central Asia.

*Etymology:* After the Bardia Forest, where it is found.

*Physical description:* Shoulder height, 11–13

feet. Massive version of the Asian elephant, whose maximum height is said to be 11 feet. Large forehead with two domes. Distinct nasal bridge. Sloping back. Thick tail.

*Tracks:* 22.5 inches across.

*Distribution:* Karnali River, Royal Bardia National Park, Nepal.

*Significant sighting:* From 1991 to 1997, John Blashford-Snell led seven expeditions to the Royal Bardia National Park in Nepal in search of these outsize elephants. He succeeded in finding and photographing two adult males, estimated to be 11 feet 3 inches at the shoulder. The larger one he nicknamed "Rajah Gaj" and the smaller one "Kancha."

*Present status:* The Bardia population is estimated to consist of 100 elephants. A DNA analysis of dung samples taken from Blashford-Snell's individuals has identified them as Asian elephants.

*Possible explanations:*

(1) An isolated, inbred variety of the Asian elephant (*Elephas maximus*).

(2) A surviving fossil elephant, *Elephas hysudricus,* known from fossils in the Siwalik Hills of northern India and western Nepal, that lived 2 million years ago.

(3) A surviving stegodon, a member of a primitive, elephant-like family of animals that died out more than 1 million years ago. *Stegodon ganesa* was 11 feet 6 inches at the shoulder but had a huge set of upper tusks that curved sideways and nearly reached the ground.

*Sources:* Peter Byrne, *Tula Hatti: The Last Great Elephant* (London: Faber and Faber, 1990); Bob Rickard and John Blashford-Snell, "The Expeditionist," *Fortean Times,* no. 70 (August-September 1993): 30–34; Nigel Hawkes, "Explorer Finds Giant Elephants in Nepal," *Times* (London), May 15, 1996; John Blashford-Snell and Rula Lenska, *Mammoth Hunt: In Search of the Giant Elephants of Nepal* (London: HarperCollins, 1996).

## Beast of Bladenboro

Apparent variety of EASTERN PUMA of North Carolina.

*Physical description:* Length, 3–4 feet, with a 14-inch tail. Shoulder height, 20 inches. Black.

*Tracks:* Catlike.

*Distribution:* Bladenboro, North Carolina.

*Significant sighting:* The Beast made its first dog kill at Clarkton, near Bladenboro, North Carolina, on December 29, 1953, and terrorized the countryside for about one week. It killed nine dogs and one pet rabbit by crushing the skulls and draining the blood. Nearly 1,000 people took part in a disorganized hunt on January 6–7 but failed to find anything.

*Sources: Fayetteville (N.C.) Observer,* January 5–9, 1954; Joseph F. Gallehugh Jr., "The Vampire Beast of Bladenboro," *North Carolina Folklore* 24 (1976): 53–58; Michael Futch, "Beast of Bladenboro Put Town on Map," *Fayetteville (N.C.) Observer,* July 23, 2000; Mark A. Hall, "The Vampire Beast of Bladenboro," *Wonders* 7, no. 1 (March 2002): 3–22.

## Beast of Bodalog

Unknown SNAKE or other mystery animal of Wales.

*Etymology:* After the Bodalog farm where the beast was seen.

*Behavior:* Nocturnal. Aquatic. Kills sheep by biting them just below the neck close to the sternum.

*Distribution:* River near Rhayader, Powys, Wales.

*Significant sighting:* By mid-October 1988, a mystery animal had killed thirty-five sheep on the Bodalog farm near Rhayader. It emerged from a nearby river at night, attacked the sheep, then returned to the river each time.

*Present status:* Only one series of reports.

*Possible explanations:*

(1) A feral Domestic dog (*Canis familiaris*) leaves a much messier carcass.

(2) A European otter (*Lutra lutra*) will not kill a sheep. Its primary food is fishes, supplemented with crustaceans, birds, small mammals, and frogs.

(3) American minks (*Mustela vison*) have been naturalized in parts of Britain. They are known to kill rabbits, cats, and dogs, but sheep would be too large to tackle.

(4) An unknown species of giant mink, perhaps greater than 2 feet long, that would be large enough to kill a sheep.

(5) Britain's only venomous snake is the European adder (*Vipera berus*), which only feeds on small animals and stays away from water.

(6) Neither of the nonvenomous snakes—the Grass snake (*Natrix natrix*) or the European smooth-snake (*Coronella austriaca*)—are in the habit of attacking sheep.

(7) A large aquatic snake not native to Wales, released by or escaped from a local pet owner.

*Sources:* *Daily Mail* (London), October 10, 1988; Karl Shuker, "A Water Vampire," *Fate* 43 (March 1990): 86–88.

## Beast of Bodmin Moor

BRITISH BIG CAT of Cornwall.

*Variant name:* Beast of Bolventor.

*Physical description:* Leopard-sized. Black.

*Behavior:* Kills livestock.

*Distribution:* Bodmin Moor, Cornwall, England.

*Significant sightings:* A large, catlike creature attacked Jane Fuller on October 26, 1993, when she was walking her Labrador dog on Bodmin Moor. She was temporarily stunned by a blow to the head but escaped. Later, two sheep were found dead in an adjoining field; one was decapitated, the other disemboweled.

An investigation from January 12 to July 1, 1995, by the British Ministry of Agriculture, Fisheries, and Food failed to turn up any conclusive evidence of a resident large cat. The inquiry looked at seventy-seven reports of puma-like or leopardlike cats and livestock killings recorded in the Bodmin Moor area between January 1994 and June 1995. The investigators, Simon Baker and Charles Wilson, determined that videos taken of the Beast showed nothing more than black domestic cats; a film showing an alleged panther cub's face close up was only a cat's, since its pupils contracted to a slit, not to a circle like a leopard's would have; the tracks were only cat-sized, except for one that was attributable to a dog; and the kills were probably perpetrated by a medley of dogs, foxes, crows, and badgers.

A skull found in the River Fowey near St. Cleer, Cornwall, on July 24, 1995, turned out to be from an imported leopard-skin rug.

Another investigation in December 1997 was prompted by bite marks on livestock, droppings, and new photos—one of which was taken through binoculars near St. Austell, Cornwall, and apparently shows a pregnant adult female puma and a cub.

*Sources:* Karl Shuker, "The Beast of Bodmin, and a Lesson in 'Skull-duggery!'" *Strange Magazine,* no. 16 (Fall 1995): 29–30; Paul Sieveking, "Not as Simple as ABC," *Fortean Times,* no. 83 (October-November 1995): 44–45; "The Beast of Bodmin Is Caught on Film," *Cryptozoology Review* 2, no. 3 (Winter-Spring 1998): 8–9; Chris Moiser, *Mystery Cats of Devon and Cornwall* (Launceston, England: Bossiney Books, 2001).

## Beast of 'Busco

Giant TURTLE of Indiana.

*Etymology:* Short for Churubusco, Indiana.

*Variant names:* Oscar, Phantom Churubusco turtle.

*Physical description:* Turtle said to be as big as a dining room table or a car top. Weight, 100–500 pounds.

*Distribution:* Fulks Lake, near Churubusco, Indiana; Black Oak Swamp, near Hammond, Indiana.

*Significant sightings:* Oscar Fulk saw a huge turtle in Fulks Lake in 1898. It was seen again in 1914 and then in July 1948 when Ora Blue and Charley Wilson glimpsed it while fishing. Gail Harris, on whose farm the turtle was spotted, launched a major effort to catch the animal in March 1949, employing scuba divers, deep-sea gear, a female sea turtle, a sump pump, and a dredging crane to drain the lake. The Fort Wayne newspapers played up the story, and thousands of people trampled across the farm looking for the turtle. On October 13, about 200 people got their wish as the turtle leaped from the water to try to catch a duck used as a lure. But by December, the draining efforts were failing, and Harris fell ill with appendicitis and called the search off. A documentary film about the event, called *The Hunt for Oscar,* was made in 1994 by Terry Doran.

When a swamp was drained near Black Oak, Indiana, in July 1950, a huge turtle with a head as big as a human's was seen swimming around a drain leading into the Little Calumet River.

*Possible explanation:* The Alligator snapping turtle (*Macroclemys temminckii*) reportedly grows to a maximum weight of 400 pounds. It has a huge head with hooked upper and lower beaks, prominent dorsal keels, and an extra row of scutes at the side of the carapace. It lives almost exclusively in the Mississippi River drainage areas of Mississippi, Louisiana, Arkansas, and Missouri. It may occasionally migrate further afield.

*Sources: Indianapolis Star Magazine,* January 1, 1950; *Indianapolis News,* July 15, 1950; Churubusco.Net: Turtle Days, http://members.aol.com/iga1/tdays1.htm.

## Beast of Exmoor

BRITISH BIG CAT of southwestern England.

*Physical description:* Large, black cat or dog. Length, 3 feet–4 feet 6 inches. Shoulder height, 2 feet 6 inches. White markings on the head and neck. Squat head. Short neck. Powerful, muscular body. Short legs.

*Behavior:* Nocturnal. Moves rapidly from cover to cover. Kills sheep by breaking the neck at the second vertebra or crushing the skull.

*Tracks:* Large, doglike prints, 4 inches across. Smaller tracks may be a female's.

*Distribution:* Exmoor, in the counties of Somerset and Devon, England.

*Significant sightings:* Attacks on livestock gained prominence in Devon in the spring of 1983, though scattered reports of a black animal in the area go back to 1982. Eric Lay, of Drewstone Farm near South Molton, thought he had lost at least forty lambs over the previous few months. Local police called in the Royal Marines, which held stakeouts in early May and June 1983 as part of Operation Beastie. They were able to observe the animal through night-vision equipment. Reports of both large cats and

dogs were logged. By late June, there were eighty-six kills, but these dropped off in July.

Two boys, Wayne Adams and Marcus White, saw the Beast on May 29, 1983, at Willingford Farm on Exmoor. It was jet black with some white markings and powerfully built. Though its head looked like a German shepherd dog's, the animal moved like a cat. The same night, a sheep was killed at Ash Mill.

Trevor Beer saw a black cat measuring 4 feet 6 inches in the summer of 1984 at a cache of deer carcasses on Exmoor that he had discovered earlier in the year. It ran swiftly and had powerful forelegs.

In January 1987, Trevor Beer discovered nine lynxlike pawprints 3 inches in diameter at Muddiford, Devon. In August, he took nine photos from a distance of about 100 yards of a black cat, 4 feet 6 inches in length, that stalked and killed a rabbit on Exmoor.

In 1990, Lars Thomas led an expedition (Operation Exmoor) to investigate sightings. At the site of a sheep kill, he found a tuft of hair that was identified as belonging to a puma.

*Possible explanations:*

(1) Trevor Beer proposed that feral Domestic dogs (*Canis familiaris*) were killing livestock, but large cats of some kind were also in the neighborhood. He noted that about 20 percent of the sightings involved a fawn-colored cat.

(2) Large, feral Domestic cats (*Felis silvestris catus*) were suggested at first by Nigel Brierly, perhaps representing a hybrid strain that has attained puma-sized proportions.

(3) A black Puma (*Puma concolor*) was Brierly's later conclusion, though melanism is virtually unknown in this strictly American species.

(4) An unknown species of indigenous big cat, suggested by Di Francis.

(5) A Eurasian lynx (*Lynx lynx*) explanation was favored by Frank Turk after lynx hairs were identified at a sheep kill in 1986. Lynxes became extinct in Britain during prehistoric times. In May 2001, a specimen later nicknamed "Lara" was captured in Cricklewood, North London, following a reported big-cat sighting. It was believed to be

*The BEAST OF EXMOOR photographed in August 1987 by Trevor Beer. (Trevor Beer/Fortean Picture Library)*

an escaped or abandoned pet. There are reports of lynxes on the loose elsewhere on the island, which has led to speculation about a relict lynx population.

*Sources:* Hope L. Bourne, *Living on Exmoor* (London: Galley Press, 1963); Bob Rickard, "The Exmoor Beast and Others," *Fortean Times,* no. 40 (Summer 1983): 52–61; "The Beast of Exmoor," *ISC Newsletter* 2, no. 3 (Fall 1983): 7–8; Trevor Beer, *The Beast of Exmoor: Fact or Legend?* (Barnstaple, England: Countryside Productions, 1984); "Once More with Felines," *Fortean Times,* no. 44 (Summer 1985): 28–31; Graham McEwan, *Mystery Animals of Britain and Ireland* (London: Robert Hale, 1986), pp. 30–36; Nigel Brierly, *They Stalk by Night: The Big Cats of Exmoor and the South-West* (Bishops Nympton, England: Yeo Valley Productions, 1989); Karl Shuker, *Mystery Cats of the World* (London: Robert Hale, 1989), pp. 44–51; Chris Moiser, *Mystery Cats of Devon and Cornwall* (Launceston, England: Bossiney Books, 2001).

## Beast of Gévaudan

An enigmatic DOG, wolf, or HYENA of south-central France.

*Etymology:* Gévaudan was the old name for an area that roughly corresponds to the modern department of Lozère, France.

*Physical description:* Bigger than a wolf. Reddish color. Large head. Small, pointed, upright ears. Muzzle like a greyhound dog's. Wide, gray chest. Black streaks on the back. Hind legs are longer than forelegs. Cropped tail.

*Behavior:* Active in the daytime. Said to be able to leap a distance of 28 feet and stand on its hind legs on occasion. Cry is like a horse neighing. Seemingly impervious to bullets. Completely ignores sheep but is wary of cows' horns. Kills and eats women and children by knocking them down and then biting into their throats and faces, sometimes decapitating them.

*Tracks:* Doglike. Long and clawed.

*Habitat:* Montane forest.

*Distribution:* The mountainous region of Languedoc, especially near Mont Mouchet, Montagne de la Margeride, Lozère Department, France.

The BEAST OF GÉVAUDAN, *from an eighteenth-century print. (Fortean Picture Library)*

*Significant sightings:* In June 1764, a young girl was tending cows in the Fôret de Mercoire near Langogne, Lozère Department, when she saw what looked like an enormous wolf running toward her. Her dogs panicked, and the animal injured her badly, but the cattle drove the Beast off with their horns. The first fatality was a fourteen-year-old shepherdess named Jeanne Boulet, who was mauled on June 30. Eleven other fatal attacks on women and children took place through the end of November, when an army unit stationed in Languedoc was called in to hunt down the Beast.

On December 24, 1764, a seven-year-old boy was killed by a similar wolflike animal, as were a shepherd and two young girls before the end of the year. On January 12, 1765, the Beast attacked a group of children near Vileret d'Apcher and seized an eight-year-old boy, but the others drove it away by jabbing it with a blade attached to a stick and throwing stones.

After the army under Captain Duhamel, aided by a host of volunteers, failed to catch the animal even though they slaughtered about 100 wolves, King Louis XV in February called in a famous Norman wolf hunter named Denneval, who fared no better. The king's harquebusier Antoine, Sieur de Beauterne, was sent to Gévaudan in late July.

A girl of the village of Vachelerie, near Paul-hac-en-Margeride, disappeared on the evening of September 8, 1765. After a shepherd found her cap, Beauterne and some gamekeepers found torn and bloodstained clothing and finally the naked body of the girl, with fang marks on her throat and one thigh eaten to the bone.

By mid-September, seventy-three people had been killed over a crescent-shaped area stretching about 31 miles long. Then, Beauterne killed an animal on September 20, 1765, near the

Royal Abbey of Chazes in Auvergne. It seems to have been a large wolf (5 feet 7 inches long and 130 pounds) with a white throat. It was autopsied, taken to a taxidermist in Clermont-Ferrand, then preserved at the Muséum d'Histoire Naturelle in Paris until it was lost. The attacks ceased for nearly three months.

On December 2, 1765, two young children were attacked near Mont Mouchet, and more children were killed in February and March 1766. The region again appealed for royal aid, but since the Beast was officially dead, the request was ignored.

When another little girl was killed at Nozerolles on June 18, 1767, the marquis d'Apcher and twelve hunters set off to track the beast. One of the hunters, Jean Chastel, shot a reddish animal on June 19, after which the depredations finally stopped. The corpse of the animal was crudely stuffed, then displayed in the region for two weeks, after which it was sent to Paris and examined by the naturalist the comte de Buffon. The animal was preserved at the Muséum d'Histoire Naturelle in Paris until 1819 and evidently identified at the time as a striped hyena.

The official tally of deaths attributed to the Beast of Gévaudan is 100, most of them women and children.

*Present status:* A French film about the Beast, *Le pacte des loups,* was released in 2001.

*Possible explanations:*

(1) A Gray wolf (*Canis lupus*) that turned to maneating, though this is completely uncharacteristic of the species and cannot explain the decapitations. Except for rare attacks by rabid animals, there is virtually no evidence for attacks by wolves on humans throughout the twentieth century.

(2) Domestic dog (*Canis familiaris*) × wolf hybrid.

(3) The Striped hyena (*Hyaena hyaena*) of Africa, like the animal killed in 1767, has a blunt muzzle, pointed nose, striped body, and shaggy mane from head to tail. Primarily interested in carrion, it can kill prey up to the size of an adult donkey. It is seldom swift enough to catch alert wild animals but is said to occasionally snatch unprotected human babies. Nonetheless, it is shier than

its relative the Spotted hyena (*Crocuta crocuta*).

(4) A baboon of some kind, since one was rumored to have been killed in the area.

(5) A Wolverine (*Gulo gulo*), suggested by Francis Petter.

(6) A sterile lion × tiger hybrid, either a Liger (male lion × female tiger) or Tigon (male tiger × female lion).

(7) Jean-Jacques Barloy has suggested that Protestant hunters deliberately unleashed huge dogs (or a hyena) on the Catholic peasantry after the first animal was killed in 1765. There was an intense Protestant-Jesuit rivalry in the area at the time.

(8) A serial killer wearing an animal skin, perhaps even one of the brothers of Jean Chastel, who may have faked killing a hyena to cover up the murders.

*Sources:* Abel Chevalley, *La Bête du Gévaudan* (Paris: Gallimard, 1936); Marie Moreau-Bellecroix, *La Bête du Gévaudan* (Paris: Éditions Alsatia, 1945); Andrew E. Rothovius, "Who or What Was the Beast of Gévaudan?" *Fate* 14 (September 1961): 32–37; Xavier Pic, *La bête qui mangeait le monde en pays de Gévaudan et d'Auvergne* (Mende, France: Chaptal, 1968); Jacques Delperrié de Bayac, *Du sang dans la montagne* (Paris: Fayard, 1970); C. H. D. Clarke, "The Beast of Gévaudan," *Natural History* 80 (April 1971): 44–51, 66–73; Gérard Ménatory, *La Bête du Gévaudan: Histoire, légende, réalité* (Mende, France: Chaptal, 1976); Jean-Jacques Barloy, "La Bête du Gévaudan soumise á l'ordinateur," *Science et Vie* 131 (June 1980): 54–59, 172; Félix Buffière, *La Bête du Gévaudan: Une grande énigme d'histoire* (Toulouse, France: Félix Buffière, 1987); Richard H. Thompson, *Wolf-Hunting in France in the Reign of Louis XV: The Beast of the Gévaudan* (Lewiston, N.Y.: Edward Mellen Press, 1991); Michel Louis, *La Bête du Gévaudan: L'innocence des loups* (Paris: Perrin, 1992); Andrew D. Gable, "The Beast of Gévaudan and Other 'Maulers,'" *Cryptozoology Review* 1, no. 3 (Winter-Spring 1997): 19–22; Franz Jullien, "La deuxième mort de la Bête de Gévaudan," *Annales du Muséum du Havre,* no. 59 (August 1998): 1–9; Michel Meurger, "A

Hyena for the Gévaudan: Testimonial Reports and Cultural Stereotypes," *Fortean Studies* 4 (1998): 227–229; Geneviève Carbone, "La Bête du Gévaudan," *Sciences et Avenir,* no. 123 (July-August 2000), on line at http://www.sciences-et-avenir.com/hs_123/page16.html.

## Beebe's Abyssal Fishes

Deep-sea FISHES observed by William Beebe in a bathysphere in the North Atlantic Ocean off Nonsuch Island, Bermuda, between 1930 and 1934 and never seen since. These include an ABYSSAL RAINBOW GAR, *BATHYSPHAERA INTACTA,* FIVE-LINED CONSTELLATION FISH, PALLID SAILFIN, and a THREE-STARRED ANGLERFISH. Beebe's bathysphere dives incorporated the first direct observations of abyssal fishes in their natural environment.

*Sources:* William Beebe, *Half Mile Down* (New York: Harcourt, Brace, 1934); Richard Ellis, *Deep Atlantic* (New York: Alfred A. Knopf, 1996), pp. 212–214; Robert D. Ballard, *The Eternal Darkness: A Personal History of Deep-Sea Exploration* (Princeton, N.J.: Princeton University Press, 2000), pp. 25–30.

## Beebe's Manta

Unknown species of FISH in the South Pacific Ocean.

*Scientific name:* Manta sp. nov., assigned by Gunter Sehm in 1996.

*Physical description:* Diamond-shaped body configuration. Dark-brown back, faintly mottled. Two broad, brilliantly white, distinctly V-shaped bands extend halfway down the back from each side of the head. Wingspan, 10 feet, approximately 1.5 times the body length. Wing tips are white, at least on the underside. Conspicuous horns. Short tail.

*Distribution:* Galápagos Islands; New Caledonia; Tabuaeran Atoll, Kiribati; off Baja California, Mexico; Great Barrier Reef, Queensland, Australia.

*Significant sightings:* Naturalist William Beebe observed a white-banded manta ray off Isla Genovesa in the Galápagos Islands on April 27, 1923. The fish collided with his vessel, *Noma,* then sped away on the surface.

*BEEBE'S MANTA, a striped ray of the Pacific Ocean. (William M. Rebsamen)*

A documentary for German television, called *Sharks: Hunters of the Seas* and broadcast on December 28, 1989, featured a thirty-second clip of a manta with white, symmetrical, V-shaped bands.

A British Broadcasting Corporation (BBC) television program titled *Holiday Guide to Australia,* broadcast on November 7, 1999, included an aerial view of a swimming manta ray with a pair of white, longitudinal bands on its wings, filmed over the Great Barrier Reef.

*Possible explanation:* The pigmentation of the uniformly dark Giant manta (*Manta birostris*) is easily rubbed off, but this only results in blotching. Sometimes, the animal is seen with white shoulder patches but not distinct banding. It has a wingspan up to 26 feet and is found in circumtropical waters.

*Sources:* William Beebe, *Galápagos, World's End* (New York: G. P. Putnam's Sons, 1924), p. 312; Gunter G. Sehm, "On a Possible Unknown Species of Giant Devil Ray, *Manta* sp.," *Cryptozoology* 12 (1996): 19–29; Karl Shuker, "There in Black and White!" *Fortean Times,* no. 131 (February 2000): 18–19.

## Behemoth

Amphibious animal of the Bible; *see* SEMIMYTHICAL BEASTS.

*Etymology:* Plural form of the Hebrew *behemah* ("beast"), inclusive of all wild and domesticated animals.

*Physical description:* Robust body. Nose

"pierces through snares." Long, strong tail "like a cedar."

*Behavior:* Herbivorous. Hearty drinker ("can draw up Jordan into his mouth").

*Habitat:* Forested rivers or swamps.

*Possible explanations:*

(1) A surviving sauropod dinosaur similar to the MOKELE-MBEMBE of Central Africa, suggested by Roy Mackal.

(2) The Hippopotamus (*Hippopotamus amphibius*) was first advocated by Samuel Bochart in 1663. Its small tail is a problem, but it is amphibious, robust, and herbivorous.

(3) The Nile crocodile (*Crocodylus niloticus*) has a strong tail, but it is a carnivore.

(4) The African elephant (*Loxodonta africana*) has been suggested by Georg Kaspar Kirchmayer and Sylvia K. Sikes.

*Sources:* Bible, Old Testament (Job 40:15–24); Samuel Bochart, *Hierozoicon, sive, bipartitum opus De animalibus Sacrae Scripturae* (London: John Martin and Jacob Allestry, 1663), vol. 2, chap. 15; Edmund Goldsmid, ed., *Un-natural History, or Myths of Ancient Science: Being a Collection of Curious Tracts on the Basilisk, Unicorn, Phoenix, Behemoth or Leviathan, Dragon, Giant Spider, Tarantula, Chameleons, Satyrs, Homines Caudati, &c.* (Edinburgh: Edmund Goldsmid, 1886); Marvin H. Pope, ed., *Job* (Garden City, N.Y.: Doubleday, 1965), p. 266; Roy P. Mackal, *A Living Dinosaur? In Search of Mokele-Mbembe* (Leiden, the Netherlands: E. J. Brill, 1987), pp. 5–7.

## Beithir

Large water SNAKE of Scotland.

*Etymology:* Gaelic and Irish, "serpent," "beast," or "bear," often with a supernatural connotation.

*Physical description:* Length, 9–10 feet.

*Behavior:* Active in summer.

*Habitat:* Lakes, caves.

*Distribution:* Around Loch a' Mhuillidh, Highland, Scotland.

*Possible explanations:*

(1) The Grass snake (*Natrix natrix*) is a greenish-olive snake with a yellowish collar around the neck. It occasionally grows up to 6 feet 6 inches long in Southern Europe, but in England and Wales, it generally attains a length of only 5 feet 9 inches. It likely inhabited the lowlands of Scotland at one time. It favors areas near lakes or streams.

(2) The European eel (*Anguilla anguilla*) rarely grows longer than 4 feet. The fishes spawn in the Atlantic Ocean, and the larvae transform into elvers on their migration route to freshwater streams and rivers in Europe, where they live for many years.

*Sources:* John Gregorson Campbell, *Superstitions of the Highlands and Islands of Scotland* (Glasgow, Scotland: J. MacLehose and Sons, 1900), pp. 223–224; Karl Shuker, "Sideshow," *Strange Magazine*, no. 15 (Spring 1995): 32.

## Bennu Bird

The sacred BIRD of Egypt that escorted souls to heaven. Found in texts of the Fifth Dynasty of the Old Kingdom (2498–2345 B.C.). It was a symbol of Osiris and resurrection and thus a possible source for the Greek PHOENIX.

*Etymology:* The ancient Egyptian *bn.w* or *benu,* for both "purple heron" and "date palm."

*Physical description:* Giant, heronlike bird. Taller than a man. White plumage. Twin red and gold plumes (or tufts) on head. Pointed bill. Slender, curved neck. Long tail feathers. Long legs.

*Behavior:* Gregarious.

*Distribution:* Egypt; Arabia.

*Significant sightings:* A large stork or heron is shown on a painted bas-relief on the inner wall of a tomb of an officer in the household of Pharaoh Khufu (2589–2566 B.C.), the builder of the Great Pyramid.

Enormous, conical bird nests, about 15 feet tall, were discovered along the coast of the Gulf of Suez, Egypt, by James Burton around 1822. The local Arabs told him they were built by a large, storklike bird that lived in the area until recently.

*Possible explanations:*

(1) The Gray heron (*Ardea cinerea*) is 3 feet 2 inches long and gray, with a black shoulder patch and black crest. It often stands

The BENNU BIRD. *Detail from the Papyrus of Anhai (British Museum no. 10,472). (Fortean Picture Library)*

with its neck in an S curve and is a common sight in lakes, rivers, and marshes of North Africa and the Arabian coast.

(2) The Goliath heron (*Ardea goliath*) is the largest living heron, nearly 5 feet long. It is blue-gray above, with a brown shoulder patch and a marbled pattern on its neck but no crest. It favors large lakes and swamps where it can wade deeper than other birds. The neck is folded when resting, but it can suddenly straighten, shooting the head forward. The bird is found on both sides of the Red Sea.

(3) The Purple heron (*Ardea purpurea*) is about 3 feet long, with a snakelike neck held in a distinctive kink. It is purplish-brown with a long, black crest and is found throughout Africa and the Middle East.

(4) A surviving Giant heron (*Ardea bennuides*), an extinct bird that was larger than the goliath heron. Fossil bones have been found on the island of Umm Al Nar near Abu Dhabi in the United Arab Emirates. Radiocarbon dating indicates the birds lived in the third millennium B.C., which encompasses the Egyptian First to the Eleventh Dynasties. Ella Hoch gave the bird its scientific name based on the possibility that it had inspired the Bennu bird symbol.

*Sources:* Bonomi, "On a Gigantic Bird Sculptured on the Tomb of an Officer of the Household of Pharaoh," *American Journal of Science* 49 (1845): 403–405; A. Wiedemann, "Die Phönix-Sage im alten Aegypten," *Zeitschrift für Ägyptische Sprache und Altertumskunde* 16 (1878): 89–106; *The Book of the Dead: The Papyrus of Ani in the British Museum,* trans. E. A. Wallis Budge [1895] (New York: Dover, 1967), pp. 280–282, 339; Karl Shuker, *In Search of Prehistoric Survivors* (London: Blandford, 1995), pp. 73–74.

## Bergman's Bear

Distinct variety or subspecies of brown BEAR of East Asia.

*Physical description:* Black. Short fur. Exceedingly large. Weight, 1,100–2,500 pounds. Large skull.

*Tracks:* Size, 14.5 inches by 10 inches.

*Distribution:* Southern Kamchatka Peninsula, Siberia.

*Present status:* This giant variety is likely extinct. The last known specimen, a pelt from Ust'-Kamchatsk, was examined by Swedish zoologist Sten Bergman in 1920.

*Possible explanation:* Brown bears (*Ursus arctos*) vary considerably in appearance; this may just have been a regional variant of the Siberian brown bear (*U. a. beringianus*) rather than a subspecies.

*Sources:* Sten Bergman, "Observations on the Kamchatkan Bear," *Journal of Mammalogy* 17 (1936): 115–120; Terry Domico, *Bears of the World* (New York: Facts on File, 1988), pp. 50–51; Igor A. Revenko, "Status and Distribution of Brown Bears in Kamchatka, Russian Far East," *Proceedings of the Tenth International Conference on Bear Research and Management,* Fairbanks, Alaska, 1995; Andrew D. Gable, "Bergman's Bear," December 19, 2000, http://www.cryptozoology.com/cryptids/godbear.php.

## Beruang Rambai

Unknown PRIMATE of Southeast Asia.

*Etymology:* Land Dayak (Austronesian), "long-haired bear," the common name for the sun bear.

*Variant name:* Bali djakai (Lawangan/Austronesian, "demon").

*Physical description:* Robust body. Shoulder height, 4 feet. Height standing erect, 6 feet. Covered in black hair. Bullet-shaped head. Bull neck. Hair on arms and thighs is 3 inches long. Thick legs.

*Behavior:* Walks on all fours. Stands on its hind legs occasionally. Beats its chest.

*Tracks:* Both humanlike and bearlike.

*Distribution:* Central Kalimantan, Indonesia, as well as in neighboring Sarawak State, Malaysia, both on the island of Borneo.

*Significant sightings:* In the 1930s, Leonard Clark ran across a Bali djakai at a water hole in the Borneo mountains. It picked up a helmet left behind, detected the scent of Clark and his guide, beat its chest, and disappeared into the bush.

Gathorne Gathorne-Hardy, earl of Cranbrook, collected descriptions of the Beruang rambai in the 1960s and concluded it was neither bear nor orangutan.

*Possible explanations:*

(1) The Sun bear (*Helarctos malayanus*) is a logical candidate, based on the name alone, though its hair is short.

(2) Misidentified Orangutan (*Pongo pygmaeus*).

*Sources:* Leonard Clark, *A Wanderer till I Die* (New York: Funk and Wagnalls, 1937), pp. 174, 188–195; Odette Tchernine, *The Yeti* (London: Neville Spearman, 1970), pp. 77–78; Jeffrey A. McNeely and Paul Spencer Wachtel, *Soul of the Tiger* (New York: Doubleday, 1988), p. 259.

## Big Bird

Large BIRD of North America, similar to the legendary THUNDERBIRD.

*Etymology:* Descriptive, though partially inspired by the character "Big Bird" on the U.S. television series *Sesame Street* (1969– ).

*Variant names:* GIANT OWL, Giasticutus (in the Ozark Mountains), MOTHMAN, PIASA, Tacuache (Spanish, "opossum"), THUNDERBIRD, THUNDERBIRD (PENNSYLVANIA).

*Physical description:* Length, 3–8 feet standing upright. Black, gray, or brown plumage. Head and short neck, either feathered and eaglelike or bald and vulturelike. Long, curved beak. Wingspan, 8–30 feet. Narrow wings. White wing tips. Short legs.

*Behavior:* Soars with wings level, using sluggish or graceful wing beats. Possibly migrates from the Pacific Northwest to the southern United States in the winter; its northward return in spring coincides with the rainy season. In the Midwest, it may migrate to the Ozark Mountains in July and fly north to Wisconsin in April. Call is "whoo whoo whoo." Feeds on

*BIG BIRD photographed in the Big Thicket, Hardin County, Texas. (James Crocker/Fortean Picture Library)*

live mammals and carrion. Nests on cliffs. Attempts to abduct human children have been reported.

*Tracks:* Three-toed. Length, 12 inches. Width, 7 inches. Baseball-sized droppings.

*Habitat:* Mountain ranges along most of its migration path.

*Distribution:* A partial list of places where Big birds have been reported follows:

*Alberta, Canada*—Lake Louise.

*Arkansas*—Blytheville.

*Florida*—Matheson Hammock Park, Sand Key, Tamiami Trail.

*Illinois*—Alton, Bloomington, Caledonia, Covell, Downs, Freeport, Glendale, Keeneyville, Lawndale, Lincoln, Odin, Shelbyville Lake, Tremont, Waynesville.

*Kentucky*—Johnson County, Lee County, Rabbit Hash, Stanford.

*Massachusetts*—Easton Center.

*Mississippi*—Tippah County.

*Missouri*—Overland, Richmond Heights, St. Louis, West Plains.

*New Jersey*—Carteret, Great Notch.

*New York*—Elizabethtown, Hudson River, New Rochelle, Rome.

*Ohio*—Gallipolis, Lowell, Nelsonville.

*Oklahoma*—Red Hills.

*Ontario, Canada*—Ramore.

*Oregon*—Hillsboro.

*Puerto Rico*—Bayamón, Naranjito.

*Texas*—Amarillo, Bethel, Brownsville, Catfish Creek, Donna, Harlingen, Laredo, Los Fresnos, Montalba, Nueces, Olmito, Palestine, Possum Kingdom Dam, Poteet, Rio Grande City, Robstown, San Antonio, San Benito.

*Utah*—Salt Lake City.

*West Virginia*—Bergoo, Oceana, Point Pleasant, Webster Springs.

*Wyoming*—Glendo.

*Significant sightings:* Eagles have occasionally been reported to carry off children in their talons, though even the largest can only lift a few pounds. The most often cited cases were in Valais, Switzerland, in 1838 when five-year-old Marie Delex was carried off and eaten by a Golden eagle (*Aquila chrysaetos*); and in the fall of 1868 in Tippah County, Mississippi, when eight-year-old Jemmie Kenney was grabbed and dropped from a height sufficient to kill him. Another incident may have occurred in January 1895 near Bergoo, West Virginia, when ten-year-old Landy Junkins disappeared in the woods and locals began reporting a huge eagle that was nesting on nearby Snaggle Tooth Knob.

Numerous reports of a bird the size of a Piper Cub airplane came from St. Louis, Missouri, and adjoining areas of Illinois in April 1948. On April 26, St. Louis chiropractor Kristine Dolezal saw it nearly collide with a plane, but the animal flapped its grayish-black wings and flew off into the clouds. She could discern ridges across the wings when they were outspread.

On February 27, 1954, Gladie M. Bills and her daughter saw what she at first thought were six jets moving in circles, diving, and playing around at a high altitude near Hillsboro, Oregon. She looked at them through a telescope and saw they were birds with glossy white wings.

David St. Albans saw a large, black bird flying over a cornfield in Keeneyville, Illinois, in July 1968. It had a tuft of white feathers at the base of its neck, but the head and neck were bare.

On January 1, 1976, a black bird more than 5 feet long, with dark-red eyes and a thick, 6-inch beak, was seen standing in a plowed field 100 yards away by two children near Harlingen, Texas. Sightings continued in the Rio Grande Valley for two months. On January 7, Alvérico Guajardo went out to see what had collided with his trailer home near Brownsville and saw a 4-foot-tall, winged creature with a long beak and covered in black feathers; it shrieked as Guajardo ran next door. On February 24, three schoolteachers driving to work near San Antonio saw a bird with a 15- to 20-foot wingspan gliding above their cars. They said it looked like a pteranodon, an extinct flying reptile. Further reports took place in December 1976.

On July 25, 1977, in Lawndale, Illinois, ten-year-old Marlon Lowe was picked up by his shirt by one of two large birds that came soaring in from the south. He screamed and punched at the bird until it dropped him after carrying him 30–40 feet. His parents and two other friends ran outside and saw the birds as they flew away. Marlon later picked out photos of California condors as the bird that attacked him. Over the next two weeks, there were at least eight other reports of similar birds in central Illinois.

Paramedic James Thompson saw a pterodactyl-like bird, with a 5- to 6-foot wingspan, gliding through the air early in the morning of September 14, 1982, east of Los Fresnos, Texas. It had a hump on its back and a pouch on its neck.

Reynaldo Ortega saw a giant bird standing on the roof of his house in Naranjito, Puerto Rico, on April 23, 1995. It was black and eaglelike, 3–4 feet tall, with a thick neck and piercing eyes. He thought it had a wolflike muzzle instead of a beak.

*Possible explanations:*

(1) The Turkey vulture (*Cathartes aura*), though all New World vultures and condors are incapable of gripping prey with their feet. This bird is widespread in the southeastern United States all year and is common through most of the rest of the country in the summer. Its wingspan is nearly 6 feet. Length, more than 2 feet. It has a distinctive, bare red head.

(2) The Black vulture (*Coragyps atratus*) is widespread in the southeastern United States and has been slowly expanding its range to the northeast. Its wingspan is nearly 5 feet. Length, 2 feet. The bare head is gray.

(3) The King vulture (*Sarcoramphus papa*) is a rain forest carrion feeder found from southern Mexico to Argentina. It has a brightly colored bald head, broad wings, and a short tail. It is also thought to prey on small reptiles and young mammals, though it lacks the strength to carry them. Length, 2 feet 6 inches.

(4) The Andean condor (*Vultur gryphus*), the world's largest bird of prey, is over 4 feet long with a wingspan up to 10 feet 6 inches,

though the average is 9 feet 3 inches for males. It weighs 23–25 pounds. The color is shiny black with white patches on the wings, a white ring on the neck, and a bare, gray-red head. It soars effortlessly without flapping its wings. Carrion is its normal diet, supplemented with seabird eggs. It will occasionally attack calves, fawns, or beached whales. It is often seen along the South American Pacific coast but returns to the Andes Mountains to roost. In 1992, some female Andean condors were introduced in Los Padres National Forest, California, as a test release for California condors, but they were all recaptured later.

(5) The California condor (*Gymnogyps californianus*) is the largest U.S. vulture, reaching a length of 4 feet, a wingspan of 9 feet 4 inches, and a weight of 20–25 pounds. Unsubstantiated wingspans up to 11 feet 3 inches have been claimed. The bird is black with white wing linings and has a naked, red-orange head that changes color with its mood. In 1987, the few remaining wild birds were caught for a captive breeding program; reintroduction began in 1992 in remote sites of Los Padres National Forest, California. Pleistocene fossil remains of this bird have been found in New York and Florida, as well as Arizona and New Mexico. There is evidence that these condors returned to the Southwest sporadically as early as the 1700s in response to the introduction of large herds of cattle, horses, and sheep that replaced the extinct Pleistocene megafauna as a source of carrion.

(6) The Golden eagle (*Aquila chrysaetos*) is the largest eagle in the United States and Europe, with a wingspan of 7 feet. Though it winters in eastern states, it is fairly scarce. It soars with wings upcurved and takes prey (small mammals) opportunistically with outstretched talons. It has a golden-bronze nape. There is some evidence in New Mexico and Oregon that it has attacked calves weighing over 200 pounds. The white head of the Bald eagle (*Haliaeetus leucocephalus*) makes it almost too recognizable to be misidentified. Both eagles grow larger in northern latitudes and higher altitudes.

(7) The Harpy eagle (*Harpia harpyja*) is the dominant bird of prey in Central and South America. Its massive feet can pick up monkeys, sloths, opossums, and snakes. It has a crest of dark feathers on its head and a wingspan of nearly 10 feet. Length, more than 3 feet. The Monkey-eating eagle (*Pithecophaga jefferyi*) of the Philippines is smaller. The rare Solitary eagle (*Harpyhaliaetus solitarios*) of Mexico is 2 feet 6 inches long but has never been reported north of Sonora.

(8) The Crowned hawk-eagle (*Stephanoaetus coronatus*) of Africa preys on small antelopes. It has a wingspan up to 6 feet 9 inches. There is some evidence from Zimbabwe and Zambia that it will occasionally attack a child.

(9) Steller's sea eagle (*Haliaeetus pelagicus*) averages an 8-foot wingspan and nests on the Asiatic side of the Bering Sea. It has a huge, orange-yellow bill and white shoulders and makes occasional visits to Alaska. The White-tailed eagle (*Haliaeetus albicilla*) is smaller, less distinctive, and an even rarer visitant to the Aleutians.

(10) The Common black-hawk (*Buteogallus anthracinus*) has a wingspan greater than 4 feet and is found in cottonwood groves in Arizona.

(11) The Crested caracara (*Caracara cheriway*), a black-crested, white-necked falcon, is fairly common in southern Texas and southern Arizona. Its wingspan is more than 4 feet.

(12) The Griffon vulture (*Gyps fulvus*) is a large, brown-winged carrion feeder with a wingspan of 9 feet and a length of nearly 4 feet. It breeds in Spain, several locations in North Africa, the Balkans, Greece, Turkey, the Caucasus, Israel, and eastward to Central Asia.

(13) The White-backed vulture (*Gyps africanus*), with a wingspan of 7 feet 10 inches, is Africa's commonest large vulture. A juvenile specimen was apparently responsible for a Big bird report in Ohio in 1972.

(14) The Great blue heron (*Ardea herodias*) is common in North American wetlands and reaches a length of nearly 4 feet.

(15) The Southern ground hornbill (*Bucorvus cafer*), the largest hornbill, is native to Central and South Africa. It attains a length of 3 feet 6 inches.

(16) The Marabou stork (*Leptoptilos crumeniferus*) grows to nearly 5 feet in length and is identifiable by its huge throat wattle and massive, wedge-shaped bill. Its range is limited to sub-Saharan Africa.

(17) The Sandhill crane (*Grus canadensis*) is about the size of a great blue heron but is gray, mottled with rust stains. It has a wingspan up to 6 feet 5 inches.

(18) The Whooping crane (*Grus americana*) is a rare and unusual sight, as it is limited to about 100 birds wintering in coastal Texas. Its length is 4 feet 4 inches, with a wingspan of 7 feet 3 inches. Males are white, with black primary feathers. Its migration path is from northern Alberta to south Texas.

(19) The Black-footed albatross (*Phoebastria nigripes*) is rarely seen over land although a fair number of these birds are seen over the Pacific Ocean in the spring and summer. Mostly dark gray, with a wingspan of over 7 feet.

(20) The Wandering albatross (*Diomedea exulans*) has a wingspan around 9 feet 9 inches, but this is primarily an Antarctic bird, with Peru as its farthest extension north. It rarely travels any distance inland.

(21) The Wood stork (*Mycteria americana*) is found in Florida and the Gulf Coast. White with a dark head and neck, it reaches a length of 3 feet 4 inches and has a wingspan over 5 feet.

(22) The Jabiru (*Jabiru mycteria*) is a tropical, white, black-headed stork with a roseate neck that is only occasionally seen in Texas. Its length is 4 feet 4 inches.

(23) The American white pelican (*Pelecanus erythrorhynchos*) has a wingspan up to 9 feet. Its black flight feathers are distinctive when it is soaring.

(24) A surviving Teratorn (Teratornithinae), a member of a subfamily of predatory vultures that resembled reptiles in some ways. Their jaws were designed to swallow living prey, though their talons were not designed for seizing. They probably used their sharp, hooked beaks to catch animals. The largest known flying bird, *Argentavis magnificens,* weighed 158 pounds, stood 5–6 feet tall, and had a wingspan of 23–25 feet. It lived in Argentina in the Late Miocene, 8–5 million years ago. In North America, *Teratornis merriami* weighed about 36 pounds and had an 11 foot 6 inch–12 foot 6 inch wingspan, while *T. incredibilis* of Nevada and California lived in the Pleistocene and had a wingspan of 17–19 feet.

(25) A surviving pterosaur, a fossil FLYING REPTILE that supposedly died out at the end of the Cretaceous period, 65 million years ago.

(26) A surviving La Brea condor (*Breagyps clarki*), a slightly smaller bird than the California condor with a long, slender beak, known from Pleistocene fossils in Nevada and southern California.

(27) An unknown species of giant bat, suggested by Mark A. Hall for the Rio Grande Valley sightings of 1976.

*Sources:* Felix-Archimede Pouchet, *The Universe: Or, The Wonders of Creation, the Infinitely Great and the Infinitely Little* (Portland, Maine: H. Hallett, 1883), pp. 236—239; "A Modern Roc: West Virginia Mountaineers Terrorized by a Giant Bird," *St. Louis (Mo.) Globe-Democrat,* February 25, 1895, p. 7; Vance Randolph, *We Always Lie to Strangers* (New York: Columbia University Press, 1951), pp. 63—66; Gladie M. Bills, "Bird Like UFO's," *Fate* 7 (December 1954): 128—129; Jerome Clark and Loren Coleman, "Winged Weirdies," *Fate* 25 (March 1972): 80—89; Maurice Kildare, "Winged Terror of the Oklahoma Hills," *True Frontier,* October 1972, pp. 29—30, 50—53; Jerome Clark, "Unidentified Flapping Objects," *Oui,* October 1976, pp. 94—100, 105—106; *Bloomington (Ill.) Daily Pantograph,* July 27, 1977, p. A-11; *Chicago Daily News,* July 27, 1977, p. 18; Jerome Clark and Loren Coleman, *Creatures of the Outer Edge* (New York: Warner, 1978), pp. 165—188, 190—194, 225—227; Mark A. Hall, *Thunderbirds! The Living Legend of Giant Birds* (Bloomington,

Minn.: Mark A. Hall, 1988); Magdalena del Amo-Freixedo, "Current Happenings on Puerto Rico," *Flying Saucer Review* 36, no. 4 (Winter 1991): 19; Jorge J. Martín, "Tambien animales imposibles: ¿Que ocurre en Puerto Rico?" *Evidencia OVNI*, no. 6 (1995): 32—33; Gerald Musinsky, The Thunderbird: Living Fossil or Living Folklore, 1997, http://members.aol.com/_ht_a/mokele/cryptozoologicalrealms/html_3.2/english/reflections/fossil.html.

## Big Grey Man

Paranormal ENTITY of Scotland, similar to a TRUE GIANT hominid.

*Variant names:* Fear liath mór, Ferla mór, Ferlas mhór, Ferlie more, Fomor (all Gaelic).

*Physical description:* Height, 10–20 feet. Gray or olive colored, or covered with short brown hair. Pointed ears. Broad shoulders. Long, waving arms.

*Behavior:* Often appears during a fog or mist. Stands erect. Makes odd crunching noises or a high-pitched humming sound. Occasionally accompanied by ghostly music or voices. Follows hikers. Creates an icy feeling in the air—a cold, physical presence that induces fear, panic, depression, or apathy.

*Tracks:* Rare and unreliable.

*Distribution:* The summit of Ben Macdhui in the Cairngorm Mountains, Grampian, Scotland.

*Significant sightings:* In 1891, Norman Collie was returning from the summit when he heard the crunch of footsteps behind him. For every step he took, he heard another crunch, as if someone had a stride three or four times the size of his own. He ran downhill the last few miles.

In 1942, Sydney Scroggie was camping out at the Shelter Stone by the Garbh Uisge on Ben Macdhui when he saw a tall, stately figure taking deliberate steps across the burns flowing into Loch Avon.

On December 2, 1952, James Alan Rennie photographed a series of tracks in the snow in a straight line on the mountain. Each print was about 19 inches long and 14 inches wide, with a stride of 7 feet. At one point, the tracks jumped a road over a distance of 30 feet.

*Possible explanations:*

(1) The Brocken spectre is the giant shadow of an observer cast on a wall of white mist and often surrounded by one or more concentric rings of color (the "glory") centered on the figure's head. However, this optical effect does not seem to be what hikers have been reporting.

(2) The high altitude, isolation, and meteorological conditions on Ben Macdhui may produce hallucinations. The warm Föhn wind of southern Germany is said to produce headaches, nausea, aching joints, fatigue, irritability, apathy, and depression. A similar effect might take place in the Cairngorms, causing confusion, stress, and disorientation.

*Sources:* Affleck Gray, *The Big Grey Man of Ben MacDhui* (Aberdeen, Scotland: Impulse, 1970); Ronald J. Willis, "Ben MacDhui: The Haunted Mountain," *INFO Journal*, no. 15 (May 1975): 2—5; F. W. Holiday, *The Goblin Universe* (St. Paul, Minn.: Llewellyn, 1986), pp. 152—154; Karl Shuker, "The Big Grey Man," *Fate* 43 (May 1990): 58—68; Andy Roberts, "The Big Grey Man of Ben Macdhui and Other Mountain Panics," *Fortean Studies* 5 (1998): 152—171.

## Big Wally

FRESHWATER MONSTER of Oregon.

*Etymology:* After the lake.

*Physical description:* Varied. Native American legends describe a manatee-like animal. A doubtful 1885 tale by a prospector involves a 100-foot-long monster with a hippo's head. Other sightings seem to be of a large sturgeon or an animal like OGOPOGO.

*Distribution:* Wallowa Lake, Oregon.

*Significant sighting:* On June 30, 1982, Marjorie Cranmer and Kirk Marks observed a 50-foot creature creating waves along the northeastern shore. It had seven dark-colored humps.

*Possible explanation:* White sturgeon (*Acipenser transmontanus*) are not known in this lake, but they grow to 20 feet in length.

*Sources:* A. W. Nelson, *Those Who Came First* (LaGrande, Oreg.: A. W. Nelson, 1934), p. 17; Vance Orchard, *Just Rambling around*

*Blue Mountain Country* (Walla Walla, Wash.: Robert Bennett, 1981); Mike Dash, "The Reporting of a Lake Monster," *Fortean Times,* no. 44 (Summer 1985): 42—43; John Kirk, *In the Domain of Lake Monsters* (Toronto, Canada: Key Porter Books, 1998), pp. 157—159.

## Bigfoot

GIANT HOMINID of western North America. By extension, the term is also applied to similar HO-MINIDS observed elsewhere.

*Etymology:* Named in 1958 when a series of huge tracks was found near Bluff Creek in northern California. Coined by newspaper columnist Andrew Genzoli, in the *Humboldt (Calif.) Times,* October 5, 1958. Plural is usually *Bigfoot,* sometimes *Bigfeet.*

*Scientific names: Paranthropus eldurrelli,* proposed by Gordon R. Strasenburgh Jr. in 1971; *Gigantopithecus canadensis, Australopithecus canadensis,* or *Gigantanthropus canadensis,* all proposed by Grover Krantz in 1985.

*Variant names:* JACKO, MATAH KAGMI, Mountain devil, PATTY, SASQUATCH, Tuni-ka (Tanana/Na-Dené). *See also* CANNIBAL GIANT.

*Physical description:* Bulky, robust body. Height, 6–9 feet, with an average of 7 feet 10 inches. Average weight estimated at 660 pounds. Shaggy body hair, ranging from dark brown or black to light brown and gray. Color variation does not seem related to height or age. Small, pointed head. Sloping forehead. Flat face. Heavy browridge with upcurled fringe of hair. Facial hair except around nose, mouth, and ears. Deep-set eyes. Broad and flat nose. Wide mouth. Short, thick neck. Huge shoulders and chest. Females have large, hairy breasts. Arms are thick and long in proportion to height.

*Behavior:* Primarily nocturnal. Walks upright with a long stride and long arm swing, leaning forward slightly with its knees bent. Not afraid of walking in water, perhaps even using water-ways as travel paths. Top running speed may be as much as 35–40 miles per hour. Inactive in cold weather. Solitary, though family groups have occasionally been reported. Calls are high-pitched whistles, screams, and howls, including: "eeek-eeek-eeek," "sooka-sooka-sooka," "gob-

BIGFOOT *sculpture carved from a redwood stump in the 1960s by Jim McClarin in front of the Willow Creek (California) Chamber of Commerce. (Bill Lewinson/ PhotoArt by Burro)*

uh-gob-uh," "ugh-ugh-ugh," and "uhu-uhu-uhu." A strong, putrid odor often reported. Omnivorous (rodents, deer, roots, larvae, car-rion, berries, grasses, clams, fishes, and vegeta-bles). Searches for rodents by digging up rocks and piling them up. Splits rotted logs in search of grubs. May also pursue and kill deer. Kid-nappings of humans, usually females, have been reported. Sometimes throws rocks at people. Shows curiosity about human activity. No ap-parent use of fire or tools. The population in the Pacific Northwest has been estimated as 1,500–2,000 adult individuals.

*Tracks:* Five-toed human print 4–27 inches long, with an average length of 14–18 inches. The width ranges from 3 to 13.5 inches at the ball of the foot, with an average of 7.2 inches. Heels are 1.5–9 inches wide, with an average of 4.8 inches. Toes are slightly curled and in a straight line like peas in a pod. Big toe is not ap-preciably larger than or separated from the oth-ers. A substantial ridge of soil or sand separates the toes from the ball of the foot. The foot is narrow in the middle (sometimes described as an "hourglass" shape), and the impression of the heel is deeper in the inner rather than the outer side. Flat arches. Transverse, midsole dermal ridges (dermatoglyphics) and sweat pores often present. Stride measures 4–6 feet. Tracks point straight ahead, with feet turning neither inward nor outward. Tracks are found in remote areas, and the movements indicated by the tracks (meandering, zigzagging) are typical for a wild animal. The morphology of the tracks, including small details, is uniform enough over a wide geographic area to suggest authenticity. Feces and hair samples have also been recovered.

*Habitat:* Montane forests.

*Distribution:* From northern California, Oregon, Washington, and Idaho north through British Columbia and southwestern Alberta to Alaska.

*Significant sightings:* In the fall of 1869, a hunter saw a male and female Bigfoot near Orestimba Creek in Stanislas County, California. The male was covered with dark-brown and cinnamon hair, stood 5 feet tall, made whistling noises, and disrupted the hunter's campfire.

When the steamer *Capilano* put into an Indian village at Bishop's Cove, British Columbia, in March 1907, the crew was assailed by terrorized villagers who wanted to escape from a 5-foot-tall monkey covered with long hair that came on the beach at night to dig clams and howl.

A group of miners claimed to have taken a shot at a huge, apelike creature near a mine in a canyon on the east side of Mount St. Helens, Washington, in July 1924. Later, one of them, named Fred Beck, shot another ape in the back three times, causing it to fall off a cliff. At night, a group of the apes assaulted the miners in their cabin for five hours, pounding on the walls and hurling rocks. The cabin had no windows, so the miners couldn't see what was attacking them. The next day, they packed up and returned to Kelso and told their story. Large tracks were found in the canyon, which was thereafter named Ape Canyon.

Prospector Albert Ostman claimed he was kidnapped in the summer of 1924 near Toba Inlet, British Columbia, and lived six days with a Bigfoot family consisting of an older male and female and two younger ones, also male and female. The adult male was between 7 and 8 feet tall and had carried Ostman in his sleeping bag for three hours to a remote valley. Ostman tried to escape, but the old male blocked his way. While he was there, Ostman made many observations about their lifestyle and habits. He finally tricked the adult male into swallowing the contents of his snuff tin, and in the ensuing confusion, he made his escape back to the coast.

In October 1941, Jeannie Chapman and her three young children fled from a Bigfoot that came toward their isolated cabin near Ruby Creek, British Columbia. Later, 16-inch, humanlike footprints were found circling the house. The creature had apparently entered a woodshed and opened up a 55-gallon barrel of salt fish.

William Roe was climbing Mica Mountain southwest of Tete Jaune Cache, British Columbia, in October 1955 when he saw what he at first thought was a grizzly bear about 75 yards away. Soon, he realized it was a huge female, 6 feet tall and 3 feet wide, completely covered with dark-brown, silver-tipped hair. Its arms reached almost to its knees, and it had breasts. It squatted down to eat the leaves from some bushes, but when it saw him, it stood up and walked away cautiously. Roe leveled his rifle to shoot it but thought it was too human-looking to kill.

Gerald Crew and others found large numbers of giant tracks around their road construction camp in high country near Willow Creek, California, several times between August 1958 and February 1959. The tracks descended an incline of 75°, and the average stride was more than 4 feet. Occasionally, the track maker would disturb heavy fuel drums and steel culverts, and once it moved a 700-pound wheel belonging to earth-moving equipment. Recently, some doubt has been cast on these events; it's been alleged that the footprints were hoaxed by the construction crew contractor and foreman, Ray and Wilbur Wallace. However, the Wallaces had difficulty keeping contract workers because the tracks terrified the crew.

On October 20, 1967, Roger Patterson and Bob Gimlin took 952 frames of 16-millimeter film showing a female Bigfoot walking away from them near Bluff Creek, California. *See* PATTY.

Numerous 18-inch tracks of a Bigfoot with an anatomically accurate clubfoot deformity (Talipes equino-varus) in its right foot were found near Bossburg, Washington, in October 1969.

Joe Medeiros and Dick Brown saw an 8- to 9-foot Bigfoot standing under an oak tree at a trailer park near The Dalles, Oregon, on June 2, 1971. Brown observed it through a telescopic rifle sight and said there was a crest on its head. It had muscular shoulders and walked with an exaggerated swinging of its arms. The next day, 20-inch-long tracks were found in the crushed grass.

On October 21, 1972, Alan Berry made a high-quality audio recording of Bigfoot calls at an altitude of 8,500 feet in the Sierra Nevada, California. The vocalization includes a wide variety of whistles and sounds, some quite humanlike. A rough transcription might read: "Gob-uh-gob-uh-gob, ugh, muy tail." A pitch-frequency analysis undertaken in 1977 by R. Lynn Kirlin and Lasse Hertel indicated that there was more than one speaker, that the animals were probably larger than adult male humans, and that their larynxes must be significantly longer than a human's in order to produce the sounds.

On June 10, 1982, U.S. Forest Service Patrolman Paul Freeman saw a Bigfoot about 8 feet tall at relatively close range in the Umatilla National Forest, Washington. On the day of the sighting, plaster casts were made of tracks the creature left. On June 16, two different sets of prints were found a few miles away at Elk Wallow by Forest Service biologist Rodney Johnson and U.S. Border Patrol tracker Joel Hardin, who made casts. All of the prints show the impressions of dermal ridges and sweat pores, features that are consistent with the friction skin on the soles of higher primates. Although Grover Krantz considered the tracks genuine, Johnson and Hardin felt they were hoaxed, since they were too shallow, followed an unnaturally straight line, had an oddly uniform stride and

pressure whether going uphill or down, appeared and disappeared abruptly, and showed abnormally pronounced dermal ridges.

In August 1987, Agnes Perkins and Charlotte White were driving west along the Trans-Canada Highway when they saw a man on the side of the road ahead. As they got closer, they realized it was a 7-foot Bigfoot covered with black hair. It climbed up the steep embankment on two legs. They had seen it for about forty-five seconds.

Early in the morning of May 23, 1988, Susan Ray Adams and Scott Stoness encountered what they thought was a bear as they were going to the public washrooms at the Crandell Lake Campground in Waterton Lakes National Park in Alberta. It snorted at them, and they ran for their car. They turned on the car headlights and could see that the animal was walking around on two legs. Another couple had seen the same thing and were similarly terrified. The animal was 8 feet tall and taking huge strides, arms swinging.

At Bella Coola, British Columbia, on November 11, 1989, Jimmy Nelson, his mother, and a friend noticed a terrible odor and saw a 7- to 8-foot creature approaching the back porch where some deer meat was hanging. It returned the following night, and the boys chased it toward a nearby creek.

A five-second video recording of a Bigfoot taken by a television film crew on August 28, 1995, in the Jedediah Smith Redwoods State Park near Crescent City, California, is interesting, if inconclusive. It shows a massive, hairy, black creature with a distinctly erect penis.

Psychologist Matthew A. Johnson was hiking with his family on July 1, 2000, near Oregon Caves National Monument, Oregon, when they smelled something skunky and heard something making "whoa whoa whoa" sounds. Johnson got a brief glimpse of a Bigfoot while he was off the trail by himself.

The body imprint of what might have been a Bigfoot's forearm, hip, thigh, and heel was found September 22, 2000, in the Skookum Meadows area of Gifford Pinchot Forest, Washington, by a Bigfoot Field Researchers Organization (BFRO) expedition led by LeRoy Fish,

Derek Randles, and Richard Noll. More than 200 pounds of plaster were required to produce a cast of the entire impression.

*Artifacts:* Numerous stone carvings referred to as "anthropoid ape heads" have been found in the Columbia River valley of Oregon and Washington. One was obtained from the Wakemap mound near The Dalles, Oregon, which would date it between 1500 B.C. and A.D. 500. They are about 6–7 inches long, carved from basalt, and show a being with a flat face, large eyes, browridges, splayed nostrils, and full lips; some have folds of loose flesh below the chin, and one appears to have a sagittal crest. Some clearly represent human faces, mountain sheep, or seals, but at least seven are apelike and could represent Bigfoot. Myron Eells, "The Stone Age of Oregon," *Annual Report of the Smithsonian Institution,* 1885, pp. 283–295; Alfred Russel Wallace, "Remarkable Ancient Sculptures from North-West America," *Nature* 43 (1891): 396; James Terry, *Sculptured Anthropoid Ape Heads Found in or near the Valley of the John Day River* (New York: J. J. Little, 1891); Frederick W. Skiff, *Adventures in Americana* (Portland, Oreg.: Metropolitan, 1935), p. 186; Roderick Sprague, "Carved Stone Heads of the Columbia River and Sasquatch," in Marjorie Halpin and Michael M. Ames, eds., *Manlike Monsters on Trial: Early Records and Modern Evidence* (Vancouver, Canada: University of British Columbia Press, 1980), pp. 229–234.

*Possible explanations:*

(1) An upright Brown bear (*Ursus arctos*) or Black bear (*Ursus americanus*) might briefly be misidentified. However, bears have short hind legs, sloping shoulders, and visible ears. Bigfoot prints do not look anything like bear tracks, in which the first toe is the shortest and the third toe the largest. Bear foreprints and hindprints overlap, the big toes are on the inside of the stride, and their feet turn inward. Black bear hind feet are about 6–7 inches long and 3–5.5 inches wide. Brown bear hind feet range from 10–11 inches long and 6–6.5 inches wide for grizzlies to 16 inches long and 10 inches wide for Kodiaks. Distortion from overlapping prints would make these seem larger.

*Artist's conception of* BIGFOOT. *(William M. Rebsamen)*

(2) Hoaxes of both sightings and tracks have definitely occurred. However, the long stride of many of the tracks would be difficult for one individual to fake, and the dermal ridges and sweat pores are unlikely touches for a hoaxer (especially prior to the publicity they were given after the discovery of this feature). Tracks are often found serendipitously in remote places, where the likelihood of their being found is equally remote. The depth of some prints would also require a hoaxer to exert as much as 450 pounds of pressure in compact soil.

(3) An evolved *Gigantopithecus blacki,* proposed by Grover Krantz. This huge-jawed Pleistocene ape lived as recently as 500,000 years ago in southern China and Vietnam, while a smaller species, *G. giganteus,* dates to 9–6 million years ago in the Siwalik Hills of India and Pakistan. It is known only from jaw

fragments and isolated teeth. It had a massive jaw and low-crowned, flat molars with thick enamel caps adapted for chewing coarse vegetation. Its estimated height was 9–10 feet tall and weight was 900–1,200 pounds. However, no weight-bearing bones have been recovered, and it is possible that the animal's teeth and jaws were disproportionate to its body size. Ivan Sanderson considered an evolved *Gigantopithecus* the best candidate for his proposed category of NEO-GIANTS.

(4) A surviving, robust form of *Homo heidelbergensis,* an archaic human known from Middle Pleistocene fossils in Europe and Africa, suggested by Will Duncan. Named from a mandible discovered in 1906 in a gravel pit at Mauer near Heidelberg, Germany, this human is thought to have lived around 500,000 years ago.

(5) A surviving *Paranthropus robustus,* suggested by Gordon Strasenburgh. The youngest known remains of this early, exclusively African hominid were found at Swartkrans, South Africa, and are 850,000 years old. However, reasonable estimates of its size, based on postcranial bones, range from 95 to 145 pounds, making this an unlikely Bigfoot candidate.

(6) Loren Coleman has suggested a surviving *Meganthropus,* a little-known hominid genus described from two partial mandibles with large teeth found in Java in 1939 and 1941. A handful of other fragmentary finds have been included in this taxon, but there is no consensus on its status. Many regard this animal as belonging to *Homo erectus,* though some consider it pathologically oversized.

(7) A Neanderthal (*Homo neanderthalensis*) population is unlikely, as these distinctively cold-adapted hominids are not found outside Europe or West Asia any later than 30,000 years ago. However, if the Central Asian ALMAS represents an extant Neanderthaloid group, individuals could have migrated across Beringia in time to populate North America. But its small stature (the average for males was 5 feet 6 inches tall) seems to rule out this species.

(8) A surviving *Homo erectus* has been pro-
posed by Ray Crowe. Following an appearance in East Africa 2 million years ago, *erectus* hominids spread into Asia and possibly into Europe. Their subsequent evolutionary history is unclear, though recent evidence suggests their persistence in Zhoukoudien, China, until as recently as 250,000 years ago. Few postcranial bones offer any glimpse of *H. erectus* stature, though femurs from East Africa indicate an average height of 5 feet 7 inches for adult males— much too small for the robust Bigfoot. *H. erectus* also seems to have preferred open, arid environments.

(9) A homegrown variety of North American primate is extremely unlikely. The oldest primatelike mammal, *Purgatorius,* appeared in the West at the end of the Mesozoic and continued through the Early Paleocene, about 66–64 million years ago. Other North American protoprimates were the Plesiadapoidea and Carpolestidae from the Early Eocene; they had snouted faces and semi-grasping feet, and some species were as large as woodchucks. Fossils are mostly known from the Rocky Mountain region, which at that time consisted of lowland tropical forest. Recent molecular evidence indicates that these were more closely related to Flying lemurs (Order Dermoptera) than true primates. Also known from the beginning of the Eocene (about 55 million years ago) are the first members of the modern primates, the lemurlike Notharctidae and the tarsier-like Omomyidae, some of which apparently were as large as medium-sized monkeys. Forests shrunk with gradual cooling toward the end of the Eocene, and these arboreal species died off or migrated to South America. The genus *Ekgmowechashala* lingered until the Late Oligocene (28 million years ago) in Oregon and South Dakota. After this, there is no evidence of any primate occupation in North America until modern humans arrived, which archaeologists are now reluctantly accepting occurred as early as 40,000 years ago.

*Sources: Antioch (Calif.) Ledger,* October 18, 1870; Theodore Roosevelt, *The Wilderness*

*Hunter* (New York: G. P. Putnam's Sons, 1893); *Vancouver (B.C.) Province,* March 8, 1907; C. P. Lyons, *Milestones on the Mighty Fraser* (Victoria, B.C., Canada: J. M. Dent, 1950), pp. 28—30; "Giant Footprints at Ruby Creek Took Railway Fence in Stride," *Agassiz-Harrison (B.C.) Advance,* September 12, 1957, reprinted in *INFO Journal,* no. 64 (October 1991): 22—24, 41; Belle Rendall, *Healing Waters: History of the Harrison Hot Springs and Port Douglas Area* (Harrison Hot Springs, B.C., Canada: Belle Rendall, 1958), pp. 30—32; Ivan T. Sanderson, "The Strange Story of America's Abominable Snowman," *True,* December 1959, pp. 40—43, 122—126; Ivan T. Sanderson, *Abominable Snowmen: Legend Come to Life* (Philadelphia: Chilton, 1961), pp. 22—147; Roger Patterson, *Do Abominable Snowmen of America Really Exist?* (Yakima, Wash.: Northwest Research Association, 1966); John Green, *On the Track of the Sasquatch* (Agassiz, B.C., Canada: Cheam, 1968); Loren E. Coleman and Mark A. Hall, "Some Bigfoot Traditions of the North American Tribes," *INFO Journal,* no. 7 (Fall 1970): 2—10; Gordon R. Strasenburgh Jr., *Paranthropus: Once and Future Brother* (Arlington, Va.: Gordon R. Strasenburgh Jr., 1971); Don Hunter and René Dahinden, *Sasquatch* (Toronto, Canada: McClelland and Stewart, 1973); John Napier, *Bigfoot: The Yeti and Sasquatch in Myth and Reality* (New York: E. P. Dutton, 1973); Peter Byrne, *The Search for Bigfoot* (Washington, D.C.: Acropolis, 1975); B. Ann Slate and Alan Berry, *Bigfoot* (New York: Bantam, 1976); John Green, *Sasquatch: The Apes Among Us* (Seattle, Wash.: Hancock House, 1978); Roderick Sprague and Grover S. Krantz, eds., *The Scientist Looks at the Sasquatch,* 2d ed. (Moscow: University of Idaho Press, 1979); Marjorie H. Halpin and Michael M. Ames, eds., *Manlike Monsters on Trial: Early Records and Modern Evidence* (Vancouver, Canada: University of British Columbia Press, 1980); Janet and Colin Bord, *The Bigfoot Casebook* (Harrisburg, Pa.: Stackpole, 1982); Grover S. Krantz, "Anatomy and Dermatoglyphics of Three Sasquatch Footprints," *Cryptozoology* 2 (1983): 53—81; René Dahinden, "Whose Dermal Ridges?" *Cryptozoology* 3 (1984): 128—131; Grover S. Krantz, "A Species Named from Footprints," *Northwest Anthropological Research Notes* 19 (1986): 93—99; Donald Baird, "Sasquatch Footprints: A Proposed Method of Fabrication," *Cryptozoology* 8 (1989): 43—46; Thomas Steenburg, *The Sasquatch in Alberta* (Calgary, Alta., Canada: Western Publishers, 1990); Grover S. Krantz, *Big Footprints* (Boulder, Colo.: Johnson, 1992); Robert Michael Pyle, *Where Bigfoot Walks* (Boston: Houghton Mifflin, 1995); Loren Coleman, "Was the First 'Bigfoot' a Hoax?" *The Anomalist,* no. 2 (Spring 1995): 8—27; Mark A. Hall, *The Yeti, Bigfoot, and True Giants* (Minneapolis, Minn.: Mark A. Hall, 1997); Wolf H. Fanrenbach, "Sasquatch: Size, Scaling and Statistics," *Cryptozoology* 13 (1997—1998): 47—75; Grover S. Krantz, *Bigfoot Sasquatch Evidence* (Blaine, Wash.: Hancock House, 1999); Thomas Steenburg, *In Search of Giants: Bigfoot Sasquatch Encounters* (Surrey, B.C., Canada: Hancock House, 2000); "Bigfoot Leaves His Mark," *Fortean Times,* no. 142 (February 2001): 12; Tatsha Robertson, "No Ifs, Butts about Bigfoot," *Boston Globe,* July 14, 2001; Mike Quast, *Big Footage: A History of Claims for the Sasquatch on Film* (Moorhead, Minn.: Mike Quast, 2001); Will Duncan, "What Is Living in the Woods, and Why It Isn't Gigantopithecus," *Crypto Hominology Special,* no. 1 (April 7, 2001), pp. 44—49, at http://www.strangeark.com/crypto/ Cryptohominids.pdf; Bigfoot Field Researchers Organization, http://www.bfro.net; The Skookum Cast, http://www.bfro.net/news/ bodycast/index.html; Benjamin Radford, "Bigfoot at 50: Evaluating a Half-Century of Bigfoot Evidence," *Skeptical Inquirer* 26 (March-April 2002): 29—34; David J. Daegling, "Cripplefoot Hobbled," *Skeptical Inquirer* 26 (March-April 2002): 35—38.

## Bili Ape

Unknown PRIMATE of Central Africa.

*Etymology:* From the Bili Forest.

*Physical description:* Chimpanzee-like.

*Habitat:* Dense rain forest.

*Distribution:* Bili Forest, northeast Democratic Republic of the Congo.

*Significant sightings:* An ape skull of unknown type was found in the area around 1900 by an unnamed explorer.

In March 2001, an expedition to the Bili Forest by National Geographic Radio turned up ape feces, a ground nest, and a large footprint in the mud near a stream. Primatologist Richard Wrangham concluded they were chimpanzee traces, possibly an unknown variety.

*Possible explanation:* Unknown variety of Chimpanzee (*Pan troglodytes*), perhaps related to the KOOLOO-KAMBA.

*Source:* Karl Shuker, "Bemused in Bili," *Fortean Times,* no. 148 (August 2001): 18.

## Bilungi

WILDMAN of Central Africa.

*Physical description:* Height, 6 feet. Covered with brown hair. Powerful chest.

*Distribution:* Near Lac Tumba, Democratic Republic of the Congo.

*Source:* Bernard Heuvelmans, *Les bêtes humaines d'Afrique* (Paris: Plon, 1980), pp. 590–591.

## BIRDS (Unknown)

Birds (Class Aves) are warm-blooded animals that have no teeth, are covered with feathers, and are wonderfully adapted for true flight. Zoologists have long recognized that birds evolved from reptiles, but with the relatively recent discovery in China that some theropod dinosaurs had feathers (*Sinosauropteryx* and *Caudipteryx*), it seems likely that early birds (such as the well-known *Archaeopteryx* of the Late Jurassic, 140 million years ago) emerged from these DINOSAURS. Feathers are complex organs requiring many different genes for their construction, and consequently, it makes sense that they evolved only once. But the feathered dinosaurs did not fly; they apparently developed feathers either as insulation to maintain body temperature, for sexual display, or possibly as an aid in jumping or gliding. When these animals acquired a strong breastbone to anchor powerful flight muscles, modified their forearms into wings, reduced their tailbones to a stump, and reengineered the rest of their skeletons into an aerodynamically sound structure, they became birds.

There are still many gaps in the avian fossil record. Unfortunately, cryptozoology may not be able to help fill them. None of the sixty-one mystery birds in this section are explainable by the survival of anything other than recent taxa, except possibly BIG BIRD or the THUNDERBIRD, which some believe may involve an extant teratorn from 8 million years ago. Flightlessness, found in such birds as the moa, is usually a late adaptation by a bird that was capable of flight but had few natural predators. The DODO, DU, KOAU, MIHIRUNG PARINGMAL, RÉUNION SOLITAIRE, ROA-ROA, and VORONPATRA are flightless.

The largest living bird is the flightless Ostrich (*Struthio camelus*); males have been recorded up to 9 feet in height and weighing 345 pounds. The heaviest flying birds are the Kori bustard (*Ardeotis kori*) of Africa and the Great bustard (*Otis tarda*) of Europe and Asia, both of which can weigh more than 40 pounds. The Wandering albatross (*Diomedea exulans*) has the largest wingspan of any living bird; a specimen caught in the Tasman Sea in 1965 had a wingspan of 11 feet 11 inches.

The seventy-four families of passerine birds, also known as perching birds, contain more than half of the world's bird species.

Sixteen of the entries are birds that, though known largely from myth and legend, might be explainable by real species, either living or extinct. These include the giant KAHA, PIASA, ROC, and SIMURGH and the smaller CALADRIUS and PHOENIX.

Nine entries are birds that have become extinct recently but may have lingered past their official extinction dates, such as the CAROLINA PARAKEET, GREAT AUK, or IVORY-BILLED WOODPECKER.

The remainder are birds about which there is simply insufficient information to classify or to verify as distinct species, such as the GOODENOUGH ISLAND BIRD or the PERUVIAN WATTLELESS GUAN.

### Mystery Birds
*Africa*
BAGGE'S BLACK BIRD; BENNU BIRD; DENMAN'S BIRD; DODO; KIGEZI TURACO; KIKIYAON;

KONDLO; LE GUAT'S GIANT; MAKALALA; MARSABIT SWIFT; MATHEWS RANGE STARLING; NGOIMA; PHOENIX; RÉUNION SOLITAIRE; ROC; SENEGAL STONE PARTRIDGE; SUDD GALLINULE; VORONPATRA

### Asia
ALOVOT; ANDAMAN WOOD OWL; ANKA; DEVIL BIRD; DOUBLE-BANDED ARGUS; DRAGON BIRD; FILIPINO SECRETARY BIRD; KAHA; PHOENIX (CHINESE); PINK-HEADED DUCK; SIMURGH; STELLER'S SEA RAVEN; SUMATRAN HUMMING-BIRD; WHISKERED SWIFT; ZIZ

### Australasia and Oceania
BIRDS OF PARADISE (UNRECOGNIZED); DU; GABRIEL FEATHER; GOODENOUGH ISLAND BIRD; HUIA; KOAU; MIHIRUNG PARINGMAL; NGANI-VATU; POUA; ROA-ROA; SASA

### Central and South America
GLAUCOUS MACAW; PERUVIAN WATTLELESS GUAN; RED JAMAICAN PARROT

### Europe
BOOBRIE; CALADRIUS; GREAT AUK; KUNGSTORN; SLAGUGGLA; STYMPHALIAN BIRD

### North America
BIG BIRD; CAROLINA PARAKEET; GIANT OWL; IVORY-BILLED WOODPECKER; PASSENGER PIGEON; PIASA; THUNDERBIRD; THUNDERBIRD (PENNSYLVANIA)

## Birds of Paradise (Unrecognized)
Distinctive species of tropical BIRDS of the Family Paradisaeidae in Papua New Guinea, known only from isolated specimens obtained without precise location data during the heyday of indiscriminate plume hunting. Six varieties were dismissed as hybrids in 1930 by Erwin Stresemann but may constitute distinct, and possibly extinct, species.

*Scientific names:* Bensbach's bird of paradise, *Janthothorax bensbachi;* Duivenbode's riflebird, *Parypheporus duivenbodei;* Elliot's sicklebill, *Epimachus ellioti;* Rothschild's lobe-billed bird of paradise, *Loborhamphus nobilis;* Ruys's bird of paradise, *Neoparadisea ruysi;* and *Pseudastrapia lobata.* See also GOODENOUGH ISLAND BIRD.

*Sources:* Errol Fuller, *The Lost Birds of Paradise* (Shrewsbury, England: Swan Hill Press, 1995); Errol Fuller, *Extinct Birds* (Ithaca, N.Y.: Cornell University Press, 2001), pp. 380—382.

## Bir-Sindic
Unknown PRIMATE of Southeast Asia.

*Variant names:* Iu-wun (in Myanmar), Olo-banda.

*Distribution:* Assam State, India; Myanmar.

*Possible explanation:* An isolated mainland population of the Orangutan (*Pongo pygmaeus*), which is now limited to the islands of Borneo and Sumatra. Orangutan fossils from around 2 million years ago have been found in Laos, Vietnam, and southern China, as well as the islands of Sumatra, Java, and Borneo.

*Source:* Bernard Heuvelmans, "Annotated Checklist of Apparently Unknown Animals with Which Cryptozoology Is Concerned," *Cryptozoology* 5 (1986): 1—26.

## Bis-Cobra
Unknown LIZARD of Central Asia.

*Behavior:* Its bite is said to be as venomous as twenty cobras. Unlike other lizards, it spits its venom.

*Distribution:* Northern India.

*Possible explanations:*

(1) A harmless gecko (*Eublepharis hardwickii*) of east India and Bangladesh. The related Leopard gecko (*E. macularius*) also has a reputation in Pakistan as a venomous reptile.

(2) A composite animal, a hybrid of poisonous snakes and harmless lizards.

*Source:* Sherman A. Minton Jr. and Madge Rutherford Minton, *Venomous Reptiles* (New York: Scribner, 1969).

## Black Dog
Canine ENTITY of Europe and North America. Distinguished from the ALIEN BIG DOG by its paranormal qualities.

*Variant names:* BARGUEST, BLACK SHUCK, Blue dog, BRAY ROAD BEAST, el Cadejo (in Costa Rica), Capelthwaite (in Cumbria), Cappel, Choin dubh (Gaelic), Church grim, CÙ SÌTH, Dando dog (in Cornwall), Fairy hound, Farbhann (in the Hebrides), Farvann, GABRIEL HOUND, GALLY-TROT, Girt dog, Gurt dog (in Somerset), GWYLLGI, HAIRY JACK, Hooter, Kludde (in Belgium), Long dog, MIRRII, MODDEY DHOE, Muckle black tyke, Owd Rugusan, Padfoot (in Leeds), POOKA, SCARFE, SHAG DOG, SNARLY YOW, Spectral hound, LE TCHAN DE BOUÔLÉ, TRASH, VARMINT, WISH HOUND.

*Physical description:* As large as a calf or collie dog. Black, like a Labrador retriever; often described as jet-black or coal-black. Shaggy coat. Occasionally said to be headless. Large, red or green, glowing eyes. Foaming or slavering mouth. Long teeth.

*Behavior:* Nocturnal. Often malevolent or menacing. Screams, growls, or howls. Bad or fiery breath reported frequently. Occasionally acts as a guide or protector to travelers. Tends to follow or run alongside people. Often apparitional in nature—seemingly real, but when a witness tries to touch or strike it, nothing solid is felt. Can appear or disappear suddenly. Sometimes grows bigger or shrinks before it disappears. Guards churchyards and treasure. Said to be an omen of death.

*Habitat:* Most often reported along roads or country lanes; also graveyards, fields, barrows, and downs. An association with waterways has also been noted.

*Distribution:* In Europe, especially common in Great Britain but also reported from Ireland, France, Belgium, Italy, Croatia, Germany, Austria, Poland. In the United States, there have been reports from Maryland, Mississippi, Missouri, Pennsylvania, and Tennessee. Other reports come from Canada, especially Nova Scotia; Costa Rica; Argentina; and Australia.

*Significant sightings:* A fearsome Black dog appeared inside a church in Bungay, Suffolk, England, on August 4, 1577, accompanied by "fearful flashes of fire" during a violent thunderstorm. It rushed down the aisle, killed two people and injured a few others, then appeared 7

*Title page of a pamphlet describing the appearance of a* BLACK DOG *in a church at Bungay, Suffolk, England, in 1577. (Fortean Picture Library)*

miles away at the church in Blythburgh, where its claws left burn marks on the church door.

In 1928, a Trinity College student was fishing in a river in County Londonderry, Northern Ireland, when a Black dog with blazing red eyes came toward him in the shallow water. Terrified, he climbed a tree, and the animal looked up at him and snarled as it passed.

In 1949, a waterworker near Keresley, Warwick, England, was confronted early one morning by a huge Black dog sitting on its haunches. Its glowing eyes watched him as he edged around it and ran away.

In the winter of 1959 or 1960, a twelve-year-old boy and his mother saw a Black dog with a huge head peering into their window on Sharpe

Street in South Baltimore, Maryland. Its eyes were oval-shaped and bright red or yellowish. Later, the boy went outside but could find no tracks in the snow.

On April 19, 1972, British coastguardsman Graham Grant was on watch at Gorleston, near the harbor entrance to Great Yarmouth, Norfolk, England, when he saw a large, black hound on the beach. It alternatively ran, then stopped and looked around, and after a short time it vanished. Grant said there was nowhere it might have hidden.

On April 30, 1976, a black-and-brown dog was seen in Abingdon, Massachusetts, feeding on a Shetland pony it had killed.

On October 31, 1984, a Mr. Lee was driving toward Molland, Devon, England, when he saw a huge, black great dane run toward the road at him. As Lee braked to a stop, the animal walked up to the hood of the car, looked at him, and vanished.

Victoria Rice-Heaps encountered a huge Black dog with glowing red eyes as she was driving past Hodsock Priory near Worksop, Nottinghamshire, England, early in the morning of May 11, 1991. It was about 18 inches taller than a great dane and seemed to be dragging something across the road.

*Possible explanations:*

(1) Black feral Domestic dogs (*Canis familiaris*); the glowing red eyes might be an indication of opacity caused by cataracts, which make the eyes shine red in reflected light.

(2) The odds are overwhelmingly in favor of the Black dog being a paranormal—rather than a biological—entity, more related to ghosts than to dogs.

(3) BRITISH BIG CATS, seen under imperfect conditions, may have contributed to Black dog folklore. However, the shaggy coat, the tendency to follow humans, and noisy movement argue against a cat.

*Sources:* Abraham Fleming, *A Straunge and Terrible Wunder Wrought Very Late in the Parish Church of Bongay* (London: Francis Godley, 1577); Robert Hunt, *Popular Romances of the West of England* (London: J. C. Hotten, 1865), pp. 220—223; Frank Hamel, *Human Animals* (New York: Frederick A. Stokes, 1917), pp.

238—246; John Symonds Udal, *Dorsetshire Folk-Lore* (Hertford, England: S. Austin, 1922), p. 167; Ethel H. Rudkin, "The Black Dog," *Folklore* 49 (1938): 111—131; Pierre van Paassen, *Days of Our Years* (London: William Heinemann, 1939), pp. 237—240; Helen Creighton, "Folklore of Lunenburg County, Nova Scotia," *Bulletin of the National Museum of Canada,* no. 117 (1950): 41; Alasdair Alpin MacGregor, *The Ghost Book* (London: Robert Hale, 1955), pp. 55—81; Robert J. Fugate, "The Devil Is a Black Dog," *Fate* 9 (January 1956): 22—24; Theo Brown, "The Black Dog," *Folklore* 69 (1958): 175—192; Theo Brown, "The Black Dog in Devon," *Transactions of the Devonshire Association* 91 (1959): 38—44; Ruth L. Tongue, *Somerset Folklore* (London: Folk-Lore Society, 1965), pp. 107—110; Patricia Dale-Green, *Dog* (London: Rupert Hart-Davis, 1966), pp. 50—84, 107—108, 183—193; Ruth E. Saint Leger-Gordon, *Witchcraft and Folklore of Dartmoor* (New York: Bell, 1973), pp. 26—41, 188; Diarmuid A. MacManus, *The Middle Kingdom* (Gerrards Cross, England: Colin Smythe, 1973), pp. 66—76, 133—137; Katharine M. Briggs, *A Dictionary of Fairies* (London: Allen Lane, 1976), pp. 16—17, 25, 62, 72, 74—75, 85, 89—90, 97—98, 140—141, 183, 207—208, 209, 216, 225—226, 282, 301, 321, 370, 412, 440; Ivan Bunn, "Black Dogs and Water," *Fortean Times,* no. 17 (August 1976): 12—13; "Killer Dog," *Fate* 29 (September 1976): 8—12; John Michell and Robert Rickard, *Phenomena: A Book of Wonders* (1977); Janet and Colin Bord, *Alien Animals* (Harrisburg, Pa.: Stackpole, 1981), pp. 77—111; Graham J. McEwan, *Mystery Animals of Britain and Ireland* (London: Robert Hale, 1986), pp. 119—149; Christopher Reeve, *A Straunge and Terrible Wunder: The Story of the Black Dog of Bungay* (Bungay, England: Morrow, 1988); Karl Shuker, "Red Eye Glow: A New Explanation," *Strange Magazine,* no. 8 (Fall 1991): 39; David McGrory, "On the Sniff," *Fortean Times,* no. 83 (October-November 1995): 42—43; Christopher Kiernan Coleman, *Strange Tales of the Dark and Bloody Ground* (Nashville, Tenn.: Rutledge Hill, 1998), pp. 31—34; Mark Chorvinsky,

"Phantom Dogs in Maryland," *Strange Magazine,* no. 19 (Spring 1998): 6—9, 52—53; "Wild Thing: Argentinian Werewolf on the Prowl," *Fortean Times,* no. 146 (June 2001): 21; Victoria Rice-Heaps, "Black Shuck Seen," *Fortean Times,* no. 154 (February 2002): 52—53; Simon Sherwood, Apparitions of Black Dogs, http://moebius.psy.ed.ac.uk/~simon/homepage/blackdog.htm.

## Black Fish (Venomous)

Unknown FISH of the Middle East.

*Physical description:* Small and black.

*Behavior:* Lethal, swift-acting bite.

*Distribution:* Shatt al Arab River, Iran and Iraq.

*Significant sightings:* Said to have killed twenty-eight people before 1975.

*Possible explanations:*

(1) Unknown relative of the Blackline fang-blenny (*Meiacanthus nigrolineatus*), a blue-and-yellow species with a black stripe along its dorsal fin, suggested by Karl Shuker. Found in the Red Sea and the Gulf of Aden, this fish has a nonfatal, venomous bite and is just under 4 inches long.

(2) The Stinging catfish (*Heteropneustes fossilis*), a black fish about 12 inches long with poisonous spines in its pectoral fins, has been introduced into the Shatt al Arab from the Indian subcontinent. Its bite is not dangerous, but its fins are.

(3) A Giant slender moray (*Thyrsoidea macrura*) from the Red Sea. This fish grows to 12 feet long, and its bites have never proven fatal.

*Sources:* Roger A. Caras, *Dangerous to Man,* rev. ed. (New York: Holt, Rinehart and Winston, 1975); Karl Shuker, "Fins, Fangs and Poison," *Fortean Times,* no. 93 (December 1996): 44.

## Black Lion

Melanistic big CAT of South Africa and West Asia.

*Physical description:* Completely or partially black Lion (*Panthera leo*).

*Distribution:* Kruger National Park, South Africa; the Zagros Mountains, Iran.

*Significant sightings:* In the 1880s, a dark brown lion was killed by soldiers of the Luristan Regiment. It was seen by Sir Henry Layard in Esfahan, Iran.

In 1975, a partially black lion cub was born at the Glasgow Zoo. It had a black chest and one black leg.

*Possible explanation:* Black lion morphs have never been verified.

*Sources:* Henry Layard, *Early Adventures in Persia, Susiana, and Babylonia* (London: John Murray, 1887); W. L. Speight, "Mystery Monsters of Africa," *Empire Review* 71 (1940): 223—228; June Kay, *Okavango* (London: Hutchinson, 1962); Karl Shuker, *Mystery Cats of the World* (London: Robert Hale, 1989), pp. 132—133.

## Black Malayan Tapir

Melanistic variety of the Malayan tapir, a HOOFED MAMMAL, of Southeast Asia.

*Scientific name: Tapirus indicus* var. *brevetianus,* given by K. Kuiper in 1926.

*Physical description:* Malayan tapir that lacks the distinctive white saddle on its back and haunches.

*Distribution:* Near Babat, Sumatera Selatan Province, Sumatra, Indonesia.

*Present status:* Only two specimens are known, both collected in 1924 by K. Brevet of the Royal Dutch-Indian Army. They both died before they could be crossbred with normal tapirs.

*Possible explanation:* Either unusual melanistic morphs or an unverified variety of the Malayan tapir (*Tapirus indicus*).

*Source:* K. Kuiper, "On a Black Variety of the Malay Tapir (*Tapirus indicus*)," *Proceedings of the Zoological Society of London,* July 1926, pp. 425—426.

## Black Sea Snake

Mystery SNAKE of West Asia.

*Physical description:* Dark brown above, white below. Length, 82—98 feet. Snakelike head.

*Behavior:* Wriggles like a snake in the water. Floats by rolling into a ball.

*Distribution:* Crimean shore of the Black Sea, Ukraine.

*Significant sighting:* For forty minutes in the spring of 1952, Vsevolod Ivanov watched a huge snake swimming in Sordolik Bay of the Black Sea near Planerskoye, Crimean Republic, Ukraine.

*Sources:* Vasilii Khristoforovich Kondaraki, *Universal'noe opisanie Kryma* (St. Petersburg, Russia, 1875), pt. 7, p. 35; Vsevolod V. Ivanov, *Perepiska s A. M. Gor'kim* (Moscow: Sov. Pisatel', 1969); Maya Bakova, "Black Sea Serpents," *Fortean Times,* no. 51 (Winter 1988—1989): 59.

## Black Shuck

BLACK DOG of southern England.

*Etymology:* Possibly from the Old English *scucca* ("demon") or from "shag" or "shaggy" after its tousled coat.

*Variant names:* Old Shock (in Suffolk), Old Shuck (in Norfolk), Shuck, Shucky dog.

*Physical description:* Size of a calf. Shaggy black dog with glowing eyes. Some writers say it has only one eye.

*Behavior:* Appears before bad weather. Accompanied by the sound of chains. Walks behind people, growling. Follows cyclists. Said to throw people down and break their legs.

*Distribution:* Norfolk and Suffolk, England.

*Significant sighting:* John Harries was followed by a Black dog in November 1945 as he cycled from East Dereham, Norfolk, to the Royal Air Force (RAF) station at Swanton Morley. Whenever he stopped, the dog would stop, and it kept pace with him even at 20 miles per hour. When he got to the base, it vanished.

*Sources:* Morley Adams, *In the Footsteps of Borrow and Fitzgerald* (London: Jerrold, 1914); John Harries, *The Ghost Hunter's Road Book* (London: Frederick Muller, 1968); Ivan Bunn, "Black Shuck: Encounters, Legends and Ambiguities," *Lantern,* no. 18 (Summer 1977): 3–6, and no. 19 (Autumn 1977): 4.

## Black Tiger

Melanistic big CAT of the Indian subcontinent and Southeast Asia.

*Variant name:* Bear tiger.

*Physical description:* The normal tiger stripes are visible over a darkened ground color.

*Distribution:* Kerala, Orissa, Assam, and Manipur States, and Mizoram Union Territory, India; Chittagong Division, Bangladesh; Bhamo District, Myanmar.

*Significant sightings:* In 1772, a Black tiger was killed in Kerala State, India. A portrait of it was painted by noted British artist John Forbes.

In March 1846, a Black tiger that had killed a local villager was shot by a poisoned arrow in the Chittagong Hill District, Bangladesh. The stripes showed distinctly against a lighter black ground.

On September 11, 1895, S. Capper and C. J. Maltby spotted a Black tiger through a telescope in the Cardamom Hills, Kerala State, India.

A Black tiger with no evidence of striping was shot in 1915 near Dibrugarh, Assam State, India.

In the early 1970s, a dark tiger cub was born to normal parents in the Oklahoma City Zoo. It had a normal ground color, but it also had smoky black pigmentation on its shoulders, pelvis, and legs. Had it not been killed by its mother shortly after its birth, it might have turned completely melanistic.

Beginning in 1975 and 1976, a number of sightings of Black tigers occurred in Similipal Tiger Reserve, near Baripada, Orissa State, eastern India. On July 21, 1993, a boy killed in self-defense a young, melanistic tigress in the village of Podagad west of the reserve. The tiger's black ventral stripes had expanded and coalesced over the tawny ground color, indicating a pseudo-melanistic morph.

*Present status:* Many reports of all-black Tigers (*Panthera tigris*) exist, but no specimen or skin showing true melanism has ever been submitted for formal description. Melanism usually occurs in tropical species such as the leopard and jaguar, so a black tropical tiger morph would not be considered genetically unusual.

*Possible explanations:*

(1) Misidentified black Leopard (*Panthera pardus*), such as the 12-foot black animal captured alive in September 1934 near Dibrugarh, Assam. However, most Black tiger observations have been in close quarters or when the animal was dead.

(2) Tigers seen in shadow or covered in charcoal, ash, or blood. However, the dark stripes have been reported in most cases.

(3) A genuine but rare melanistic tiger morph.

*Sources:* C. J. Buckland, "A Black Tiger," *Journal of the Bombay Natural History Society* 4 (1889): 149; T. A. Hauxwell, "Possible Occurrence of a Black Tiger," *Journal of the Bombay Natural History Society* 22, no. 4 (1913): 88—89; Karl Shuker, *Mystery Cats of the World* (London: Robert Hale, 1989), pp. 101—107; Karl Shuker, "Melanism, Mystery Cats, and the Movies," *Strange Magazine,* no. 19 (Spring 1998): 23, 54—55; Karl Shuker, "Black Is Black . . . Isn't It?" *Fortean Times,* no. 109 (April 1998): 44; Lala A. K. Singh, *Born Black: The Melanistic Tiger in India* (New Delhi: World Wide Fund for Nature—India, 1999).

## B'lian

WILDMAN of Southeast Asia.

*Distribution:* Southern peninsular Malaysia.

*Sources:* Boris F. Porshnev, *Sovremennoe sostoianie voprosa o relikhtovykh hominoidakh* (Moscow: Viniti, 1963); Bernard Heuvelmans, "Annotated Checklist of Apparently Unknown Animals with Which Cryptozoology Is Concerned," *Cryptozoology* 5 (1986): 1–26.

## Blood-Sweating Horse

Unknown horse (a HOOFED MAMMAL) of East Asia.

*Behavior:* Bleeds from the shoulder when running at full speed. Said to be able to travel an incredible distance in one day.

*Distribution:* Tian Shan Range, Xinjiang Uygur Autonomous Region, China; formerly ranged in Uzbekistan and Tajikistan.

*Significant sightings:* An emperor of the Han Dynasty (206 B.C.–A.D. 200) sent armies to catch this horse along the Silk Road trade route.

In August 2000, Japanese horse researcher Hayato Shimizu took a photo of a horse in the western Tian Shan that shows blood running from its shoulders.

*Possible explanations:*

(1) The bleeding is caused by some sort of parasitic worm.

(2) The horse's arteries stand out under exertion, producing a reddish flush.

(3) The horse actually bleeds under extreme exertion.

(4) An unknown horse breed, similar to the recently discovered Nangchen and Riwoche horses of Tibet.

*Source:* "Expert Sees Legendary Asian Horse 'Sweat Blood,'" *Japan Times,* April 15, 2001.

## Blue Horse

Odd blue-colored horse (a HOOFED MAMMAL) of South Africa.

*Physical description:* Shoulder height, just under 5 feet. Smooth, blue-mauve skin. Completely hairless. Buff-colored face. One large, beige patch on its back. Tail like a pig's.

*Behavior:* Can be broken in for riding. Performs well in harness.

*Distribution:* South Africa.

*Significant sighting:* In South Africa in 1860, a man named Lashmar spotted a blue-colored horse in a herd of QUAGGAS (*Equus quagga*). He captured it and sent it to Cape Town, where it was sold and sent to London in 1863. It was ridden as a fox-hunting horse on Lord Stamford's estate, examined by Charles Spooner at the Royal Veterinary College in London, then sold in February 1868 to a Mr. Moffat for exhibition at the Crystal Palace. By then its original blue coloration had faded to gray.

*Possible explanation:* A mutant form of a gene controlling hair development could produce hairlessness, according to Karl Shuker. Presence of the pigment eumelanin in combination with others might result in a blue color. Where the horse came from or why it was accompanying a herd of quaggas is unknown.

*Sources:* C. O. G. Napier, "The Blue Horse," *Land and Water,* February 22, 1868, app., p. 80; Karl Shuker, "A Horse of a Different Color," *Fate* 47 (May 1994): 66—69.

## Blue Men of the Minch

MERBEING of Scotland.

*Physical description:* Blue skin. Long, gray face.

*Behavior:* Swims alongside ships to lure sailors into the water but can be overwhelmed by a skilled rhymer or riddler. Able to conjure storms.

*Habitat:* Underwater caverns.

*Distribution:* The Minch, the strait between the Isle of Lewis in the Hebrides and Scotland, especially off the Shiant Islands.

*Possible explanations:*

(1) A personification of dangerous waters.

(2) A folk memory of Tuareg slaves from North Africa taken to Scotland in the ninth century by Norse pirates and slave traders. These nomadic people are still known as Blue Men today because of their indigo robes. The wide-ranging Vikings did apparently visit North Africa and may have even engaged in some slave trading there. However, equating the two Blue Men groups seems a stretch.

*Sources:* John Gregorson Campbell, *Superstitions of the Highlands and Islands of Scotland* (Glasgow, Scotland: J. MacLehose and Sons, 1900), pp. 199—202; Donald Alexander Mackenzie, *Scottish Folk-Lore and Folk Life* (London: Blackie, 1935); Gwen Benwell and Arthur Waugh, *Sea Enchantress* (London: Hutchinson, 1961), pp. 173—174.

## Blue Tiger

Bluish big CAT of East Asia.

*Physical description:* Well-defined black stripes over a grayish-blue ground color. Deep blue on chest and ribcage.

*Distribution:* Rongcheng area, Fujian Province, China.

*Significant sighting:* In September 1910, hunter and missionary Harry R. Caldwell saw a tiger with a bluish-gray ground color and deep blue underparts near Rongcheng. He had it in his rifle sights but could not pull the trigger because there were two young boys nearby who might be endangered.

*Present status:* Other individuals may exist in the area if this variety entered the gene pool.

*Possible explanation:* Karl Shuker suggests a Tiger (*Panthera tigris*) may have possessed a recessive melanistic mutant allele and a recessive dilute mutant allele to produce a morph called a "blue dilution."

*Sources:* Harry R. Caldwell, *Blue Tiger* (New York: Abingdon, 1924); Roy Chapman Andrews, "The Trail of the Blue Tiger," *True,* January 1950, reprinted in *North American BioFortean Review,* no. 6 (May 2001): 80—91, http://www.strangeark.com/nabr/NABR6.pdf; Karl Shuker, *Mystery Cats of the World* (London: Robert Hale, 1989), pp. 100—101.

## Blue-Nosed Frog

Unknown AMPHIBIAN of Central Africa.

*Physical description:* Frog with a blue spot on its snout that glows in the dark.

*Distribution:* Northern Cameroon.

*Significant sighting:* Jonathan Downes found several of these tree frogs for sale at an animal fair in Newton Abbot, Devon, England, in July 1997. Later on, he discovered that there was no such frog known to science.

*Source:* Karl Shuker, "The Frog with the Luminous Nose," *Strange Magazine,* no. 19 (Spring 1998): 23.

## Bobo

SEA MONSTER of the North Pacific Ocean.

*Etymology:* Possibly from the Portuguese *bobo* ("silly") because of the reaction to eyewitness reports. Portuguese-speaking fishermen were common in the area in the 1940s.

*Distribution:* Cape San Martin and Monterey Bay, California.

*Significant sighting:* On November 7, 1946, a monster with the face of a gorilla appeared off Cape San Martin. Apparently it had been seen for the previous ten years.

*Sources:* "No Such Animal," *Doubt,* no. 17 (1947): 260; Randall A. Reinstedt, *Mysterious Sea Monsters of California's Central Coast* (Carmel, Calif.: Ghost Town Publications, 1993), pp. 26—27.

## Bokyboky

Mystery CIVET of Madagascar.

*Etymology:* Malagasy (Austronesian) word.

*Variant names:* Vontira, Votsotsoke.

*Physical description:* Size of a cat. Broad face. Large ears.

*Behavior:* Kills rats and snakes by sticking its tail down their burrows and farting.

*Distribution:* Southwestern Madagascar.

*Possible explanation:* The Narrow-striped mongoose (*Mungotictus decemlineata*) holds its bushy tail erect when alarmed.

*Source:* David A. Burney and Ramilisonina, "The *Kilopilopitsofy, Kidoky,* and *Bokyboky:* Accounts of Strange Animals from Belo-sur-Mer, Madagascar, and the Megafaunal 'Extinction Window,'" *American Anthropologist* 100 (1998): 957–966.

## Booaa

Unknown HYENA of West Africa.

*Etymology:* After the cry it makes.

*Physical description:* Large hyena.

*Distribution:* Senegal.

*Possible explanation:* May be a western range extension of the hyena-like NANDI BEAR of East Africa.

*Source:* Karl Shuker, "Death Birds and Dragonets: In Search of Forgotten Monsters," *Fate* 46 (November 1993): 66—74.

## Boobrie

Unknown water BIRD of Scotland, often confused with the WATER HORSE.

*Physical description:* Like the Common loon (*Gavia immer*). White streak on neck and breast. Eaglelike bill is 18 inches or longer. Neck is nearly 3 feet long. Short, black legs. Webbed, clawed feet.

*Behavior:* Call is like the roar of an angry bull. Said to feed on lambs and otters.

*Distribution:* Argyll and Bute, Scotland.

*Possible explanations:*

(1) Unknown species of Loon (Family Gaviidae).

(2) The Yellow-billed loon (*Gavia adamsii*) is an occasional visitor to Scotland. It has a striking white bill, but it is not at all eaglelike. First-winter birds have much paler and whiter coloration about the neck than other loons.

*Sources:* John Francis Campbell, *Popular Tales of the West Highlands* (Edinburgh: Edmonston and Douglas, 1860—1862), vol. 4, p. 308; Katharine M. Briggs, *A Dictionary of Fairies* (London: Allen Lane, 1976), p. 34.

## Booger

Local name for various cryptids in mountainous areas of the United States, including HAIRY BIPEDS, DEVIL MONKEYS, NORTH AMERICAN APES, EASTERN PUMAS, and ALIEN BIG DOGS.

*Etymology:* Originally "ghost" or "haunt" but by the late 1960s, it also meant "monster" or "animal" in the southern United States.

*Variant names:* Booger dog, Booger man.

*Sources:* Frederic G. Cassidy, ed., *Dictionary of American Regional English* (Cambridge, Mass.: Harvard University Press, 1985), vol. 1, pp. 333—334; Christopher Kiernan Coleman, *Strange Tales of the Dark and Bloody Ground* (Nashville, Tenn.: Rutledge Hill, 1998), pp. 63—65.

## Bornean Tiger

Unrecognized big CAT of Southeast Asia.

*Distribution:* Borneo, Indonesia.

*Significant sightings:* In the late 1990s, Erik Meijaard collected scattered evidence from the north, east, and interior of Borneo that tigers existed there in recent times. Tiger skins, skulls, and teeth are in the possession of some of the indigenous peoples, and sightings have occurred as recently as 1995.

*Possible explanations:* Either a remnant population of Tigers (*Panthera tigris*) has existed on the island since the Pleistocene or, at some point in the recent past, it was introduced, perhaps by the sultans of Sarawak, Sabah, or Brunei.

*Sources:* Douchan Gersi, *Dans la jungle de Bornéo* (Paris: Éditions G.P., 1975); Erik Meijaard, "The Bornean Tiger: Speculation on Its Existence," *Cat News,* no. 30 (Spring 1999): 12—15.

## Bothrodon pridii

Supposed giant SNAKE of South America.

*Physical description:* Based on the misidentification of an alleged 2.5-inch-long fossil poison fang found in the Gran Chaco area of South America in the 1920s and presumed to be from an unknown snake. In 1939, the object was positively identified as a prong from the shell of the Chiragra spider conch (*Lambis chirarga*).

*Sources:* John Graham Kerr, "*Bothrodon pridii,* an Extinct Serpent of Gigantic Dimensions," *Proceedings of the Royal Society of Edinburgh* 46 (1926): 314—315; David Heppell, "Gigantic Serpent Really a Gastropod!" *Conchologists' Newsletter,* no. 16 (March 1966): 108—109.

## Bozho

FRESHWATER MONSTER of Wisconsin.

*Etymology:* Potawatomi (Algonquian), "hello." May also be a shortened form of the name of the Algonquian trickster figure Man-abozho.

*Physical description:* Serpentine. Long head and neck. Large eyes. Long tongue.

*Distribution:* Lake Mendota, Wisconsin.

*Significant sightings:* On June 27, 1883, Billy Dunn and his wife encountered a huge, green snake with light spots that had to be beaten back from their rowboat with an oar and a hatchet.

In the autumn of 1917, a fisherman saw a head and neck 100 feet off Picnic Point.

*Sources:* "A True Snake Story," Madison *Wisconsin State Journal,* June 28, 1883; "Western Lake Resorts Have Each a Water Monster," *Chicago Tribune,* July 24, 1892; Charles E. Brown, *Sea Serpents: Wisconsin Occurrences of These Weird Water Monsters* (Madison: Wisconsin Folklore Society, 1942).

## Brachystomos

WILDMAN of East Africa.

*Etymology:* Greek, "narrow mouth."

*Physical description:* Narrow throat.

*Behavior:* Drinks with the aid of straws.

*Distribution:* East Africa.

*Possible explanation:* Bernard Heuvelmans suggests that this description is based on the distortion of the lips as practiced by the Luba-Kasai people of the Democratic Republic of the Congo and the Kyabé of Chad.

*Sources:* Pomponius Mela, *De chorographia,* III. 9; Bernard Heuvelmans, *Les bêtes humaines d'Afrique* (Paris: Plon, 1980), pp. 152—153, 163.

## Bray Road Beast

BLACK DOG or WEREWOLF of Wisconsin.

*Etymology:* After rural Bray Road, the scene of many encounters.

*Physical description:* Height, 5 feet 7 inches. Weight, 150 pounds. Brownish-silver hair or fur. Glowing, yellowish eyes. Pointed ears. Wolflike muzzle. Fangs. Wide chest and shoulders. Muscular forelegs. Fingers with claws. Hind legs oddly shaped and longer than a dog's. Tail like a husky or German shepherd dog.

*Behavior:* Walks uncertainly on two feet. Runs on all fours. Growls. Holds food with palms facing up. Chases people.

*Tracks:* Doglike, 4 inches wide, and 4–5 inches long.

*Distribution:* East of Elkhorn, Wisconsin.

*Significant sightings:* In the fall of 1989, Lorianne Endrizzi saw a wolflike creature kneeling by the side of Bray Road. It seemed to be eating something.

Farmer Scott Bray encountered a "strange-looking dog" that left tracks in his cow pasture in September or October 1989.

On October 31, 1991, Doristine Gipson was attacked by a large animal as she was stopped along the road. The creature hit the car trunk as she drove away.

Along Bray Road on the night of August 13, 1999, a woman and her family saw what at first looked like a deer, but it was about 5 feet tall and had glowing, red eyes. It approached the car steadily to within 50 feet before they drove away.

*Possible explanations:*

(1) Hoaxes.

(2) Black bears (*Ursus americanus*) are not normally found in the area.

(3) Coyotes (*Canis latrans*) are too small.

(4) The Gray wolf (*Canis lupus*) is not as ro-bustly built, never walks upright, and is not known in the area.

(5) A hybrid dog × wolf. These hybrids have become fashionable in the United States, where some estimates place the number at 600,000. Second-generation hybrids tend to be strongly territorial, prone to roaming, and shy of people. Wolf × German shepherd hybrids tend to be less stable than wolf × mala-mute or husky crosses.

(6) A paranormal ENTITY.

(7) Escaped baboon of some type.

*Sources:* Scarlett Sankey, "The Bray Road Beast: Wisconsin Werewolf Investigation," *Strange Magazine,* no. 10 (Fall-Winter 1992): 19—21, 44—46; Loren Coleman, "The Wisconsin Werewolf," *Fortean Times,* no. 108 (March 1998): 47; Richard Hendricks, Weird Wisconsin: The Bray Road Beast, http://www.weird-wi.com/brayroad/.

## Brenin Llwyd

Mythical giant ENTITY of Wales.

*Etymology:* Welsh, "gray king."

*Variant names:* Gray king, Monarch of the mist.

*Behavior:* Accompanied by clouds or mist. Stalks hikers. Feared as a child stealer.

*Distribution:* Snowdonia, Cadair Idris, and the Cambrian Mountains in Gwynedd and Powys, northern Wales.

*Present status:* Best known now as a character in Susan Cooper's fantasy novel *The Grey King* (1975).

*Source:* Marie Trevelyan, *Folklore and Folk-Stories of Wales* (London: Elliot Stock, 1909), p. 69.

## Brentford Griffin

Dubious flying GRIFFIN of England.

*Physical description:* The size of a dog. Dark color. Long muzzle. Wings. Four legs.

*Habitat:* Said to live on an island in the Thames River.

*Distribution:* Brentford, Greater London, England.

*Significant sightings:* In June or July 1984, Kevin Chippendale was walking down Braemar Road, close to the Griffin Pub, in Brentford when he glimpsed what looked like a dog with wings flying across the street ahead of him. He saw the same thing at the same spot in late February 1985. Other alleged witnesses came forward, but the media soon lost interest.

*Present status:* In 1995, novelist Robert Rankin finally admitted that he was responsible for some, if not all, of the reports and publicity as a joke.

*Sources:* Andrew Collins, *The Brentford Griffin* (Wickford, England: Earthquest Books, 1985); Stuart Coolie, "To Brentford and Back," *Fortean Times,* no. 80 (April-May 1995): 28—29.

## British Big Cat

ALIEN BIG CAT of the British Isles.

*Variant names:* BEAST OF BODMIN MOOR, BEAST OF EXMOOR, NOTTINGHAM LION, and SURREY PUMA. Many other nicknames have been bestowed by the media, among them: Ashley leopard (Kent), Ayrshire puma (Scotland), Beast of Ballymeana (Antrim, Northern Ireland), Beast of Barnet (Hertfordshire), Beast of Basingstoke (Hampshire), Beast of Beacon Hill (Sussex), Beast of Bennachie (Scotland), Beast of Bin (Grampians), Beast of Blagdon (Somerset), Beast of Bont (Ceredigion, Wales), Beast of Broadoak (Gloucestershire), Beast of Broomhill (Yorkshire), Beast of Bucks, Beast of Carsington (Derbyshire), Beast of Chiswick (London), Beast of Essex, Beast of Inkberrow (West Midlands), Beast of Margam (Wales), Beast of Milton Keynes (Buckinghamshire), Beast of Otmoor (Oxfordshire), Beast of the Borders (Shropshire), Beast of Tonmawr (Wales), Beast of Tweseldown (Hampshire), Black beast (Gloucestershire), Black beast of Moray (Scotland), Brechfa beast (Carmarthenshire, Wales), Cadmore cat (Gloucestershire), Cannich puma (Highland, Scotland), Catmose cat (Rutland), Chiltern puma (Buckinghamshire), Durham puma, Eccles cheetah (Norfolk), Fen tiger (Cambridgeshire), Highland puma (Scotland), Lindsey leopard (Lin-

*British big cat carrying a rabbit in its mouth, photographed by Selwyn Jolly at Morvah, Cornwall, in 1988. It measured 3 feet from head to hindquarters. (Fortean Picture Library)*

colnshire), Mendips monster (Somerset), Monster of the M25 (Hertfordshire), Munstead monster (Surrey), Norfolk gnasher, Peak panther (Derbyshire), Penistone panther (Yorkshire), Penwith puma (Cornwall), Powys beast (Wales), Rossshire lioness (Scotland), Skerray beast (Highland, Scotland), Terror of Tedburn (Devon), Tilford lynx (Surrey), and Wildcat of the Wolds (Humberside).

*Physical description:* Puma- or lionlike cat. Ranges in size from a terrier to a great dane. Length, 3–5 feet. Shoulder height, 18 inches–2 feet 6 inches. Brown, rusty, gray, or sandy but often pure black. Sometimes striped or spotted. Flat face. Ears are sometimes tufted. Short, or long, powerful legs. Large paws. Pointed tail is of variable length, sometimes with a white tip.

Specific descriptions vary widely. Witnesses have also referred to similarities with a cheetah, monkey, tiger, partially striped cat, bear, or German shepherd dog. There may be two varieties in Scotland, both black but in two different sizes. *See* Kellas Cat.

*Behavior:* Snarls, howls, roars, or screams. Kills and sometimes eats livestock and game. Dogs are terrified of it.

*Tracks:* Cat- or doglike. Up to 5 inches long, which is larger than a German shepherd's. Some of the tracks show claws, which rules out an arboreal habitat and most often indicates a dog; however, many cats keep their claws extruded to facilitate movement and balance when sprinting, leaping, or walking on certain types of terrain.

*Habitat:* Fields, gardens, woods, hills, streets.

*Distribution:* In many parts of Great Britain but especially Hampshire, Surrey, and East and West Sussex in the southeast; Devon and Cornwall in the southwest; Lancashire, Cheshire, Nottinghamshire, and North Yorkshire; Ceredigion in Wales; and Renfrewshire, Ayrshire, Strathclyde, Highland, and East Lothian in Scotland.

*Significant sightings:* Sheep kills in May 1810 at Ennerdale Water, Cumbria, England; in 1905 at Great Badminton, South Gloucestershire, England; and in January 1927 in Inverness, Highland, Scotland, might have been depredations by big cats. *See* ALIEN BIG DOG.

On July 18, 1963, a truck driver named David Back stopped in Oxleas Wood, Shooters Hill, Bexley, Greater London, to help what he thought was an injured dog. However, the animal, which had long legs and a curled tail, jumped up and ran off into the woods. Later, police reported that a cheetahlike animal had jumped over the hood of a squad car. A search covered some 850 acres and turned up some clawed tracks.

On September 20, 1976, Alec Jamieson of Skegness, Lincolnshire, England, saw a sandy-colored cat about 5 feet long that left 2.5-inch × 3-inch tracks.

Donald Mackenzie and his son were hunting foxes December 12, 1977, when they saw and wounded a large cat swimming the River Naver near Bettyhill, Highland, Scotland. It had red eyes, a dark coat, and a white chest. They chased it in their Land Rover. It outpaced them, though Mackenzie managed to shoot it again.

Using a sheep's head as bait, Ted Noble captured a female puma alive near Cannich, Highland, Scotland, October 29, 1980. However, cat sightings and livestock killings persisted afterward, indicating there were other predators in the area. In fact, the animal captured by Noble turned out to be elderly, lame, well-groomed, overweight, and so tame it purred the next day for visitors at the Highland Wildlife Park, where it was taken. Some suspected that an illegal exotic-pet owner had pulled a fast one by planting an unwanted puma in Noble's trap.

In November 1981, a large gray cat was seen by many witnesses at Tonmawr, Neath Port Talbot, Wales. Steven Joyce managed to get some photos of a large cat at a distance and two smaller ones at close range. Di Francis visited the area later, taking a photo of a 4-foot black cat and finding prints 5 inches long.

On June 16, 1999, a large, orange-and-yellow cat with black stripes reared up on its hind legs and attacked a forklift truck driven by Raymond Cibor near Armthorpe, Doncaster, England. Cibor said it looked like a tiger, but police found tracks later identified as a dog's.

In September 1999, video footage of a large black cat was captured by a closed-circuit security camera at a brick-making plant near Telford, Shropshire, England. Security guards had seen a 6-foot panther twice before.

Shortly after midnight one night in December 1999, Alastair Skinner was driving near Rogie Falls, Highland, Scotland, when he got within 5 feet of what looked like a black panther with a long tail.

A 5-foot cat allegedly clawed and bit eleven-year-old Joshua Hopkins near Trellech, Carmarthenshire, Wales, on August 23, 2000. His wounds were treated at a nearby medical center.

Elaine Ainslie saw a black panther–like cat while she was walking her dog in a field in Orniston, East Lothian, Scotland, on July 19, 2001. It had a long tail and was very muscular. She picked up her dog and backed away until it was out of sight.

Vicar Kenneth Wakefield observed a glossy black panther crossing a road near Launceston, Cornwall, on September 25, 2001. It was about 6 feet long and 3 feet high at the shoulder.

*Possible explanations:* Although classed as a big cat (with pumas, leopards, and lions), the British big cat might be explained in some instances by small cats (wildcats, lynxes, or feral domestics). When seen at a distance, sizes are difficult to estimate. Because descriptions vary so widely, a multicausal explanation seems likely.

(1) The Scottish wildcat (*Felis silvestris grampia*) was common in England, Wales, and Scotland until the end of the fifteenth century. It was exterminated everywhere in Britain by the 1860s except in Scotland, where after World War I, it began to increase in numbers. It grows up to 3 feet 6 inches in length, slightly larger than a domestic cat. The coat is a gray-brown or tabby color, with white on the throat. Its head is broader, teeth sharper, limbs longer, and tail shorter than a domestic cat's. It has a bushy, blunt-ended tail with a well-defined pattern of black stripes. The average weight of males is

11 pounds. Primarily nocturnal, the wildcat feeds mostly on rodents, as well as rabbits and birds. It inhabits woodlands (especially deciduous or mixed), scrubland, seacoasts, and rocky areas with low human density. Some mystery cats could be surviving populations of this wildcat in pockets of England and Wales.

(2) Feral Domestic cats (*Felis silvestris catus*) undoubtedly account for many sightings; among them are the smaller gray cats of Tonmawr, Wales. Feral cats do not grow appreciably larger in the wild than in domesticity. The largest recorded weight is 42 pounds. An odor of brussels sprouts characteristic of feral cats has been noted in some sightings.

(3) Hybrid feral domestic cats × Scottish or European wildcats (*Felis silvestris silvestris*). Mating is common between these closely related species. Colin Matheson has suggested that a wildcat strain exists among feral cats in parts of Wales. Between 1873 and 1904, the Scottish wildcat was experimentally crossed with several domestic breeds, but the hybrids proved too wild for domestication. Color or size variations are not necessarily evidence of a hybrid, which tend to be smaller, with tapered tails, fused black banding, and white markings. *See* KELLAS CAT.

(4) Hybrid feral domestic cats × escaped Jungle cats (*Felis chaus*), suggested by Karl Shuker. Such hybrids are bred in the United States and have been foundation registered since 1995. Called a chausie, this breed is known for its nearly 6-foot vertical leap, large size (14–18 inches at the shoulder), tufted ears, speed, and a weight of 20 pounds or more.

(5) At least four introduced specimens of the Leopard cat (*Felis bengalensis*) were shot or found dead in England between 1984 and 1994. This Asian cat can also mate with domestic cats to produce spotted hybrids, which have been bred in the United States since 1963 as the Bengal variety.

(6) Four types of escaped or released exotic pets that are now naturalized and breeding in the wild could be responsible for mystery cat sightings: a tawny or gray Puma (*Puma concolor*), 5 feet long with a 2-foot tail, weight 200 pounds, small head, neck relatively long, short ears, large paws—even a hypothetical melanistic (black) variety; a Eurasian lynx (*Lynx lynx*), 3 feet 4 inches long with a 4.5–11-inch tail, long limbs, black-tufted ears, golden eyes, yellowish-gray to reddish-brown color; a Leopard (*Panthera pardus*), 6 feet 6 inches long with a 3-foot tail, spots or melanistic coat, elongate and muscular body; and a female Lion (*Panthera leo*), identifiable by a black tuft at the end of the tail. This could especially be true after the 1976 Dangerous Wild Animals Act required special licenses for exotic pets. (The 1981 Wildlife and Countryside Act offered stricter penalties.) Irresponsible owners may have released the animals rather than pay for a license. The Clouded leopard (*Neofelis nebulosa*) and Cheetah (*Acinonyx jubatus*) are other exotic possibilities.

(7) Big cats (pumas and panthers) kill by sinking their claws into the victim's head or hindquarters (usually deer), while breaking of the neck is used for stronger adversaries; consumption of the victim's abdomen, lack of skeletal damage, and location of the carcass in a secluded spot are also characteristic. Small cats (lynxes and wildcats) subsist primarily on rabbits and rodents, rarely attacking larger prey.

(8) Feral Domestic dogs (*Canis familiaris*) or Red foxes (*Vulpes vulpes*). When seen from a distance, many dog breeds can appear catlike—especially those with small heads, rounded ears, and short legs. Packs of stray dogs can quickly leave a sheep devoid of flesh. Large droppings found at a deer kill in Scotland in 1998 contained fox DNA. Massive pawprints (5 inches or more) are more likely indicative of a dog (unless the cat's feet are vastly out of proportion to its body length).

(9) Skulls that have been found turned out to be from a leopard-skin rug (discovered behind a hedge on Dartmoor in January 1988) and from a wall-mounted tiger trophy (found on Exmoor in 1993).

(10) Escaped Wolverines (*Gulo gulo*) are occasionally found in portions of Wales and southern England. The world's largest weasel, it can grow to 4 feet long and 14–17 inches at the shoulder.

(11) A surviving pumalike Pleistocene felid, such as the lion-sized, short-tailed Scimitar-toothed cat (*Homotherium*). However, it is unlikely that such an animal could persist virtually unnoticed by hunters and livestock owners for thousands of years when the smaller Scottish wildcat was nearly exterminated.

(12) An unknown species of big cat, proposed by Di Francis, though it would have to account for a nearly impossibly wide range of colors, anatomy, and behaviors.

(13) A surviving indigenous variety of Eurasian lynx (*Lynx lynx*) that did not die out at the end of the Pleistocene.

(14) Other escaped or released exotic animals have accounted for cat reports, including a Binturong (*Arctictis binturong*), Spotted hyena (*Crocuta crocuta*), and Eurasian badger (*Meles meles*).

*Sources:* William Cobbett, *Rural Rides* (London: J. M. Dent, 1912), vol. 1, pp. 286—287; "If You Go Down to the Woods Today," *INFO Journal*, no. 13 (May 1974): 3—18; Robert J. M. Rickard, "The 'Surrey Puma' and Friends: More Mystery Animals," *Fortean Times*, no. 14 (January 1976): 3—9; Mike Tomkies, *My Wilderness Wildcats* (London: Macdonald and Jane's, 1977); Bob Rickard, "The Scottish 'Lioness,'" *Fortean Times*, no. 26 (Summer 1978): 43—44; Bob Rickard, "The Scottish Lions," *Fortean Times*, no. 32 (Summer 1980): 23—26; Janet and Colin Bord, "Strange Creatures in Powys," *Fortean Times*, no. 34 (Winter 1981): 18—20; "Scottish Puma: Saga or Farce?" *Fortean Times*, no. 34 (Winter 1981): 24—25, 36; Di Francis, *Cat Country: The Quest for the British Big Cat* (Newton Abbot, England: David and Charles, 1983); "Once More with Felines," *Fortean Times*, no. 44 (Summer 1985): 28—31; "The Black Beasts of Moray," *Fortean Times*, no. 45 (Winter 1985): 10—12; Andy Roberts, *Cat Flaps: Northern Mystery Cats* (Brighouse, England: Brigantia, 1986); D. D. French, L. K. Corbett, and N. Easterbee, "Morphological Discriminants of Scottish Wildcats (*Felis silvestris*), Domestic Cats (*F. catus*) and Their Hybrids," *Journal of Zoology* 214 (1988): 235—259; Nigel Brierly, *They Stalk by Night: The Big Cats of Exmoor and the South-West* (South Molton, England: Nigel Brierly, 1989); Karl Shuker, *Mystery Cats of the World* (London: Robert Hale, 1989), pp. 33—69; James Wallis, "British Big Cats," *Fortean Times*, no. 54 (Summer 1990): 30—31; "Mystery Moggies," *Fortean Times*, no. 59 (September 1991): 18—20; Michael Goss, "Alien Big Cat Sightings in Britain: A Possible Rumour Legend?" *Folklore* 103 (1992): 184—202; Mike Dash, "Mystery Moggies," *Fortean Times*, no. 64 (August-September 1992): 44—45; Karl Shuker, "The Lovecats," *Fortean Times*, no. 68 (April-May 1993): 50—51; Richard Halstead and Paul Sieveking, "An ABC of British ABCs," *Fortean Times*, no. 73 (February-March 1994): 41—44; Paul Sieveking, "Beasts in Our Midst," *Fortean Times*, no. 80 (April-May 1995): 37—43; Karl Shuker, "Who's Afraid of the Big Bad Wolverine?" *Fortean Times*, no. 85 (February-March 1996): 36—37; Jonathan Downes, *The Smaller Mystery Carnivores of the Westcountry* (Exwick, England: CFZ Publications, 1996); Paul Sieveking, "Cool Cats," *Fortean Times*, no. 88 (July 1996): 28—31; Paul Sieveking, "Watch Out: Big Cat About," *Fortean Times*, no. 101 (August 1997): 23—26; Paul Sieveking, "Cats in the Hats," *Fortean Times*, no. 111 (June 1998): 14—15; Paul Sieveking, "Nothing More than Felines," *Fortean Times*, no. 121 (April 1999): 20—21; Paul Sieveking, "Where the Wild Things Are," *Fortean Times*, no. 133 (April 2000): 18—19; "Alien Big Cat Attacks Boy," *Fortean Times*, no. 140 (December 2000): 6; Paul Sieveking, "Millennium Moggy Survey," *Fortean Times*, no. 146 (June 2001): 16—17; Chris Moiser, *Mystery Cats of Devon and Cornwall* (Launceston, England: Bossiney Books, 2001); Sarah Hartwell, Domestic × Wild Hybrids in the Wild, 2001, http://members.aol.com/jshartwell/hybrids.htm; Jim Gilchrist, "Beasts on the Prowl," *The Scotsman*,

February 2, 2002; Scott Weidensaul, *The Ghost with Trembling Wings* (New York: North Point Press, 2002), pp. 128—150; British Big Cat Society, http://www.britishbigcats.com.

## Brosnie

FRESHWATER MONSTER of Russia.

*Etymology:* After the lake.

*Physical description:* Serpentine. Length, 13–16 feet. Fish- or snakelike head. Large eyes. Enormous tail.

*Distribution:* Lake Brosno (250 miles northwest of Moscow), Tver' Region, European Russia.

*Significant sightings:* A monster tradition dates back to 1854.

In late 1996, tourists from Moscow snapped a photo of the monster after their seven-year-old son shouted that he had seen a dragon in the lake.

*Possible explanation:* The Sturgeon (*Acipenser sturio*) often grows to 11 feet 6 inches long, with outsize specimens reaching 19 feet. Some individuals are thought to live as long as 100 years. It is found in the Baltic Sea and spawns in rivers that drain into it. The Tver' Region would be a bit remote for this species.

*Sources:* Nikolai Pavlov, "Russia's 'Nessie' Frightens Villagers," Reuters, December 14, 1996; "A Russian Lake Monster," *Cryptozoology Review* 2, no. 1 (Summer 1997): 4; Karl Shuker, "Freshwater Monsters: The Next Generation," *Fate* 51 (February 1998): 18—21.

## Buffalo Lion

Maneless big CAT of East Africa.

*Etymology:* After its preference for large prey such as buffalo.

*Variant names:* River lion; Tsavolion.

*Physical description:* Male lion without a mane. Length, 9 feet 8 inches, including tail. Weight, around 400 pounds.

*Behavior:* Solitary. Adept at attacking large prey.

*Distribution:* Kenya.

*Significant sightings:* The famous pair of maneless, man-eating lions of Tsavo, Kenya, were responsible for killing 140 railway workers during a nine-month period in 1898. Now on exhibit at the Field Museum in Chicago, these

*The maneless, man-eating lions that terrorized Tsavo, Kenya, in 1898. (Field Museum of Natural History, Chicago)*

large males were shot by Chief Engineer John Henry Patterson in December.

In 1998, two maneless male lions were photographed in Tsavo National Park after bringing down a buffalo cow.

*Possible explanations:*

(1) Manelessness among male Lions (*Panthera leo*) could be due to hormonal problems or genetic defects. The condition may even constitute a form of natural selection in response to the preference by big-game hunters for lions with impressive manes.

(2) A surviving Cave lion (*P. l. spelaea*), a Middle Pleistocene felid from Southern Europe, although it was much heavier, ranging from 550 to 1,100 pounds.

*Sources:* John Henry Patterson, *The Man-Eaters of Tsavo and Other East African Adventures* (London: Macmillan, 1907); Peter von Buol, "'Buffalo Lions': A Feline Missing Link?" *Swara: The Magazine of the East African Wildlife Society* 23, no. 2 (July-December 2000): 20—25; Philip Caputo, *Ghosts of Tsavo: Stalking the Mystery Lions of Tsavo* (Washington, D.C.: National Geographic, 2002).

## Bukwas

CANNIBAL GIANT of western North America.

*Etymology:* Kwakiutl (Wakashan), "man of the woods" or "ape."

*Variant names:* Boks or Puks (Bella Coola/Salishan), Bowis (Tsimshian/Penutian), Pi'kis (Nass-Gitksian/Penutian), Pokwas, Pukmis (Nootka/Wakashan), Pukwubis (Makah/Wakashan).

*Physical description:* Height, about 5 feet. Covered with long hair. Face hairless and protruding. Thick browridges. Splayed nostrils. Pointed ears. No chin. Strong chest. Long arms.

*Behavior:* Walks with a stooping gait. Shrieks and whistles, especially at night. Has a bad odor. Eats clams. Has no fear of fire. Travels by canoe. Sometimes described as the spirit of a drowned person or a transformed otter.

*Distribution:* British Columbia and Washington State.

*Significant sighting:* Represented on carved, wooden masks used for ritual purposes. One mask was collected around 1914 from Nass-Gitksian Indians in northern British Columbia and is in Harvard's Peabody Museum. It features browridges, splayed nostrils, a jutting jaw without a chin, and thick lips.

*Sources:* Franz Boas and George Hunt, "Kwakiutl Texts," *Memoirs of the American Museum of Natural History* 5 (1902): 250—270; Thomas F. McIlwraith, "Certain Beliefs of the Bella Coola Indians Concerning Animals," *Archaeological Reports of the Ontario Department of Education* 35 (1924—1925): 17—27; Philip Drucker, "The Northern and Central Nootkan Tribes," *Bulletin of the Bureau of American Ethnology* 144 (1951): 152—153, 325; Alice Henson Ernst, *The Wolf Ritual of the Northwest Coast* (Eugene: University of Oregon Press, 1952), pp. 16—17, 34; Bruce Rigsby, "Some Pacific Northwest Native Language Names for the Sasquatch Phenomenon," *Northwest Anthropological Research Notes* 5, no. 2 (1971): 153—156; Edwin L. Wade, "The Monkey from Alaska: The Curious Case of an Enigmatic Mask from Bigfoot Country," *Harvard Magazine*, November-December 1978, pp. 48—51; Philip W. Davis and Ross Saunders, *Bella Coola Texts* (Victoria, Canada: British Columbia Provincial Museum, 1980), pp. 192—199; Marjorie M. Halpin, "The Tsimshian Monkey Mask and Sasquatch," in Marjorie Halpin and Michael M. Ames, eds., *Manlike Monsters on Trial* (Vancouver, Canada: University of British Columbia, 1980), pp. 211—228; Grant R. Keddie, "On Creating Un-Humans," in Vladimir Markotic and Grover Krantz, eds., *The Sasquatch and Other Unknown Hominoids* (Calgary, Alta., Canada: Western Publishers, 1984), pp. 22—29; John E. Roth, *American Elves* (Jefferson, N.C.: McFarland, 1997), p. 183.

## Bulgarian Lynx

The Eurasian lynx (*Lynx lynx*) is thought to have become extinct in Bulgaria in the 1940s, but unconfirmed reports indicate it persists in some areas. Increases in ungulate populations and reintroductions of this CAT in other parts of Europe may encourage its return.

*Sources:* Nikolai Spassov, "Cryptozoology: Its Scope and Progress," *Cryptozoology* 5 (1986): 120—124; Kristin Nowell and Peter Jackson, "Eurasian Lynx," from *Wild Cats: Status Survey and Conservation Action Plan*, IUCN, 1996, at http://lynx.uio.no/jon/lynx/eulynx1.htm.

## Bunyip

Mystery MARSUPIAL of Australia.

*Etymology:* Probably derived from the Australian BANIB. A "monster of Aboriginal legend, supposed to haunt water-holes; any freak or impostor," according to G. A. Wilkes, *Dictionary of Australian Colloquialisms*, 3d ed. (Sydney, Australia: Sydney University Press, 1990). The form *Bahnyip* appeared in the *Sydney Gazette* in 1812. Bernard Heuvelmans thought the word derived from *Bunjil,* the supreme being of the Victorian Aborigines. The name is widely used in Victoria and New South Wales and was first heard by whites in the Sydney area. By 1852, the word had become a synonym for "impostor" or "humbug" in Sydney. The term *bunyip aristocracy* refers to snobbish Australian conservatives.

*Variant names:* BANIB, Bunnyar (in Western Australia), Bunyup, Burley beast, Dongus (in New South Wales), Gu-ru-ngaty (Thurawal/Australian, New South Wales), Kajanprati, Katenpai, Kianpraty (in Victoria), Kine praty, Kinepràtia, KUDDIMUDRA, Mirree-ulla (Wiradhuri/Australian, New South Wales), MOCHEL

MOCHEL, MOOLGEWANKE, Munni munni (in Queensland), Toor-roo-don (in Victoria), Tumbata (in Victoria), TUNATPAN, WAA-WEE, Wangul (in Western Australia), Wouwai (near Lake Macquarie, New South Wales).

*Physical description:* According to Tony Healy and Paul Cropper, about 60 percent of the sightings resemble seals or swimming dogs, and 20 percent are long-necked creatures with small heads. (The remainder are too ambiguous to categorize.)

*Seal-dog variety*—Seal-like. Length, 4–6 feet. Shaggy, black or brown hair. Round head and whiskers like a seal's, otter's, or bulldog's. Shining eyes. Prominent ears. No tail.

*Long-necked variety*—Length, 5–15 feet. Black or brown fur. Head like a horse's or an emu's. Large ears. Small tusks. Elongated, maned neck about 3 feet long, with many folds of skin. Four legs. Three toes. Horselike tail.

*Behavior:* Amphibious. Nocturnal. Swims swiftly with fins or flippers. Loud, roaring call. Eats crayfish. Lays eggs in platypus nests in underwater burrows. Said by the Aborigines to be a guardian water spirit that eats women and children and causes sickness.

*Tracks:* Three-toed. Emulike.

*Habitat:* Lakes, rivers, and swamps.

*Distribution:* Traditions range throughout the continent, with sightings centered in Victoria, southern New South Wales, and eastern South Australia.

*Significant sightings:* In June 1801, mineralogist Joseph Charles Bailly of the French *Le Géographe* Expedition reported hearing the bellow of some large animal in the Swan River, Western Australia.

Hamilton Hume and James Meehan found skulls and bone fragments of amphibious animals the day after they discovered Lake Bathurst, New South Wales, in April 1818.

The earliest sightings by a colonist were at Lake Bathurst by Edward Smith Hall (later a founder of the Bank of New South Wales), who saw both the seal-dog and the long-necked varieties. In November 1821, Hall saw a black Bunyip with a bulldog's head thrashing in the water for five minutes. In December 1822, he was drying himself off after bathing in the eastern end of the lake when he saw a 3-foot, black head and neck gliding along the surface for about 300 yards. Some of the reports in the lake of animals with bulldoglike heads that made noises like a porpoise were possibly prompted by seals, which are known to have migrated to the nearby Mulware River in 1947.

Employees of George Holder (or Hopper) saw two horselike Bunyips in Paika Lake, New South Wales, in the 1840s.

Atholl Fletcher found a fresh skull along the lower Murrumbidgee River, New South Wales, in 1846. The top of the cranium, the front of the snout, and the lower jaw were missing. The cranium was about 9 inches long. The eye sockets were abnormally close to the upper jaw. No other bones were present. It was first examined by James Grant, who considered it to be a fetal skull of an unknown animal. William S. Macleay in Sydney also considered it to be from a young animal, possibly a fetus; comparing it to a fetal mare's skull, he thought it most likely belonged to a deformed colt. Based on an illustration, Sir Richard Owen in England pronounced it a calf's skull. It was taken to the Australian Museum in Sydney but has long since vanished. The Aboriginal name for the animal was said to be Katenpai.

Geologist E. J. Dunn observed several animals swimming in the flooded Murrumbidgee River near Gundagai in 1850.

A naturalist named Stocqueler reported "freshwater seals" in the Goulburn and Murray Rivers, New South Wales, in 1857.

Horsemen saw a whitish, dog-sized animal in 1886 along the River Molonglo, Australian Capital Territory.

On September 8, 1949, L. Keegan and his wife reported they had seen a 4-foot animal with shaggy ears several times over the past two weeks in the Lauriston Reservoir, Victoria. They claimed it used its ears in swimming through the water at tremendous speed.

In the 1960s, Jack Mitchell collected many reports by Aborigines, farmers, and tourists of a seal-dog Bunyip in the Macquarie River between Wellington and Warren, New South Wales.

A roaring animal that splattered mud around the bank of the Nerang River was heard near Gilston, Queensland, in 1965.

*Present status:* Widely believed in as a partially supernatural being by the Aborigines of southeastern Australia at the time of white settlement. There are few modern sightings, and most are vague or fanciful. The long-necked variety of Bunyip has not been reported since the nineteenth century and may be extinct.

*Possible explanations:*

(1) Australian fur seals (*Arctocephalus pusillus doriferus*) or Australian sea lions (*Neophoca cinerea*) that stray inland through the river systems might explain some sightings of the seal-dog Bunyip. In the nineteenth century, these were known to travel many miles up the Murray, Shoalhaven, and Murrumbidgee Rivers. Elephant seals (*Mirounga leonina*) were also known along the coast. Either of these animals seen unexpectedly in an unusual habitat could be misidentified.

(2) An unknown form of freshwater seal endemic to southeastern Australia.

(3) Booming calls of the Brown bittern (*Botaurus poiciloptilus*) of Victoria and New South Wales have been attributed to the Bunyip. One of its nicknames is the "bunyip bird."

(4) The Musk duck (*Biziura lobata*) was responsible for one report in Sydney in 1960.

(5) Some reports may have involved large Murray cod (*Maccullochella peelii peelii*), which grow to more than 5 feet.

(6) The Saltwater crocodile (*Crocodylus porosus*), the largest living reptile, is found in northern Australia, but it may have been known to Aborigines in the south in precolonial times, forming the basis for a Bunyip legend. Mature males average 14–16 feet long and are generally dark, with lighter tan or gray areas.

(7) An Australian version of the long-necked FRESHWATER MONSTER.

(8) Aboriginal legends of surviving Quaternary marsupials. Two candidates are the terrestrial, herbivorous, tapir-snouted *Palorchestes,* suggested by Tim Flannery and Michael Archer, said to have been the size of a bull, or *Diprotodon optatum,* the largest known marsupial, about 10 feet long with a 3-foot skull, suggested by C. W. Anderson

and Karl Shuker. Neither were amphibious, however.

(9) An unknown species of otterlike marsupial.

*Sources:* "The Bunyip, or *Kine pratie," Sydney Morning Herald,* January 21, 1847, p. 2; William H. Hovell, "The Apocryphal Animal of the Interior of New South Wales," *Sydney Morning Herald,* February 9, 1847; William Sharp Macleay, "On the Skull Now Exhibited at the Colonial Museum of Sydney, As That of the 'Bunyip,'" *Sydney Morning Herald,* July 14, 1847; William Westgarth, *Australia Felix* (Edinburgh: Oliver and Boyd, 1848); Ronald C. Gunn, "On the 'Bunyip' of Australia Felix," *Tasmanian Journal of Natural Science* 3 (1849): 147—149; John Morgan, *The Life and Adventures of William Buckley* (Hobart, Tasm., Australia: A. Macdougall, 1852), pp. 48, 108—109; *Moreton Bay (Queensl.) Free Press,* April 15, 1857, p. 3; Charles Gould, "Large Aquatic Animals," *Papers and Proceedings of the Royal Society of Tasmania,* 1872, pp. 32—41; Robert Brough Smyth, *The Aborigines of Victoria* (Melbourne, Australia: Government Printer, 1878); William Hardy Wilson, *The Cow Pasture Road* (William Hardy Wilson, 1920), p. 19; C. W. Anderson, "The Largest Marsupial," *Australian Museum Magazine* 2 (1924): 113—116; John Gale, *Canberra: History and Legends* (Queanbeyan, N.S.W., Australia: A. M. Fallick, 1927); Charles Fenner, *Bunyips and Billabongs* (Sydney, Australia: Angus and Robertson, 1933); Gilbert Whitley, "Mystery Animals of Australia," *Australian Museum Magazine* 7 (1940): 132—139; Charles Barrett, *The Bunyip and Other Mythical Monsters and Legends* (Melbourne, Australia: Reed and Harris, 1946), pp. 7—30; Alan Marshall, "Bunyips Never Whistle," *Melbourne Argus Magazine,* December 14, 1951; K.G. Dugan, "Darwin and *Diprotodon:* The Wellington Cave Fossils and the Law of Succession," *Proceedings of the Linnaean Society of New South Wales* 104 (1980): 265—272; Patricia Vickers-Rich and Gerard Van Tets, eds., *Kadimakara: Extinct Vertebrates of Australia* (Lilydale, Vic., Australia: Pioneer Design Studio, 1985), pp. 17, 234—244; W. S. Ramson, ed., *The Australian National Dictionary* (Melbourne, Australia: Oxford University Press, 1988), pp.

109—110; Christopher Smith, "A Second Look at the Bunyip," *INFO Journal,* no. 64 (October 1991): 11—13, 37; Tony Healy and Paul Cropper, *Out of the Shadows: Mystery Animals of Australia* (Chippendale, N.S.W., Australia: Ironbark, 1994), pp. 161—180; Malcolm Smith, *Bunyips and Bigfoots* (Alexandria, N.S.W., Australia: Millennium Books, 1996), pp. 1—24; Robert Holden and Nicholas Holden, *Bunyips: Australia's Folklore of Fear* (Canberra: National Library of Australia, 2001).

## Bu-Rin

Giant SNAKE of Southeast Asia.

*Physical description:* Length, 40–50 feet.

*Behavior:* Aquatic. Aggressive. Attacks swimmers and small boats.

*Distribution:* Near Putao, Myanmar.

*Source:* Alan Rabinowitz, *Beyond the Last Village: A Journey of Discovery in Asia's Forbidden Wilderness* (Washington, D.C.: Island Press, 2001), p. 116.

## Buru

Unknown LIZARD of Central Asia.

*Etymology:* Apatani and Nisi (Sino-Tibetan) word, possibly from its call.

*Physical description:* Roundish, elongated body. Length, 11–14 feet. Mottled blue-black above. Broad white band on the underside. Head, 20 inches. One account gives it three plates on the head, one on the top and on each side. Eyes are close behind a flat-tipped snout. Flat teeth, except for a single pair of large, pointed teeth in both the upper and lower jaws. Forked tongue. Neck, 3 feet. Three lines of short spines run down its back and sides. Back, 18 inches wide. One account said it has legs 20 inches long with clawed feet, while another only gave it paired lateral flanges. Round, tapering tail 3–5 feet long and fringed at the base.

*Behavior:* Completely aquatic. Raises its head out of the water occasionally. Basks in the sun on the bank in the summer. Remains in the mud when the swamps dry up. Makes a hoarse, bellowing noise. Does not eat fishes. Young are born alive in the water. Can grab a man with its tail and drag him underwater.

*Distribution:* Swamps and lakes near Ziro in the Apatani Valley, Arunachal Pradesh Union Territory, India; 50 miles to the southwest in the Dafla hills, Arunachal Pradesh Union Territory, India.

*Significant sightings:* In 1945 and 1946, James Phillip Mills and Charles Stonor collected descriptions of the Buru from the Apatani people, who are said to have killed the last of them in their area when they were draining swamps for rice cultivation.

In 1948, Ralph Izzard and Charles Stonor visited a swamp in the Dafla hills near Chemgeng in the hopes of finding a living Buru but returned with conflicting stories from the Nisi people.

*Present status:* It may still be possible to find skeletal remains of the animals in the Apatani Valley, since the precise kill spots are still known.

*Possible explanations:*

(1) A surviving dinosaur of some type, suggested by Ralph Izzard.

(2) An unknown species of Monitor lizard (*Varanus* sp.), suggested by Roy Mackal.

(3) An unknown species of Crocodile (Order Crocodylia), suggested by Tim Dinsdale.

(4) A large, swamp-dwelling Lungfish (Order Lepidosireniformes) would explain the Buru's ability to keep submerged in mud, according to Karl Shuker. The body structure also matches a lungfish more than a reptile. Its bellow might be caused by its ventilating air.

(5) An unknown species of Bonytongue fish similar to the Pirarucu (*Arapaima gigas*) of South America, which also has an air bladder fashioned into a lung.

*Sources:* Christopher von Fürer-Haimendorf, "The Valley of the Unknown," *Illustrated London News* 121 (November 8, 1947): 526—530; Ralph Izzard, *The Hunt for the Buru* (London: Hodder and Stoughton, 1951); Desmond Doig, "Bhutan," *National Geographic* 120 (September 1961): 384, 391—392; Tim Dinsdale, *The Leviathans* (London: Routledge and Kegan Paul, 1966), pp. 105—110; Roy P. Mackal, *Searching for Hidden Animals* (Garden City, N.Y.: Doubleday, 1980), pp. 79—98; Karl Shuker, *Extraordinary Animals Worldwide* (London: Robert Hale, 1991), pp. 54—61.

# Caddy

SEA MONSTER of the coast of British Columbia, Canada.

*Etymology:* Name popularized if not coined October 11, 1933, by *Victoria (B.C.) Daily Times* editor Archie H. Wills after repeated sightings in Cadboro Bay, British Columbia. Short form of *Cadborosaurus,* coined at the same time.

*Variant names:* Amy, Cadborosaurus, Edizgiganteus (after Ediz Hook Light, Washington), HAIETLUK, Klamahsosaurus (on Texada Island), Penda (after Pender Island).

*Scientific name: Cadborosaurus willsi,* proposed by Edward L. Bousfield and Paul H. LeBlond in 1995.

*Physical description:* Serpentine body that forms many humps or loops. Length, 16—100 feet. Diameter, 2 feet 6 inches—8 feet. Light brown to black. Small head resembles a sheep, horse, giraffe, or camel. Eyes in the front of the head. Small ears or horns. Pointed tongue. Two rows of fishlike teeth. Mane or fur sometimes reported. Neck is 3—12 feet long, about as thick as an arm. One pair of front flippers. Back sometimes appears serrated, sometimes smooth. Flat tail is fluked or formed from fused back flippers.

*Behavior:* Does not appear to undulate when it swims. Fast swimming speed, clocked at 40 knots. Breathes in short pants. Makes whalelike grunts and hisses. Feeds on herring, salmon, and ducks.

*Distribution:* British Columbia seacoast, especially around Cadboro Bay and the Strait of Georgia.

*Significant sightings:* A crew member of the ship *Columbia* under American fur trader Capt. Robert Gray was the first to report a Caddy sighting in 1791.

Osmond Fergusson watched a 25-foot animal with a long neck near the Queen Charlotte Islands, British Columbia, on June 26, 1897.

In September 1905 or 1906, Philip H. Welch saw a brown animal with a 6- to 8-foot neck from a distance of 100 yards away in Johnstone Strait. It had two bumps on its head that were 5 inches high and rounded on top.

F. W. Kemp and his wife and son watched an 80-foot maned animal while they were sitting on the Chatham Island beach, British Columbia, on August 10, 1932.

On September 23, 1933, Dorothea Hooper and a neighbor observed a serpentine animal with a serrated back cavorting in Cadboro Bay about 400 yards distant. It created a commotion in the water as it swam out to sea.

Maj. W. H. Langley and his wife were sailing in Haro Strait on October 1, 1933, when they heard a loud grunt off Chatham Island. They saw the back of a huge, dark-green creature with serrated markings on the top and sides.

Charles F. Eagles sketched a 60-foot animal that he saw in Oak Bay on October 14, 1933. It had crocodile-like spines on its neck.

On December 3, 1933, Justice of the Peace G. F. Parkyn of Bedwell Harbour was one of twelve people watching from Pender Island as an animal with a large, horselike head and neck gulped down a duck that had just been shot by Cyril Andrews.

In 1936, E. J. Stephenson and his wife and son watched a yellow-and-bluish, 90-foot-long, 3-foot-thick animal crawling over a reef into a lagoon on Saturna Island.

A 10- to 12-foot carcass of apparently a young Caddy was removed from the stomach of a sperm whale, photographed, and displayed for a while at Naden Harbour whaling station in 1937. The photo shows it stretched out on packing cases. It was about 10 feet long, with a camel-like head, traces of flippers, and a paddling tail. The carcass was allegedly shipped off to the Field Museum in Chicago, but there is no record of its arrival.

A Canadian naval officer was fishing in an open boat off Esquimalt Harbour in November 1950 when a 30-foot Caddy appeared and created a heavy wash. It swam with an undulating motion using large flippers on either side. It snapped its teeth together once before it dived after twenty-five seconds.

On February 12, 1953, R. D. Cockburn, C. P. Crawford, and Ron Loach saw an animal with three humps off Qualicum Beach for five minutes. Two other men got into a boat and rowed within 20 feet, but it submerged and reappeared 100 yards away. Its head was dog-shaped and had two horns.

In late November 1959, David Miller and Alfred Webb came within 30 feet of an animal with a 10-foot neck sticking straight up out of the water off Discovery Island. It had coarse brown fur, red eyes, and small ears.

A 16-inch-long juvenile Caddy was caught in a net by William Hagelund in 1968 off De Courcy Island, but it was thrown back. It had spiny teeth, a saw-toothed ridge of plates along its backbone, and a bilobate tail. A soft, yellow fuzz covered its undersides.

Mechanical engineer Jim M. Thompson was fishing off Spanish Banks, Vancouver, in January 1984 when an 18- to 22-foot serpentine animal surfaced about 100 feet away. It had a giraffelike head with small stubby horns and floppy ears.

In May 1992, music professor John Celona saw a multihumped animal about 25 feet long while sailing.

Students Damian Grant and Ryan Green were swimming across Telegraph Bay in May 1994 when they saw a 20-foot animal with two humps.

*Possible explanations:*

(1) The Northern sea lion (*Eumetopias jubatus*) can appear serpentine in the water but only grows to about 10 feet 6 inches long.

(2) The Northern elephant seal (*Mirounga angustirostris*) is found in British Columbian waters in the nonbreeding season, but it only measures up to 16 feet long and does not have an elongated neck.

(3) A surviving basilosaurid type of archaic whale, suggested by Roy Mackal and Karl Shuker. Some basilosaurids were serpentine, grew up to 80 feet long, and lived in the Late Eocene, about 42 million years ago. They had a tail fluke, but it's unknown whether it was used primarily for propulsion or steering. They are mainly known from the eastern United States and Egypt but may have been worldwide in distribution.

(4) An evolved plesiosaur, suggested by Edward Bousfield and Paul LeBlond. This group of long-necked marine reptiles swam with paddlelike limbs and had a body length that varied from 6 to 46 feet. Plesiosaur fossils are found continuously from the Middle Triassic, 238 million years ago, to the Late Cretaceous, 65 million years ago.

(5) A decaying Basking shark (*Cetorhinus maximus*) might account for the 1937 Naden Harbour carcass. These sharks take on a remarkably plesiosaur-like appearance due to the differential decomposition rates of their gill slits and lower tail fluke. A 30-foot carcass found in November 1934 by Hugo Sandstrom on Henry Island turned out to be a Basking shark.

(6) Some kind of decapod (crayfish or lobster) has been suggested by Aaron Bauer and Anthony Russell as an explanation for Hagelund's juvenile Caddy capture in 1968.

*Sources:* "Yachtsmen Tell of Huge Sea Serpent off Victoria," *Victoria (B.C.) Daily Times,* October 5, 1933, p. 1; "The Loch Ness Monster Paralleled in Canada," *Illustrated London News* 184 (January 6, 1934): 8; "A Canadian 'Monster,'" *Illustrated London News* 185 (December 15, 1934): 1011; Ray Gardner, "Caddy, King of the Coast," *Maclean's Magazine* 63 (June 15, 1950): 24, 42—43; D. Mattison, "An 1897 Sea Serpent Sighting in the Queen Charlotte Islands," *B.C. Historical News* 17, no. 2 (1964): 15; Paul H. LeBlond and John Sibert, *Observations of Large*

*Unidentified Marine Animals in British Columbia and Adjacent Waters* (Vancouver, Canada: University of British Columbia, Institute of Oceanography, June 1973); William A. Hagelund, *Whalers No More: A History of Whaling on the West Coast* (Madeira Park, B.C., Canada: Harbour, 1987); Frederic C. Howay, ed., *Voyages of the "Columbia" to the Northwest Coast, 1787—1790 and 1790—1793* (Portland: Oregon Historical Society, 1990), p. 249; Penny Park, "Beast from the Deep Puzzles Zoologists," *New Scientist* 137 (January 23, 1993): 16; Jessica Maxwell, "Seeing Serpents," *Pacific Northwest* 27 (April 1993): 30—34; Mike Dash, "The Dragons of Vancouver," *Fortean Times,* no. 70 (August-September 1993): 46—48; Edward L. Bousfield and Paul H. LeBlond, "An Account of *Cadborosaurus willsi,* New Genus, New Species, a Large Aquatic Reptile from the Pacific Coast of North America," *Amphipacifica* 1, suppl. 1 (1995): 3—25; Paul H. LeBlond and Edward L. Bousfield, *Cadborosaurus: Survivor from the Deep* (Victoria, B.C., Canada: Horsdal and Schubart, 1995); Aaron M. Bauer and Anthony P. Russell, "A Living Plesiosaur? A Critical Assessment of the Description of *Cadborosaurus willsi,*" *Cryptozoology* 12 (1996): 1—18; Darren Naish, "Another Caddy Carcass?" *Cryptozoology Review* 2, no. 1 (Summer 1997): 26—29; Paul H. LeBlond, "Caddy: An Update," *Crypto Dracontology Special,* no. 1 (November 2001): 55—59.

## Cait Sìth

Fairy CAT of Scotland.

*Etymology:* Gaelic, "fairy cat."

*Variant names:* Big ears, Cat sìth, Cath paluc.

*Physical description:* Size of a dog. Black with a white spot on its breast.

*Behavior:* Arches its back and bristles when angered.

*Distribution:* Highland, Scotland.

*Possible explanation:* Folk tradition about the KELLAS CAT.

*Sources:* John Gregorson Campbell, *Superstitions of the Highlands and Islands of Scotland* (Glasgow, Scotland: J. MacLehose and Sons,

1900), p. 32; James MacKillop, *Oxford Dictionary of Celtic Mythology* (New York: Oxford University Press, 1998), pp. 78, 81.

## Caitetu-Mundé

Unknown peccary-like HOOFED MAMMAL of South America.

*Etymology:* The Collared peccary is known as the Caitetu in Brazil, possibly from a Tupí word. *Mundé* is said to be a Tupí word for animal trap and is added to the names of animals that are hunted for game, such as *coatimundi* for the South American coati (*Nasua nasua*).

*Physical description:* Smaller than the White-lipped peccary (*Tayassu pecari*) and larger than the Collared peccary (*T. tajucu*). Length, about 3 feet. Shoulder height, about 20 inches.

*Behavior:* Lives in pairs or family groups of four.

*Distribution:* Rio Aripuanã, Mato Grosso State, Brazil.

*Source:* Karl Shuker, "New Beasts from Brazil?" *Fortean Times,* no. 139 (November 2000): 22.

## Caladrius

Mythical BIRD of Northern Europe.

*Variant name:* Charadrius (possibly from the Greek *charadrai,* "clefts," where the bird is said to live).

*Physical description:* Completely white plumage. Yellow bill. Swanlike neck. Yellow legs.

*Behavior:* Can detect human illness. Takes the disease on itself, flies toward the sun, vomits, and disperses the illness into the air. Its dung is said to cure blindness.

*Habitat:* Rivers.

*Distribution:* Northern Europe or West Asia.

*Possible explanations:* Birds with prominent white plumage have been proposed, among them:

(1) The Ringed plover (*Charadrius hiaticula*) is mostly white on the underside with an orange bill. It is found primarily on coasts and estuaries.

(2) The Common crane (*Grus grus*) is grayish with a long neck. It breeds in

northern Europe and Asia and winters in Southern Europe and Africa.

(3) A white parrot (Family Psittacidae) of some kind.

(4) The Great white egret (*Egretta alba*) has a black bill. It winters in Southern Europe.

(5) The Lapwing (*Vanella vanella*) has a short black bill and is only white on the underside, with green upperparts. It looks black and white in flight or from a distance.

(6) The Woodcock (*Scolopas rusticola*) is widespread in the British Isles and much of Europe. Its bill is grayish, and its plumage looks more brown than white.

(7) The White wagtail (*Motacill alba*), suggested by T. H. White, has a gray back, rump, and flanks.

(8) The Northern fulmar (*Fulmaris glacialis*) is a light gray-and-white seabird with a yellow bill. It spends a lot of time gliding along coastal cliffs.

(9) A seagull of some type, especially the Herring gull (*Larus argentatus*) or Yellow-legged gull (*L. cachinnans*), which have yellow bills.

*Sources:* Pierre de Beauvais, *A Medieval Book of Beasts,* trans. Guy R. Mermier (Lewiston, N.Y.: Edwin Mellon, 1992), pp. 27—28; George Claridge Druce, "The Caladrius and Its Legend, Sculptured upon the Twelfth-Century Doorway of Alne Church, Yorkshire," *Archaeological Journal* 69 (1912): 381—416; T. H. White, ed., *The Bestiary: A Book of Beasts* (New York: G. P. Putnam's, 1960), pp. 115—116.

## Calchona

WILDMAN of South America.

*Etymology:* Spanish, "ghost" or "bogey."

*Variant name:* Chilludo.

*Physical description:* A large man. Covered with long, sheeplike wool. Bearded.

*Behavior:* Nocturnal. Scares horses and travelers.

*Habitat:* Hills and fields.

*Distribution:* Western Neuquén Province, Argentina; Chile.

*Sources:* Zorobabel Rodriguez, *Diccionario de chilenismos* (Santiago de Chile: El

Independiente, 1875); Gregorio Alvarez, *El tronco de oro: Folklore del Neuquén* (Buenos Aires: Editorial "Pehuén," 1968), p. 121.

## Camahueto

SEA MONSTER of South America.

*Etymology:* Mapudungun (Araucanian), "sea elephant."

*Physical description:* Horse- or calflike head with one or two horns, which can be regenerated if lost. Sharp teeth. Strong claws.

*Behavior:* Born in freshwater; adults migrate to the sea. Said to bore holes in the cliffs and reefs. Eats fishes as well as humans.

*Distribution:* Isla de Chiloé, Los Lagos Region, Chile.

*Source:* Julio Vicuña-Cifuentes, *Mitos y supersticiones recogidos de la tradicion oral Chilena* (Santiago de Chile: Universitaria, 1915), pp. 32—33.

## Camazotz

Giant BAT of Central and South America.

*Etymology:* Zapoteco (Oto-Manguean), "death bat" or "snatch bat."

*Variant names:* Chonchon (in Peru and Chile), H'ik'al (Tzotzil, "black-man"), Soucouyant (in Trinidad), Tin tin (in Ecuador), Zotzilaha chamalcan (Mayan).

*Physical description:* Batlike head. Large knife- or leaflike protuberance on the nose. Sometimes depicted solely as a flying head.

*Behavior:* Nocturnal. Call an "eek eek" or "tui-tui-tui." In Mayan lore, kills dying men on their way to the center of the earth.

*Distribution:* Southern Mexico to northern Argentina.

*Possible explanations:*

(1) Much Latin American bat-demon mythology can be traced to the Common vampire bat (*Desmodus rotundus*), which feeds entirely on the blood of vertebrates–especially cattle and horses but sometimes on humans. It silently approaches an animal, lands on it, makes a tiny cut in the skin, and laps up the blood flow. It runs and hops on all fours as well as flies.

(2) The False vampire bat (*Vampyrum spectrum*) has an elongated face and a small noseleaf, unlike *Desmodus*. It is also much larger, with a wingspan of 3 feet.

(3) Spear-nosed bats (Subfamily Phyllostominae) have large noseleaves and are common throughout Central and South America.

(4) Surviving GIANT VAMPIRE BAT (*Desmodus draculae*), a Pleistocene bat known from fossils in southeastern Brazil.

*Sources: Popol Vuh: The Mayan Book of the Dawn of Life*, trans. Dennis Tedlock (New York: Simon and Schuster, 1996), pp. 71, 125, 275; Eduard Seler, "The Bat God of the Maya Race," *Bulletin of the Bureau of American Ethnology* 28 (1904): 231—241; Sarah Blaffer Hrdy, *The Black-Man of Zinacantán: A Central American Legend* (Austin: University of Texas Press, 1972); Elizabeth P. Benson, "Bats in South American Folklore and Ancient Art," *Andean Past* 1 (1987): 165—190; Elizabeth P. Benson, "The Maya and the Bat," *Latin American Indian Literatures Journal* 4 (1988): 118—120; Andrew D. Gable, "Two Possible Cryptids from Precolumbian Mesoamerica," *Cryptozoology Review* 2, no. 1 (Summer 1997): 17—25.

## Camelops

A genus of large, North American, camel-like HOOFED MAMMALS thought to have died out 10,000 years ago.

*Physical description:* Camel with a long neck, long legs, and probably a single hump.

*Distribution:* Utah.

*Significant sighting:* In 1926, Hector Lee and other high school students on a field trip discovered an unfossilized *Camelops* skull in a cave near Tabernacle Crater southwest of Fillmore, Utah. A strip of dried ligament was still attached.

*Possible explanations:*

(1) Oliver Hay was of the opinion that the skull was preserved exceptionally well because it had been covered by a 3- to 4-foot layer of fine dust in the cave. He thought it dated from the Aftonian interglacial period, no later than 700,000 years ago.

(2) Alfred Romer was convinced that *Camelops* persisted into recent times. A radiocarbon date of 11,075 years ago was obtained from the skull in 1979, supporting this theory.

*Sources:* Alfred Sherwood Romer, "A 'Fossil' Camel Recently Living in Utah," *Science* 68 (1928): 19—20; Oliver P. Hay, "An Extinct Camel from Utah," *Science* 68 (1928): 299—300; Alfred Sherwood Romer, "A Fresh Skull of an Extinct American Camel," *Journal of Geology* 37 (1929): 261—267; Michael E. Nelson and James H. Madsen Jr., "The Hay-Romer Camel Debate: Fifty Years Later," *University of Wyoming Contributions to Geology* 18 (1979): 47—50.

## Campchurch

UNICORN of Southeast Asia.

*Physical description:* Size of a deer. Horn, 3 feet 6 inches long. Two webbed feet like a duck's.

*Behavior:* Amphibious. Eats fishes. Horn contains an antitoxin.

*Distribution:* Strait of Malacca, Indonesia.

*Present status:* A similar sea-unicorn was noted in the sixteenth century as living off the southeastern coast of Africa.

*Sources:* Garcia ab Huerto, *Aromatum et simplicium aliquot medicamentorum* (Antwerp, Belgium, 1567), I. 14; André Thevet, *La Cosmographie universelle* (Paris: P. L'Huilier, 1575), vol. 2, p. 2.

## Canadian Alligator

Large CROCODILIAN of western Canada.

*Variant name:* PITT LAKE LIZARD.

*Physical description:* Length, usually 5—10 feet, with a maximum of 20 feet. Relatively smooth, dark skin. Horns or ears are sometimes reported. Long snout. Jaws 12 inches long. Four legs, 10 inches long.

*Behavior:* Aquatic but seen on land occasionally.

*Tracks:* Webbed.

*Distribution:* Pitt Lake, Kootenay Lake, Chilliwack Lake, Cultus Lake, Nitinat Lake, and the Fraser River, in British Columbia.

*Significant sightings:* On October 10, 1900, George Goudereau saw an animal like a 12-foot alligator crawl out of Crawford Bay on Kootenay Lake and root for food in a garbage heap. Later, a trail of large, webbed tracks was found.

In 1915, Charles Flood, Green Hicks, and Donald Macrae found some black, alligator-like lizards in a small mud lake south of Hope, British Columbia.

*Possible explanation:* An unknown species of cold-adapted crocodilian. The American alligator (*Alligator mississippiensis*) is the most northerly American crocodilian and is found as far north as the North Carolina coast. It was reported in southern Virginia in colonial times. Crocodilians depend on their environment to provide body warmth, and their hatchlings are more susceptible to chilling than adults. In fact, eggs incubated at temperatures lower than 88°F will tend to produce only female offspring and ultimately threaten the viability of the population. Nonetheless, both the American and the Chinese alligators (*A. sinensis*) dig burrows into which they can retreat during cold spells. They can also survive in lakes that are frozen by keeping their nostrils above the surface as their metabolism and body temperature drop. In warmer times, at least three species of crocodilians lived in Canada: *Leidyosuchus canadensis* and *Stangerochampsa* in Alberta during the Late Cretaceous, 65 million years ago, and *Borealosuchus acutidentatus* in Saskatchewan during the Paleocene, 60 million years ago.

*Sources:* Ivan T. Sanderson, *Abominable Snowmen: Legend Come to Life* (Philadelphia: Chilton, 1961), pp. 39—41; John Kirk, *In the Domain of Lake Monsters* (Toronto, Canada: Key Porter Books, 1998), pp. 176, 185—186; Chad Arment and Brad LaGrange, "Canadian 'Black Alligators': A Preliminary Look," *North American BioFortean Review* 1, no. 1 (April 1999): 6—12, http://www.strangeark.com/nabr/NABR1.pdf.

## Canavar

FRESHWATER MONSTER of West Asia.

*Etymology:* Turkish, "monster."

*Variant name:* Vanna.

*Physical description:* Length, 24—50 feet. Width, 3—6 feet. White with a black stripe along its back. Hairy head with horns. Three erect spines or fins.

*Distribution:* Van Gölü (Lake Van), Van Province, Turkey.

*Significant sightings:* Provincial deputy governor Bestami Alkan observed a black, dinosaur-like animal with triangular spikes on its back in 1995.

On June 10, 1997, Unal Kozak took video footage of the creature. It shows a dark object moving through the water near the shore before submerging. Enlarged, the object resembles a dark-brown hump, possibly showing an eye on one side.

*Sources:* Mustafa Y. Nutku and Unal Kozak, *Van Gölü Canavari* (Van, Turkey: Y.Y.U. Matbaasi, 1996); Karl Shuker, "Teggie and the Turk," *Strange M8agazine,* no. 17 (Summer 1996): 25—27; CNN, "Sea Monster or Monster Hoax?" June 12, 1997, http://www.cnn.com/WORLD/9706/12/fringe/turkey.monster/.

## Cannibal Giant

North American Indian tribes often had legends of ENTITIES similar to GIANT HOMINIDS or WILDMEN. Largely mythical and partially historical, these tales in some cases may be based on a traditional knowledge of BIGFOOT, HAIRY BIPEDS, or NORTH AMERICAN APES. Descriptions vary, but most of these creatures are said to be large and hairy; they live in remote areas and are said to eat people. Some of their behaviors are clearly fanciful, such as their ability to cause unconsciousness, their knack for trickery, or their penchant for driving people crazy. Sometimes, they are said to have stiff legs, a spike on their toes, or no odor at all–attributes not matching Bigfoot characteristics very well. "Stick Indians" generally refers to any group that lives in a wilderness area (and thus could refer to other tribes as well as unknown hominids) or that throws sticks at people. "Stone giants" were the ancient, stone-clad beings in Iroquoian mythology who were generally unpleasant and acted to mislead or kill humans; they represented both the forces of evil and the hardships of winter.

CANNIBAL GIANT *dance performed by the Chilkat Dancers, Haines, Alaska. (Photo by Martin Cordes; reprinted with permission from Allie Cordes. From a postcard in the author's collection)*

*Variant names:* ÄLÄKWIS, ATAHSAIA, ATCHEN, BUKWAS, CHIHALENCHI, CHIYE TANKA, DSONO-QUA, DZOAVITS, EL-ISH-KAS, ESTI CAPCAKI, GE-NOŜGWA, GETʼQUN, GILYUK, GOUGOU, GUGWÉ, GYEDM GYILILIX, HAITLÓ LAUX, HECAITOMIXW, KAIGYET, KASHEHOTAPALO, KECLEH-KUDLEH, KE-LÓ-SUMSH, KHOT-SA-POHL, KIWÁKWE, KOOSH-TAA-KAA, LA LA, LENGHEE, LOO POO OIʼYES, MADUKARAHAT, MATLOX, MESINGW, MIITIIPI, MISAABE, NAKANI, NATLISKELIGUTEN, NUMUZOʼHO, OEH, OH-MAH, OKEE, OLAYOME, PA-SNU-TA, PIAMUPITS, RUGARU, SASQUATCH, SEE-ATCO, SKADEGAMUTC, SKOOKUM, SMAYʼIL, SNANAIK, SNE-NAH, SOʼYOKO, STEETATHL, STEN-WYKEN, STE-YE-HAHʼ, Stick Indian, Stone giant, STRENDU, TAH-TAH-KLEʼ-AH, TENATCO, THAMEK-WIS, Timber giant, TOKÉ-MUSSI, TORNAIT, TOR-NIT, TʼOYLONA, TSADJATKO, TSAMEKES, TSOʼAPITTSE, TSULKALU, WAHTEETA, Wauk-wauk, WINDIGO, XUDELE, YAHYAHAAS, YIʼ DYIʼ TAY.

*Sources:* Marvin A. Rapp, "Legend of the Stone Giants," *New York Folklore Quarterly* 12 (1956): 280—282; Wayne Suttles, "On the Cultural Track of the Sasquatch," *Northwest Anthropological Research Notes* 6 (Spring 1972): 65—90; Joseph Bruchac, *Stone Giants and Flying Heads* (Trumansburg, N.Y.: Crossing, 1979); Wayne Suttles, "Sasquatch: The Testimony of Tradition," in Marjorie M. Halpin and Michael M. Ames, eds., *Manlike Monsters on Trial* (Vancouver, Canada: University of British Columbia Press, 1980), pp. 245—254; Raymond D. Fogelson, "Windigo Goes South: Stoneclad among the Cherokees," in Marjorie M. Halpin and Michael M. Ames, eds., *Manlike Monsters on Trial* (Vancouver, Canada: University of British Columbia Press, 1980), pp. 132—151; Grant R. Keddie, "On Creating Un-humans," in Vladimir Markotic and Grover S. Krantz, eds., *The Sasquatch and Other Unknown Hominoids* (Calgary, Alta., Canada: Western, 1984), pp. 22—29; Loren Coleman and Mark A. Hall, "From 'Atshen' to Giants in North America," in Vladimir Markotic and Grover S. Krantz, eds., *The Sasquatch and Other Unknown Hominoids* (Calgary, Alta., Canada: Western, 1984), pp. 30—43; Kyle Mizokami, Bigfoot-Like Figures in North American Folklore and Tradition, http://www.rain. org/campinternet/bigfoot/bigfoot-folklore.html.

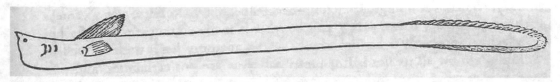

CAPTAIN HANNA'S FISH, *an odd fish caught off New Harbor, Maine, in 1880 by Captain S. W. Hanna. (U.S. Fish Commission)*

## Captain Hanna's Fish

Odd, eellike FISH of the North Atlantic Ocean.

*Physical description:* Serpentine. Length, 25 feet; 10 inches thick at the largest part. Darkish slate color on top, grayish-white below. Flat head that extends over a small mouth with sharp teeth. Prominent gill slits. Two small, rayed pectoral fins and a triangular rayed dorsal fin behind the head. A caudal fin extends around the tail.

*Significant sighting:* In August 1880, Capt. S. W. Hanna caught a fish of this description off New Harbor, Maine, but discarded it because it had torn his net.

*Distribution:* Gulf of Maine.

*Possible explanations:*

(1) Unknown species of elongated shark, perhaps a large form of the Frilled shark (*Chlamydoselachus anguineus*), suggested by Bernard Heuvelmans. The Frilled shark grows to over 6 feet in length and lives in deep waters near the sea floor, primarily in the eastern Atlantic, though three individuals have been recorded in the western Atlantic. It is the only shark to have a mouth that opens at the tip of the snout. However, sharks do not have rayed fins. Larger varieties may exist. David Stead examined the skull and vertebrae of a 12-foot fish that washed up in the harbor at Sydney, Australia, in August 1907 and identified it as a Frilled shark, though double the length of the largest known specimens.

(2) Unknown species of bony fish, suggested by Ben Roesch, based on its lack of pelvic fins, and the position of the dorsal fin.

*Sources:* S. W. Hanna, "Description of an Eel-Like Creature Taken in a Net at New Harbor, Maine, in 1880," *Bulletin of the U.S. Fish Commission* 3 (1883): 407—410; David G. Stead, *Sharks and Rays of Australian Seas* (Sydney, Australia: Angus and Robertson, 1963).

## Carabuncle

FRESHWATER MONSTER of Ireland, as well as a mysterious South American animal.

*Etymology:* From the Latin *carbunculus* ("gem").

*Physical description:* Serpentine. Said to have a shining, precious stone or a pearl hanging from its head that glitters like silver in the night.

*Behavior:* Nocturnal.

*Distribution:* Lough Geal, on Mount Brandon, County Kerry, Ireland; the Straits of Magellan, Argentina; Paraguay.

*Sources:* Gonzalo Fernándo de Oviedo y Valdés, *Natural History of the West Indies,* trans. Sterling A. Stoudemire [1526] (Chapel Hill: University of North Carolina Press, 1959); Martín del Barco Centenera, *The Argentine and the Conquest of the River Plate* [1602] (Buenos Aires: Instituto Cultural Walter Owen, 1965); Charles Smith, *The Antient and Present State of the County of Kerry* (Dublin: Charles Smith, 1756), p. 124; Henry Hart, "Notes on the Plants of Some of the Mountain Ranges of Ireland," *Proceedings of the Royal Irish Academy, Science,* ser. 2, 4 (1884): 211, 220; Nathaniel Colgan, "Field Notes on the Folklore of Irish Plants and Animals," *Irish Naturalist* 23 (March 1914): 53—64.

## Caribbean Crowing Snake

Unknown SNAKE of the West Indies.

*Physical description:* Length, 4 feet. Thick body. Dull ochre color with dark spots. Pale red pyramidal crest like a rooster. Scarlet wattles.

*Behavior:* Crows like a rooster. Eats poultry.

*Distribution:* Eastern portion of Jamaica; Haiti.

*Significant sightings:* In 1829, a medical doctor saw a crested snake, dead and slightly decomposed, in Jamaica.

A snake with wattles was shot in Jamaica on March 30, 1850, by the son of Jasper Cargill.

*Possible explanation:* Vocalizing snakes are not unknown. *See* CROWING CRESTED COBRA.

*Source:* Philip Henry Gosse, *The Romance of Natural History, Second Series* (London: J. Nisbet, 1861), pp. 211—219.

## Caribbean Monk Seal

Nondescript SEAL of the West Indies, presumed extinct since 1952.

*Scientific name: Monachus tropicalis,* given by Gray in 1850.

*Physical description:* Length, 7—8 feet. Brown on the back with a gray tinge.

*Behavior:* Approachable and unaggressive.

*Distribution:* Caribbean Sea, off Haiti and Jamaica. It formerly extended throughout the northern and western Caribbean and the Gulf of Mexico.

*Significant sightings:* Sixteen out of ninety-three Haitian and Jamaican fishermen interviewed in 1997 claimed to have seen at least one monk seal in the previous two years.

*Present status:* Five major surveys of former monk seal habitats have been conducted by trained naturalists since 1950, with no definite evidence of the animal's survival past 1952.

*Possible explanations:*

(1) A misidentified California sea lion (*Zalophus californianus*) that escaped from captivity.

(2) The Hooded seal (*Cystophora cristata*) occasionally strays as far south as Florida.

(3) The Harbor seal (*Phoca vitulina*) and Harp seal (*Phoca groenlandica*) are even rarer visitants to the Caribbean.

*Source:* I. L. Boyd and M. P. Stanfield, "Circumstantial Evidence for the Presence of Monk Seals in the West Indies," *Oryx* 32 (1998): 310—316.

## Carolina Parakeet

Small BIRD of the Parrot family (Psittacidae) in the southeastern United States, presumed extinct since 1918.

*Scientific name: Conuropsis carolinensis,* given by Tommaso Salvadori in 1891.

*Physical description:* Bright green plumage. Yellow head. Orange forehead and cheeks.

*Habitat:* Wetlands.

*Distribution:* Santee River, South Carolina; Okefenokee Swamp, Georgia; Okeechobee County, Florida.

*Significant sightings:* Henry Redding reported a flock of thirty parakeets near Fort Drum Creek, Florida, in 1920.

In 1926, Charles E. Doe saw three pairs of parakeets at Grapevine Hammock in Okeechobee County, Florida. He took some of their eggs, which have been preserved.

The National Audubon Society bird wardens for the Santee Swamp area in South Carolina reported the presence of green parakeets with yellow heads on several occasions in the 1930s. In 1933 and 1934, George M. Melamphy saw as

The CAROLINA PARAKEET (Conuropsis carolinensis), *presumed extinct since 1918. (© 2002 ArtToday.com, Inc., an IMSI Company)*

many as nine together at a time, feeding on sun-flower seeds. Ornithologist Alexander Sprunt Jr. claimed to have seen a juvenile fly swiftly by in the fall of 1936. Sprunt's companion, Robert Porter Allen, had come to believe by 1949 that they had seen mourning doves or released exotic parrots.

Orsen Stemville took a color film of some type of parakeet in the Okefenokee Swamp, Georgia, in 1937.

In 1938, a woodsman named Shokes saw two yellow-headed parakeets circling above him as a juvenile flew up to join them near Wadmacaun Creek, South Carolina. The Santee habitat was destroyed during the completion of the Santee-Cooper Hydroelectric Project in 1936—38.

*Present status:* The last wild specimens were shot in April 1904 at Lake Okeechobee, Florida. The last captive specimen died at the Cincinnati Zoo in February 1918.

*Possible explanation:* Nonnative green para-keets escaped from pet owners or zoos.

*Sources:* M. S. Curtler, "Carolina Parakeet Not Extinct?" *Animals* 7 (November 23, 1965): 532; James C. Greenway Jr., *Extinct and Vanishing Birds of the World* (New York: Dover, 1967); Christopher Cokinos, *Hope Is the Thing with Feathers* (New York: Jeremy P. Tarcher, 2001), pp. 5—58; Errol Fuller, *Extinct Birds* (Ithaca, N.Y.: Cornell University Press, 2001), pp. 239—243.

## Carugua

Unknown PRIMATE of South America.

*Behavior:* Kills cattle by pulling their tongues out.

*Distribution:* Ybytymí area, Paraguay.

*Significant sighting:* In the 1940s, in nearly eight months, about 100 cattle were found dead in the Ybytymí area with no wounds except for their tongues being torn out. The events re-curred in 1952 or 1953 on a different ranch.

*Source:* Bernard Heuvelmans, *On the Track of Unknown Animals* (New York: Hill and Wang, 1958), p. 308.

## Caspian Tiger

Big CAT of West Asia, presumed extinct since the 1970s.

*Scientific name: Panthera tigris virgata.*

*Variant name:* Hyrcanian tiger.

*Distribution:* Talysh, Azerbaijan; the Cudi Mountains of Turkey and Iran.

*Significant sightings:* The Caspian subspecies of tiger formerly had a vast range from Afghanistan, Turkestan, and Kazakhstan through the Caucasus to Iran and Turkey. The last time a living specimen was seen was in Afghanistan in 1967. A fresh skin had been pur-chased by a druggist in eastern Turkey some-time in the 1970s, but it has not been properly examined and may have been an illegally ob-tained Bengal pelt. However, hunters in Azer-baijan and Turkey still report hearing and see-ing it.

*Sources:* Nikolai Spassov, "Cryptozoology: Its Scope and Progress," *Cryptozoology* 5 (1986): 120—124; Karl Shuker, "Tracking a Turkish Tiger," *Fortean Times,* no. 146 (June 2001): 46.

## Cassie

A MULTIHUMPED SEA MONSTER of the North Atlantic Ocean.

*Etymology:* After Casco Bay, in imitation of other named water monsters. Coined in 1986 by Loren Coleman.

*Distribution:* Casco Bay and other points along the coast of the Gulf of Maine, including Penobscot and Portland.

*Significant sightings:* Future naval commodore Edward Preble was serving as an ensign on the warship *Protector* in June 1779 when he saw a large serpent lying on the surface of a bay along the Maine coast. Commander John Foster Williams ordered Preble to launch a longboat in an attempt to shoot the animal, which appeared to be 100—150 feet long and as thick as a barrel. As the boat approached, the serpent raised its head 10 feet above the water. A shot was fired, but the snake swam away quickly.

On July 12, 1818, a sea monster was seen in the harbor of Portland, Maine, in full view of a number of observers at Weeks's wharf.

In the summer of 1836, Captain Black of the

schooner *Fox* spotted an animal in the inlet between the mainland and Mount Desert Rock, Maine. It held its snakelike head 2—3 feet above the water.

Maj. Gen. H. C. Merriam was sailing with his sons opposite Wood Island Light, Maine, on August 5, 1905, when they saw a mottled brown "monster serpent" with its head 4 feet above the water slowly moving toward their becalmed boat. The animal then circled them from a distance of 300 yards at about 12 miles an hour. Its head was like a snake's, the neck was 15—18 inches in diameter, and the total length was 60 feet or more. It remained visible for ten minutes, then reappeared a short time later after the wind started up.

On August 20, 1910, the fishing steamer *Bonita* passed an 80-foot animal in Casco Bay. It was black with large white spots.

*Present status:* Few sightings since the 1950s.

*Sources:* "Distinguished Visiter," *Boston Weekly Messenger* 7 (July 23, 1818): 651; James Fenimore Cooper, *Lives of Distinguished American Naval Officers* (Philadelphia: Carey and Hart, 1846), pp. 180—182; Van Campen Heilner, *Salt Water Fishing* (Philadelphia: Penn, 1937), app.; Loren Coleman, "Casco Bay's Sea Serpent," *Portland Monthly,* May 1986; J. P. O'Neill, *The Great New England Sea Serpent* (Camden, Maine: Down East, 1999).

## Cat-Headed Snake

Mystery SNAKE of Central Europe.

*Physical description:* Length, 7 feet. Gray and black color. Head like a cat's. Ridged back.

*Distribution:* The Swiss Alps.

*Significant sighting:* In April 1711, Jean and Thomas Tinner killed a large snake with a catlike head at Hauwelen on the Frumsemberg mountain, Switzerland.

*Source:* Johann Jakob Scheuchzer, *Helvetica* (Leiden, the Netherlands: Petri Vander Aa, 1723).

## Catoblepas

Mammal of North Africa; *see* SEMIMYTHICAL BEASTS.

*Etymology:* Greek, "that which looks downward."

*Variant names:* Catoplepe, Gorgon.

*Physical description:* Body like a bull's. Oversized head so heavy that it always hangs close to the ground. Bristled, scraggly head-hair from crown to nose. Scaly back. Small wings, according to later sources.

*Behavior:* Sluggish. Has an overpowering breath (odor). Its gaze can kill (though Pliny may have gotten this characteristic mixed up with that of the BASILISK).

*Distribution:* The Nile River in Egypt, east to Ethiopia.

*Possible explanation:* The Black wildebeest or White-tailed gnu (*Connochaetes gnou*) is a stocky, dark-brown antelope with front-facing horns and a strange tuft of long hair on the muzzle that acts as a scent dispenser for its face glands. It is found in South Africa. Its fighting posture is to kneel on the ground and hook its opponent from below with its sharply angled horns. The related Blue wildebeest or Brindled gnu (*C. taurinus*), of South and East Africa, ranged much father north, where Pliny was more likely to have heard about it, possibly persisting on the North African coast into historical times.

*Sources:* Pliny the Elder, *Natural History,* VIII. 32; Ælian, *De natura animalium,* VIII. 105; Edward Topsell, *The Historie of Foure-Footed Beastes* (London: William Iaggard, 1607); Peter Costello, *The Magic Zoo* (New York: St. Martin's, 1979), pp. 111—114.

## CATS (Unknown)

The Cat family (Felidae) arose from primitive carnivores in Europe about 30 million years ago, in the Oligocene. Common characteristics include extraordinarily acute hearing, eyesight adapted for night vision, a sensitive nose, whiskers, long canine teeth for seizing, large carnassial teeth for tearing, a spotted or striped coat, flexible vertebrae, and retractile claws (except for the cheetah).

The largest living cat is the male Tiger (*Panthera tigris*), which in Siberia attains an average length of 10 feet 4 inches from nose to tail tip.

Claims of tigers up to 13 feet or longer are sometimes based on stretched skins or faulty measurements. Weights up to 700 pounds have been reported.

Felid taxonomy remains somewhat controversial. Some authorities put modern cats into three groups: the Subfamily Pantherinae, incorporating leopards, jaguars, lions, tigers, lynxes, and bobcats; the Subfamily Felinae, which includes everything in the genus *Felis;* and an undetermined category for the cheetah. Others place all living cats into the Subfamily Felinae.

Genetic variations in coat coloration have been one reason why cat classification has been problematic. The general purpose for spots, stripes, and splotches on the coat is to provide camouflage and make it more difficult for the animal to be identified when it is stalking its prey. Each species has a basic set of markings, and individual cats often sport a unique pattern. Significant variations in coat pigmentation (albino, melanistic, chinchilla, agouti, and so on) are produced by the mutant alleles of six major genes. If enough variants occur in an isolated population to allow it to interbreed successfully and preserve those characteristics, it becomes a subspecies.

The occurrence of melanism (black pigment) in some mystery cats is perplexing. Melanism is the commonest coat variation in wild cats and occurs in thirteen different species. Black leopards are most often found in the forested part of their range in tropical Asia and less often in Africa. At one time, they were considered a distinct species, but both normal and melanistic individuals can be found in the same litter. The American puma exhibits very little tendency toward melanism, making reports of black EASTERN PUMAS especially anomalous.

Of the sixty-two mystery cats in this list, seventeen probably represent little-known color morphs of known species, such as the BLUE TIGER and the KING CHEETAH. Hybridization between various cat species in the wild is rare but possible under certain circumstances in areas where two species overlap. Eight cat cryptids may represent recurring hybrids, such as the KELLAS CAT and the TINICUM CAT. Ten others, such as the BORNEAN TIGER and the ARIZONA JAGUAR, are unverified extensions of the distribution of known species.

Twelve cats in the list may be undescribed species new to science; these include the SPOTTED LION and the NAYARIT RUFFED CAT. Five may represent surviving species known from fossils, especially the saber-toothed cats *Machairodus* and *Smilodon* (among them the TIGRE DE MONTAGNE, WATER LION, and WATER TIGER). Four versions of ALIEN BIG CATS are included; their origins are undoubtedly multicausal, but some have been tempted to think a new species is involved.

Four on the list do not seem to be cats at all, while the WINGED CAT is apparently an extreme form of a disease afflicting domestic cats.

## Mystery Cats

*Africa*

BLACK LION; BUFFALO LION; DARK LEOPARD; KING CHEETAH; KITANGA; MALAGASY LION; MNGWA; QATTARA CHEETAH; SPOTTED LION; TIGRE DE MONTAGNE; WATER LION; WOBO; WOOLLY CHEETAH

*Asia*

BLACK TIGER; BLUE TIGER; BORNEAN TIGER; CASPIAN TIGER; CAUCASIAN BLACK CAT; CIGAU; DOGLAS; HARIMAU JALUR; SAT-KALAUK; SEAH MELANG PÀA; SHING MUN TIGER; STRIPELESS TIGER; SUNDANESE HORNED CAT; YAMAMAYA

*Australasia and Oceania*

AUSTRALIAN BIG CAT

*Central and South America*

ANOMALOUS JAGUAR; COLUMBUS'S APE-FACED CAT; JAGUARETÉ; MITLA; ONÇA-CANGUÇÚ; PERUVIAN JUNGLE LION; PERUVIAN JUNGLE WILDCAT; RAINBOW TIGER; SHIASHIA-YAWÁ; SIEMEL'S MYSTERY CAT; SPECKLED JAGUAR; STRIPED JAGUAR; TAPIR TIGER; WARACABRA TIGER; WATER TIGER; YANA PUMA

*Europe*

ALIEN BIG CAT; BRITISH BIG CAT; BULGARIAN LYNX; CAIT SITH; ÎLE DU LEVANT WILDCAT; IRISH WILDCAT; ISTURITZ SCIMITAR CAT; KELLAS CAT

## North America
ARIZONA JAGUAR; CUITLAMIZTLI; EASTERN
PUMA; MANED AMERICAN LION; NAYARIT
RUFFED CAT; ONZA; OZARK HOWLER; SANTER;
TINICUM CAT

## Various
WINGED CAT

## Caucasian Black Cat
Mystery CAT of West Asia.

*Scientific name:* Felis daemon, given by K. A.
Satunin in 1904.

*Physical description:* Length, 22—30 inches.
Black with a reddish tinge to reddish-brown.
Thin scattering of white hairs all over. Lighter
below. Black stripes on the flanks. Tail, 13—15
inches long.

*Distribution:* Armenia and Azerbaijan, south
of the Caucasus Mountains.

*Significant sighting:* Described by K. A. Satunin on the basis of two mounted specimens,
skins, and skulls in the Leningrad Academy of
Sciences.

*Present status:* No recent reports.

*Possible explanations:*

(1) Melanistic morph of the Caucasian
wildcat (*Felis silvestris caucasica*).

(2) Feral Domestic cat (*Felis silvestris catus*),
according to S. I. Ognev and Reginald
Pocock.

(3) Feral cat × Caucasian wildcat hybrid,
suggested by Karl Shuker. Possibly similar
to the KELLAS CAT of Scotland.

*Sources:* Konstantin A. Satunin, "The Black
Wild Cat of Transcaucasia," *Proceedings of the
Zoological Society of London,* 1904, pp.
163—164; F. B. Aliev, "The Caucasian Black
Cat, *Felis silvestris caucasica* Satunin 1905,"
*Säugetierkundliche Mitteilungen* 22 (1973):
142—145; Karl Shuker, *Mystery Cats of the World*
(London: Robert Hale, 1989), pp. 80—82.

## Cax-Vinic
WILDMAN of Mexico.

*Etymology:* Mayan, possibly from *c'as* or *kaax*
("bush" or "wild") + *vinic* ("man").

*Variant names:* Cangodrilo, Fantasma hu-
mano (Spanish, "human phantom"), Hombre
oso (Spanish, "bear-man").

*Physical description:* Covered in black or
brown hair. Glowing eyes.

*Behavior:* Nocturnal. Emits a loud, threatening cry.

*Distribution:* Sierra Madre, Chiapas State,
Mexico. Rumors exist of other Mexican wildmen from Chihuahua to Veracruz States.

*Significant sighting:* W. C. Slater reported
finding humanlike tracks ("the size of a small
woman's hand") in the snow at an altitude of
6,500 feet on Volcán Popocatépetl, near Mexico
City, in the 1930s. Local people attributed them
to the "men of the snows."

*Sources:* W. C. Slater (letter), "The
'Abominable Snowmen': Footprints in
Mexico," *Times* (London), August 2, 1937,
p. 6; Ivan T. Sanderson, *Abominable Snowmen:
Legend Come to Life* (Philadelphia: Chilton,
1961), pp. 157—158; Scott Corrales,
"Paranormal Manimals in Latin America,"
http://www.strangemag.com/paranormalanim.
html.

## Cecil
FRESHWATER MONSTER of Nevada.

*Physical description:* Length, 45—50 feet.

*Behavior:* Swift swimmer.

*Distribution:* Walker Lake, Nevada.

*Significant sighting:* In April 1956, two witnesses saw an animal that was able to pace their
car at 35 miles per hour.

*Source:* J. K. Parrish, "Our Country's
Mysterious Monsters," *Old West,* Fall 1969,
pp. 25, 37—38.

## Centaur
SEMIMYTHICAL BEAST of Southern Europe and
the Middle East.

*Etymology:* Greek *kéntauros,* derived from
"those who round up bulls."

*Physical description:* Head, arms, and trunk of
a human. Legs and body of a horse.

*Behavior:* One group was fierce, sensuous,
rude, and barbarous; they were destroyed by
Herakles and the Lapithae in a symbolic fight

When great Attempts are undergone,
Ioyne Strength *and* Wisedome, *both in one.*

SAPIENTIA ✥ VIRIBVS IVNGENDAS

ILLVSTR. XLI.                                        Book. 2

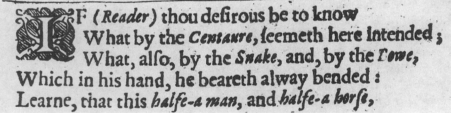

IF (*Reader*) thou desirous be to know
What by the *Centaure*, seemeth here intended ;
What, also, by the *Snake*, and, by the *Bowe*,
Which in his hand, he beareth alway bended :
Learne, that this *halfe-a man*, and *halfe-a horse,*

*The* CENTAUR *as a symbol of strength and wisdom. From George Wither,* A Collection of Emblemes, Ancient and Moderne *(London: Augustine Matthews, 1635). (Fortean Picture Library)*

between humans and beasts. Another group was wise and friendly and included Chiron and Photius, the companions of heroes.

*Distribution:* Coastal mountains of Thessaly, Greece, between Mount Ossa and Mount Pilion; possibly Arabia and elsewhere.

*Significant sightings:* Pliny claims to have seen a dead Centaur, preserved in honey, taken to Rome from Arabia via Egypt during the reign of Emperor Claudius (A.D. 41—54). Phlegon of Tralles saw it about sixty years later and wrote that it had a fierce face and hairy arms and fingers. Its human torso merged smoothly with its horse's body, and its hooves were firm. The entire body had turned dark brown.

John Farrell and Margaret Johnson were driving along a country road near Drogheda, County Louth, Ireland, in the spring of 1966 when their way was blocked for two minutes by a horse with a man's face.

*Possible explanations:*

(1) Early depictions show Centaurs as hairy giants. They are considered to be the personification of a wild mountain tribe of horsemen who herded cattle. The Thessalians were noted for their riding ability.

(2) Hobby-horse dancers in ancient Greek rituals, suggested by Robert Graves.

(3) Pliny's Centaur may have been a manufactured composite assembled from mummified human and pony parts.

*Sources:* Lucretius, *The Nature of the Universe,* trans. R. E. Latham (Baltimore, Md.: Penguin, 1951), pp. 198—199 (bk. 5); Pliny the Elder, *Natural History* (New York: Penguin, 1991), p. 80 (VII. 33); John Cuthbert Lawson, *Modern Greek Folklore and Ancient Greek Religion* (Cambridge: University Press, 1910); Georges Dumézil, *Le problème des centaures* (Paris: P. Geuthner, 1929); *New Larousse Encyclopedia of Mythology* (New York: Putnam, 1968), pp. 161—162; Graham J. McEwan, *Mystery Animals of Britain and Ireland* (London: Robert Hale, 1986), pp. 165—166; Adrienne Mayor, *The First Fossil Hunters: Paleontology in Greek and Roman Times* (Princeton, N.J.: Princeton University Press, 2000), pp. 228—243.

# CEPHALOPODS (Unknown)

Squids, octopuses, cuttlefish, the fossil ammonites, and the chambered nautilus are all members of the Class Cephalopoda, marine INVERTEBRATES with large eyes, a head surrounded by muscular tentacles, and a chitinous beak like a parrot's. They are considered the most highly evolved invertebrates. Octopuses have eight tentacles, squids and cuttlefish have ten, and the nautilus has sixty to ninety. All move by taking in water and expelling it forcibly through a siphon; octopuses can also crawl along the ocean bottom.

The first cephalopods were the nautiloids, which appeared in the Ordovician, some 450 million years ago. The ammonites became abundant from the Devonian to the Cretaceous, 350—65 million years ago. Cuttlefish, squids, and octopuses have only vestigial internal shells or no shells at all; consequently, they have left virtually no fossils behind. Their closest fossil relatives were the belemnites, common in Mesozoic seas, that had conical shells and eight tentacles equipped with hooks.

The largest living invertebrate is the Giant squid (*Architeuthis* sp.). Although the top size for this animal is a matter of some controversy (see KRAKEN), the 2.2-ton specimen that washed ashore at Thimble Tickle, Newfoundland, on November 2, 1878, measured 20 feet from beak to tail, and one of its tentacles measured 35 feet, giving it a total length of 55 feet. It also had the largest eye of any known animal, living or extinct, with a diameter of about 15.75 inches. The largest known octopus is the Giant Pacific octopus (*Enteroctopus dofleini*), which can exceed a radial spread of 20 feet.

There are ten cephalopod cryptids in this list, three squids and seven octopuses. Two of the octopuses have supposedly been found in freshwater environments, and one of the squids was reported living in a toxic oil-emulsion pit. Either habitat would be a first for any of the cephalopods, which are exclusively marine. Six in the list involve animals of considerable size.

## Mystery Cephalopods

CUERO; FRESHWATER OCTOPUS; GIANT BRITISH OCTOPUS; GIANT MEDITERRANEAN OCTOPUS;

GIGANTIC OCTOPUS; GIGANTIC PACIFIC OCTO-PUS; KRAKEN; LUSCA; OIL PIT SQUID; SEA MONK

## CETACEANS (Unknown)

Whales, dolphins, and porpoises belong to the Order Cetacea, a group of mammals that have completely adopted an aquatic existence. They have streamlined, hairless bodies with two front flippers, no hind legs, a muscular tail for propulsion, and a blowhole at the top of the head for breathing. There are two main types: The Toothed whales (Odontocetes) include dolphins, porpoises, sperm whales, the beluga, the narwhal, and beaked whales; the toothless Baleen whales (Mysticetes) include the rorquals, right whales, and the gray whale.

Recent molecular studies have shown that whales are most closely related to modern cattle, deer, and pigs (HOOFED MAMMALS of the Order Artiodactyla) and may have evolved from them in the Early Eocene, about 52 million years ago. Alternatively, they may have emerged from the mesonychids, an extinct group of archaic ungulates that ranged in size from a weasel to a bear. The earliest cetacean fossils are the freshwater pakicetids, known mostly from teeth and skulls found on the Indian subcontinent. The ambulocetids are the oldest marine whales, found in Pakistan in the Eocene, 50—45 million years ago. They had four large legs used for swimming, looked more like crocodiles than whales, and lived offshore, although they apparently swam into river estuaries to drink fresh water. Next came the protocetids, which lived in fully marine tropical and subtropical environments around the world during the Middle and Late Eocene, 48—42 million years ago.

Most important to cryptozoology are the basilosaurids, the last of the archaic whales. These emerged in the Late Eocene, about 42 million years ago, and are characterized by long, flexible vertebrae; reduced but functional hind limbs; a muscular, fluked tail; and a serpentine body that grew up to 80 feet long. Basilosaurid fossils are found in Louisiana, Alabama, Mississippi, and Egypt. The type fossil, *Basilosaurus cetoides,* was misidentified as a reptile in 1843 when it was first described; the genus name

means "king reptile." The nineteenth-century anatomist Richard Owen later renamed it *Zeuglodon* ("yoked tooth"), and it is often cited as such in some cryptozoological literature. However, the rules of nomenclature defer to the earlier name even if it's inaccurate. Because it looks remarkably like some SEA MONSTERS reported in modern times, a surviving basilosaurid is frequently suggested as a candidate.

Baleen whales first turned up in the Late Eocene, 40 million years ago, while toothed whales emerged in the Early Oligocene, 35 million years ago. Although the toothed whales were originally marine, some have returned to freshwater.

The largest mammal ever recorded is the Blue whale (*Balaenoptera musculus*). A female measuring 110 feet 2.5 inches was brought into the Cia Argentina de Pesca shore station in Grytviken, South Georgia, in 1909.

Of the 17 cetaceans in this section, all but 2 are toothed whales, and of these, 9 are probable dolphins. The 2 lone baleen cryptids are the MAGENTA WHALE and the SCRAG WHALE. Their geographic breakdown is: Antarctic 3, Atlantic Ocean 4, Indian Ocean 2, Mediterranean 1, Pacific Ocean 6, and South America 1.

**Mystery Cetaceans**
ALULA WHALE; ANTARCTIC KILLER WHALE; ANTARCTIC LONG-FINNED WHALE; DIMORPHIC BEAKED WHALE; GREEK DOLPHIN; HIGH-FINNED SPERM WHALE; ILIGAN DOLPHIN; MAGENTA WHALE; PALMYRA FISH; PINK DOLPHIN; RHINOCEROS DOLPHIN; SAWTOOTH DOLPHIN; SCOTT'S DOLPHIN; SCRAG WHALE; SENEGAL DOLPHIN; SOUTHERN NARWHAL; WHITE-FLIPPERED BEAKED WHALE

## Chagljevi

Unknown DOG of Eastern Europe.

*Physical description:* Doglike. The size of a puppy.

*Behavior:* Nocturnal. Afraid of humans.

*Distribution:* Montenegro Republic, Yugoslavia.

*Possible explanation:* The Golden jackal (*Canis aureus*) is still found in southeastern Europe, as far west as Italy and as far north as Austria. It stands 15—20 inches at the shoulder.

*Source:* Marcus Scibanicus, "Strange Creatures from Slavic Folklore," *North American BioFortean Review* 3, no. 2 (October 2001): 56—63, http://www.strangeark.com/nabr/NABR7.pdf.

## Challenger Deep Flatfish

A flat abyssal FISH of the North Pacific Ocean.

*Etymology:* Named after the Challenger Deep in the Mariana Trench.

*Physical description:* Solelike. Two distinct eyes.

*Habitat:* Visits or inhabits abyssal oceanic depths where no light penetrates.

*Distribution:* The Mariana Trench, east of Guam.

*Significant sighting:* On January 23, 1960, the bathyscaph *Trieste,* piloted by Jacques Piccard and Donald Walsh, reached a record depth of 35,800 feet (7 miles) in the Challenger Deep. As they touched down on the bottom, a flatfish with two distinct eyes swam away to avoid them.

*Possible explanation:* Torben Wolff suggested the animal was a Sea cucumber, perhaps the cushion-shaped *Galatheathauria aspera,* which has an oval shape. Eyes would be of absolutely no use at this depth.

*Source:* Jacques Piccard and R. S. Dietz, *Seven Miles Down* (New York: G. P. Putnam's Sons, 1961).

## Champ

FRESHWATER MONSTER of Lake Champlain in Vermont, New York, and Québec.

*Scientific name: Champtanystropheus,* proposed by Dennis Hall.

*Etymology:* After the lake.

*Variant names:* Champy, CHAOUSAROU, Sammy, Tatoskok (Abenaki/Algonquian)

*Physical description:* Reports from the nineteenth century to the 1960s generally describe an enormous serpent. Fiery eyes. Possibly hooded. Glistening scales. Fishlike tail. Spouts water.

Reports from the 1960s onward are more like those of the classic Loch Ness—like freshwater LONGNECK. Length, 15—50 feet. Dark brown or black color. Rough skin. Height out of the water, 3—8 feet. Horse- or snakelike head with two horns or ears. Visible teeth. Long, upright neck, 12 inches thick and 4—5 feet long. Maned. One to ten humps reported, with two or three most frequently observed.

*Behavior* (post-1960 observations): Most frequently seen in the summer between 7:00 and 8:00 P.M. in clear weather with a calm surface. Elusive. Moves by vertical undulations. Leaves a well-defined wake. Aggressive and noisy, though the noise from motors seems to frighten it. Probably feeds on fishes.

*Distribution:* Sightings have been scattered throughout the length of Lake Champlain. Clusters of sightings seem to occur at Rouses Point, Plattsburgh, and Bulwagga Bay near Port Henry in New York, as well as off Burlington in Vermont.

*Significant sightings:* The earliest sighting may be a dubious report from July 1819 of a 187-foot monster witnessed by Captain Crum from his scow in Bulwagga Bay, New York.

In early July 1873, a crew laying track for the New York & Canada Railroad along the shore near Dresden, New York, saw a serpent with an enormous head approaching them from across the lake. The men started to retreat but saw the animal turn and swim rapidly away. It seemed to be covered with bright, silvery scales, and it spurted water about 20 feet into the air. Its tail resembled that of a fish. A few days afterward, others saw the animal and farmers complained of missing livestock. On August 9, a party of monster hunters organized by the *Whitehall Times* allegedly trapped the serpent in Axehelve Bay and shot it from the decks of a steamboat they had commandeered, the *Molyneaux.* On September 7, railway workers eager for the $50,000 reward that P. T. Barnum had recently offered thought they had found the missing carcass, but it turned out to be a log.

On July 30, 1883, Sheriff Nathan H. Mooney saw a huge serpent 25—35 feet long with a flat, triangular head in Cumberland Bay, New York. It stood out about 5 feet above the

*Two versions of* CHAMP, *a lake monster in Lake Champlain, New York and Vermont. (Richard Svensson/Fortean Picture Library)*

water. Sightings continued throughout the summer.

In 1945, Charles Langlois and his wife, of Rutland, Vermont, got close to the animal in a rowboat.

Orville Wells watched a 20-foot animal with a long neck and two humps in Treadwell Bay, New York, in 1976.

On July 5, 1977, Sandra Mansi and her family were picnicking by the lake when they saw the head and neck of a "dinosaur" some 100—160 feet offshore near St. Albans, Vermont. She managed to take a color Instamatic photograph of the animal before leaving hurriedly in the car. The photo has held up under scrutiny and apparently shows a gray-black object at least 15—20 feet long at the waterline. It has a long neck, a small head, and a hump. B. Roy Frieden of the University of Arizona's Optical Sciences Center in 1981 determined that the photo was not a montage and appeared to show a separate set of surface waves coming from the object that are independent from the waves from the rest of the lake. A 1982 analysis of wave patterns in the photo by oceanographer Paul H. LeBlond gave an estimate ranging from 16 to 56 feet for the waterline length of the object.

Jim Kennard and Joseph Zarzynski picked up a target using towed side-scan sonar on June 3, 1979, in Whallon Bay, New York. The object was moving at a depth of 175 feet. However, a school of fishes was not ruled out.

On July 28, 1984, Michael Shea, Bette Morris, and about sixty other people watched Champ for ten to fifteen minutes from the vessel *The Spirit of Ethan Allen* off Appletree Point, Burlington, Vermont. It was approximately 30 feet long and had three to five humps.

On August 10, 1988, Martin Klein, Joseph Zarzynski, and others aboard an air-sea rescue vessel between Westport, New York, and Basin Harbor, Vermont, saw an animate object thrashing on the surface of the lake.

On July 6, 2000, Dennis Jay Hall obtained about forty-five minutes of digital video of two long-necked animals in shallow water just south of the mouth of Otter Creek, Vermont. He has several videos of single animals taken on several other occasions, one as recently as October 6, 2000, in Button Bay, Vermont.

*Possible explanations:*

(1) Newspaper hoaxes, especially in the nineteenth century.

(2) Wave effects created by passing watercraft.

(3) Floating logs.

(4) The Lake sturgeon (*Acipenser fulvescens*) is still found in Lake Champlain. This fish can grow to 7—9 feet in length, though most are a bit smaller. The lake supported a small commercial fishery that harvested 50—200 sturgeons annually in the late nineteenth and early twentieth centuries. The annual harvest declined rapidly in the late 1940s, and the fishery finally closed in 1967. In 1998, the Vermont Department of Fish and Wildlife began a project to assess and ultimately restore a viable lake sturgeon population in Lake Champlain.

(5) A stray Harbor seal (*Phoca vitulina*), known to colonize small lakes and rivers in northern Canada, may account for some early sightings. In February 1810, a 4-foot seal was found crawling on the ice of Lake Champlain south of Burlington. Two other specimens were found in 1846 and 1876.

(6) An evolved plesiosaur has been theorized by J. Richard Greenwell and Karl Shuker. Long-necked plesiosaurs such as *Elasmosaurus* had a large body, short tail, four limbs modified into paddles, a long neck with a small head, and a maximum known length of 46 feet. Their primary food was probably fishes. Plesiosaur fossils are found continuously from the Middle Triassic (238 million years ago) to the end of the Cretaceous (65 million years ago), though there was a smaller extinction at the end of the Jurassic (144 million years ago) that resulted in a reduction in diversity. Fossils have been found in abundance in marine sediments in England and Kansas, but all continents including Antarctica have yielded some remains. They were exclusively marine; consequently, a variety that could subsist in a freshwater environment would have had to undergo significant modifications.

(7) A surviving archaic basilosaurid whale has been suggested by Roy Mackal and Gary Mangiacopra. These predecessors of modern cetaceans lived in the Late Eocene, about 42 million years ago, and had serpentine bodies that grew up to 80 feet long.

(8) *Tanystropheus longobardicus,* a diapsid reptile from the Middle Triassic, 230 million years ago, has been suggested by Dennis Jay Hall, although it has a much longer neck and smaller body than Champ appears to have. Young specimens have relatively short necks, which apparently grew quickly as the animal reached adulthood. Its long neck was more than twice the length of its body and tail, and it apparently attained a total length of 10 feet. Found in marine sediments in Central Europe, *Tanystropheus* may have been a coastal swimmer that fed on fishes. In the 1970s, Hall discovered a 12-inch reptile with a forked tongue in a marshy area bordering Lake Champlain. It was sent to the University of Vermont, where it was subsequently lost. He later ran across a drawing of *Tanystropheus* and thought it was very similar. A smaller relative from the Late Triassic, *Tanytrachelos,* has been found in Virginia.

*Sources:* Leon Dean, "Champlain Ace in the Hole," *Vermont Life* 13 (Summer 1959): 19; "Monster Time Again," *Vermont Life* 16 (Spring 1962): 49; Marjorie L. Porter, "The Champlain Monster," *Vermont Life* 24 (Summer 1970): 47—50; Gary S. Mangiacopra, "Lake Champlain: America's Loch Ness," *Of Sea and Shore* 9, no. 1 (Spring 1978): 21—26, and no. 2 (Summer 1978): 89—92; *New York Times,* Science Times section, June 30, 1981; "People," *Time* 118 (July 13, 1981): 64; Joseph W. Zarzynski, "'Champ': A Personal Update," *Pursuit,* no. 54 (1981): 51—53, 58; Paul H. LeBlond, "An Estimate of the Dimensions of the Lake Champlain Monster from the Length of Adjacent Wind Waves in the Mansi Photograph," *Cryptozoology* 1 (1982): 54—61; Michel Meurger and Claude Gagnon, *Lake Monster Traditions: A Cross-Cultural Analysis* (London: Fortean Tomes, 1988), pp. 39—40; Joseph W. Zarzynski, *Champ: Beyond the Legend* (Wilton, Vt.: M-Z Information, 1988); Yasushi Kojo, "Some Ecological Notes on Reported Large, Unknown Animals in Lake Champlain," *Cryptozoology* 10 (1991): 42—54; Jerome Clark, *Encyclopedia of Strange and Unexplained Physical Phenomena* (Detroit, Mich.: Gale Research, 1993), pp. 45—50; *USA Today,* September 8, 1993; Joseph A. Citro, *Green Mountain Ghosts, Ghouls and Unsolved Mysteries* (Montpelier: Vermont Life, 1994), pp. 103—125; Loren Coleman, "Lake Monsters' Fate Sealed?" *Fortean Times,* no. 88 (July 1996): 40; Dennis Jay Hall, *Champ Quest 2000 the Ultimate Search: Field Guide and Almanac for Lake Champlain* (Jericho, Vt.: Essence of Vermont, 2000); Dennis Jay Hall, Champ Quest: The Ultimate Search, http://www.champquest.com.

## Chan

FRESHWATER MONSTER of Mexico.

*Physical description:* Like a sauropod dinosaur.

*Distribution:* Lago La Alberca and neighboring lakes in the Valle de Santiago, Guanajuato State, Mexico.

*Present status:* Photographs are likely hoaxes. Every September, the locals, who consider the monster a god, offer it gifts.

*Sources:* Leopoldo Bolaños, June 15, 1998, accessed in 2001, http://www.fortunecity.com/roswell/daniken/62/invest.html; John Kirk, *In the Domain of Lake Monsters* (Toronto, Canada: Key Porter Books, 1998), pp. 209—211.

## Chaousarou

FRESHWATER MONSTER of Lake Champlain in Vermont, New York, and Québec.

*Etymology:* Huron (Iroquoian) word.

*Significant sighting:* Samuel de Champlain saw a creature in the lake in July 1609. It was only 5 feet long and had a double row of sharp

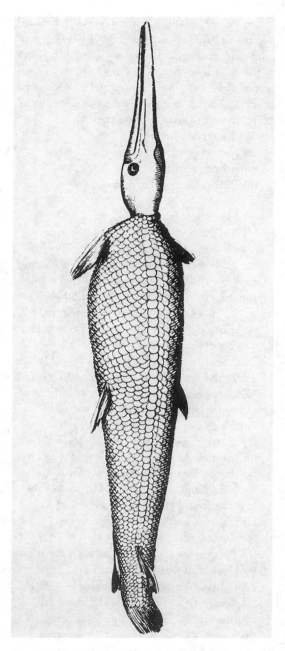

*The CHAOUSAROU, described by Samuel de Champlain in 1609. From François Du Creux,* Historiae canadensis *(Paris: Cramoisy and Mabre-Cramoisy, 1664). (Fortean Picture Library)*

teeth and silvery-gray scales.

*Possible explanations:*

(1) The Longnose gar (*Lepisosteus osseus*) has a long, narrow snout but only one row of

teeth on its upper jaw. It grows to 6 feet in length and is olive-brown above and white below. It is currently found in Lake Champlain.

(2) The Alligator gar (*Atractosteus spatula*) has a double row of teeth, but it is currently found no farther north than the Ohio River.

(3) The Lake sturgeon (*Acipenser fulvescens*) is olive-gray above and white below, with scutes along the back and the sides.

*Source:* Samuel de Champlain, *Les voyages faits au Grand Fleuve Sainct Laurens* [1613], in *The Works of Samuel de Champlain,* Henry Percival Biggar, ed. (Toronto, Canada: Champlain Society, 1925), vol. 2, p. 91.

## Chemisit

Unknown animal of East Africa that the British call the NANDI BEAR.

*Etymology:* Kalenjin (Nilo-Saharan) word meaning "devil"; however, at one level of meaning, it is regarded as an animal, not a spirit.

*Variant names:* Chemoiset, Chemosit, Chimisit, Gononet.

*Physical description:* Tawny or reddish color. Sometimes striped. Face like an ape's.

*Behavior:* Nocturnal. Stands on its hind legs sometimes. Makes a peculiar moaning cry or blood-curdling roar. Said to break into native huts at night, kill the occupants, and eat their brains. Used as a threat by mothers to make their children obey.

*Tracks:* Round and bearlike.

*Distribution:* Western Kenya.

*Significant sighting:* In the 1960s, engineer Angus McDonald was sleeping in a hut near Kipkabus, Kenya, when he was awakened by a shriek as a large animal jumped in the window and chased him around the hut for about five minutes. It seemed to be about 7 feet tall with an ape's face, and it was able to run on both two feet and all fours, leaving round tracks. The El-geyo tribesmen identified it as a Chemosit.

*Sources:* Alfred C. Hollis, *The Nandi: Their Language and Folk-Lore* (Oxford: Clarendon, 1909), p. 41; Geoffrey Williams, "An Unknown Animal on the Uasingishu," *Journal of the East Africa and Uganda Natural History*

*Society,* no. 4 (1912): 123—125; I. Q. Orchardson (letter), *Journal of the East Africa and Uganda Natural History Society,* no. 28 (1927): 19, 23; Charles R. S. Pitman, *A Game Warden among His Charges* (London: Nisbet, 1931), p. 291; Odette Tchernine, *The Yeti* (London: Neville Spearman, 1970), pp. 69—72; Martin Pickford, "Another African Chalicothere," *Nature* 253 (1975): 85.

## Chessie

SEA MONSTER that frequents the Chesapeake Bay, Maryland.

*Etymology:* After the bay. First named in July 1977 by a reporter for the *Richmond (Va.) Times-Dispatch.*

*Variant names:* Chesapeake Chessie, Potomac Patty.

*Physical description:* Serpentine or eel-like in general shape. Length, 12—35 feet. Smooth. Diameter, 8—10 inches. Dark gray to black. Three or four humps. Turtlelike head and neck held 3 feet above the surface. No fins.

*Behavior:* Sometimes reported swimming with horizontal undulations; at other times, it is seen swimming with vertical undulations, while the head and neck are kept steady. Passive toward observers.

*Tracks:* Described as reptilian or snakelike.

*Distribution:* Chesapeake Bay, Potomac River, and Rappahannock River in Maryland and Virginia. Favored spots are Love Point on Kent Island, Eastern Bay, and the mouth of the Potomac River.

*Significant sightings:* An animal suspiciously like a large turtle, 12 feet long with a large shell and four fins, was seen by the crews of two schooners on July 26, 1840, near North Point, Maryland.

An engineer performing helicopter test flights out of Aberdeen Proving Grounds saw an enormous, eel-like animal in the Bush River in 1963.

In 1965, Pam Peters saw a serpentine animal in the South River off Hillsmere Shores, Maryland.

In July 1977, Gregg Hupka took a fuzzy photograph of an animal at the mouth of the Potomac River.

G. F. Green III and his family saw a 25-foot, humped animal while they were water-skiing on June 22, 1980. It sank when they approached it.

On May 21, 1982, Robert and Karen Frew videotaped a 30- to 35-foot sea monster, slightly under 1 foot in diameter, off Love Point, Maryland. They observed the animal from about 200 feet away as it repeatedly broke the surface of the water and moved against the current at about 4—5 knots. George Zug, Clyde Roper, and five other scientists from the Smithsonian Institution could not identify the object during a special meeting on August 20 to evaluate the tape, calling it "animate but unidentifiable."

Clyde Taylor and his daughter Carol were walking along Cloverdale Community Beach on Kent Island, Maryland, on July 16, 1982, when they saw a 30-foot serpentine animal in 3 feet of water moving in vertical undulations toward the shore. Carol ran toward it, and it dove out of sight. Clyde, a commercial artist, drew a series of pictures of it.

*Possible explanations:*

(1) Floating logs.

(2) The Harbor porpoise (*Phocoena phocoena*) only grows to slightly more than 6 feet in length.

(3) Sea turtles may be responsible for some sightings, especially the 9-foot Atlantic loggerhead (*Caretta caretta*), the 4-foot Green turtle (*Chelonia mydas*), and the 8-foot Leatherback turtle (*Dermochelys coriacea*), all of which have been found in the Chesapeake Bay.

(4) The American eel (*Anguilla rostrata*) only grows to about 5 feet. It has a slender, snakelike body but does not hold its head and neck above the surface.

(5) The Northern water snake (*Nerodia sipedon sipedon*) has dark dorsal crossbands and reaches a length of 4 feet 6 inches.

(6) A surviving archaic basilosaurid whale has been suggested by Roy Mackal for similar SEA MONSTERS.

(7) An out-of-place Anaconda (*Eunectes murinus*) was suggested by John Meriner, though these are tropical freshwater snakes.

(8) Moray eels (Family Muraenidae) and Sea snakes (Family Hydrophiidae) are similarly tropical or subtropical. Morays rarely swim on the surface.

(9) The name Chessie has recently been appropriated by a Florida manatee (*Trichechus manatus latirostris*) captured in the Chesapeake Bay in the fall of 1994 and returned to Florida. Subsequently monitored by the U.S. National Biological Service, Chessie was tracked migrating all the way north to Rhode Island in 1995. This or another manatee might have been responsible for earlier sightings.

*Sources:* "Nessie Junior," *Washington Post,* August 18, 1978; Gary S. Mangiacopra, "The Great Unknowns into the 20th Century," *Of Sea and Shore* 11, no. 4 (Winter 1980—81): 259—261; Russ Robinson, "Chessie May Have Made Video Debut," *Baltimore Sun,* July 11, 1982; "Chesapeake Bay Monster Filmed on Videotape," *ISC Newsletter* 1, no. 2 (Summer 1982): 9—10; "Chessie Videotape Analysis Inconclusive," *ISC Newsletter* 2, no. 1 (Spring 1983): 9; Michael T. Shoemaker, "The Day They Caught 'Chessie,'" *Strange Magazine,* no. 3 (1988): 30—31; Michael Bright, *There Are Giants in the Sea* (London: Robson, 1989), pp. 64—78; Michael A. Frizzell, "The Chesapeake Bay Serpent," *Crypto Dracontology Special,* no. 1 (November 2001): 129—137.

## Cheval Marin

SEA MONSTER of the coastal waters of Canada and West Africa.

*Etymology:* French, "sea horse."

*Physical description:* Horselike head. Clawed forearms. Fishlike, scaly tail.

*Behavior:* Neighs like a horse.

*Distribution:* Île Brion and Rivière-St.-Jean, Québec, Canada; West Africa.

*Possible explanations:*

(1) Explorer Jacques Cartier saw two Walruses (*Odobenus rosmarus*) on the Île Brion in 1534 and fish-shaped, horselike animals in a river that may have been the modern Rivière-St.-Jean off the St. Lawrence. The French naturalist Louis Nicolas conflated the two stories and mixed in Native American legends of the HORSE'S

HEAD to describe a composite animal.

(2) Early reports from French Africa may have confused the Hippopotamus (*Hippopotamus amphibius*) and the West African manatee (*Trichechus senegalensis*).

(3) A SEA MONSTER resembling Heuvelmans's MERHORSE.

*Sources:* Marc Lescarbot, *History of New France* [1609], trans. Henry Percival Biggar (Toronto, Canada: Champlain Society, 1907—1914), vol. 7, p. 73; Gabriel Sagard, *Le grand voyage du pays des Hurons* [1632], ed. Marcel Trudel (Montreal, Canada: Hurtubise HMH, 1976); Girolamo Merolla, *A Voyage to Congo* [1682], in Awnsham Churchill, ed., *A Collection of Voyages and Travels* (London: A. and J. Churchill, 1704), vol. 1, pp. 651—756; Henry Percival Biggar, *The Voyages of Jacques Cartier* (Ottawa: F. A. Acland, 1924); Michel Meurger and Claude Gagnon, *Lake Monster Traditions: A Cross-Cultural Analysis* (London: Fortean Tomes, 1988), pp. 211—216.

## Chick-Charney

Legendary FLYING HUMANOID of the West Indies.

*Variant name:* Chiccharnie.

*Physical description:* Half bird, half human. Huge eyes. Beak instead of a nose. Three toes.

*Behavior:* Makes nests by tying together the tops of trees. Harmful to humans if laughed at.

*Habitat:* In kapok or pine trees.

*Distribution:* Andros Island, Bahamas.

*Sources:* Curt Rowlett, "Chick-Charney: Bird-man of the Bahamas," *Strange Magazine*, no. 7 (April 1991): 34; Richard Alsopp, ed., *Dictionary of Caribbean English Usage* (Oxford: Oxford University Press, 1996), p. 149.

## Chihalenchi

CANNIBAL GIANT of the western United States.

*Etymology:* Miwok (Penutian) word.

*Physical description:* Covered with hair.

*Habitat:* Limestone caverns.

*Distribution:* Sierra Nevada, California.

*Source:* Craig D. Bates, "California," in Colin F. Taylor, ed., *Native American Myths and Legends* (New York: Smithmark, 1994), pp. 72—73.

## Chimera

Composite monster of West Asia; *see* SEMI-MYTHICAL BEASTS.

*Etymology:* From the Greek *chimaira* ("monster"). Originally a "she-goat," the feminine form of *chímaros*, "he-goat one winter old."

*Variant names:* Chimaera, Khimaira.

*Physical description:* An assemblage of anatomical parts borrowed from totally unrelated animals. Head of a lion. Body of a goat. Serpentine tail. According to Hesiod, an animal with three heads (goat, lion, snake).

*Behavior:* Said to breathe fire. Lustful.

*Distribution:* Mount Olimpos (Tahtali or Yanartas), Olimpos—Bey Mountains National Park, Antalya Province, south-central Turkey.

*Significant sighting:* Wreaked havoc in ancient Lycia. Killed by the Greek hero Bellerophon, mounted on the winged horse Pegasus. This may have provided some inspiration for the legend of St. George slaying a DRAGON (BRITISH).

*Possible explanations:*

(1) A fantastic product of the imagination.

(2) Based on a volcanic crevice on Mount Olimpos that still vents flammable methane gas and burns both day and night.

(3) Lions are said to have lived at the top of the Mount Olimpos, goats were pastured in the midheights, and snakes lived at the bottom.

(4) A personification of a storm cloud.

*Sources:* Homer, *Iliad*, VI. 179; Hesiod, *Theogony*, 319ff; Jorge Luis Borges, *The Book of Imaginary Beings* (New York: E. P. Dutton, 1969), pp. 62—63.

## Chiparemai

WILDMAN of South America.

*Etymology:* Kalihna (Carib) word.

*Variant names:* Ewaipahoma, Rayas (Spanish, "sting rays").

*Physical description:* Headless human or a human with either a doglike head or a head that sags below the shoulders. Eyes in shoulders. Mouth in chest. Long hair down the shoulders.

*Behavior:* Uses bow and arrows.

*Distribution:* Río Orinoco delta, Venezuela.

*Present status:* Sometimes erroneously cited in support of MONO GRANDE reports.

*Possible explanations:*

(1) The West Indian manatee (*Trichechus manatus*) looks superficially human and has sparse facial hair.

(2) An Indian tribe that wears elaborate head coverings that make them look headless and uses body art in the form of a face to decorate the torso, as suggested by Victor Von Hagen.

(3) A deformed Indian, possibly a hunchback.

*Sources:* Walter Raleigh, *Discovery of the Large, Rich, and Beautiful Empire of Guiana* [1596], and Lawrence Kemys, *A Relation of the Second Voyage to Guiana* [1596], in Richard Hakluyt, ed., *The Principal Navigations, Voyages, Traffiques & Discoveries of the English Nation* [1598] (New York: Macmillan, 1910), vol. 10, pp. 406—407, 437—438, 465; Alexander von Humboldt, *Personal Narrative of a Journey to the Equinoctial Regions of the New Continent, during the Years 1799—1804* [1825] (New York: Penguin, 1995), pp. 203—204, 216; Victor Wolfgang Von Hagen, *South America Called Them* (New York: Alfred A. Knopf, 1945), p. 119.

## Chipekwe

Dinosaur-like animal of Central Africa, similar to the EMELA-NTOUKA.

*Etymology:* Bemba (Bantu), "monster." Around Lake Tanganyika, Chipekwe is a common name for the Giant tigerfish (*Hydrocynus goliath*), a 4-foot fish that can grow to 150 pounds. The word may actually refer to any dangerous animal, from venomous snakes to man-eating crocodiles, in several Bantu languages.

*Variant names:* Chepekwe, Chimpekwe, Mbilintu ("frightful unknown monster"), Mfuku.

*Physical description:* Smooth, dark body. Size ranges from smaller than a hippo to as large as an elephant. Has a single smooth, white, ivory horn or tusk.

*Behavior:* Nocturnal. Amphibious. Aggressive. Kills and eats hippopotamuses and rhinoceroses.

*Tracks:* Like hippopotamus or crocodile tracks but 2 feet 6 inches—3 feet long.

*Habitat:* Swamps, lakes.

*Distribution:* Lake Bangweulu, Kafue Flats, Luapula River, Lukulu River, Lake Mweru, and Lake Shiwa Ngandu in Zambia; Dililo Swamps in the Democratic Republic of the Congo; Lake Tanganyika in Tanzania; Lago Dilolo and the Kasai River in Angola.

*Significant sightings:* Joseph Menges and Hans Schomburgk independently reported the existence of an unknown animal–half elephant, half dragon–in Zambian swamps at the end of the nineteenth century. Schomburgk noted the absence of hippos in parts of Lake Bangweulu as the direct result of an unknown amphibious predator.

The Aushi people have a tradition of the death of a Chipekwe, which they killed with their harpoons in the deep waters of the Luapula River, Zambia. It had a smooth, dark body, and a single ivory horn on its snout.

Native Commissioner Robert Young shot at a large animal in Lake Shiwa Ngandu, Zambia, that dived and left a wake like a steamboat.

Retired magistrate H. Croad was camped by a small lake in Zambia when he heard a splashing noise in the middle of the night. The next morning, he found large footprints of an animal that he could not recognize.

Around 1906, settler R. M. Green was told by natives that a hippo had its throat torn out by a Chipekwe on the Lukulu River, Zambia.

In October 1919, a Belgian railway manager named Lepage was charged by a dinosaur-like monster with "tusks like horns," a pointed snout, and a scaly hump. It is said to have later stampeded through the village of Fungurume, Katanga Region, Democratic Republic of the Congo. The story was followed by another report by a Belgian big-game hunter named Gapelle who is said to have tracked a similar beast for 12 miles in the Congo and finally sighted it. It had a kangaroo-like tail and a horn on its snout. An expedition by the Smithsonian Institution was allegedly in search of the beast when three members of the party were killed in a railway accident. However, Wentworth Gray exposed these stories as hoaxes in 1920, pointing out that Gapelle was an anagram for Lepage and that the Smithsonian (though the accident had occurred on November 28) was not hunting dinosaurs.

In 1932, Franz W. Grobler reported he had seen a photo of a Chipekwe standing on the back of a hippo that it had killed in Lago Dilolo, Angola. However, the photo was a crude hoax by someone who superimposed a Komodo dragon onto the photo of a dead hippo. It was based on a dubious report by a Swede, J. C. Johanson, who on February 16, 1932, took a photo of a 38-foot-long, lizardlike monster in the Kasai Valley on the border of Angola and Democratic Republic of the Congo.

In May 1954, Alan Brignall saw a small head and long neck rise up out of the water about 25 yards from the shore of Lake Bangweulu, Zambia. The head had a distinct brow, blunt nose, and visible jawline, and it moved from side to side. After a few seconds, it sank down vertically and disappeared.

*Possible explanations:*

(1) A ceratopsian dinosaur, suggested by Karl Shuker, although this suborder is not known from Africa. Better-known species include *Monoclonius, Psittacosaurus, Protoceratops,* and *Triceratops.* Usually, these dinosaurs had a large frill or flange around the head along with facial horns.
(2) A surviving saber-toothed cat, suggested by Bernard Heuvelmans.
(3) An unknown species of aquatic rhinoceros, suggested by Denis David Lyell.
(4) Old, solitary, exceptionally aggressive male Hippopotamuses (*Hippopotamus amphibius*) that attack boats and other hippos, according to Alain Chevillat.
(5) William Hichens suggested a surviving chalicothere, a family of fossil ungulates that lived 25 million years ago in the Miocene and survived in East Africa until 12,000 years ago. However, they were ground-based terrestrial browsers and not amphibious.

*Sources:* Carl Hagenbeck, *Beasts and Men* (London: Longmans, Green, 1909), pp. 96—97; Hans Schomburgk, *Wild und Wilde im Herzen Afrikas* (Berlin: E. Fleischel, 1910), pp. 219—220; "A Tale from Africa: Semper Aliquid Novi," *Times* (London), November 17, 1919; "Dragon of the Prime: Congo Monster Sighted," *Times* (London), December 12, 1919; C. G. James, "Congo Swamp Mystery," *Daily Mail* (London), December 26, 1919, p. 2; Wentworth D. Gray, "The Brontosaurus," *Times* (London), February 23, 1920; Victor Forbin, "Les patientes recherches des savants, leurs travaux laborieux ont permis de reconstituer les squelettes des animaux de l'époque Tertiare," *Sciences et Voyages,* May 27, 1920, pp. 206—208; John G. Millais, *Far Away Up the Nile* (London: Longmans, Green, 1924), pp. 61—67; "Mystery Animal of African Swamps?" *Cape Argus* (Cape Town), July 5, 1932, p. 16; "Meet the Mystery Monster," *Cape Argus* (Cape Town), July 7, 1932, p. 13; "Chepekwe Does Exist: Native Stories of Mystery Monster," *Cape Argus* (Cape Town), July 9, 1932, p. 16; Joseph E. Hughes, *Eighteen Years on Lake Bangweulu* (London: The Field, 1933), pp. 146—148; Hans Schomburgk, *Meine Freunde im Busch* (Berlin: Freiheitsverlag, 1936), pp. 35, 313, 374—375; William Hichens, "African Mystery Beasts," *Discovery* 18 (1937): 369—373; F. B. Macrae, "More African Mysteries," *National Review* 111 (1938): 791—796; Frank W. Lane, *Nature Parade* (London: Jarrolds, 1955), p. 265; Bernard Heuvelmans, *On the Track of Unknown Animals* (New York: Hill and Wang, 1958), pp. 434—441, 450—455, 475—478; Bernard Heuvelmans, *Les derniers dragons d'Afrique* (Paris: Plon, 1978), pp. 115—134, 191—202, 207—213, 218—221, 231—233, 303—305, 383—386; Dwight Smith and Gary S. Mangiacopra, "Carl Hagenbeck and the Rhodesian Dinosaurs," *Strange Magazine,* no. 6 (1990): 50—52; Ulrich Magin, "Living Dinosaurs in Africa: Early German Reports," *Strange Magazine,* no. 6 (1990): 11; Karl Shuker, *In Search of Prehistoric Survivors* (London: Blandford, 1995), pp. 18—20, 28—30.

## Chiye Tanka

CANNIBAL GIANT of the north-central United States.

*Etymology:* Lakota (Siouan), "big elder brother."

*Variant name:* Chiha tanka (Dakota/Siouan).

*Distribution:* North Dakota; South Dakota; Montana.

*Sources:* Mary Eastman, *Dacotah* (New York: John A. Wiley, 1849), pp. 208—211; Henry Rowe Schoolcraft, *Historical and Statistical Information Respecting the History, Condition, and Prospects of the Indian Tribes of the United States* (Philadelphia: Lippincott, Grambo, 1851—1857), vol. 3, p. 232; J. Owen Dorsey, "Teton Folk-Lore," *American Anthropologist* 2 (1889): 143, 155.

## Chollier's Ape

Unknown PRIMATE of West Africa.

*Etymology:* After its only observer.

*Physical description:* Size of a man. Dense black hair.

*Distribution:* Southeastern Mali near the Burkina Faso border.

*Significant sighting:* In 1938, Louis Chollier was on safari south of Tiou Tiou, Mali, when his party came across a huge, black animal with long hair that looked like a bear or an ape. It disappeared into the rocks before they could pursue it.

*Possible explanation:* An escaped Chimpanzee (*Pan troglodytes*) or a remnant population of chimps left over from the species' former range.

*Source:* Bernard Heuvelmans, *Les bêtes humaines d'Afrique* (Paris: Plon, 1980), pp. 560—562.

## Christina

FRESHWATER MONSTER of Alberta, Canada.

*Etymology:* After the lake.

*Physical description:* Length, 30—40 feet. Horselike head. Hairy neck, 3 feet long. Large eyes.

*Distribution:* Christina Lake, Alberta (near Conklin).

*Significant sighting:* The animal was seen by a fisherman in June 1984.

*Sources: Toronto Sunday Sun,* September 2, 1984; Stephen Lequire, "Sunday Showcase," *Calgary Sunday Sun,* July 28, 1985.

## Chuchunaa

GIANT HOMINID of eastern Siberia.

*Etymology:* Related to Yakut (Turkic) word for "fugitive" or "outcast."

*Variant names:* Abas, Abasy, Chuchuna, Kuchena, Kuchuna, Mulen (Evenki/Altaic, "bandit"), Mulena, Siberian snowman.

*Physical description:* Heavily built. Height, usually 6 feet 6 inches—7 feet, though smaller sizes are reported. Long, matted head-hair. Lively expression. Big, black face. Protruding browridge. Small forehead, nose, and eyes. Broad chin. Heavy beard. Wide shoulders on relatively narrow body. Long arms.

*Behavior:* Travels singly or in small groups. May migrate south in the summer. Swift runner. Excellent swimmer. Incapable of speech but utters a piercing, modulated whistle. Hunts and eats reindeer and mountain goats. Catches or steals fishes from nets. Picks wild berries. Lives in caves. Allegedly raids settlements to steal food. Normally shy but occasionally picks fights with hunters or herders. Wears deerskin clothing, boots, and headband. Uses a knife, spear, and fire-steel, as well as a bow and feathered arrows in a quiver. Throws stones when otherwise unarmed.

*Distribution:* From the east bank of the Lena River, Sakha Republic, to the Chukotskiy Peninsula, Chukot Autonomous Province, Siberia, with concentrations in the Verkhoyansk and Poloustnaya Ranges and the upland massifs east of the Yana and Indigirka Rivers, Sakha Republic; as far south as the Khrebet Dzhugdzhur Range, Khabarovsk Territory, Siberian Russia.

*Significant sightings:* Many Mulen were said to have been killed during the Russian Civil War, 1918—1921, when refugees moved into previously uninhabited areas.

In the 1920s, Tatyana Zakharova and other Evenk villagers came across a Chuchunaa while gathering berries near Khoboyuto Creek. It was also picking and eating berries, but it stood up to a full height of nearly 7 feet when it saw them and ran away swiftly. The Chuchunaa was dressed in deerskin, had long arms, a small forehead, and jutting chin.

In czarist times and during World War II, many Chuchunaa were said to have been rounded up and killed, their corpses buried secretly.

*Possible explanations:*

(1) Neanderthals (*Homo neanderthalensis*), cold-adapted hominids of Europe and West Asia that flourished 70,000—35,000 years ago, have been proposed by Myra Shackley, although the nearest fossils are some ambiguous teeth found in the Middle Paleolithic layers of Denisova Cave in the Altai Mountains bordering Kazakhstan. The Chuchunaa and Mulen may actually represent a more advanced archaic human, since their toolkit seems more developed.

(2) *Homo gardarensis,* proposed by Mark Hall, although this medieval skeleton from Greenland is now universally considered to be a Norseman with acromegaly, a skull deformity caused by a malfunctioning pituitary gland.

(3) Paleoasiatic aborigines who have retreated into wilderness areas, suggested by Russian scientist S. Nikolaev.

(4) Completely mythical beings of the Siberian nomadic tribes, suggested by G. V. Ksenofontov.

*Sources:* P. L. Dravert, "Dikie lyudi muleny i chuchuna," *Budushchaya Sibir,* no. 6 (1933): 40—43; Georgii U. Ergis, ed., *Istoricheskie predaniia i rasskazy Yakutov,* vol. 1 (Moscow: Izd-vo Akademii Nauk SSSR, 1960); Bernard Heuvelmans and Boris F. Porshnev, *L'homme de Néanderthal est toujours vivant* (Paris: Plon, 1974), pp. 143—146; Vladimir Pushkarev, "Nevye svidetel'stra," *Tekhnika Molodezhi,* 1978, no. 6, pp. 48—52; "Sighting the Yeti's Relatives," *Nature* 271 (1978): 603; Myra Shackley, "The Case for Neanderthal Survival: Fact, Fiction, or Faction?" *Antiquity* 56 (1982): 31—41; Myra Shackley, *Still Living? Yeti, Sasquatch and the Neanderthal Enigma* (New York: Thames and Hudson, 1983), pp. 134—139; Gavriil V. Ksenofontov, *Uraangkhai-sakhalar* (Yakutsk, Siberia: Natsional'noe Izd-vo Respubliki Sakha, 1992); Dmitri Bayanov, *In the Footsteps of the Russian Snowman* (Moscow: Crypto-Logos, 1996), pp. 123—125, 129—130.

## Chu-Mung

Variant name for the YETI of Central Asia.

*Etymology:* Lepcha (Sino-Tibetan), "spirit of the glaciers."

*Variant names:* Hlo-mung, Thlo-mung ("mountain spirit").

*Physical description:* Covered in long, dark hair.

*Behavior:* Solitary. Climbs trees sometimes.

*Habitat:* High mountains.

*Distribution:* Himalaya Mountains of eastern Nepal; West Bengal and Sikkim States, India.

*Sources:* Charles Stonor, *The Sherpa and the Snowman* (London: Hollis and Carter, 1955), pp. 11—12; René de Nebesky-Wojkowitz, *Oracles and Demons of Tibet* (The Hague, the Netherlands: Mouton, 1956), p. 344; Halfdan Siiger, "'The Abominable Snowman': Himalayan Religion and Folklore from the Lepchas of Sikkim," in James F. Fisher, ed., *Himalayan Anthropology: The Indo-Tibetan Interface* (The Hague, the Netherlands: Mouton, 1979).

## Chunucklas

FRESHWATER MONSTER of British Columbia, Canada.

*Etymology:* Salishan word.

*Variant name:* Shunuklas.

*Physical description:* Black. Head like a snake's. Ears or fins behind the head. Neck, 12—15 feet long.

*Distribution:* Harrison Lake and Harrison River, British Columbia.

*Significant sightings:* In 1908, Hans Oluk and the crew of a tugboat saw the head and neck of an animal in Harrison Lake.

On August 24, 1936, Maggie Mills and others watched an animal with a long neck and tail swimming in Harrison River at the mouth of Old Jim's Slough.

*Sources: Chilliwack (B.C.) Progress,* August 26, 1936; *Vancouver Daily Province,* October 18, 1976; Mary Moon, *Ogopogo* (Vancouver, Canada: J. J. Douglas, 1977), p. 150; John Kirk, *In the Domain of Lake Monsters* (Toronto, Canada: Key Porter Books, 1998), pp. 176—180.

# Chupacabras

Paranormal ENTITY of the West Indies, South America, and parts of North America.

*Etymology:* Spanish, "goat-sucker."

*Variant names:* Canguro ("kangaroo"), Ciguapas (in Dominican Republic), Comecogollos ("banana-tree eater"), Conejo ("rabbit"), Gallinejo (contraction of *gallina,* "chicken," and *conejo*), Goatsucker, Maboya (Taíno/Arawakan, "evil spirit"), Moca vampire, Sacalenguas ("draw tongue," in El Salvador).

*Physical description:* Height, 4—5 feet. Covered in short, gray fur. Said to have a chameleon-like ability to change color. Skin appears to have darkened spots. Large, round head. Huge, lidless, fiery-red eyes run up to the temples and spread to the sides; white sclera not present. Ears small or absent. Two small nostrils. Lipless mouth. Sharp, protruding fangs. Pointy spikes run from the head down the spine; these may double as wings. Attached to the spikes are fleshy membranes that vary in color from blue to red or purple. Thin arms with three webbed fingers. Red claws. Muscular but thin hind legs. Three clawed toes. No tail.

*Behavior:* Nocturnal. Moves awkwardly with arms outstretched or by hopping. Said to be able to jump over trees. Kills goats, chickens, sheep, and other farm animals and drinks their blood. Animal victims generally have two or three circular puncture wounds about 0.25—0.5 inches in diameter in the neck or lower jaw; often, one of the wounds punctures the cerebellum. In some cases, organs are said to have been removed.

*Tracks:* Leaves a trail of slime and rancid meat. Tracks vary; some near Miami were 5 inches long by 4.5 inches wide.

*Distribution:* Puerto Rico; Dominican Republic; Sinaloa, Chihuahua, Coahuila, Nayarit, Veracruz, and Jalisco States, Mexico; Donna and San Antonio, Texas; Miami area, Florida; Tucson, Arizona; New York City; Cambridge, Massachusetts; Boaco and Tolapa, Nicaragua; Costa Rica; El Salvador; Guatemala; Calama, Chile; Varginha and Sorocaba, São Paulo State, Brazil; Touloes, Castelo Branco District, Portugal; Valmaseda, in Spain's Basque Country.

*Significant sightings:* In Moca, Puerto Rico, in March 1975, something was killing cows, goats, pigs, and geese and draining all their blood. Deep stab or puncture wounds were found on the carcasses, causing the perpetrator to be christened the Moca vampire. Killer snakes and birds were blamed, and the Puerto Rico Agricultural Commission called on the police for a full investigation. On March 25, Juan Muñiz became the first human to be attacked by what he described as a "horrible creature covered in feathers," forcing him to hide behind some bushes. In April, other towns on the island reported animal killings and attacks by a weird bird or doglike animal or unidentified flying object (UFO) aliens, but reports died out after a few more weeks.

In March 1991, another rash of pig, goose, and chicken killings erupted in Lares, Puerto Rico. Residents reported an apelike creature, while officials blamed feral dogs. In June 1991, livestock predation at Aguada was accompanied by reports of banana trees being ripped apart by something the island press dubbed Comecogollos, a hairy BIGFOOT with glowing eyes.

Eight sheep were found dead with three puncture wounds and drained of blood in Orocovis, Puerto Rico, in early March 1995.

In the second week of August 1995, Madelyne Tolentino of Canóvanas, Puerto Rico, was one of the first to see the Chupacabras responsible for a series of about 150 animal deaths in the area. It was about 4 feet tall, dark gray, with skinny arms and legs and apparent burn marks on its abdomen. It seemed to have feathers along its spine.

Mrs. Bernardo Gómez of Caguas, Puerto Rico, on November 15, 1995, saw a hairy, red-eyed beast rip open a bedroom window, destroy a stuffed teddy bear, and leave a puddle of slime and rancid meat on the windowsill.

Residents of San Germán, Puerto Rico, chased a Chupacabras away November 16, 1995, just as it was about to kill three fighting roosters. It had large, almond-shaped eyes, an oval face, and small hands protruding from its shoulders.

On November 28, 1995, a hand- or footprint was found after an attack at Vega Baja, Puerto Rico. It showed 6 fingers or toes.

*The* CHUPACABRAS, *a paranormal entity responsible for killing livestock in Puerto Rico and Mexico in the 1990s. (John Sibbick and Fortean Times/Fortean Picture Library)*

Osvaldo Rosado of Guánica, Puerto Rico, claimed that on December 23, 1995, he was grabbed from behind by a gorilla-like animal that gave him a bear hug so tight that wounds appeared on his abdomen.

A Chupacabras killed a pair of sheep at Canóvanas, Puerto Rico, on January 8, 1996. José Febo saw it sitting in a tamarind tree; when it saw him, it jumped down and ran off like a gazelle.

On March 9, 1996, Ovidio Méndez of Aguas Buenas, Puerto Rico, was burying a dead and mutilated chicken when he saw a creature 4 feet tall and walking on two legs. It had large fangs, red eyes, pointed ears, and clawlike hands.

About sixty-nine animals—goats and fowl—were killed in Sweetwater, Florida, in March 1996. Teide Carballo saw a dark-brown, monkeylike creature walking on two legs like a hunchback. Dade County officials attributed the attacks to wild dogs.

From March through May 1996, numerous Chupacabras reports were made in Mexico. Teodora Ayala Reyes in a village in Sinaloa State and José Angel Pulido in Tlajomulco de Zuñiga, Jalisco State, both claimed to be cut or bitten by a Chupacabras. By late May, there were forty-six attacks on 300 animals and four people. Mexican Chupacabras were said to be more rodent-like and only 3 feet tall.

Violeta Colorado's dogs cornered a strange animal in Zapotal, Mexico, on May 9, 1996. It hissed weirdly and escaped. The same night, nine sheep were killed nearby and their blood drained.

Fifty animals were found drained of blood on a farm near Utuando, Puerto Rico, on November 20, 1997. Twin triangular perforations appeared on their stomachs.

On May 3, 2000, in Concepción, Chile, Liliana Romero Castillo was awakened by barking dogs and looked out to see a 7-foot, winged humanoid in her garden. At 6 P.M. the next day, her children found a dead, bloodless dog with two puncture marks at its throat. The Chilean military police removed it shortly afterward.

Some 200 sheep were killed in the area around Calama, Antofagasta Region, Chile, in the first twenty days in May 2000. A half-human, half-animal shape had been seen. The events were blamed on National Aeronautics and Space Administration (NASA) genetic experiments that got out of control.

On the night of August 28, 2001, a couple were returning home from a church meeting in Calama, Chile, when they saw a small, hairy, gray-and-white figure by the side of the road coming out of the bushes. It was apparently suspended a few inches off the ground. It sped across the road in an oddly rigid manner.

On January 12, 2002, two teenagers near Calama, Chile, reported a 6-foot-tall, dog-headed, football-shaped monster that hopped menacingly toward them. It had three fingers and toes and a tail 2 inches thick.

*Present status:* Modern reports began quietly around 1974, peaked in 1995 and 1996, and began to decline in frequency in 1997, with a reoccurrence in Chile beginning in 2000.

*Possible explanations:*

(1) Folklore in the making, spread by contagion as stories are made up, misinterpreted, repeated, and embellished.

(2) Ritual killings by practitioners of Santería or youthful delinquents.

(3) Paranormal origins prevail in the popular press: aliens from an unidentified flying object (UFO), genetic experiments gone awry, or black magic.

(4) A freshwater MERBEING, suggested by Loren Coleman, based apparently on its clumsy gait or webbed fingers and toes.

(5) The Long-tailed weasel (*Mustela frenata*) feeds on small mammals, but it is not native to Puerto Rico, which has no large wild mammals except for monkeys (see below) and some mongooses introduced to control rats on sugarcane plantations.

(6) Feral Domestic dogs (*Canis familiaris*) could be responsible for some of the livestock depredations.

(7) Predation by escaped Rhesus monkeys (*Macaca mulatta*) that had been brought to offshore islands for research purposes, suggested by Juan A. Rivero of the University of Puerto Rico at Mayaguez. Fugitive monkeys have spread through the island as a result of an escape at a facility at

La Paraguera in the 1970s. However, these macaques usually eat insects, shoots, fruit, and seeds, with the occasional small animal; even a troop would not attack a goat.

(8) Humberto Cota Gil blamed the False vampire bat (*Vampyrum spectrum*) for animal attacks in Sinaloa, Mexico, although this bat does not range this far north and limits its prey to birds, other bats, and small rodents.

(9) The Band-winged nightjar (*Caprimulgus longirostris*), a night-flying, insect-eating bird related to the whippoorwill that ranges from Venezuela to Argentina, was said to have contributed to the sightings because its name in Spanish is *chotacabras*.

*Sources:* Salvador Freixedo, *Defendámonos de los dioses* (Madrid: Editorial Algar, 1984); Scott Corrales, *The Chupacabras Diaries: An Unofficial Chronicle of Puerto Rico's Paranormal Predator* (Derrick City, Pa.: Samizdat, 1996); Dudley Althaus, "'Goatsucker' Spreading Fear across Mexico," *Houston Chronicle*, May 12, 1996; Gregory McNamee and Luis Alberto Urrea, "Hellmonkeys from Beyond," *Tucson Weekly*, May 30-June 5, 1996, http://www.tucsonweekly.com/tw/05-30-96/cover.htm; Scott Corrales, "How Many Goats Can a Goatsucker Suck?" *Fortean Times*, no. 89 (September 1996): 34—38; Rafael A. Lara Palmeros, "Chupacabras: Puerto Rico's Paranormal Predator," *INFO Journal*, no. 76 (Autumn 1996): 12—18; Tito Armstrong, The Chupacabra Home Page, 1996—1997, http://www.princeton.edu/~accion/chupa.html; Scott Corrales, *Chupacabras and Other Mysteries* (Murfreesboro, Tenn.: Greenleaf, 1997); Virgilio Sánchez-Ocejo, *Miami Chupacabras* (Miami, Fla.: Pharaoh Production, 1997); Scott Corrales, "Night of the Chupacabras," *Inexplicata*, no. 2 (Winter 1998), at http://www.inexplicata.com/issue2/night_of_the_chupacabras.html; Jonathan Downes, *Only Fools and Goatsuckers* (Exeter, England: Center for Fortean Zoology, 1999); Scott Corrales, *Chupacabras Rising: The Paranormal Predator Returns* (Derrick City, Pa.: Scott Corrales, 2000); Mark Pilkington, "Chupacabras Fever," *Fortean Times*, no. 140 (December 2000): 22—23; Thomas E. Bullard, "Chupacabras in Perspective," *International UFO Reporter* 25, no. 4 (Winter 2000-2001): 3—9, 26—30; Virgilio Sánchez-Ocejo, "On the Trail of the Chupacabras," *Inexplicata*, no. 8 (Spring 2001), at http://www.inexplicata.com/issue8/on_the_trail.htm; Loren Coleman, *Mothman and Other Curious Encounters* (New York: Paraview, 2002), pp. 104—110; "The Hopping Horror," *Fortean Times*, no. 158 (June 2002): 16.

## Chuti

HYENA-like animal, depicted in traditional art of Central Asia.

*Physical description:* Canine head. Tigerlike stripes. Has four forward claws and one rear claw.

*Behavior:* Kills cattle. Leaves humans alone.

*Distribution:* Choyang and Iswa Valleys, Makalu-Barun National Park, Nepal.

*Possible explanations:*

(1) The Striped hyena (*Hyaena hyaena*) has well-defined stripes on its body and limbs and a massive head. It is common in open country and less abundant in the forested regions of Nepal.

(2) Confusion with the similarly named hominid- or bearlike DZU-TEH.

*Source:* Hamish MacInnes, *Look behind the Ranges* (London: Hodder and Stoughton, 1979), pp. 216, 225.

## Cigau

Mystery CAT of Southeast Asia.

*Physical description:* Said to be half tiger, half ape. Smaller but more heavily built than the tiger. Unpatterned yellow or tan fur. Neck ruff. Short tail.

*Behavior:* Aggressive. Said to attack humans without provocation.

*Distribution:* Wilderness east of Mount Kerinci and south toward Bangko, Sumatra, Indonesia.

*Source:* Karl Shuker, "Blue Tigers, Black Tigers, and Other Asian Mystery Cats," *Cat World*, no. 214 (December 1995): 24—25.

## CIVETS AND MONGOOSES
### (Unknown)

Civets and mongooses are carnivores, more primitive than DOGS or CATS, belonging to the Family Viverridae. They share a common ancestor and probably originated in Africa 30 million years ago in the Oligocene.

*Civets* are medium-sized carnivores of Africa and Asia with long bodies and relatively short legs. Most species have stripes, spots, or bands on their bodies, and their tails are often ringed with contrasting colors. Their claws are retractile. Most have perianal (not anal) glands that produce a strong-smelling substance; in some species, the odor is sufficiently potent to ward off predators. The secretion, called civet, is used as a perfume base and medicine.

*Mystery civets:* A civet in Ceram, Indonesia, apparently with an 18-inch tail covered in fur with dark rings, was found by Tyson Hughes in 1986. Locals describe the animal as half dog, half cat. There are no known civets in Ceram.

Two unusual civet specimens have been reported by Dr. Pham Nhat in Lào Cai Province, Vietnam.

Sightings of a wild dog—like animal in Java, Indonesia, called the Anjing hutan and thought to be the Dhole (*Cuon alpinus*), have been shown in some instances to be a surviving population of the Sulawesi palm civet (*Macrogalidia musschenbroeki*), thought extinct since the 1940s.

*Mongooses* typically have pointed heads, long tails, and thick hair, except on the lower legs. Indian mongooses of the genus *Herpestes* are used by snake charmers to entertain tourists with cobra shows. Though not completely immune to cobra venom, the mongoose's speed, agility, and erectile fur usually make the snake miss its mark.

*Mystery mongooses:* An odd, gray, musky, ring-tailed animal with a blunt snout that was killed by a dog near Pablo, Montana, in 1980 may have been an escaped pet mongoose of some kind. However, importation of these animals is tightly controlled. There are no known indigenous viverrids in North America.

The Malagasy cryptid BOKYBOKY is probably the Narrow-striped mongoose (*Mungotictus decemlineata*).

*Sources:* Karl Shuker, "Menagerie of Mystery," *Strange Magazine*, no. 16 (Fall 1995): 28—33, 48—49; Karl Shuker, "A Surfeit of Civets?" *Fortean Times*, no. 102 (September 1997): 17; Chad Arment and Brad LaGrange, "Crypto-Varmints," *North American BioFortean Review* 2, no. 3 (December 2000): 18—20, http://www.strangeark.com/nabr/NABR5.pdf.

### Clear Lake Catfish

Odd FISH of California.

*Physical description:* Coelacanth-like fish with leglike lobed fins. Doglike head. Barbels. Distinctive scales. Horizontally oriented tail.

*Distribution:* Clear Lake, California.

*Significant sighting:* In October 1993, Lyle Dyslin caught a strange-looking, dog-headed fish; after taking photographs of it, he released it back into the lake because it reminded him of his dachshund.

*Possible explanation:* Malformed catfish or, less likely, a dramatic new species. Clear Lake is known for its large Channel catfish (*Ictalurus punctatus*), which put up a struggle when hooked. Catfishes are not native to California but began to be introduced sometime in the 1870s.

*Sources: San Francisco Examiner,* October 3, 1993; Karl Shuker, "Sounds Fishy to Me!" *Strange Magazine,* no. 17 (Summer 1996): 23.

### Coelacanth (Unrecorded Populations)

The Coelacanth is the only surviving member of a class of lobefin FISHES that dates back 400 million years, to the Early Devonian. No fossil Coelacanths have been found that are more recent than the Cretaceous period, 65 million years ago.

*Etymology:* Former genus name, from the Greek *koilos* ("hollow") + *akantha* ("spine").

*Scientific names: Latimeria chalumnae,* given by J. L. B. Smith in 1939 for the Indian Ocean species; *L. menadoensis,* given in 1999 for the Celebes Sea species.

*Variant names:* Ikan fomar (in Java), Patuki (Rapa Nui/Austronesian).

*Physical description:* Length, 5 feet. Weight, up to 150 pounds. Slate-blue with white flecks. Thick, armorlike scales, lined with serrated rows of hardened, toothpick-pointed denticles. Rays of the first dorsal fin are arranged like a fan, but the second dorsal, pectoral, and pelvic fins are lobed. Three-lobed tail. A pressure-sensitive lateral line senses the proximity of other fishes and surrounding structures.

*Behavior:* Congregates in submarine caves in the daytime. Sculls in an upside-down position with paired pectoral and pelvic fins moving in unison.

*Distribution:* Now known from the Comoro Islands and nearby African waters and off Sulawesi in Indonesia. Evidence for their presence in the South Pacific, the Gulf of Mexico, the coast of India, and Java is tenuous yet intriguing.

*Significant sightings:*

*Off Easter Island*–The Easter Islanders have legends of a Coelacanth-like fish called the Patuki, which has leglike fins.

*In the Gulf of Mexico*–A Tampa, Florida, souvenir seller bought a bucketful of Coelacanth scales, now lost, from a local fisherman in 1949. Silver ornaments in the shape of a Coelacanth, apparently dating from the seventeenth to eighteenth centuries, were found in Bilbao and Toledo, Spain, in 1964 and 1965; initially thought to have originated in the Spanish colonies of North America, they were shown, in 2001, to be recent manufactures using the Comoran coelacanth as a model.

*Off the coast of India*–An eighteenth-century Indian miniature painting shows a Muslim holy man standing beside a Coelacanth-like fish with armorlike scales.

*Java*–George Serres caught a 25-pound specimen in 1995 off southwestern Java, though documentation was later lost. Local people know about the fish, which they call Ikan fomar.

*Possible explanations:*

(1) The Spanish silver ornaments may have been based on a fossil Coelacanth, though its three-dimensional shape conforms well with the living fish.

(2) The Indian miniature painting could portray the Climbing perch (*Anabas testudineus*), found in India and Southeast Asia, which is famous for its ability to survive several days out of water.

*Sources:* Francis Mazière, *Mysteries of Easter*

*Island* (London: Collins, 1969); B. Brentjes, "Eine Vorentdeckung des Questenflossers in Indien?" *Naturwissenschaftliche Rundschau* 25 (1972): 312—313; Hans Fricke, "Quastie im Baskenland?" *Tauchen,* no. 10 (October 1989): 64—67; Michel Raynal and Gary S. Mangiacopra, "Out-of-Place Coelacanths," *Fortean Studies* 2 (1995): 153—165; Karl Shuker, "Long May the Coelacanth Reign as King of the Sea in Indonesia," *Strange Magazine,* no. 20 (December 1998): 36—37; Hans Fricke and Raphaël Plante, "Silver Coelacanths from Spain Are Not Proofs of a Pre-scientific Discovery," *Environmental Biology of Fishes* 61 (August 2001): 461—463.

## Coje Ya Menia

WATER LION of Central Africa.

*Etymology:* Mbundu-Loanda (Bantu), "water lion."

*Physical description:* Slightly smaller than a hippopotamus. Has tusks or large canine teeth.

*Behavior:* Nocturnal. Amphibious. Moves to smaller rivers and swamps during the rainy season. Makes a loud, rumbling roar. Kills hippopotamuses but does not eat them.

*Tracks:* Like an elephant's but with toes overprinted on the impression of the sole.

*Habitat:* Rivers and swamps.

*Distribution:* Upper Cuango and Cuanza Rivers, Angola, and smaller tributaries.

*Significant sighting:* In the 1930s, a Portuguese truck driver heard that a Coje ya menia had killed a hippopotamus along the Cuango River the night before. He went off with some trackers and for several hours followed the trail of the hippo and another smaller animal. They found the dead, uneaten hippo ripped to shreds in an area where the grass and shrubs had been crushed down.

*Possible explanations:*

(1) A surviving aquatic saber-toothed cat, first suggested by Ingo Krumbiegel in 1947.

(2) An unknown monitor lizard.

(3) A surviving dinosaur of some type.

*Sources:* Ilse von Nolde, "Der *Coje ya menia:* Ein sagenhaftes Tier Westafrikas," *Deutsche Kolonialzeitung* 51, no. 4 (1939): 123—124;

Ingo Krumbiegel, "Was ist der 'Löwe des Wassers'?" *Kosmos* 42 (1947): 143—146; Martin Wilfarth, "Leben heute noch Saurier?" *Prisma*, October 1949, pp. 279—282; Bernard Heuvelmans, *Les derniers dragons d'Afrique* (Paris: Plon, 1978), pp. 239—241, 319—323, 326, 329.

## Coleman Frog

Giant AMPHIBIAN of New Brunswick, Canada.

*Physical description:* Like a huge bullfrog. Length, 27 inches. Weight, 42 pounds.

*Behavior:* Consumes baked beans, June bugs, whiskey, and buttermilk toddies. Said to have been used to tow canoes and race against tomcats.

*Distribution:* Killarney Lake, New Brunswick.

*Present status:* Said to have been dynamited from the lake in 1885. On display for many years at Fred B. Coleman's Barker House Hotel, the lone specimen was donated to the York-Sudbury Historical Museum, Fredericton, in 1959.

*Possible explanations:*

(1) Bullfrogs (*Rana catesbeiana*) only grow to 8 inches long. The record weight is 1 pound 4 ounces.

(2) A fake, made to advertise a patent medicine for relieving sore throats. A 1988 report by the Canadian Conservation Institute refers to the artifact as consisting of canvas, wax, and paint. A letter refers to the exhibit as "an amusing example of a colossal fake and deception."

*Sources:* Gerald L. Wood, *The Guinness Book of Animal Facts and Feats* (Enfield, England: Guinness Superlatives, 1982), p. 119; Joe Nickell, *Real-Life X-Files* (Lexington: University Press of Kentucky, 2001), pp. 157—159.

## Colossal Claude

SEA MONSTER of Oregon.

*Variant name:* Marvin.

*Physical description:* Length, 15—40 feet. Round, tan body. Snake- or horselike head. Neck, 8 feet long. Long tail.

*Behavior:* Raids fishing lines.

*Distribution:* Mouth of the Columbia River, Oregon, and the neighboring seacoast.

*Significant sightings:* In 1934, L. A. Larson, mate of the Columbia River lightship, saw a 40-foot animal at the mouth of the river. Crew members studied it for some time with binoculars.

In 1963, the Shell Oil Company during an oil search off the Oregon coast recorded a videotape that shows a 15-foot animal with barnacled ridges swimming in 180 feet of water.

*Source:* Peter Cairns, "Colossal Claude and the Sea Monsters," Portland *Oregonian*, September 24, 1967.

## Colovia

Unknown SNAKE of Southern Europe.

*Physical description:* Serpentine. Length, 11 feet. Scaly.

*Distribution:* Sicily, Italy.

*Significant sighting:* A snakelike animal was tracked down and killed in a marsh near Siracusa, Sicily, in December 1933. It was destroyed because local superstition held that its appearance presaged disaster.

*Possible explanation:* Escaped python or boa (Family Boidae).

*Source:* *Times* (London), December 27 and 29, 1933.

## Columbus's Ape-Faced Cat

Odd CAT or other mammal of Central America, seen on Columbus's fourth voyage.

*Physical description:* Large cat. Apelike face. Prehensile tail.

*Behavior:* Aggressive. Killed a peccary with only its foreleg and thick tail.

*Distribution:* Costa Rica.

*Significant sighting:* Christopher Columbus sent an armed party into the interior of Costa Rica in 1502, and this odd-looking animal was captured.

*Possible explanations:*

(1) The Central American spider monkey (*Ateles geoffroyi*) has a prehensile tail and a catlike body 1—2 feet long. The Costa Rican variety is brown to silvery above, lighter below. It growls and acts threatening when disturbed, but whether this fruit-eating

monkey, even if provoked, could dispatch a pig is doubtful.

(2) The Kinkajou (*Potos flavus*), a relative of the raccoon, is even more catlike, also has a prehensile tail, and is common in Central American forests. However, it is extremely docile.

(3) An unknown species of carnivorous cat, now extinct, suggested by Herbert Wendt.

*Sources:* Letter from Christopher Columbus to the King and Queen of Spain, July 7, 1503, in J. M. Cohen, trans., *The Four Voyages of Christopher Columbus* (New York: Penguin, 1969), p. 298; Herbert Wendt, *Out of Noah's Ark* (Boston: Houghton Mifflin, 1959).

## Columbus's Serpent

Mystery CROCODILIAN of the West Indies.

*Physical description:* Described as a *sierpe* ("snake"). Length, 5 feet.

*Habitat:* Freshwater lagoon.

*Distribution:* Crooked Island, Bahamas.

*Significant sighting:* On October 21, 1492, Christopher Columbus's men killed a "snake" on the northwest coast of Crooked Island in the Bahamas. Martin Alonso Pinzón of the *Pinta* killed another one the next day. He planned to skin one and return it to Spain; however, no such skin has been preserved.

*Possible explanations:*

(1) Long thought to have been an Iguana (Family Iguanidae), although these large lizards rarely enter water and there is no fossil evidence for them on the island.

(2) When remains of a village site visited by Columbus was excavated in 1987, the femur from an American crocodile (*Crocodylus acutus*) was discovered. Its size indicated it belonged to an animal about 4 feet long. Though it may not have been the animal killed by Columbus's men, it shows that crocodiles used to live on the island.

*Sources:* Bartolomé de las Casas, "Digest of Columbus's Log-book on His First Voyage" [1492], in J. M. Cohen, trans., *The Four Voyages of Christopher Columbus* (New York: Penguin, 1969), pp. 70–72; Karl Shuker,

"Close Encounters of the Cryptozoological Kind," *Fate* 53 (May 2000): 26–29.

## Con Rít

MULTIFINNED SEA MONSTER of the China Sea.

*Etymology:* Vietnamese (Austroasiatic) name for a millipede with a toxic bite.

*Physical description:* Length, 60 feet. Dark brown above, light yellow below. Body composed of armored segments 2 feet long and 3 feet wide. A pair of thin appendages, 2 feet 4 inches long, is attached to each segment.

*Distribution:* Halong Bay, Vietnam.

*Significant sighting:* Tran Van Con and other Vietnamese found a carcass washed ashore at Hong Gai, Vietnam, around 1883. The head was gone, but the remainder was formed of odd segmented joints that rang like sheet metal when hit with a stick. It smelled so badly that it was towed out to sea.

*Possible explanations:*

(1) The backbone of a whale, though the vertebral structure should have been obvious and described in a different way.

(2) The caudal vertebrae of an Oarfish (*Regalecus glesne*). However, its bones are shaped differently, and this fish generally only grows to 36 feet.

(3) Surviving archaic basilosaurid whale, similar to those in Bernard Heuvelmans's MULTIFINNED SEA MONSTER category, which he theorized had armored plates. However, it's now known that basilosaurids were not armored.

(4) A surviving Sea scorpion (Class Eurypterida), a group of arthropods that flourished from the Ordovician to the Permian periods, 500–250 million years ago, had an abdomen divided into twelve segments, but no appendages were attached to them. In addition, they actually lived in brackish or freshwater instead of the open sea, and the largest one, a species of *Pterygotus*, only reached 9 feet in length.

(5) A giant crustacean of an unknown type, proposed by Karl Shuker. The carcass represents only the exoskeleton and limbs. However, the largest known living

crustacean is the Japanese spider crab (*Macrocheira kaempferi*), which has a claw span of 10–12 feet but a body size not much over 1 foot—nowhere near the size of the Con rít.

*Sources:* Abel Gruvel, *L'Indochine: Ses richesses marines et fluviales* (Paris, 1925), p. 123; "Le grand serpent de mer," *La Nature Supplément* 53 (November 14, 1925): 153; A. G. L. Jourdan, "A propos du Serpent de mer," *La Nature Supplément* 53 (December 12, 1925): 185–186; Bernard Heuvelmans, *In the Wake of the Sea-Serpents* (New York: Hill and Wang, 1968), pp. 416–421; Karl Shuker, *In Search of Prehistoric Survivors* (London: Blandford, 1995), pp. 126–127.

## Coromandel Man

WILDMAN of Australasia.

*Etymology:* From the mountain range.

*Variant names:* Forest taniwah, Hairy moehau, Matau, Moehau monster, Toangina, Tu-uhourangi.

*Physical description:* Covered with red or silver hair.

*Habitat:* Caves.

*Distribution:* Coromandel Range, Waikato River, and Tongariro National Park, North Island, New Zealand; Cameron Mountains, Milford Wilderness, and Lake Wakatipu, South Island, New Zealand.

*Significant sightings:* In 1878, gold prospectors on Martha Hill in Waihi reported large, long-haired man-beasts carrying stone knives, hand axes, and wooden clubs.

Large, five-toed, humanlike footprints were found embedded in mud along a creek in 1903 by miners in the Karangahake Gorge.

In early February 1952, hunters Douglas Tainvhana and Roy Norman got a fleeting glimpse of a hairy man running along a track on the Coromandel Peninsula.

In 1963, Carl McNeil saw an apelike creature running along a track bed on the Coromandel Peninsula.

Trevor Silcox was hunting wild pig with a companion in the Coromandel Range in 1972 when they spotted a 6-foot, naked man covered with dark hair moving through the scrub. Four tracks measuring 14 inches long and 7 inches wide were found.

*Possible explanation:* A surviving remnant of postulated pre-Maori inhabitants of New Zealand.

*Sources:* Craig Heinselman, "Hairy Maeroero," *Crypto* 4, no. 1 (January 2001): 23–26; Rex Gilroy, *Giants from the Dreamtime: The Yowie in Myth and Reality* (Katoomba, N.S.W., Australia: URU, 2001).

## Cressie

FRESHWATER MONSTER of Newfoundland, Canada.

*Etymology:* After the lake.

*Variant names:* Cressiteras anguilloida (quasi-scientific name), Haoot tuwedyee (possibly Beothuk/Algonquian, "swimming demon"), Woodum haoot ("pond devil").

*Physical description:* Serpentine. Length, 15 feet. Black. Rounded hump. No fins or flukes.

*Behavior:* Swims with a rolling motion.

*Distribution:* Crescent Lake, Newfoundland.

*Significant sightings:* Around June 7, 1960, Bruce Anthony and three other loggers watched an object that looked like an overturned boat swim and cross a nearby sandbar.

On September 5, 1991, Pierce Rideout saw an unusual wave and then a black, 15-foot animal swimming with a rolling motion about 150 yards offshore.

*Possible explanation:* An oversized American eel (*Anguilla rostrata*), which normally grows to less than 5 feet. Eels usually spawn in the ocean, but they have never been seen in Tommy's Arm Brook, Crescent Lake's only outlet to the sea. However, the lake's deeper water is saline, which might allow the eels to stay in the lake to spawn.

*Sources:* John Braddock, "Monsters of the Maritimes," *Atlantic Advocate* 58 (January 1968): 12–17; Cressie, the Monster of Crescent Lake, http://www.nfcap.nf.ca/central/RobertsArm/attract/cressie.html.

## Cretan Pterosaur

FLYING REPTILE of Southern Europe.

*Physical description:* Batlike. Pelican-like beak. Fingerlike protrusions on the wings. Large claws.

*Distribution:* Crete.

*Significant sighting:* A flying reptile was seen in the summer of 1986 in the Asteroussia Mountains, near Pirgos, Crete, by three youths who were hunting. It was large and dark gray, with membranous wings that made a flapping noise as it flew over them slowly.

*Source:* Thanassis Vembos, "A Prehistoric Flying Reptile?" *Strange Magazine,* no. 2 (1988): 29.

## CROCODILIANS (Unknown)

Crocodiles, alligators, and gavials are the three families of living Crocodilians (Order Crocodylia) containing twenty-two species. Modern crocodilians have a full secondary palate that allows them to breathe and eat at the same time, just like a mammal. They are semiaquatic ambush predators that hunt in water but go on land to bask in the sun. Although they generally will eat whatever they can catch (including outboard motors and gas tanks), crocs prefer large fishes, birds, turtles, and mammals.

The oldest known modern crocodilian was *Hylaeochampa vectiana* from the Early Cretaceous, 110 million years ago, from the Isle of Wight, England. One of the largest crocodiles ever was the 30- to 50-foot *Deinosuchus,* found in the Late Cretaceous of North America; its skull alone measures 6 feet 6 inches in length. Another was the 40-foot, 10-ton *Sarcosuchus* from the Ténéré Desert, Niger, which had a massive growth on the end of its snout, was covered with bony plates, and thrived 110–90 million years ago. The order diversified successfully by the end of the Mesozoic and flourished in the Cenozoic. Different species spread as far north as Sweden in Europe (*Thoracosaurus macrorhynchus*) and Alberta and Saskatchewan in North America (*Leidyosuchus canadensis* and *Borealosuchus acutidentatus*).

The largest living crocodilian is the Saltwater crocodile (*Crocodylus porosus*), found from South-east Asia to northern Australia and the Solomon Islands. The average length for mature males is 14–16 feet, but outsize specimens greater than 23 feet long have been reliably reported.

Of the nine animals in this list, three (LIPATA, MAHAMBA, and the SULAWESI LAKE CROCODILE) may be surviving fossil species. The others are distribution anomalies, a mundane explanation for a lake monster (MIGO), and an odd color variant (PINK ALLIGATOR).

### Mystery Crocodilians

CANADIAN ALLIGATOR; COLUMBUS'S SERPENT; KIPUMBUBU; LIPATA; MAHAMBA; MIGO; PINK ALLIGATOR; PSKOV CROCODILE; SULAWESI LAKE CROCODILE

## Crowing Crested Cobra

Mystery SNAKE of East and Central Africa.

*Variant names:* Bubu (on the Lower Zambezi River in Mozambique), Hongo (Ngindo/ Bantu), Inkhomi (Ngoni/Bantu and Nyakyusa-Ngonde/Bantu, "the killer"), Kovoko (Nyamwezi/Bantu), Mbobo (Rungwa/Bantu), N'gokwiki (Gbaya/Ubangi), Ngoshe (Bemba/ Bantu), Songo (Yao/Bantu, "strikes down at the head").

*Physical description:* Cobralike snake. Length, up to 20 feet. Buff-brown or grayish-black. Bright red, forward-projecting crest on its head. Scarlet face. The male has a pair of red facial wattles. The dorsal vertebra of one specimen had articulating surfaces of $8 \times 9$ millimeters.

*Behavior:* Arboreal. May also be aquatic. Extremely vicious. The male makes a loud sound like a rooster crowing. The female makes a hen-like clucking sound. Both male and female emit a warning cry of "chu-chu-chu-chu." Feeds on maggots from rotting flesh; it supposedly kills animals so that maggots will grow on the carcasses. Also eats hyraxes. Attacks humans by lunging down from a tree toward the head or face. The venom is extremely toxic, resulting in death almost instantaneously.

*Habitat:* Trees, hills, rocks.

*Distribution:* KwaZulu-Natal Province, South Africa; Mozambique; Zimbabwe; Malawi; Zambia; Tanzania; Central African Republic.

*Significant sightings:* From a witch doctor in Malawi, J. O. Shircore obtained a plate of bone from the crest (with bits of skin attached), some neck bones, and several vertebrae from at least two different specimens of this snake.

In May 1959, John Knott accidentally ran over a 7-foot black snake in his Land Rover in the Lake Kariba area of Zimbabwe. It had a symmetrical crest on its head that could be erected by raising five bony structures.

*Possible explanations:*

(1) A nonexistent composite of several different snakes, suggested by Charles R. S. Pitman, who also proposed that the crowing was not done by the snake but by its victims.

(2) The Gaboon adder (*Bitis gabonica*) has a pair of hornlike scales on its snout, and its head is pale brown with a dark central line. It is now endangered and found only in coastal Natal and eastern Zimbabwe. Pitman noted in the 1930s that in Kawambwa, Zambia, people thought the animal had a crest and made a crowing noise.

(3) The Rhinoceros viper (*Bitis nasicornis*) of West and Central Africa has a flat, triangular-shaped head with two or three hornlike projections. Its brilliant color patterns vary among individuals.

(4) The Black mamba (*Dendroaspis polylepis*), Africa's most feared snake, is found from Kenya to Mozambique. It sometimes carries molted skin on its head, which makes it look crested. It is also rumored to lunge down at people from trees.

(5) An unknown species of venomous snake with a crest or frill.

(6) The Puff adder (*Bitis arietans*) of South Africa is known to emit a bell-like note; the Indian cobra (*Naja naja*) is said to purr or hiss; and the Bornean cave racer (*Elaphe taeniurae grabowskyi*) makes an eerie, meowing sound. However, these snakes have no vocal cords, so they must produce the sounds using other frictive organs.

*Sources:* Horace Waller, *The Last Journals of David Livingstone* (London: John Murray, 1875), vol. 2, p. 344; Charles R. S. Pitman, *A Report on a Faunal Survey of Northern Rhodesia* (Livingstone, Zambia: Government Printer, 1934); William Hichens, "African Mystery Beasts," *Discovery* 18 (December 1937): 369–373; J. O. Shircore, "Two Notes on the Crowing Crested Cobra," *African Affairs* 43 (1944): 183–186; John Knott, "'Crowing Snake,'" *African Wildlife* 16 (September 1962): 170; Charles Cordier, "Animaux inconnus du Congo," *Zoo* 38 (April 1973): 185–191; Karl Shuker, *Extraordinary Animals Worldwide* (London: Robert Hale, 1991), pp. 31–37.

## Cù Sìth

BLACK DOG of Scotland.

*Etymology:* Gaelic, "fairy dog."

*Physical description:* Size of a yearling bullock. Usually dark green, sometimes white. Shaggy. Paws as wide as a man's hand. Long tail is coiled up or plaited.

*Behavior:* Gives three loud bays.

*Tracks:* Sometimes found in snow or mud.

*Distribution:* Scotland and the Hebrides.

*Sources:* John Gregorson Campbell, *Superstitions of the Highlands and Islands of Scotland* (Glasgow, Scotland: J. MacLehose, 1900), pp. 30–32; Alasdair Alpin MacGregor, *The Ghost Book* (London: Robert Hale, 1955), pp. 55–81; James MacKillop, *Oxford Dictionary of Celtic Mythology* (New York: Oxford University Press, 1998), p. 122.

## Cuero

FRESHWATER MONSTER of South America.

*Etymology:* Spanish, "cowhide," from its rough skin.

*Variant names:* El Bien peinado ("well-groomed one"), Cuero unudo, Hide, Huecú, Lafquen trilque, Manta ("blanket"), TRELQUE-HUECUVE.

*Physical description:* Dark color. Rough skin. Usually just a hump or long neck is seen. Sometimes said to have four eyes on the head and numerous eyes on the perimeter of its body.

*Behavior:* Most active in the evening. Can walk on land. Creates large wakes.

*Distribution:* Lago Lacar and Lago Nahuel Huapí, Neuquén Province, Argentina; other lakes in the region, including Chile; also said to be marine.

*Possible explanations:*

(1) Jorge Luis Borges characterized the animal as a FRESHWATER OCTOPUS.

(2) Karl Shuker has suggested an unknown species of large, freshwater jellyfish, perhaps related to the Moon jelly (*Aurelia aurita*) found near the coast and inland in warm and tropical waters. It ranges in size from 2 to 16 inches.

(3) An evolved Sea scorpion (Class Eurypterida), which flourished from the Ordovician to the Permian periods, 500–250 million years ago, proposed by Mark Hall. One species, *Pterygotus buffaloensis,* attained a length of 9 feet and was the largest known arthropod. Sea scorpions were roughly cylindrical and had distinct latitudinal scales.

*Sources:* Hartley Burr Alexander, *Latin American Mythology* [1920] (New York: Cooper Square, 1964), p. 328; Maurice Burton, "Muck and Monsters," *Illustrated London News* 237 (1960): 570; Ulrich Dunkel, *Abenteuer mit Seeschlangen* (Stuttgart, Germany: Kreuz-Verlag, 1961); Gregorio Alvarez, *El tronco de oro: Folklore del Neuquén* (Buenos Aires: Editorial "Pehuén," 1968), pp. 120–121; Harold Osborne, *South American Mythology* (Feltham, England: Paul Hamlyn, 1968), p. 116; Jorge Luis Borges, *The Book of Imaginary Beings* (New York: E. P. Dutton, 1969), pp. 100–101; Félix Coluccio, *Diccionario de creencias y supersticiones* (Buenos Aires: Corregidor, 1983), p. 522; Mark A. Hall, *Natural Mysteries,* 2d ed. (Minneapolis, Minn.: Mark A. Hall, 1991), pp. 59–64.

## Cuino

Supposed HOOFED MAMMAL of Mexico.

*Physical description:* Piglike animal, thought in the nineteenth century to be a hybrid Domestic pig sow (*Sus scrofa scrofa*) × and a polled Domestic ram (*Ovis aries*). Curly hair is black, white, black and white, or brown and white.

*Distribution:* Oaxaca State, Mexico.

*Present status:* The name is now a Spanish synonym for the Yucatan miniature breed of pig.

*Possible explanation:* W. B. Tegetmeier examined the skull of an alleged Cuino in 1902 but found it to be a full-blooded pig.

*Source:* Karl Shuker, *Mysteries of Planet Earth* (London: Carlton, 1999), p. 19.

## Cuitlamiztli

Unknown CAT or HYENA of Mexico.

*Etymology:* Nahuatl (Uto-Aztecan), "glutton cat."

*Physical description:* Pumalike animal.

*Behavior:* Attacks and eats deer ravenously, hence its name.

*Distribution:* Mexico.

*Significant sighting:* In the Aztec ruler's menagerie at Teotihuacan, Mexico, Bernal Díaz reported seeing in 1520 a wolflike animal, which some have equated with the Cuitlamiztli.

*Possible explanations:*

(1) The ONZA, though this cat is now considered a regional variation of the Puma (*Puma concolor*).

(2) Surviving *Chasmaporthetes ossifragus,* the only fossil hyaenid found in North America, suggested by Karl Shuker to explain Díaz's animal. It was a mobile hunter that lived 2 million–10,000 years ago, from the Pliocene to the Pleistocene. However, Díaz's description is so vague that it might even refer to a Coyote (*Canis latrans*).

*Sources:* Bernardino de Sahagún, *Florentine Codex: General History of the Things of New Spain* [1577?], Arthur J. O. Anderson and Charles E. Dibble, trans. (Salt Lake City: University of Utah Press, 1950–1982), pt. 12, p. 6; Bernal Díaz del Castillo, *The Conquest of New Spain* [1632], J. M. Cohen, trans. (New York: Penguin, 1963), p. 229; Karl Shuker, *Mystery Cats of the World* (London: Robert Hale, 1989), p. 187.

## Curinquéan

TRUE GIANT hominid of South America.

*Physical description:* Height, 11 feet. Covered with hair.

*Behavior:* Adorns lips and nostrils with nuggets of gold.

*Distribution:* Brazil.

*Source:* Simão de Vasconcellos, *Noticias curiosas, y necessarias das cousas do Brasil* (Lisbon: I. da Costa, 1668).

## Curupira

LITTLE PEOPLE of South America.

*Etymology:* From the Guaraní (Tupí) *curu-mim* ("boy") + *pira* ("body"). *Kuru* in Aché means "short" or "small."

*Variant names:* Caá-porá ("mountain lord"), Caiçara (for the female), Caipora, Caypóré, Coropira, Corubira (Bakairí/Carib), Kaaguerre, Kaapore, Korupira (Tupí/Guaraní), Kurupi (Guaraní), Kurú-piré (Guaraní), Yurupari (Tucano/Tucanoan).

*Physical description:* Height, 3–4 feet. Covered with hair. Red or yellow skin. Large head like a chimpanzee. Red head-hair. Shaggy mane around the neck. Flattened nose. Large mouth. Green or blue teeth. Large feet, said to point backwards. Crooked toes.

*Behavior:* Arboreal. Poor swimmer. Emits a birdlike whistle. Eats bananas. Said to smoke a pipe. Lives in hollow trees. Said to abduct children and rape women. Can shape-shift. Protects trees, forests, and game. Rides a pig or deer.

*Tracks:* Apelike prints.

*Habitat:* Forests, hills, ravines, mountains.

*Distribution:* Pará, Amazonas, and Pernambuco States in northern Brazil; Paraná, Rio Grande do Sul, and Goiás States in southern Brazil; Misiones Department in Paraguay; Chaco Province, Argentina.

*Present status:* Caipora has become a minor god in the Candomblé religion.

*Possible explanation:* Surviving *Protopithecus,* a Late Pleistocene spider monkey known from fossils in eastern Brazil.

*Sources:* Charles Carter Blake, "Note on Stone Celts, from Chiriqui," *Transactions of the Ethnological Society of London,* new ser., 2 (1863): 166–170; Herbert H. Smith, *Brazil: The Amazons and the Coast* (New York: Charles Scribner's Sons, 1879), pp. 560–569; Daniel G. Brinton, "The Dwarf Tribe of the Upper Amazon," *American Anthropologist* 11 (1898): 277–279; Juan B. Ambrosetti, *Supersticiones y leyendas* (Buenos Aires: La Cultura Argentina, 1917), pp. 89–92; Luís da Câmara Cascudo, *Dicionário do folclore Brasileiro* (Rio de Janeiro: Instituto Nacional do Livro, 1962), vol. 1, pp. 166–168, 261–262; Napoleão Figueiredo and Anaíza Vergolino e Silva, *Festas de santo e encantados* (Belém, Brazil: Academia Paraense de Letras, 1972); Maria Thereza Cunha de Giacomo, *Curupira: Lenda indigena* (São Paulo, Brazil: Melhoramentos, 1975); Karl Shuker, "On the Trail of the Curupira," *Fortean Times,* no. 102 (September 1997): 17; John E. Roth, *American Elves* (Jefferson, N.C.: McFarland, 1997), pp. 50–54, 83–89, 94–95, 107.

## Cyclops

One-eyed GIANT HOMINID of Southern Europe.

*Etymology:* Greek, "round eye."

*Variant names:* Arimaspean (Scythian/Indo-European), Kyklops, Monopthalmos ("one-eyed person"), Polyphemus, Triamates.

*Physical description:* Giant. One eye in the center of the forehead.

*Behavior:* Eats humans.

*Distribution:* Sicily; Crete; India; Africa.

*Present status:* While investigating the ancient voyage of Odysseus, Tim Severin discovered that there was a modern tradition of giants that once lived on Crete.

*Possible explanation:* Inspired by observations of fossil elephant skulls in the Mediterranean, where the nasal opening is mistaken for an eye socket.

*Sources:* Homer, *Odyssey,* I. 69; Hesiod, *Theogony* 139–146, 501–506; Euripides, *The Cyclops;* Herodotus, *The Histories,* ed. John Marincola (New York: Penguin, 1996), pp. 198, 221–222, 225 (III. 116, IV. 13–15, 27); Othenio Abel, *Der Tiere der Vorwelt* (Berlin: Teubner, 1914); Othenio Abel, *Das Reich der Tiere: Tiere der Vorzeit in ihrem Lebensraum* (Berlin: Deutscher Verlag, 1939); Willy Ley, *The Lungfish, the Dodo, and the Unicorn* (New York: Viking, 1948), pp. 47–51; William Elgin Swinton, *Giants Past and Present* (London:

Robert Hale, 1966), pp. 21–22; Justin Glenn, "The Polyphemus Myth: Its Origin and Interpretation," *Greece and Rome* 25 (1978): 141–155; Tim Severin, *The Ulysses Voyage: Sea Search for the Odyssey* (London: Hutchinson, 1987), pp. 88–98; Erich Thenius and Norbert Vávra, *Fossilien im Volksglauben und im Alltag: Bedeutung und Verwendung vorzeitlicher Tier- und Pflanzenreste von der Steinzeit bis heute* (Frankfurt am Main, Germany: Kramer, 1996), pp. 19–21.

# D

## Daisy Dog

Doglike ENTITY of Cornwall, England.

*Etymology:* From the cross-shaped plot of daisies on the dog's grave.

*Physical description:* Size of a cat. Laughing face. Pug nose. Feathery ears. Plumed tail draped over on its back.

*Behavior:* Its bite is said to be fatal.

*Distribution:* Cornwall.

*Significant sighting:* Much feared by Cornish fishermen in the nineteenth century and perhaps earlier.

*Possible explanation:* Said to be the ghosts of Pekinese dogs that were sent by the Chinese emperor to Queen Elizabeth I in the late sixteenth century. The dog's keeper, a Chinese princess, was killed by mutinous sailors, and the dogs were thrown overboard. The bodies were later found and buried. No record of this event exists outside of folklore. However, the similarity of the Daisy dog to the real Pekinese (a breed that did not officially reach England until October 1860) is remarkable.

*Sources:* Ruth L. Tongue, *Forgotten Folk Tales of the English Counties* (London: Routledge and Kegan Paul, 1970); Karl Shuker, *Extraordinary Animals Worldwide* (London: Robert Hale, 1991), pp. 49–53.

## Dakuwaqa

Giant FISH of the South Pacific Ocean.

*Etymology:* Fijian (Austronesian) word.

*Variant name:* Dakuwaqua.

*Physical description:* Shark with light spots. Length, 35 feet. Turtle-shaped head. Short neck, 2 feet in diameter. Enormous dorsal fin. Fluked tail like a whale's.

*Behavior:* Said to attack canoes when hungry.

*Distribution:* Koro Sea, off Vanua Levu, Fiji Islands.

*Significant sighting:* Rev. A. J. Small saw the animal in 1912 when he was on board an 8-ton cutter.

*Possible explanation:* The Whale shark (*Rhincodon typus*) is known in these waters and fits the general physical description.

*Sources:* Colman Wall, "Dakuwaqa," *Transactions of the Fijian Society,* 1917, pp. 6–12, 39–46; George T. Bakker, "Dakuwaqa," *Transactions of the Fijian Society,* 1924, pp. 30–36; Paul Sieveking, "The Dakuwaqua," *Fortean Times,* no. 58 (July 1991): 28.

## Dakwa

FRESHWATER MONSTER of Tennessee.

*Etymology:* Cherokee (Iroquoian) word.

*Behavior:* Knocks over boats. Said to be able to swallow a man whole.

*Distribution:* Little Tennessee River, near the mouth of Little Toqua Creek, Monroe County, Tennessee; French Broad River, Buncombe County, North Carolina.

*Sources:* Albert S. Gatschet, "Water-Monsters of American Aborigines," *Journal of American Folklore* 12 (1899): 255–260; Don Chesnut, Eastern Cherokee Place Names, 1999, http://www.users.mis.net/~chesnut/pages/cherokeeplace.htm.

## Dard

Mystery LIZARD of Western Europe.

*Etymology:* French, "forked tongue."

*Variant name:* CAT-HEADED SNAKE.

*Physical description:* Catlike head. A mane extends down the back. Four legs. Short tail like a viper.

*Behavior:* Not venomous but bites viciously when it attacks. Hisses loudly. Said to suck the udders of cows.

*Distribution:* Vienne Department, east-central France.

*Possible explanation:* Popular folklore regarding the harmless Slow worm (*Anguis fragilis fragilis*), a legless European anguid lizard that grows to 20 inches. Some confusion may exist with the related European glass lizard (*Ophisaurus apodus*), which is longer (up to 4 feet), has vestigial hind limbs, and is only found in Eastern Europe and West Asia.

*Source:* Henri Ellenberger, "Le monde fantastique dans le folklore de la Vienne," *Nouvelle Revue des Traditions Populaires* 1 (1949): 407–435.

## Dark Leopard

Unusual big CAT of Africa and Asia.

*Scientific name:* Panthera pardus var. *melanotica,* based on the Grahamstown specimen.

*Variant names:* Damasia (Gikuyu/Bantu, in Kenya), KIBAMBANGWE, NDALAWO, SHING MUN TIGER.

*Physical description:* Leopard with dark coat patterns that are distinct from the melanistic, all-black variety.

*Distribution:* Aberdare Highlands, Kenya; Bufumbira County, southwest Uganda; Virunga Volcanos region of Rwanda; Eastern Cape Province, South Africa; Kerala State, India; Bali, Indonesia; and Hong Kong, China.

*Significant sightings:* A pseudomelanistic leopard was shot near Grahamstown, Eastern Cape Province, South Africa, in the 1880s. It had a tawny background color, with an orange gloss on the shoulders. Small spots coalesced on its back to form a solid black color from head to tail. The underparts looked like a typical leopard's (white with large spots). Its total length was 6 feet 7 inches.

Another specimen was killed in Kerala State, southwestern India, in 1912. The rosettes were fused into a solid black over the entire upper body.

G. Hamilton-Snowball shot a large, dark leopard in the Aberdare Highlands, Kenya, in the 1920s. The local Gikuyu people told him it was called a Damasia and was different from a leopard.

A mounted leopard, supposedly obtained in Belize, is on display at the Wildlife World Museum in Springfield, Missouri. Its background color is a very dark reddish-brown.

*Present status:* Melanistic Leopards (*Panthera pardus*) have an abnormally dark background color, but the rosettes are still visible under proper lighting conditions or up close. They are common in Myanmar, peninsular Malaysia, southern India, Java, southwestern China, and some parts of Nepal and Assam. Black leopards are less common in Africa, though they have been reported in Ethiopia, Kenya, Rwanda, and Cameroon.

Pseudomelanistic leopards are known from only a few specimens. The background color is a normal orange-yellow, but the rosettes are so abundant that they have fused together into a solid black color over portions of the coat. The normal background color is sometimes visible in thin, irregular yellow streaks. Though documented, the pseudomelanistic morph is not well known and might be misinterpreted as an unknown animal if it is found unexpectedly.

Brown mutants are also on record.

*Sources:* Albert Günther, "Note on a Supposed Melanotic Variety of the Leopard, from South Africa," *Proceedings of the Zoological Society of London,* March 3, 1885, pp. 243–245; Holdridge Ozro Collins, in *Bulletin of the Southern California Academy of Sciences* 14 (1915): 49–51; G. Hamilton-Snowball, "Spotted Lions," *The Field* 192 (October 9, 1948): 412; S. H. Prater, *The Book of Indian Animals* (Mumbai, India: Bombay Natural History Society, 1971), p. 68; C. A. W. Guggisberg, *Wild Cats of the World* (New York: Taplinger, 1975); Gerald L. Wood, *The Guinness Book of Animal Facts and Feats,* 3d ed. (Enfield, England: Guinness Superlatives, 1982), p. 35; Karl Shuker, *Mystery Cats of the World* (London: Robert Hale, 1989), pp. 112–115, 133–137; Bill Rebsamen, "A Mounted Cat Mystery," *North American BioFortean Review* 1, no. 1 (April 1999): 20–21, http://www.strangeark.com/nabr/NABR1.pdf.

## Das-Adder

Unknown LIZARD of South Africa.

*Etymology:* The local name for Rock hyraxes (*Procavia* spp.) is "dassie"; thus, a combination of snake and hyrax.

*Variant name:* Dassie-adder.

*Physical description:* Snakelike body. Head like a hyrax. Skin around ear openings is folded into a crest. Red and yellow stripes on the tail, which is about 2 feet long.

*Behavior:* Extremely venomous. Allegedly capable of luring prey with its irresistible gaze.

*Distribution:* Drakensberg Mountains, South Africa.

*Possible explanations:*

(1) The Rock monitor (*Varanus albigularis*) has a shorter head than the Water monitor lizard (*V. niloticus*) and grows to just over 4 feet. Its tail is long but not striped.

(2) An unknown species of monitor lizard.

*Sources:* W. L. Speight, "Mystery Monsters in Africa," *Empire Review* 71 (1940): 223–228; Karl Shuker, "Here Be Dragons," *Fate* 49 (June 1996): 31–34.

DE LOYS'S APE, *a mystery primate photographed in 1920 by Swiss geologist François de Loys in the Serranía de Parijá of Colombia and Venezuela. (Fortean Picture Library)*

## Dav

WILDMAN of West Asia.

*Etymology:* Svan (South Caucasian) variant of DEV.

*Behavior:* Eats berries. Wears animal skins.

*Distribution:* Caucasus Mountains of Georgia.

*Sources:* Douglas William Freshfield, *The Exploration of the Caucasus* (London: E. Arnold, 1896), vol. 2, pp. 191, 210; Dmitri Bayanov, *In the Footsteps of the Russian Snowman* (Moscow: Crypto-Logos, 1996), p. 14.

## De Loys's Ape

Apelike PRIMATE of South America.

*Scientific names: Ameranthropoides loysi,* proposed by George Montandon in 1929; *Ateles loysi,* proposed by Arthur Keith in 1929 to counter Montandon.

*Physical description:* Gibbonlike primate. Thick coat of long, grayish-brown hair. Height, 5 feet 1.75 inches. Oval face, with developed forehead. Triangular patch of pale pigment on the forehead. Round ridges surrounding the eye sockets. Flat nose with flared nostrils. Powerful jaws with thirty-two teeth (instead of the normal thirty-six for platyrrhine monkeys in South America). Flat chest. Broad shoulders. Sturdy arms. Monkeylike hands. Long fingers. Vestigial thumbs. Oversized clitoris. Long toes. Opposed big toe. No tail.

*Behavior:* Can apparently walk upright by holding on to bushes. Screams wildly. Angrily confronts humans and uses tree branches and its own excrement as weapons.

*Distribution:* Serranía de Parijá of Colombia and Venezuela.

*Significant sighting:* In 1920, members of Swiss geologist François de Loys's expedition killed the female of a pair of tall, tailless apes that appeared to threaten them. The incident occurred along the Río Tarra, on the border between Colombia and Venezuela. His famous photograph shows the dead animal sitting on a gasoline crate with its chin propped up by a stick. Other photos were tragically lost when de Loys's boat capsized. The skull was retained, but the expedition's cook used it as a salt container, and it disintegrated (as did the pelt).

*Possible explanations:*

(1) Thought to be a Black spider monkey (*Ateles paniscus*) by Sir Arthur Keith, who suspected that the animal had been creatively manipulated for the camera by de Loys. However, the reported size alone makes this doubtful, since these monkeys are rarely more than 3 feet 6 inches when standing on the hind legs. No tail is visible in the photo, and the animal's body is thicker and more massive. De Loys's ape is less hairy, has more powerful jaws, and has an oval face with little prognathism.

(2) The White-bellied spider monkey (*A. belzebuth*) is a strong possibility, especially if the size of the crate in the photo is less than 16 inches, instead of the 18–20 inches estimated by George Montandon in his analysis of the case. Ivan Sanderson favored this explanation, suggesting that the decomposing body had already started bloating when the photo was taken.

(3) Loren Coleman and Michel Raynal argue persuasively that the story was a hoax by Montandon, based on de Loys's photo of a spider monkey. Montandon would have done this in order to lend credence to his racist theory that primate evolution took place independently in South America, with Indians as the ultimate result.

(4) An unclassified ape or monkey known locally as the MONO GRANDE.

(5) A surviving *Protopithecus brasiliensis,* a fruit-eating spider monkey from the Late Pleistocene of eastern Brazil that was twice as large as any extant species.

*Sources:* George Montandon, "Découverte d'un singe d'apparence anthropöide en Amérique du Sud," *Journal de la Société des Américanistes de Paris* 21, no. 6 (1929): 183–186; George Montandon, "Un singe d'apparence anthropöide en Amérique du Sud," *Comptes Rendus de l'Academie de Sciènces* 188 (1929): 815–817; François de Loys, "A Gap Filled in the Pedigree of Man?" *Illustrated London News* 84 (June 15, 1929): 1040; Francis M. Ashley-Montague, "The Discovery of a New Anthropoid Ape in South America?" *Scientific Monthly* 29 (1929): 275–279; Léonce Joleaud, "Remarques sur l'evolution des primates Sud-Americains, à propos du grand singe du Venezuela," *La Revue Scientifique* 67 (1929): 269–273; Arthur Keith, "The Alleged Discovery of an Anthropoid Ape in South America," *Man* 29 (1929): 135–136; Nello Beccari, "Ameranthropoides Loysi, gli atelini e l'importanza della morfologica cerebrale nella classificazione delle scimmie," *Archivio per l'Antropologia e la Etnologia* 73 (1943): 5–114; Don Cousins, "Ape Mystery," *Wildlife* 24 (April 1982): 148–149; Michael T. Shoemaker, "The Mystery of the Mono Grande," *Strange Magazine* 7 (April 1991): 2–5, 56–60; Marc E. W. Miller and Khryztian E. Miller, "Further Investigation into Loy's 'Ape' in Venezuela," *Cryptozoology* 10 (1991): 66–71; Loren Coleman and Michel Raynal, "De Loys' Photograph: A Short Tale of Apes in Green Hell, Spider Monkeys, and *Ameranthropoides loysi* As the Tools of Racism," *The Anomalist,* no. 4 (Autumn 1996): 84–93; Letters, *The Anomalist,* no. 5 (Summer 1997): 143–153; Ángel L. Viloria, Franco Urbani, and Bernardo Urbani, "François de Loys (1892–1935) y un hallazgo desdeñado: La historia de una controversia antropológica," *Interciencia* 23 (March-April 1998): 94–100; Karl Shuker, "Monkeying around with Our Memories?" *Strange Magazine,* no. 20 (December 1998): 40– 42; Letters, *Interciencia* 24 (July-August 1999): 229–231.

## Dediéka

Mystery PRIMATE of Central Africa.

*Etymology:* Kota or Teke (Bantu) word.

*Variant names:* Dodiéka, Tschimpênso (Yombe/Bantu).

*Physical description:* Like a chimpanzee but larger and with a black face. Sagittal crest.

*Distribution:* Republic of the Congo; Gabon.

*Possible explanations:*

(1) Originally thought to be a Chimpanzee (*Pan troglodytes*) × Gorilla (*Gorilla gorilla*) hybrid or transition species.

(2) Skulls and skins are now considered to be those of either male chimpanzees or female gorillas.

*Sources:* Eduard Pechuël-Loesche, *Die Loango-Expedition* (Leipzig, Germany: P. Frohberg, 1879–1907), vol. 3, pp. 246–250; Henri Neuville, "A propos d'un crâne de gorille rapporté de la Likouala-Mossaca par le Dr. A. Durieux," *L'Anthropologie* 23 (1912): 363–396; Bernard Heuvelmans, *Les bêtes humaines d'Afrique* (Paris: Plon, 1980), pp. 418–421, 426–427.

## Deep-Sea Spider

Mystery INVERTEBRATE of the South Pacific Ocean.

*Physical description:* Triangular fore body, 0.52 inches in length. Hind body, 2 inches long, separated from the front by a narrow waist. Five pairs of jointed appendages on the fore body. The first pair seems to be used as feelers. The second pair is very long and carried high above the body. The last three pairs are walking legs.

*Significant sighting:* During the first Disturbance and Recolonization Experiment in a Manganese Nodule Area of the Deep South Pacific (DISCOL) expedition, a scorpion-like animal was photographed several times at a depth of 13,616 feet in the Peru Trench on February 12, 1989.

*Possible explanations:*
(1) The animal was tentatively identified as a Tailless whip scorpion (Order Amblypygi), which it certainly resembles. All other tailless whip scorpions are strictly terrestrial, as are modern arachnids. However, recent discoveries have shown that ancestors of the True scorpions (Order Scorpiones) were once primarily aquatic and were equipped with organs called book gills, formed of respiratory sheets like the pages of a book. These scorpions often grew large; *Praearcturus gigas* of the Early Devonian, 400 million years ago, was more than 3 feet long. Aquatic or amphibious species persisted until at least the end of the Triassic, 205 million years ago. Some scorpions might have survived by adapting to abyssal depths.

(2) Sea spiders (Subphylum Pycnogonida) are marine arthropods that are found from intertidal regions to a depth of 23,000 feet. There are at least 1,000 species, but they are little known. Some deep-sea forms grow to more than 2 feet across the legs. Often classed as an order within the chelicerate arthropods, they bear such unusual features (a long proboscis with an odd terminal mouth, a reduced abdomen, a pair of ovigers for carrying eggs, long pouches in the intestines that extend to the end of the legs) that they probably represent a sister taxon to the Chelicerata. The second pair of legs of the DISCOL spider may have been the ovigers.

*Source:* Hjalmar Thiel and Gerd Shriever, "The Enigmatic DISCOL Species: A Deep Sea Pedipalp?" *Senckenbergiana Maritima* 20 (October 1989): 171–175.

## Denman's Bird

Mystery BIRD of East Africa.

*Distribution:* Ruwenzori Range, Uganda.

*Significant sighting:* Canadian mountaineer Earl Denman watched a pair of unidentifiable birds dive swiftly and almost vertically in the high mountain air.

*Possible explanation:* Verreaux's eagle (*Aquila verreauxii*), a black raptor with a wingspan up to 8 feet that lives in the highlands of East Africa and performs spectacular aerial courtship displays. Adults are often seen in pairs, diving together to seize hyraxes and hares.

*Sources:* Earl Denman, *Animal Africa* (London: Robert Hale, 1957), p. 159; Ben S. Roesch, letter, *Strange Magazine*, no. 18 (Summer 1997): 2–3.

## Derketo

Fish-tailed fertility goddess of the Middle East, one of the first known MERBEINGS to be depicted in art.

*Etymology:* Greek word.

*Variant names:* Atargatis, Dea Syria, Derceto.

*Physical description:* Half woman, half fish.

*Distribution:* Eastern Mediterranean Sea.

*Significant sighting:* Worshiped in Phoenicia, where Lucian saw her mermaidlike image in the second century A.D.

*Sources:* Diodorus Siculus, *History*, II.; Lucian of Samosata, *De dea Syria;* Gwen Benwell and Arthur Waugh, *Sea Enchantress* (London: Hutchinson, 1961), pp. 28–29.


## Dev

WILDMAN of West and Central Asia.

*Etymology:* Tajik (Persian), "demon." The Indo-European root *dyeu-* is the basis for both the Sanskrit (Indo-Aryan) *devah* ("god") and the Avestan *daēva* ("demon").

*Variant names:* Daeva, Div, PARÉ.

*Physical description:* Height, 4 feet 10 inches. Covered with shaggy, reddish-brown or black hair. Black skin. Has horns, claws, fangs, and tail.

*Behavior:* Bipedal. Travels either singly or in pairs. Feeds on marmots and other rodents.

*Distribution:* Armenia; Northern Iran; the Pamir Mountains, Tajikistan.

*Significant sightings:* Geologist B. M. Zdorik ran across a Dev sleeping along a path high in the upper reaches of the Dondushkan River in the Pamirs, Tajikistan, in 1934. Its body was covered with yaklike fur. Zdorik and his guide panicked and fled before the creature awoke. The local people said there were families of Devs living in the Tal'bar and Safid-Dara Valleys. An adult Dev had been caught in 1933 at a flour mill a few miles from Tutkaul, where it was kept chained up for two months before it escaped.

*Sources:* Boris F. Porshnev and A. A. Shmakov, eds., *Informatsionnye materialy, Komissii po Izucheniyu Voprosa o "Snezhnom Cheloveke,"* 4 vols. (Moscow: Akademiia Nauk SSSR, 1958–1959); Ivan T. Sanderson, *Abominable Snowmen: Legend Come to Life* (Philadelphia: Chilton, 1961), pp. 310–311; Dmitri Bayanov, *In the Footsteps of the Russian Snowman* (Moscow: Crypto-Logos, 1996), pp. 78–80.


## Devil Bird

Mystery BIRD of the Indian subcontinent.

*Physical description:* Pigeon-sized bird that is rarely seen but often heard. Long tail.

*Behavior:* Nocturnal. Cry is a hideous, strangling sound, said to be heard in cemeteries.

*Distribution:* Sri Lanka.

*Significant sighting:* Mitford of the Ceylon Civil Service saw a big black bird by moonlight at Kurunegala in the nineteenth century. Its cry was like a boy being tortured and strangled. He thought it was a nightjar of some type.

*Possible explanations:* None of the following emit anything like the reported cry of the Devil bird, but many are similar in shape and plumage:

(1) The Brown wood owl (*Strix leptogrammica indranee*) was assumed by Charles Pridham and James Tennent to be the Devil bird. It measures 14–21 inches, and its call is a series of three or four short hoots. Some hold it responsible for an eerie scream.

(2) The Forest eagle owl (*Bubo nipalensis blighi*), suggested by G. M. Henry. Its call is a deep hoot, while its mating calls are said by some to consist of shrieks like those of a woman being strangled.

(3) The Sri Lanka frogmouth (*Batrachostomus moniliger*) only attains a length of 9 inches, and its call consists of liquid chuckles or soft "karoo" or "whoo" cries. It is shy and not often vocal.

(4) The Gray nightjar (*Caprimulgus indicus*), proposed by William Vincent Legge, makes restrained "chunk-chunk-chunk" calls.

(5) The Sri Lankan Changeable hawk-eagle (*Spizaetus cirrhatus ceylanensis*) has a ringing scream "kleee-klee-ek," whether perched or on the wing. It also has a rapid "ki-ki-ki-ki-ki-ki-ki-keeee," beginning short, rising in crescendo, and ending in a scream.

(6) The Mountain hawk-eagle (*Spizaetus nipalensis kelaarti*) has a noisy "klu-weet-weet" call.

(7) The Oriental honey buzzard (*Pernis ptilorhynchus ruficollis*) emits loud and high-pitched ringing notes.

(8) A composite bird created from the calls of several species.

(9) An unknown owl. A new species of owl is said to have been discovered in Sri Lanka in January 2001 by an ornithologist who had been tracking its unfamiliar call for several years.

(10) An unknown species of hawk-eagle.

(11) An unknown species of nightjar, suggested by Karl Shuker.

*Sources:* Charles Pridham, *An Historical, Political, and Statistical Account of Ceylon and Its Dependencies* (London: T. and W. Boone, 1849), pp. 737–738; James Emerson Tennent, *Sketches of the Natural History of Ceylon* (London: Longman, Green, Longman, and Roberts, 1861); William Vincent Legge, *A History of the Birds of Ceylon* (London, 1880); George Morrison Henry, *A Guide to the Birds of Ceylon* (London: Godfrey Cumberledge, 1955); Richard L. Spittel, *The Devil Birds of Ceylon,* suppl. to *Loris* 11 (December 1968): 1–14; Sálim Ali, *Indian Hill Birds* (Mumbai, India: Oxford University Press, 1987); Karl Shuker, "Horned Jackals and Devil Birds," *Fate* 42 (January 1989): 57–64; "New Bird Discovered in Sri Lanka," *BBC News Online,* February 26, 2001.


## Devil Monkey

Large, tailed PRIMATE of North America.

*Etymology:* Name coined by Mark A. Hall.

*Variant names:* Giant monkey, NALUSA FALAYA.

*Physical description:* Height, 3–8 feet. Light brown to black hair. Pointed ears. Baboonlike or doglike muzzle. Strong chest. A blaze of white fur from neck to belly. Short forelegs with claws. Muscular hind legs. Large feet. A long tail is sometimes reported, alternately described as black and bushy or hairless.

*Behavior:* Sometimes walks bipedally, at other times quadrupedally. Aggressive toward dogs and humans. Emits a wide range of hoots, calls, screams, and whistles. Said to kill livestock and small game.

*Tracks:* Three rounded toes, with regular spacing between. Length, 12–15 inches.

*Distribution:* British Columbia; Appalachian Mountain region of the United States.

*Significant sightings:* Early one morning in 1959, a monkeylike creature rushed at the car of a couple driving down a rural road near Saltville, Virginia. It chased the vehicle for a short while, grabbing at it with its front paws. Later, they found three long, deep scratches in the metal from the front door to the rear. Around the same time, the animal apparently ripped off the convertible top of a car driven by two nurses.

After midnight on June 26, 1997, Debbie Cross saw a strange animal outside her rural home near Dunkinsville, Ohio. It was 3–4 feet tall, hairy, and with long arms and a short tail. As it moved away from her, it walked on its knuckles.

A giant black monkey was spotted in rural Danville, New Hampshire, on September 9, 2001, and on at least nine other occasions over a two-week period.

*Present status:* Distinctions between NORTH AMERICAN APES, Devil monkeys, HAIRY BIPEDS, and BIGFOOT are nebulous and possibly arbitrary. In general, NORTH AMERICAN APES are tailless and primarily quadrupedal, and they resemble chimpanzees; Devil monkeys are tailed and resemble baboons; HAIRY BIPEDS cover a wide range of descriptions from apes to wildmen and even paranormal entities; BIGFOOT is a robust, tall hominid with a range that seems restricted to the Pacific Northwest. Loren Coleman even notes certain similarities with PHANTOM KANGA-ROO reports and suggests that juvenile Devil monkeys may resemble kangaroos.

*Possible explanations:*

(1) Feral pet monkeys have formed breeding colonies, especially in Florida. A group of Squirrel monkeys (*Saimiri sciureus*) live in Broward County's Hugh Taylor Birch State Park, and Rhesus monkeys (*Macaca mulatta*) became well established around Silver Springs after they were first released into the wild in Marion County during the filming of Tarzan movies in 1933. Rhesus monkeys are also known in Jack Kaye Park in Fort Lauderdale. A few of the 500 monkeys uncaged in 1992 by Hurricane Andrew from research labs, private owners, and the Miami Metrozoo are probably still living and breeding in the Everglades. Some officials believe Capuchin monkeys (*Cebus*

spp.) are forming troops in south Florida. A free-roaming group of Japanese macaques (*Macaca fuscata*) was brought to Dilley, Texas, in 1972 to save the animals from destruction in Kyoto, Japan, where they are regarded as a nuisance; in 1980, the monkeys became the property of the South Texas Primate Observatory and were confined for behavioral research at a ranch there. But in the late 1980s, their enclosure fell into disrepair, and several escaped. The monkeys have roamed the south Texas brush ever since, their population swelling to more than 600 by 1995. However, none of these colonies are in areas where Devil monkeys have been reported.

(2) Surviving *Protopithecus brasiliensis* (or related species), an extinct, fruit-eating spider monkey from the Late Pleistocene of eastern Brazil that was twice as large as any extant species, suggested by Chad Arment.

(3) Surviving *Theropithecus oswaldi*, a large baboon that lived 650,000 years ago in East Africa and is the ancestor of the modern Gelada baboon (*T. gelada*), suggested by Mark A. Hall. The male was roughly the size of a female gorilla and weighed 250 pounds. A ground dweller, this animal was too big to live in trees and could not use its long forearms for swinging.

*Sources:* Chad Arment, "Virginia Devil Monkey Reports," *North American BioFortean Review* 2, no. 1 (2000): 34–37, http://www.strangeark.com/nabr/NABR1.pdf; Chad Arment, "Devil Monkeys or Wampus Cats?" *North American BioFortean Review* 2, no. 2 (2000): 45–48, http://www.strangeark.com/nabr/NABR4.pdf; Loren Coleman, *Mysterious America,* rev. ed. (New York: Paraview, 2001), pp. 184–187.

## Devil Pig

Large, piglike HOOFED MAMMAL or MARSUPIAL of Australasia.

*Variant names:* Gazeka, Monckton's gazeka.

*Physical description:* Dark skin with patterned markings. Length, 5 feet. Shoulder height, 3 feet 6 inches or greater. Long snout. Horselike tail. Even-toed (cloven) feet.

*Distribution:* Owen Stanley Range, Papua New Guinea.

*Significant sightings:* Ancient stone carvings depicting strange animals with long, trunklike snouts were first found in 1962 in the Ambun Valley.

Huge (rhinoceros-sized) excrement was found by the crew of the HMS *Basilisk* on the northeast Papuan coast in the 1870s. Dung from feral pigs, which are the largest Papuan ungulates, is less substantial.

Two native Papuans, Private Ogi and the village constable Oina, saw two large, porcine animals on Mount Albert Edward, Papua New Guinea, on May 10, 1906. Ogi tried to shoot one, but his hands shook, and he misfired.

*Possible explanations:*

(1) A feral Domestic pig (*Sus scrofa* var. *domesticus*) is rarely larger than 2 feet 6 inches at the shoulder.

(2) The Malayan tapir (*Tapirus indicus*) is odd-toed and not found as far east as New Guinea.

(3) The Babirussa (*Babyrousa babyrussa*), found in Sulawesi, Indonesia, is not a close match.

(4) A Papuan occurrence of the Javan rhinoceros (*Rhinoceros sondaicus*) is unlikely.

(5) A Long-nosed echidna (*Zaglossus bruijni*), especially a newly hatched juvenile, might account for the Ambun sculptures.

(6) A surviving diprotodont marsupial, such as the tapirlike *Palorchestes* or the rhinoceros-like, nasal-horned *Nototherium.* Most of New Guinea's native mammals are marsupials, making these large animals viable possibilities for the Devil pig. The snouted *Palorchestes* seems particularly akin to the animal depicted in the Ambun stones. The last diprotodonts are thought to have died out in Australia between 18,000 and 6,000 years ago.

*Sources:* Alfred O. Walker, "The Rhinoceros in New Guinea," *Nature* 11 (1875): 248, 268; Adolf Bernhard Meyer, "The Rhinoceros in New Guinea," *Nature* 11 (1875): 268; Charles A. W. Monckton, *Some Experiences of a New Guinea Resident Magistrate* (London: John Lane, 1920); Charles A. W. Monckton, *Last*

*Days in New Guinea* (London: John Lane, 1922), pp. 52–56; Charles A. W. Monckton, *New Guinea Recollections* (London: John Lane, 1934), pp. 214–215; W. G. Heptner, "Über das Java-Nashorn auf Neu-Guinea," *Zeitschrift für Säugetierkunde* 25 (1960): 128–129; "A Remarkable Stone Figure from the New Guinea Highlands," *Journal of the Polynesian Society* 74 (1965): 78–79; Laurent Forge, "Un marsupial géant survit-il en Nouvelle Guinée?" *Amazone*, no. 2 (January 1983): 9–11; James I. Menzies, "Reflections on the Ambun Stones," *Science in New Guinea* 13 (1987): 170–173.

## Devil's Hoofmarks

Tracks made by a mystery mammal, possibly a RODENT, in England and elsewhere.

*Etymology:* Newspapers in 1855 reported that some people attributed the marks to Satan.

*Tracks:* Vaguely donkeylike; some appear to have been made by hooves, while others do not. Length, 3.5–4 inches. Width, 1.5–2.75 inches. Depth, 0.5–4 inches. The prints are 8–16 inches apart and directly in front of each other, rather than alternating left and right.

*Significant sightings:* On the night of February 8–9, 1855, something left a trail of thousands of prints in the snow across Devon, England, from Torquay in the south to Exeter in the north. The tracks wandered through gardens, lanes, and cemeteries in at least thirty villages, crossing roofs and jumping across walls and haystacks. Several groups of people followed them, but all failed to find a track maker, to note whether there was more than one trail, or to determine if the prints varied in shape from place to place.

Other cases of mysterious trails that have been compared to the Devon case include cloven tracks in the snow found in Inverness, Scotland, the same month, which a local naturalist declared were made either by a hare or by a European polecat (*Mustela putorius*); the donkeylike tracks discovered by the crew of James Clark Ross in May 1840 on desolate Kerguelen Island in the South Indian Ocean, which may have been made by an animal cast ashore on a shipwrecked vessel; the 2-mile-long set of semicircular tracks found in the snow on January 10, 1945, near Everberg, Belgium, by Eric Frank Russell that were 9–12 inches apart; the hoofmarks clearly impressed in the wet sand of a Devonshire coastal town that were found by a Mr. Wilson in October 1950; and small, horseshoe-shaped tracks in the snow photographed by Ruth Christiansen in January 1975 at Frederic, Wisconsin.

*Possible explanations:*

(1) Many wild guesses were made, including an escaped monkey, mouse, rat, swan, hare, deer, otter, toad, Domestic cat (*Felis silvestris catus*), escaped kangaroo, heron, and a Great bustard (*Otis tarda*).

(2) Zoologist Richard Owen suggested the tracks were made by Eurasian badgers (*Meles meles*) out looking for food on this particularly cold night. However, badger prints overlap and clearly alternate left and right.

(3) Birds may have easily left tracks in closed gardens or on top of walls, but bird prints, whether clawed or webbed, are easily recognizable.

(4) The Donkey (*Equus asinus*) does actually place its hooves in a straight line when walking and could have accounted for some of the tracks.

(5) The Red fox (*Vulpes vulpes*) may have contributed to the mystery, according to Gordon Stein, but it would have had a difficult time scaling walls or roofs.

(6) The Wood mouse (*Apodemus sylvaticus*) makes a V-shaped track when it moves along in kangaroo-like leaps of 2–3 feet, climbing bushes and trees with ease. It is quite common in English gardens. Alfred Leutscher thought of this explanation when he came across dozens of trails made by this mouse in the snow, all four feet coming together in a single track roughly conforming to the hoofmarks.

(7) Manfri Wood recounted a tale he heard about a carefully planned prank by Romanies using several hundred pairs of specially made stilts in an attempt to scare away their rivals, the Didikais (Romanies of mixed heritage) and Pikies (criminals expelled from Romany society). The scheme depended on making

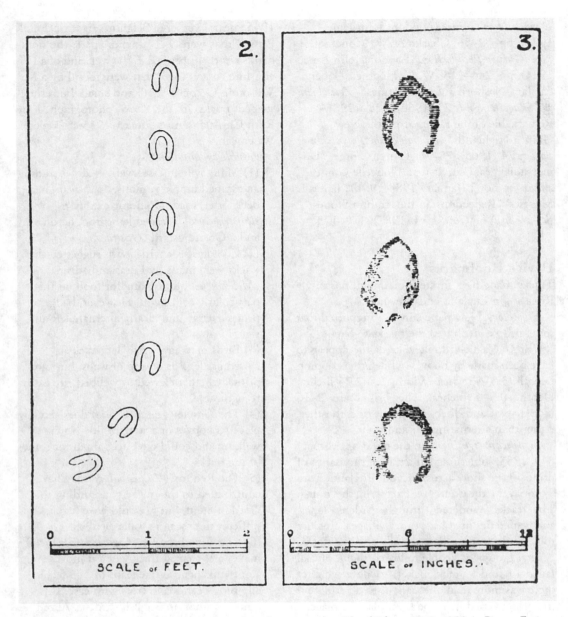

*The* DEVIL'S HOOFMARKS, *unidentified tracks made in the snow on the night of February 8–9, 1855, in Devon, England. (Illustrated London News/Fortean Picture Library)*

the tracks appear supernatural. However, the event took place in Somerset at an unspecified time in the past and involved size 27 boots at the base of the stilts.

(8) James Alan Rennie proposed that the tracks were made by freakish air currents, an interesting suggestion in the light of similar explanations for the crop circles of southern

England. He claimed to have seen a line of much larger tracks being made in such a fashion in the snow in northern Canada in 1924.

(9) Morris K. Jessup and George Lyall suggested that unidentified flying objects (UFOs) sent down "rays" or laser beams to create the tracks.

(10) Mike Dash has discovered that some apparent anomalies in this case have been overstated: There were multiple trails of varying lengths, not a continuous one of 100 miles, as one account had it. Several sources suggest that tracks were found for several days afterward. One report that the tracks led to a haystack with no marks in the snow on its surface and reappeared on the opposite side is difficult to explain, but the source was anonymous and secondhand. Some witnesses saw claw marks or rounded prints, while others found clearly defined hoofmarks. Also, in some instances the tracks were not strictly one in front of the other. All of this indicates multiple causes for the tracks.

*Sources:* James Clark Ross, *A Voyage of Discovery and Research in the Southern and Antarctic Regions* (London: John Murray, 1847), vol. 1, p. 87; "Extraordinary Occurrence," *Times* (London), February 16, 1855; [William D'Urban], "Foot-Marks on the Snow, in Devon," *Illustrated London News* 26 (February 24, 1855): 187; "Mysterious Footprints in the Snow," *Inverness Courier,* March 1, 1855; Richard Owen, "Professor Owen on the Foot-Marks in the Snow in Devon," *Illustrated London News* 26 (March 3, 1855): 214; "The Mysterious Footprints in Devonshire," *Times* (London), March 6, 1855; "The Foot-Marks in the Snow in Devon," *Illustrated London News* 26 (March 10, 1855): 238; Rupert T. Gould, *Oddities: A Book of Unexplained Facts* (New York: Frederick A. Stokes, 1928), pp. 9–22; [Eric Frank Russell], "Our Cover," *Doubt,* no. 15 (1946): 218; Theo Brown, "The Great Devon Mystery of 1855, or 'The Devil in Devon,'" *Report and Transactions of the Devonshire Association* 82 (1950): 107–112; Theo Brown, "A Further Note on the 'Great Devon Mystery,'" *Report and Transactions of the Devonshire Association* 84 (1952): 163–171; Morris K. Jessup, *The Case for the UFO* (New York: Citadel, 1955), pp. 153–160; James Alan Rennie, *Romantic Strathspey* (London: Robert Hale, 1956), pp. 81–82; Eric J. Dingwall, "Did the Devil Walk Again?" *Tomorrow* 5, no. 3 (Spring 1957); Alfred Leutscher, "The Devil's Hoof-Marks," *Animals* 6, no. 8 (April 20, 1965): 108–109; George Lyall, "Did a Laser Create the Devil's Footprints?" *Flying Saucer Review* 18, no. 1 (January-February 1972): 24–25; Manfri Frederick Wood, *In the Life of a Romany Gypsy* (London: Routledge and Kegan Paul, 1973); "Woman Photographs Strange Tracks," *A.P.R.O. Bulletin* 23, no. 8 (June 1975); "The Devil's Walk in Devon," *Fortean Times,* no. 39 (Spring 1983): 16; G. A. Household, ed., *The Devil's Footprints: The Great Devon Mystery of 1855* (Exeter, England: Devon Books, 1985); Gordon Stein, "The Devil's Footprints," *Fate* 38 (August 1985): 88–95; Mike Dash, ed., "The Devil's Hoofmarks: Source Material on the Great Devon Mystery of 1855," *Fortean Studies* 1 (1994): 71–150; Joe Nickell, *Real-Life X-Files* (Lexington: University Press of Kentucky, 2001), pp. 10–17.

## Didi

Unknown PRIMATE of South America.

*Etymology:* Possibly a Carib word.

*Variant names:* Dai-dai, Didi-aguiri, Dru-di-di, Massikruman, Quato.

*Physical description:* Height, 5 feet. Reddish-brown or black hair or fur. Thickset, powerful build. Receding forehead. Heavy brows. Large eyes. Big-lobed ears. Flared nostrils. High cheekbones. Thick lips. Jutting jaw. Opposable thumbs. Long arms. Long, slender feet. No tail.

*Behavior:* Shy. Swings arms while walking erect. Apparently lives and travels as part of a pair. Call is "hoo hoo" or a long, melancholy whistle, beginning in a high key then dying away. Builds crude brush houses from palm leaves. Throws sticks and mud. Accepts food that is left out for it. Said to be able to mate successfully with humans.

*Tracks:* Apelike. Large toe joint of the male flares out, while the female's does not.

*Distribution:* Mazaruni, Cotinga, Berbice, and Demerara Rivers in Guyana; French Guiana.

*Significant sightings:* A British prospector named Haines came across two Didi in the Konawaruk Mountains, Guyana, in 1910. They were covered in reddish-brown hair.

A guide named Miegam was traveling up the Berbice River in Guyana in 1918 with three others when they saw two hairy creatures on the riverbank. The creatures' footprints looked ape-like rather than human.

Mycologist Gary Samuels observed a 5-foot-tall Didi about 60 feet away in the Guyanese forest in 1987. It walked past on two feet, making an occasional "hoo" sound.

*Sources:* Edward Bancroft, *An Essay on the Natural History of Guyana, in South America* (London: T. Becket and P. A. De Hondt, 1769), pp. 130–131; Charles Barrington Brown, *Canoe and Camp Life in British Guiana* (London: E. Stanford, 1876), pp. 87–88, 123, 385; L. C. van Panhuys, "Are There Pygmies in French Guiana?" *Proceedings of the International Congress of Americanists* 13 (1905): 131–133; Nello Beccari, "*Ameranthropoides loysi*, gli Atelini e l'importanza della morfologia cerebrale nella classificazione delle scimmie," *Archivio per l'Antropologia e la Etnologia* 73 (1943): 1–112; Ivan T. Sanderson, *Abominable Snowmen: Legend Come to Life* (Philadelphia: Chilton, 1961), pp. 178–181; Mark A. Hall, *Living Fossils: The Survival of* Homo gardarensis, *Neandertal Man, and* Homo erectus (Minneapolis, Minn.: Mark A. Hall, 1999), pp. 50–51; Loren Coleman and Patrick Huyghe, *The Field Guide to Bigfoot, Yeti, and Other Mystery Primates Worldwide* (New York: Avon, 1999), pp. 72, 183.

## Dientudo

GIANT HOMINID of South America.

*Etymology:* Spanish, "big teeth."

*Physical description:* Half bear, half man. Huge teeth.

*Habitat:* Forests.

*Distribution:* El Gato Creek, Ringuelet, and Toloso in Buenos Aires Province, Argentina.

*Source:* Fabio Picasso, "South American Monsters and Mystery Animals," *Strange Magazine,* no. 20 (December 1998): 28–35.

## Dilali

WATER LION of Central Africa.

*Etymology:* Gbaya-Bossangoa (Ubangi), "water lion."

*Variant names:* Dilaï, Mama himé, Mamaimé (Zandé/Ubangi, "water lion").

*Physical description:* Length, 5 feet. Shoulder height, 3 feet. The size of a horse. Mane. Large tusks. Hairy legs. Claws like a lion's.

*Behavior:* Aquatic. Feeds on fish and leaves. Kills hippopotamuses and crocodiles but does not eat them. In Central African Republic, said to kill and eat humans.

*Distribution:* Southern Chad; Central African Republic.

*Significant sighting:* In 1912, Naumann of Ulm, a lieutenant in the German Imperial Defense Corps, offered a reward for the Dilali while he was stationed north of the Ouham River in Chad. He failed to find any evidence other than stories.

*Possible explanations:*

(1) The aquatic variety of the PYGMY ELEPHANT.

(2) A surviving saber-toothed cat, suggested by Bernard Heuvelmans.

*Sources:* Ingo Krumbiegel, *Von neuen und unentdeckten Tierarten* (Stuttgart, Germany: Franckh'sche Verlagshandlung, 1950), pp. 57–68; Bernard Heuvelmans, *On the Track of Unknown Animals* (New York: Hill and Wang, 1958), pp. 463–465, 468, 474; Robert Kirch, "Animaux inconnus en Afrique?" *Connaissance de la Chasse,* no. 60 (April 1981): 62–65, 92.

## Dimorphic Beaked Whale

Unclassified CETACEAN of the eastern Pacific Ocean.

*Etymology:* From its two distinct color forms.

*Scientific name: Mesoplodon* species A.

*Variant name:* Unidentified beaked whale.

*Physical description:* Length, 16–18 feet. Long, wide beak. Relatively flat head with a small but distinct melon. Low, triangular dorsal fin. Males have a broad, white swath across the body; the head and beak are reddish-brown or tan, while the back and flanks behind the swath are black-brown or chocolate-brown. Females and young are gray-brown, fading to pale gray on the underside.

*Behavior:* Usually seen traveling in tight groups at a moderate pace. Feeds on squid.

*Distribution:* Eastern Pacific Ocean from central Mexico to Peru. Favors deep water.

*Significant sightings:* Known from about sixty-five sketchy sightings at sea, as well as a series of aerial photographs taken by a helicopter in November 1999.

*Present status:* Classification is pending, until a stranded specimen can be examined.

*Possible explanations:*

(1) This animal is assumed to be an unknown species of beaked whale.

(2) Longman's beaked whale (*Indopacetus pacificus*) is known only from two skulls; there have been no live observations.

(3) Bahamonde's beaked whale (*Mesoplodon bahamondi*) is known from a single skull.

(4) The Lesser beaked whale (*Mesoplodon peruvianus*) was officially described in 1991 after a male specimen was found on a deserted beach north of Lima, Peru. Its range overlaps with the Dimorphic, but it has been thought to be smaller, at 11–12 feet long. Robert Pitman and Morgan Lynn consider that it is a good match and that Dimorphic sizes have been overestimated.

*Sources:* Robert L. Pitman, Anelio Aguayo L., and Jorge Urban R., "Observations of an Unidentified Beaked Whale (*Mesoplodon* sp.) in the Eastern Tropical Pacific," *Marine Mammal Science* 3 (1987): 345–352; Mark Carwardine, *Whales, Dolphins, and Porpoises* (New York: Dorling Kindersley, 1995), pp. 112–113; Robert L. Pitman and Morgan S. Lynn, "Biological Observations of an Unidentified Mesoplodont Whale in the Eastern Tropical Pacific and Probable Identity: *Mesoplodon peruvianus*," *Marine Mammal Science* 17 (2001): 648–657.

## Dingonek

WATER LION of East Africa.

*Etymology:* Masai and Okiek (Nilo-Saharan) word.

*Variant names:* Ndamathia (Kikuyu/Bantu), Ol-maima, Ol-umaina (Masai/Nilo-Saharan).

*Physical description:* Length, 14–18 feet. Spotted like a leopard. Covered with scales. Head like an otter's or a lioness's. Small ears. Two straight, white tusks in the upper jaw. Short neck. Back is as broad as a hippo's. Short legs. Claws. Long, broad tail.

*Behavior:* Basks on logs or on riverbanks. Slips in the water when disturbed. Swims with only its head above the water.

*Tracks:* As big as a hippo's. Clawed.

*Distribution:* Rivers in Kenya and Tanzania that feed into the east side of Lake Victoria; the Kikira River, a source of the Tana River, Kenya.

*Significant sightings:* In 1907, big-game hunter John Alfred Jordan took a shot at an animal with leopard's spots and two large fangs along the Migori River, where it flows into Lake Victoria, Kenya. He also found clawed tracks the size of a hippo's.

Around the same time, a man known to C. W. Hobley saw a large animal floating on a log in the Mara River on the border between Kenya and Tanzania. It was spotted like a leopard, covered with scales, had a head like an otter's, and was about 16 feet long.

*Possible explanation:* A surviving saber-toothed cat, suggested by Bernard Heuvelmans. The scales might be explained by clumps of wet, shiny, matted fur.

*Sources:* Edgar Beecher Bronson, *In Closed Territory* (Chicago: A. C. McClurg, 1910), pp. 112–116, 130–136; C. W. Hobley, "Further Researches into Kikuyu and Kamba Religious Beliefs and Customs," *Journal of the Royal Anthropological Institute* 41 (1911): 406–422; C. W. Hobley, "On Some Unidentified Beasts," *Journal of the East Africa and Uganda Natural History Society,* no. 6 (1912): 48–52; John Alfred Jordan, "The Brontosaurus: Hunter's Story of Tusked and Scaly Beast," *Daily Mail* (London), December 16, 1919, p. 7; John Alfred Jordan, *The Elephant Stone* (London: Nicholas Kaye, 1959), pp. 78–81; Bernard Heuvelmans, *Les derniers dragons d'Afrique* (Paris: Plon, 1978), pp. 172–181, 374–377.

## DINOSAURS (Living)

Until the 1980s, there was ongoing controversy (occasionally reflected in cryptozoological literature) over whether dinosaurs had a single ancestor or many different ones. In the current view, it appears that Richard Owen had it right in

1842 when he invented the name Dinosauria ("terrible reptiles"), based only on three known fossil genera that he thought had one common ancestor. The defining characteristic of the Dinosauria is now considered to be (along with a few other minor skeletal characteristics of the femur, humerus, ankle, and foot) a ball-and-socket joint at the hip, like the mammals, that supports the body weight and allows for an erect, bipedal gait in certain types. As a group, they flourished for 160 million years, from the Late Triassic to the end of the Cretaceous (225–65 million years ago). The most primitive dinosaur yet found is the 3-foot-long *Eoraptor,* discovered in northwestern Argentina in 1991.

Not all huge fossil reptiles were dinosaurs. The flying pterosaurs, the marine plesiosaurs and mosasaurs, the diapsids *Tanystropheus* and *Champsosaurus,* the mammal-like therapsids—none of these are classed with the dinosaurs.

The Superorder Dinosauria is subdivided into two orders, the Saurischia and the Ornithischia.

The Saurischia included carnivorous, bipedal therapods such as *Tyrannosaurus* and the herbivorous, long-necked sauropodomorphs such as *Apatosaurus.* They had in common elongated necks, long second fingers, and skeletal cavities housing air-filled sacs connected to the lungs. It was this type of dinosaur that survived extinction at the end of the Cretaceous in the form of BIRDS.

The Ornithischia included dome-headed and horned cerapods (such as *Iguanodon* and *Triceratops*) and the armored thyreophorans (such as *Stegosaurus* and *Ankylosaurus*). They shared key characteristics of the jaws and teeth that enabled them to chew plants efficiently.

*Giganotosaurus* may have been the largest carnivorous animal that ever lived on land. A theropod dinosaur from the Early Cretaceous of Argentina that was first described in 1995, it was at least 42 feet from nose to tail tip. Vertebrae from a related species yet unnamed that was discovered in Patagonia in 2000 indicate an even greater length of 45 feet. The largest *Tyrannosaurus rex* was only 40 feet long.

At 110–120 feet, the herbivorous sauropod *Seismosaurus* of the Late Jurassic of New Mexico is the longest land vertebrate yet discovered,

weighing in at 33 tons. Its tail alone was about 50 feet, and its head and neck were nearly that length. The Cretaceous sauropod *Argentinosaurus* of Patagonia may also have attained this size, though it is only known from vertebrae and limb bones. In late 1999, some vertebrae from a possibly even larger sauropod were discovered in southern Patagonia; preliminary estimates gave it a length of 167 feet.

Different species of dinosaurs went extinct throughout the Mesozoic, not just at the end of the Cretaceous. For example, more time elapsed between the death of the last *Stegosaurus* and the hatching of the first *Tyrannosaurus* than between the extinction of the last dinosaur and the birth of the first modern human.

There is no unambiguous evidence for dinosaur fossils after the Cretaceous-Tertiary boundary. Dinosaur teeth mixed with mammalian bones in Paleocene deposits have been found in the Hell Creek Formation in Montana, but it's not clear whether they had originally come from earlier, dinosaur-bearing levels. Redeposition of older fossils into younger sediments by rivers or streams is not uncommon.

Surviving dinosaurs are not a zoological impossibility, especially in areas that have been geologically stable for the past 60 million years (such as Africa). Large dinosaurs that are cold-blooded (ectothermic) would have a better time surviving in hot, equatorial regions than warm-blooded (endothermic) animals with high metabolic rates. Ectotherms also require only 10 percent of the amount of food taken in by full endotherms. However, determining dinosaur energetics and thermal biology without living models is, at best, a speculative endeavor.

The two major types of African dinosaur in this section are the MOKELE-MBEMBE, which might be a surviving sauropod, and the EMELA-NTOUKA, which some think might be a ceratopsian survivor such as *Monoclonius.* Both are known by many different local names. The others in the list are much less documented.

**Mystery Dinosaurs**

EMELA-NTOUKA; MBIELU-MBIELU-MBIELU; MOKELE-MBEMBE; PARTRIDGE CREEK BEAST; ROW; SILWANE MANZI.

*See also* DRAGONS, LIZARDS (UNKNOWN), and the WATER LION.

*South American dinosaurs:* A few rumors of huge, amphibious beasts in South America are on record, but no local Indian names have surfaced.

In 1882, an odd, 40-foot saurian was killed on the Río Beni, El Beni Department, Bolivia. It was said to have two additional, doglike heads sprouting from its back, a long neck, and scaly armor. "A Bolivian Saurian," *Scientific American* 49 (1883): 3.

The explorer Percy Fawcett mentioned dinosaur-like animals briefly on several occasions as occurring in the Río Guaporé area on the border of Bolivia and Brazil, in the Madidi region of La Paz Department in northwestern Bolivia, and in swamps around the Rio Acre in Acre State, Brazil. Percy H. Fawcett, *Exploration Fawcett* (London: Hutchinson, 1955).

In late 1907, Franz Herrmann Schmidt and Rudolph Pfleng allegedly encountered an aquatic, dinosaur-like monster, 35 feet long, in a swampy area in the forested swamps of Loreto Department, Peru. It had a tapirlike head "the size of a beer keg," a snakelike neck, and heavy, clawed flippers. Their bullets seemed to have no effect on the animal. Franz Herrmann Schmidt, "Prehistoric Monsters in Jungles of the Amazon." *New York Herald,* January 11, 1911.

In 1931, Swedish explorer Harald Westin saw a 20-foot lizard walking along the shore of the Rio Mamoré on the border of Brazil and Bolivia. It had an alligator-like head, four legs, and a body like a distended boa constrictor. Harald Westin, *Tjugu års djungel- och tropikliv* (Stockholm: Bonnier, 1933).

Leonard Clark heard rumors of an animal resembling a sauropod dinosaur from Peruvian Indians around the Río Marañón, Peru, in 1946. Leonard Clark, *The Rivers Ran East* (New York: Funk and Wagnalls, 1953).

In 1975, a Swiss businessman hired a seventy-five-year-old guide named Sebastian Bastos, who told him that the Amazonian Indians knew of animals 18 feet long that overturn canoes and kill humans. Bastos himself had survived an attack several years earlier. *Liverpool Daily Post,* January 3, 1976.

*Artifacts:*

A gold figurine from Ashanti Province in Ghana, West Africa, and now located at the University of Pennsylvania Museum seems to depict a sauropod dinosaur. It was made as a trademark representing a particular family of gold dealers and resembles an *Apatosaurus* (bulky body, four legs, long tail), except for a relatively large head that looks more like a *Tyrannosaurus.* Some researchers see it as a representation of the MOKELE-MBEMBE. Margaret Plass, *African Miniatures: The Goldweights of the Ashanti* (London: Lund Humphries, 1967); "An Iguanodon from Dahomey," *Pursuit,* no. 9 (January 1970): 15–16; Bernard Heuvelmans, *Les derniers dragons d'Afrique* (Paris: Plon, 1978), pp. 336–337.

In October and November 1924, an expedition led by archaeologist Samuel Hubbard and paleontologist Charles W. Gilmore explored the Havasu Canyon area on the Havasupai Indian Reservation west of the Grand Canyon in northern Arizona. Near where the Tobocobe Trail intersects Lee Canyon, they discovered pictographs on the red sandstone along the trail, one of which seems to show a bipedal ornithopod dinosaur. Oakland Museum, *Discoveries Relating to Prehistoric Man by the Doheny Scientific Expedition in the Hava Supai Canyon, Northern Arizona* (San Francisco, Calif.: Sunset Press, 1927); A. Hyatt Verrill, *Strange Prehistoric Animals and Their Stories* (Boston: L. C. Page, 1948).

In July 1944, German merchant Waldemar Julsrud discovered a cache of clay and stone figurines depicting dinosaurs, weird animals, humans, masks, and vessels on El Toro hill near Acámbaro, Guanajuato State, Mexico. By the mid-1950s, he had found some 33,500 separate objects, which filled his twelve-room mansion and, it is said, forced him to sleep in the bathtub. The collection is no longer open to the public, and it is suspected that only a fraction of the original number of objects exist now. Though apparently seven distinct artistic styles are represented in the collection, none are typical of artifacts found elsewhere in Mesoamerica. Most, if not all, of the dinosaur-like figures are fanciful or composite animals, though some have seen resemblances to the sauropod *Brachiosaurus,* the ornithopod *Iguanodon,* and an *Ankylosaurus.* Other figures resemble such extinct Pleistocene fauna as CAMELOPS. Radiocarbon dates for the artifacts range from 4530–1110 B.C., though in some cases, laboratories have retracted these findings upon learning of their controversial nature, referring to suspected contamination or even "regenerated light signals." William N. Russell, "Did Man Tame the Dinosaur?" *Fate* 5 (February-March 1952): 20–27; Charles C. Di Peso, "The Clay Figurines of Acambaro, Guanajuato, Mexico," *American Antiquity* 18 (1953): 388–389; William N. Russell, "Report on Acambaro," *Fate* 6 (June 1953): 31–35; Ronald J. Willis, "The Acambaro Figurines," *INFO Journal,* no. 6 (Spring 1970): 2–17; "The Julsrud Ceramic Collection in Acambaro, Mexico," *Pursuit,* no. 22 (April 1973): 41–43; Charles H. Hapgood, *Mystery in Acambaro* (Winchester, N.H.: Charles H. Hapgood, 1973; Kempton, Ill.: Adventures Unlimited, 2000); Dennis Swift, *Dinosaurs of Acambaro,* http://www.omniology.com/3-Ceramic-Dinos.html.

In 1966, Peruvian physician Javier Cabrera obtained a rock on which was a picture of a fish, seemingly carved thousands of years ago. He found where it came from and eventually amassed a collection of thousands of volcanic rocks with pictures of dinosaurs, kangaroos, mastodons, winged humanoids, telescopes, open-heart surgery, and other fantastic images. Now housed in his Museo de Piedras Grabadas in Ocucaje, near Ica, Peru, Cabrera claims they were made 1 million–250,000 years ago by an unknown culture. Others have accused Cabrera of producing the stones himself or at least turning a blind eye to local forgers. Ryan Drum, "The Cabrera Rocks," *INFO Journal,* no. 17 (May 1976): 6–11; Javier Cabrera Darquea, *El mensaje de las piedras grabadas de Ica* (Lima, Peru: INTI-Sol, 1976); David Hatcher Childress, *Lost Cities and Ancient Mysteries of South America* (Stelle, Ill.: Adventures Unlimited, 1986), pp. 29–31, 48–52; Michael D. Swords, "The Cabrera Rocks Revisited," *INFO Journal,* no. 48 (March 1986): 11–13; Robert Todd Carroll, "Ica Stones," in *Skeptic's Dictionary,* http://skepdic.com/icastones.html.

## Djinni

LITTLE PEOPLE of Southeast Asia.

*Etymology:* Arabic (Semitic), possibly a borrowing from the Latin *genius* ("guardian spirit of a man or place"). Plural, *Djinn.*

*Variant names:* Cin (Turkish), Djihin, Djin (Djinniyah for the female), Genie (English), Jann (in Iran), Jinni.

*Physical description:* Body is composed of vapor or smokeless flame. In Malay folklore, Djinni is used as a polite equivalent for an evil spirit (*hantu*).

*Behavior:* Nocturnal. Intelligent. Capable of appearing in different forms, including a black cat, goat, BLACK DOG, duck, hen, buffalo, fox, snake, or human.

*Habitat:* Ruined houses, cisterns, rivers, wells, crossroads, markets.

*Distribution:* Malaysia; Indonesia; Iran; elsewhere in the Islamic world.

*Possible explanations:*

(1) In pre-Islamic Arabia, Djinn were elemental nymphs and spirits of the desert.

(2) In Islamic metaphysics, Djinn are supernatural beings.

(3) In popular culture, Djinni is a general folkloric name for many types of indigenous spirits.

*Sources:* B. Lewis, Ch. Pellat, and J. Schacht, eds., *Encyclopedia of Islam: New Edition* (Leiden, the Netherlands: E. J. Brill, 1965), vol. 2, pp. 546–550; Jorge Luis Borges, *The Book of Imaginary Beings* (New York: E. P. Dutton, 1969), pp. 133–134.

## Dobharchú

A large OTTERlike animal of Ireland that may be responsible for some PÉIST legends or sightings of FRESHWATER MONSTERS.

*Etymology:* Irish, "otter," though in County Leitrim, the word is said to connote a mythical "king otter." The Irish word *madra usice* ("water hound") is more commonly used for the European otter.

*Variant names:* Anchu, Dhuraghoo, Dorraghow, Doyarchu, King otter, Master otter, Water hound.

*Physical description:* Like an otter but said to

be about five times as large (perhaps 10–15 feet). White pelt. Black ear tips. Black cross on its back.

*Behavior:* Aggressive and dangerous. Call is a whistling sound. Can break a rock with its snout.

*Distribution:* County Leitrim, Ireland.

*Significant sighting:* Grace (or Grainne) Connolly is said to have been killed by a Dobharchú as she washed her clothes in Glenade Lough, County Leitrim, on September 24, 1722. A gravestone in Conwall Cemetery south of Kinlough portrays the creature as doglike, with a long tail, large paws, long neck, and short head.

*Possible explanation:* The European otter (*Lutra lutra*) can grow to a length of 4 feet 6 inches, including the tail. However, it is not particularly aggressive toward humans. Formerly common throughout Europe (including Ireland), it has been virtually eliminated through much of its range since the 1950s.

*Sources:* Roderick O'Flaherty, *A Chorographical Description of West or H-Iar Connaught, Written* A.D. *1684* (Dublin: Irish Archaeological Society, 1846), pp. 19–20; Patrick Tohall, "The Dobhar-chú Tombstones of Glenade, Co. Leitrim," *Journal of the Royal Society of Antiquaries of Ireland* 78 (1948): 127–129; Dáithí Ó hÓgáin, *Myth, Legend and Romance* (London: Ryan, 1990); Karl Shuker, "In the Spotlight: The Dobhar-chú," *Strange Magazine*, no. 16 (Fall 1995): 32–33, 49; Karl Shuker, *Mysteries of Planet Earth* (London: Carlton, 1999), pp. 172–173.

## Dobsegna

Doglike animal of Australasia that resembles a THYLACINE.

*Etymology:* Dani (Papuan) word.

*Physical description:* Light-brown fur. Strong mouth. Huge jaws. Head and shoulders like a dog. Stripes on the rear portion of its body. Thin tail nearly as long as its body.

*Behavior:* Nocturnal. Hunts in packs at dawn or dusk. Feeds on small marsupials, pigs, chickens, and birds.

*Habitat:* Caves and rocky areas in the highlands.

*Distribution:* Baliem Valley, the central mountains around Wamena, and the Gunung Lorentz National Park of Irian Jaya, Indonesia; Mount Giluwe, Papua New Guinea.

*Significant sightings:* In 1993, Ned Terry investigated reports that a Thylacine-like animal existed in the Baliem Valley; the local Dani people identified his photos of a Thylacine as the Dobsegna.

In March 1997, a supposed Thylacine attacked villagers' livestock in the Jayawijaya District of Irian Jaya.

*Possible explanation:* Surviving THYLACINE, which lived in New Guinea during the Pleistocene.

*Sources:* Albert S. Le Souëf and Harry Burrell, *The Wild Animals of Australasia* (London: G. G. Harrap, 1926), p. 332; Karl Shuker, "Thylacines in New Guinea?" *Fortean Times*, no. 108 (March 1998): 16; "More Tasmanian Tigers," *Cryptozoology Review* 2, no. 3 (Winter-Spring 1998): 5–6.

## Dodo

Flightless BIRD of the Solitaire family (Raphidae), thought extinct in Mauritius since at least 1690.

*Etymology:* From the Portuguese *doudo* ("fool" or "simpleton").

*Scientific name: Raphus cucullatus,* given by Carl von Linné in 1758.

*Variant name:* Dronte.

*Physical description:* Based on seventeenth-century accounts, this bulky bird had grayish plumage with whitish tail feathers. Length, about 3 feet 3 inches. Long, featherless, hooked beak. Its rudimentary wings were incapable of flight. Four toes, instead of the usual three.

*Behavior:* Clumsy. When running to escape capture, its body wobbled and its belly scraped the ground.

*Distribution:* Mauritius, in the Indian Ocean.

*Significant sightings:* Natives of Mauritius told Lawrence G. Green in the 1930s that Dodos still existed in caves and mountains on remote parts of the island.

Reports of Dodolike birds walking at dusk along the beach in the Plain Champagne area turned up in the early 1990s, enough to con-

*The DODO (Raphus cucullatus), a flightless bird thought extinct in Mauritius since at least 1690. From H. E. Strickland,* The Dodo and Its Kindred *(London: Reeve, Benham, and Reeve, 1848). (From the original in the Special Collections of Northwestern University Library)*

vince William J. Gibbons to mount an expedition to the island in 1997.

*Present status:* The last known sighting of a living Dodo was in 1662 by Volquard Iverson.

*Possible explanations:*
(1) Surviving Dodo, though this is extremely unlikely.
(2) Surviving unknown species of Solitaire (Raphidae) related to the Dodo.
(3) A Southern giant petrel (*Macronectes giganteus*), which averages 37 inches in length.
(4) Observations after 1662 may be the Mauritius red hen (*Aphanapteryx bonasia*), which was called the Dodo after the original bird was extinct.

*Sources:* Lawrence G. Green, *Secret Africa* (London: Stanley Paul, 1936); Masauji Hachisuka, *The Dodo and Kindred Birds* (London: H. F. and G. Witherby, 1953); Willy Ley, *Exotic Zoology* (New York: Viking, 1959), pp. 334–354; A. W. Diamond, ed., *Studies of Mascarene Island Birds* (New York: Columbia University Press, 1987); Karl Shuker, "How Dead Is the Dodo?" *Fate* 42 (May 1989): 62–69; Karl Shuker, "From Dodos to Dimetrodons," *Strange Magazine,* no. 19 (Spring 1998): 22; Errol Fuller, *Extinct Birds* (Ithaca, N.Y.: Cornell University Press, 2001), pp. 194–203.

## Doglas

Unrecognized big CAT hybrid of the Indian subcontinent.

*Etymology:* Unknown; possibly Hindi (Indo-Aryan), "double-faced."

*Physical description:* Length, 8 feet. Spotted, leopardlike head. Tigerlike neck ruff. Shoulders and body covered with tiger stripes that sometimes blur into rosettes. Grayish ground color. Leopardlike tail.

*Distribution:* India.

*Significant sighting:* An apparent leopard × tiger hybrid was killed by Frederick C. Hicks in 1910, who claimed that the locals were familiar with these animals.

*Possible explanation:* An interspecies Leopard (*Panthera pardus*) × Tiger (*P. tigris*) cross is unknown in the wild, but if it occurred, it would probably result in a sterile animal.

*Sources:* Frederick C. Hicks, *Forty Years among the Wild Animals of India from Mysore to the Himalayas* (Allahabad, India: Pioneer Press, 1910); Karl Shuker, *Mystery Cats of the World* (London: Robert Hale, 1989), pp. 111–112.

# DOGS (Unknown)

The Dog family (Canidae) arose from primitive carnivores in the Eocene, about 35 million years ago, in North America. Common characteristics include elongated jaws, long legs relative to body size, five toes on the front feet and four toes on the hind feet, nonretractile claws, and an omnivorous diet. Most species are uniform in coloration, with special markings usually confined to the head and the tip of the tail. In size, canids range from the Fennec fox (*Fennecus zerda*) that weighs about 3 pounds to the Gray wolf (*Canis lupus*) that weighs up to 175 pounds.

The earliest canids were the hesperocyonines of North America, small- to medium-sized predators of the Late Eocene, 35 million years ago. They were replaced by the borophagines, a group that ranged in size from foxes to lions and was dominant from the Miocene through the Pleistocene, 25–1.5 million years ago. An early true dog (*Canis davisi*) moved across the Bering land bridge to Asia and Europe in the Miocene. About the size of a coyote, it ultimately gave rise to the foxes and wolves that colonized all of Eurasia. From 2 million to 700,000 years ago, wolves, coyotes, and foxes moved back into North America from Asia. All domestic dog breeds are descended from the gray wolf, which was apparently domesticated at different times and places as early as 12,000 years ago.

Of the twenty canids in this list, only a handful could represent new or surviving species (perhaps the HUNGARIAN REEDWOLF or the WAHEELA); most will likely turn out to be color morphs, deformed individuals, or misidentifications of known animals. Some, such as the ALIEN BIG DOG or the PHANTOM WOLF, are undoubtedly multicausal.

## Mystery Dogs

ADJULÉ; ALIEN BIG DOG; ANDEAN WOLF; BEAST OF GÉVAUDAN; CHAGLJEVI; HORNED JACKAL; HUNGARIAN REEDWOLF; ITZCUINTLIPOTZOTLI; MITLA; MLULARUKA; PHANTOM WOLF; RED WOLF; al-SALAAWA; SHAMANU; SHUNKA WARAK'IN; WAHEELA; WALDAGI; WALRUS DOG; WOLF DEER; YOKYN

## Doko

SMALL HOMINID of East Africa.

*Etymology:* Swahili (Bantu), "small." Similar to the ancient Egyptian Dongo; *see* PYGMY (CLASSICAL). In some places, it has the meaning "ignorant or stupid."

*Physical description:* Height, 4 feet. Olive-bronze skin. Straight head-hair. Flat nose. Small eyes. Thick lips.

*Behavior:* Eats fruit, roots, mice, snakes, ants, and honey. Worships a superior being called Yer. Taken as a slave by neighboring tribes.

*Distribution:* East of Lake Turkana, Kenya.

*Present status:* Possibly corresponds with one of the Kenyan cryptids, designated as hominid X5, described by Jacqueline Roumeguère-Eberhardt in 1990.

*Sources:* William Cornwallis Harris, *The Highlands of Æthiopia* (London: Longman, Brown, Green, and Longmans, 1844), vol. 3, pp. 63–66; Ludwig Krapf, *Travels, Researches and Missionary Labours, during an Eighteen Years' Residence in Eastern Africa* (London: Trübner, 1860), pp. 171–172, 302; Jacqueline Roumeguère-Eberhardt, *Les hominidés non-identifiés des forêts d'Afrique: Dossier X* (Paris: Robert Laffont, 1990).

## Domenech's Pseudo-Goat

An odd, goatlike HOOFED MAMMAL seen once in Texas.

*Physical description:* Size of a cat. White, glossy fur. Rose-colored, goatlike horns. Claws instead of hooves.

*Distribution:* Central Texas.

*Significant sighting:* Around 1850 in Fredericksburg, Texas, the French missionary Emmanuel Domenech talked to an American officer who told him that a Comanche woman kept one of these animals and that they were found wild in the woods.

*Possible explanation:* If described accurately, a taxonomic anomaly. An unusual array of ungulate mammals evolved in the Cenozoic of South America, which was an island during much of that era. Among them was a family of notoungulates known as homalodotheriids, robust forest browsers with clawlike toes instead of hooves that lived in Argentina until about 9 million years ago. None of them, however, had true horns; only the distantly related toxodonts developed dermal horns like those of rhinos. The combination of horns and claws makes it nearly impossible to find a match for this animal in the fossil record.

*Sources:* Emmanuel Domenech, *Missionary Adventures in Texas and Mexico: A Personal Narrative of Six Years' Sojourn in Those Regions* (London: Longman, Brown, Green, Longmans, and Roberts, 1858), pp. 122–123; Karl Shuker, "A Rose-Horned, Snow-Furred, Claw-Footed Controversy," *Fate* 45 (April 1992): 59–60.

## Dorsal Finner

A category of SEA MONSTER identified by Gary Mangiacopra.

*Physical description:* Length, 70–100 feet. Diameter, 9–15 feet. Smooth skin, with occasional scales or warts. Dull green to dark brown in color. Yellow, shading to lighter, on the underside. Froglike or alligator-shaped head, 15 feet long. Large green or red eyes. Jaw, 5 feet long, with 6-inch-long teeth. Round neck. One pair of frontal flippers. Large fin standing straight up on the back.

*Behavior:* Swims rapidly by vertical undulations. Churns up the water. Possibly attracted to ships.

*Distribution:* Atlantic coast of the United States.

*Significant sighting:* Captain Delory of the sloop *Mary Lane* sighted a huge, alligator-like head sticking out of the water southwest of Point Judith, Rhode Island, on August 4, 1888. He estimated its length as 70 feet as it passed close by, its bright green scales glistening.

*Present status:* Mangiacopra identified five sightings from 1878 to 1888, then seems to have dropped this category.

*Possible explanation:* Some type of unknown whale.

*Sources:* "The Sea Serpent Once More," *New York Times,* August 7, 1888, p. 5; Gary S. Mangiacopra, "The Great Unknowns of the 19th Century," *Of Sea and Shore* 8, no. 3 (Fall 1977): 175–178.

## Double-Banded Argus

Mystery BIRD of Southeast Asia.

*Scientific name: Argusianus bipunctatus,* proposed by T. W. Wood in 1871. Placed in the same genus as the Great argus of Indonesia.

*Physical description:* Has two reddish-brown bands with white dots on its primary feather, instead of one band.

*Distribution:* Java in Indonesia or Tioman Island, Johor State, Malaysia.

*Significant sighting:* Pheasant of uncertain origin, known from a single feather in the British Museum of Natural History.

*Possible explanations:*

(1) Surviving fossil pheasant of some type.

(2) An aberrant form of the Great argus (*Argusianus argus*) which lives in peninsular Malaysia and Sumatra and Borneo in Indonesia.

*Sources:* T. W. Wood, "*Argus bipunctatus,* sp. n., Described from a Single Feather," *Annals and Magazine of Natural History,* ser. 4, 8 (1871): 67–68; G. W. H. Davison, "Notes on the Extinct *Argusianus bipunctatus,*" *Bulletin of the British Ornithologists Club* 103, no. 3 (1983): 86–88; Errol Fuller, *Extinct Birds* (Ithaca, N.Y.: Cornell University Press, 2001), p. 111.

## Dover Demon

Bizarre humanoid ENTITY of Massachusetts.

*Etymology:* Name given by Loren Coleman and picked up by local newspapers.

*Physical description:* Height, 3 feet 6 inches–4 feet. Hairless, peach-colored, sandpapery skin. Large, watermelon-shaped head. Eyes, orange or green, shining, round, and lidless. No nose,

*The DOVER DEMON of April 21, 1977. From a painting by Bill Bartlett. (Loren Coleman)*

ears, or mouth. Thin neck. Spindly arms and legs. Long fingers and hands.

*Behavior:* Bipedal.

*Distribution:* Dover, Massachusetts.

*Significant sightings:* At 10:30 P.M. on April 21, 1977, Bill Bartlett was driving through Dover, Massachusetts, with two friends when he saw a strange, thin creature with glowing eyes and a large head crawling along some rocks on the far side of the road. It was visible only for a few seconds, but it terrified Bartlett. Over the next twenty-four hours, there were two other sightings by local teens: About two hours later, John Baxter saw a humanoid creature run away from him down a wooded gully, and Will Taintor and Abby Brabham spotted an orange-eyed, monkeylike apparition crouching on all fours on the night of April 22.

*Possible explanations:*

(1) A hoax by the teenagers, though the initial investigators (Loren Coleman and Walter Webb) thought the youngsters were sincere.

(2) An escaped laboratory monkey, though none was reported lost.

(3) A Red fox (*Vulpes fulva*) that had lost its hair due to sarcoptic mange.

(4) A newborn horse.

(5) A yearling Moose (*Alces alces*), suggested by Martin Kottmeyer. There is a permanent breeding population of 200–300 moose in Massachusetts, mostly west of the Connecticut River, though some have wandered close to Boston. In 1996, a young moose cow crashed through some fences around row houses near Boston College. A calf would make an unexpected sight in the darkness; however, some of the characteristics do not match, and moose were not as populous in 1977 as they are now.

(6) A MERBEING, according to Mark A. Hall, though it was seen in the woods.

(7) A paranormal apparition or unidentified flying object (UFO) entity.

*Sources:* Jerome Clark, "The Dover Humanoid," *Fate* 31 (March 1978): 50–55; Joseph A. Citro, *Passing Strange* (New York: Houghton Mifflin, 1997), pp. 136–147; Martin Kottmeyer, "Demon Moose," *The Anomalist* 6 (1998): 104–110; Loren Coleman, *Mysterious America,* rev. ed. (New York: Paraview Press, 2001), pp. 42–61.

# Dragon (Asian)

SEMIMYTHICAL BEAST of East Asia. One of the four sacred animals of Chinese mythology.

*Variant names:* Chèn (Mandarin Chinese/Sino-Tibetan), Chi lung ("wingless dragon"), Chi'ih, Féi-yu, Fu-ts'ang lung ("treasure dragon"), Jiao lung ("scaly dragon"), Kiao lung, Kioh lung, Kura-mitsu-ha (Japanese, "dark water snake"), Kura-okami (Japanese, "dragon god of the valleys"), Kura-yama-sumi (Japanese, "lord of the dark mountains"), Lóng, Long-ma (Vietnamese), Lung ("five-clawed dragon"), Lung wang ("dragon king"), Mang ("four-clawed dragon"), NAGA, Qiu lung ("horned dragon"), Riong (Korean/Altaic), Riu (Japanese), Shen lung ("spiritual dragon"), T'ao t'ieh ("glutton"), Tatsu (Japanese), Ti lung ("river dragon"), T'ien lung ("celestial dragon"), Ying lung ("winged dragon"), Yu lung ("fish dragon").

*Physical description:* A huge body with both serpentine and crocodilian characteristics. Has 117 fishlike scales. Straight horns like a deer's, through which it can hear. Flat, long head like a camel's. Has a bladderlike swelling on the top of its head. Bearded. Eyes like a rabbit's. Ears like a cow's. Tongue and neck like a snake's. The male has a luminous pearl concealed under its chin by a fold of skin. Long mane. Wings seen only in mature specimens. Belly like a frog's. Four feet, with claws like a hawk's. Footpads like a tiger's. Chinese dragons have four or five toes; Japanese dragons only have three.

*Behavior:* Can fly without wings. Has the ability to change forms. Sometimes guards treasure. Lays a brightly colored, gemlike egg. Said to have a 3,000-year growth cycle in which it first looks like a water snake, grows a carp's head and scales, develops four limbs and a long tail, sprouts a pair of horns, and finally grows wings. A benevolent creature symbolizing authority, strength, experience, wisdom, and goodness. Originally the Chinese rain god, the Dragon was associated with the Chinese emperor, ancestor worship, fertility, and pools.

*Habitat:* Wells, rivers, lakes (in China); the ocean (in Japan).

*Distribution:* China; Japan; Korea; Indonesia.

*Significant sightings:* The oldest known image of a Chinese dragon is a rock painting dating from 8000 B.C. that was found in 1993 on a cliff in southwestern Shanxi Province.

In the fourth millennium B.C., a Dragon delivered the eight mystic triagrams, Hae Pa Kua, to a legendary emperor.

The Northern Song emperor Huizong in A.D. 1110 classified all Dragons into five families—Blue Spirit Dragons, very compassionate kings; Red Spirit Dragons, the kings of lakes; Yellow Spirit Dragons, kings who receive vows favorably; White Spirit Dragons, virtuous and pure

DRAGON *sculpture in the Botanical Gardens, Saigon, Vietnam, 1930s. (From a postcard in the author's collection)*

kings; and Black Spirit Dragons, the kings of mysterious lakes.

Another official classification of Dragons divided them into Spirit Dragons that fly into heaven and Earthly Dragons that protect treasure or hide in the earth.

The Russian monk Elder Barsanuphius served with a nursing detachment during the Russo-Japanese War. Some Chinese soldiers told him that in 1902, when they were stationed at a post in the mountains 40 miles from Muling, Heilongjiang Province, they saw a winged Dragon creep out from a cave on several occasions.

*Possible explanations:*

(1) The Chinese alligator (*Alligator sinensis*) may have been the prototype for the legendary Dragon, according to Richard Carrington. Now restricted to the lower Yangtze River Valley in Anui Province, China, it may have had a much wider range in eastern China in historical times. It prefers slow-moving, freshwater rivers, streams, and swamps. Reports of individuals 9 feet long exist in Chinese historical records, but today, the animal does not exceed 6 feet. It is the most endangered of all crocodilians, thanks to rampant habitat destruction. Chinese apothecaries have traditionally sold dried alligator parts as remedies derived from Dragons.

(2) SEA MONSTERS seen sometimes in the Gulf of Tonkin would be regarded as Dragons.

(3) Dinosaur fossils in numerous areas of China and Mongolia have probably contributed to Dragon mythology. Chinese Dragon eggs in apothecary shops often turned out to be dinosaur or fossil ostrich eggs from Mongolia.

(4) Some Dragon legends may have been inspired by fossil elephants or mammoths.

(5) Monitor lizards, especially the Komodo dragon (*Varanus komodoensis*), found in Indonesia, which grows to 10 feet 6 inches long, may have inspired Dragon mythology. The largest known monitor was *Megalania prisca,* a 15- to 21-foot lizard that lived in central Australia in the Pliocene and Pleistocene (2 million–20,000 years ago). Other monitor species currently unknown to science may also have contributed to Dragon lore.

(6) Carl Sagan suggested that Dragon legends may stem from primal memories of dinosaurs passed on to us from our mammalian ancestors who were their contemporaries.

*Sources:* Nicholas Belfield Dennys, *The Folk-Lore of China* (London: Trübner, 1876), pp. 102–111; Charles Gould, *Mythical Monsters* (London: W. H. Allen, 1886), pp. 212–259; M. W. de Visser, *The Dragon in China and Japan* (Amsterdam: Johannes Müller, 1913); J. O'Matley Irwin, "Is the Chinese Dragon Based on Fact, Not Mythology?" *Scientific American* 114 (1916): 399, 410; L. Newton Hayes, *The Chinese Dragon* (Shanghai, China: Commercial Press, 1922); Ernest Ingersoll, *Dragons and Dragon Lore* (New York: Payson and Clarke, 1928); L. C. Hopkins, "The Dragon Terrestrial and the Dragon Celestial," *Journal of the Royal Asiatic Society,* 1931, pp. 791–806, and 1932, pp. 91–97; B. Gokan, "Historical Review of Discussions on the Fossil Elephants Found in Japan in the Late Yedo Period," *Chishitsugaku zasshi* 45 (1938): 773–776; Maria Leach, *Funk and Wagnalls Standard Dictionary of Folklore, Mythology and Legend* (New York: Funk and Wagnalls, 1949–1950), vol. 1, p. 323; Martin Birnbaum, "Chinese Dragons and the Bay de Halong," *Western Folklore* 11 (1952): 32–37; Richard Carrington, *Mermaids and Mastodons* (London: Chatto and Windus, 1957); Frank James Daniels, "Snake and Dragon Lore of Japan," *Folklore* 71 (1960): 145–164; Jorge Luis Borges, *The Book of Imaginary Beings* (New York: E. P. Dutton, 1969), pp. 64–66, 82–84; Carl Sagan, *The Dragons of Eden* (New York: Random House, 1977); Donald A. Mackenzie, *Myths of China and Japan* (New York: Gramercy, 1994); Karl Shuker, *Dragons: A Natural History* (New York: Simon and Schuster, 1995), pp. 86–93; Victor Afanasiev, *Elder Barsanuphius of Optina* (Platina, Calif.: St. Herman of Alaska Brotherhood, 2000).

## Dragon (British)

Snakelike monster of the British Isles; *see* SEMI-MYTHICAL BEASTS.

*Variant names:* Amphiptere, Knucker (from Old English *Nicor*), NYKUR, WELSH WINGED SNAKE, Worm, WYVERN.

*Physical description:* Serpentine. Slimy body. Black, red, yellow, or white. Red eyes. Forked tongue. Sharp teeth. Sometimes winged. Sometimes with two or four legs, other times limbless.

*Behavior:* Basks in the sun. Can fly. Spits venom. Can rejoin or regenerate severed body parts. Breathes fire. Drinks large quantities of milk. Eats livestock. Kills by crushing or strangulation. Eats humans, especially girls. Guards treasure.

*Habitat:* Rivers, pools, hills, forests, caves.

*Distribution:* England, Scotland, Wales, and the Channel Islands. A partial list of places where British dragons have been reported follows:

*Anglesey, Wales*—Penmynedd.

*Angus, Scotland*—Kirkton of Strathmartine.

*Barnsley*—Warncliff [Wantley] Lodge (near Wortley).

*Borders, Scotland*—Linton Hill.

*Buckinghamshire*—Hughenden.

*Cheshire*—Bache Pool (near Moston), Grimesditch Brook (near Lower Whitley).

*Derbyshire*—Winlatter Rock.

*Devon*—Dolbury Hill (Exe River).

*Dorset*—Kingston.

*Durham*—Bishop Auckland, Lambton Castle (River Wear), Sockburn Manor (Tees River).

*Essex*—Henham.

*Gloucestershire*—Deerhurst.

*Hampshire*—Dragon Field (near Bisterne).

*Herefordshire*—Brinsop, Mordiford, Wormbridge.

*Hertfordshire*—Brent Pelham.

*Highland, Scotland*—Ben Vair.

*Jersey, Channel Islands*—Five Oaks.

*Lincolnshire*—Anwick, Castle Carlton.

*Norfolk*—Ludham.

*North Yorkshire*—Filey, Loschy Hill (near Nunnington), Scaw Wood (near Handale), Sexhow, Slingsby.

*Northumberland*—Bamburgh Castle, Longwitton, Spindleston Hough.

*Oxfordshire*—Dragon Hill, Uffington.

*Powys, Wales*—Llandeilo Graban, Llanrhaiadr-ym-Mochnant.

*Somerset*—Aller, Carhampton, Churchstanton, Kingston St. Mary, Norton Fitzwarren, Shervage Wood (near Crowcombe).

*Suffolk*—Bures Saint Mary.

*Sussex*—Bignor Hill, Knucker Hole (near Lyminster), St. Leonard's Forest (near Horsham).

*Significant sightings:* St. George (a knight of Cappadocia in Turkey) was said to have killed a Dragon in a pond near Silene (possibly Shahhat or Suluntah in Cyrenaica or Zlitan in Tripolitania), Libya. The citizens of the town were sacrificing teenage girls to the monster in order to keep it from killing everyone and devastating the countryside. When it was the turn of the king's daughter, an itinerant knight named George stuck the Dragon with his lance. The girl then led it through the town where George killed it. Afterward, the townsfolk became Christian. The legend may have originated in sixth-century North African folktales or in the Caucasus Mountains of Georgia, but St. George was adopted as an Anglo-Saxon Christian hero in England. The tale circulated widely during the Middle Ages, eventually becoming a somewhat erotic romance. The Dragon was seen as a symbol of paganism or evil.

The Lambton Worm was a loathsome Dragon that surfaced in the River Wear, Durham, in the fourteenth century. Lord Lambton caught it on his fishing line but threw it down a nearby well when he realized what it was. For the next few years, the creature grew in size and began to terrorize the locals, consuming livestock and killing any would-be slayers. The villagers had to pacify it by keeping a trough filled with milk for it to drink. Lambton himself finally killed it but only because he had protected himself with a spike-studded coat of mail. A piece of the Dragon's hide and the milk trough were still on exhibit at the castle in the nineteenth century.

In the early fifteenth century, Sir Maurice de Berkeley is said to have killed a scaly, fire-breathing Dragon at Dragon Field near Bisterne, Hampshire.

Sir Thomas Venables is said to have shot and killed a Dragon just as it was about to eat a child in Bache Pool, near Moston, Cheshire, in the sixteenth century. A 1632 carving in the church vestry shows the crest of the Venables as a Dragon swallowing a child.

A scaly Dragon—9 feet long, black on top, reddish below, and with a white ring around its neck—was roaming St. Leonard's Forest, near Horsham, Sussex, in August 1614. It could run as fast as a man on its four feet, and it killed but did not eat several cattle, two dogs, and two people on different occasions. The animal left behind a slimy trail and spat venom.

A flying Dragon 8–9 feet long with two rows of sharp teeth and a pair of wings was seen near Henham, Essex, beginning on May 27 and 28, 1669. It was observed basking in the sun by several people, but when they returned with guns and pitchforks, it darted into Birch Wood.

*Possible explanations:*

(1) Physical characteristics borrowed from a vague knowledge of pythons, cobras, and crocodiles.

(2) An evolved *Kuehneosaurus,* a 2-foot-long, winged reptile that lived in England in the Late Triassic, 200 million years ago, proposed by Mark A. Hall. Though known fossil forms were only capable of gliding flight, Hall suggests that by the Middle Ages, it may have grown in size and developed true flight.

(3) Windsock banners used by armies to identify specific military units. There was a whistling device attached to the silk banner that made hissing noises as the banner was waved vigorously. A lighted torch was also placed in the mouth of the banner. The custom probably originated in China, but the Romans picked it up during various wars with the Persians, Scythians, and Dacians. A Dragon was the standard of a Roman cohort (one-tenth of a legion). After the Romans left Britain, the Britons and Saxons adopted the custom for their own armies. After the Battle of Hastings in 1066, the Dragon standard was adopted by the Normans and was used throughout the Hundred Years' War. The national flag of Wales is a red Dragon, Y Ddraig Goch.

(4) A completely mythical animal used in moralistic tales.

(5) A symbolic expression of the raids of the Vikings, whose longboats featured brightly painted Dragon figureheads.

(6) Tales constructed to explain monuments, carvings, and heraldic devices that depicted Dragons; alternatively, place-names that referred to them.

(7) Legends that underscore the uniqueness of a community whose lord of the manor is portrayed as a Dragon slayer or whose local farm lad has outwitted and killed a monster.

(8) The Dragon is seen by Paul Devereux and others as a symbol for the unusual forces and energies associated with sacred sites in the British landscape. These earth energies are centered on megalithic structures such as Stonehenge and are channeled into invisible streams that coincide with "leys," or alignments of roads, trackways, standing stones, and other landmarks.

*Sources:* Jacobus de Viragine, *The Golden Legend of Jacobus de Voragine* [1265], ed. Frederick S. Ellis (Hammersmith, England: Kelm Scott, 1892), vol. 1, pp. 454–455; *The Flying Serpent, or, Strange News out of Essex* (London: Peter Lillicrap, 1669?); Samuel Rudder, *A New History of Gloucestershire* (Cirencester, England: Samuel Rudder, 1779), pp. 402–403; *True and Wonderfull: A Discourse Relating a Strange and Monstrous Serpent, or Dragon, Lately Discovered and Yet Living to the Great Annoyance and Divers Slaughters Both Men and Cattel, by His Strong and Violent Poyson; in Sussex, Two Miles from Horsam, in a Woode Called St. Leonards Forrest, and Thirtie Miles from London, This Present Month of August, 1614,* in *The Harleian Miscellany* (London: Robert Dutton, 1809), vol. 3, pp. 227–231; William Eastmead, *Historia Rievallensis* (Thirsk, England: R. Peat, 1824); James Dacres Devlin, *Helps to Hereford History, Civil and Legendary* (London: J. R. Smith, 1848); William Henderson, *Notes on the Folk Lore of the Northern Counties of England and the Borders* (London: Longmans, Green, 1866), pp. 245–247; Egerton Leigh, *Ballads & Legends of Cheshire* (London: Longmans, 1867), pp. 223–227; J. O. Halliwell, "The Serpent of St. Leonard's Forest," *Sussex Archaeological Collections* 19 (1867): 190–191; Llewellyn Jewitt, "The Dragon of Wantley and the Family of Moore," *Reliquary,* new ser. 18 (1878):

193–202; H. A. Heaton, "St. George and the Dragon," *Antiquary* 35 (1899): 113–118; Cornelia Steketee Hulst, *St. George of Cappadocia in Legend and History* (London: David Nutt, 1909), pp. 12–39; John Francis Campbell, *The Celtic Dragon Myth* (Edinburgh: John Grant, 1911); H. R. Ellis Davidson, "The Hill of the Dragon," *Folklore* 61 (1950): 169–185; Richard Carrington, *Mermaids and Mastodons* (New York: Rinehart, 1957), pp. 64–77; Gwyn Williams, *Green Mountain, an Informal Guide to Cyrenaica and its Jebel Akhdar* (London: Faber and Faber, 1963); Ruth L. Tongue, *Somerset Folklore* (London: Folk-Lore Society, 1965), pp. 79, 129–131; Rosemary Dickens, *Dragon Legend of Burley Beacon and Bisterne* (Salisbury, England: Rosemary Dickens, n.d.); Whitall N. Perry, "The Dragon That Swallowed St. George," *Studies in Comparative Religion* 10 (Summer 1976): 136–172; Janet and Colin Bord, *The Secret Country* (New York: Walker, 1976), pp. 69–88; Paul Screeton, *The Lambton Worm and Other Northumbrian Dragon Legends* (London: Zodiac House, 1978); Jacqueline Simpson, "Fifty British Dragon Tales: An Analysis," *Folklore* 89 (1978): 79–93; Paul Newman, *The Hill of the Dragon* (Totowa, N.J.: Rowman and Littlefield, 1979); Peter J. Hogarth and Val Clery, *Dragons* (London: Allen Lane, 1979); Ralph Whitlock, *Here Be Dragons* (London: Allen and Unwin, 1983); Clive Harper, *The Hughenden Dragon* (High Wycombe, England: Torsdag, 1985); Mark A. Hall, *Natural Mysteries,* 2d ed. (Minneapolis, Minn.: Mark A. Hall, 1991), pp. 43–50; Carl Lofmark, *A History of the Red Dragon,* ed. G. A. Wells (Llanrwst, Wales: Gwasg Carreg Gwalch, 1995); Karl Shuker, *Dragons: A Natural History* (New York: Simon and Schuster, 1995), pp. 12–15, 58–63; Gordon Rutter, "The Lambton Worm: A Cryptozoological Folklore Story from the Past," *Cryptozoology Review* 3, no. 2 (Autumn 1998): 29–31; Dragoncrafts, http://www.dragoncrafts.co.uk.

## Dragon (European)

Snakelike animal of Europe; *see* SEMIMYTHICAL BEASTS.

# DE DRACONE.

RACO vocabulum eſt Græcum à Latinis vſurpatum, quod aliquando pro quouis ſerpente ſumitur præcipuè apud Græcos & Poëtas, ſiquidem eadem videtur vtriuſque nominis apud Græcos ratio. Draconem enim παρὰ τὸ δλέρκεϑαι, id eſt, à cernendo dictum volunt, ſicut ὄφιϲ ἀπὸ τᴕ̈ ὀπἴϲιν quod idem ſignificat; Propriè tamen draco dicitur de ſerpente

DRAGONS, *both winged and wingless. From Konrad Gesner,* Historiae animalium *(Zürich, Switz.: Christ. Froschoverum, 1551–1587). (From the original in the Special Collections of Northwestern University Library)*

*Etymology:* From the Greek *drákon* ("serpent" or "sea fish") or, more literally, "that which kills at a glance."

*Variant names:* Draco (Latin), Dragonet, Drake, Firedrake, GARGOUILLE, LINDORM, Lindwurm, Peluda, Python, TARASQUE, VOUIVRE.

*Physical description:* Serpentine. Scaly or slimy. Black, red, yellow, or white. Crest on the head. Red eyes. Small mouth. Lion's limbs. Sometimes winged. Eagle's claws. Strong tail.

*Behavior:* Leaves a putrid slime behind when it moves on land. Said to be capable of flight. Extremely venomous, toxic, or contagious. Inflicts injury with its tail. Can also kill by constriction. Said to herald the beginning of wars or other disasters. Causes floods.

*Distribution:* Throughout Europe.

*Significant sightings:* To the ancient Greeks, the Dragon was a large snake found near tombs.

In 714, the Basque hero Don Teodosio killed a bat-winged Dragon on Mount Aralar, Spain.

In either 1410 or 1420, a man was lost in a cave on Mount Pilatus, Switzerland, for five months. It was the lair of two flying Dragons, and he escaped by grabbing the tail of one as it flew away.

Ulrich Vogelsang, who sculpted the winged Dragon of Klagenfurt, Austria, in 1590, based his design on the skull of a Pleistocene Woolly rhinoceros (*Coelodonta antiquitatis*) dug up in a nearby quarry in 1335. The legend of a Lindwurm that caused floods in the River Glan is much older, however.

On July 26, 1713, a giant serpent, 17 feet 4 inches long, was killed by a forester named Zander near Wroclaw, Poland.

*Possible explanations:*

(1) Windsock banners carried by medieval armies. At the Battle of Liegnitz in 1241, Kaidu's Mongol army carried Dragon banners that flamed and fumed. *See also* DRAGON (BRITISH).

(2) Viking or Byzantine ships in the shape of Dragons may have popularized the myth.

(3) Such astronomical events as comets or meteors were thought to be flying Dragons.

(4) A union of the more disagreeable aspects of the Egyptian gods Isis, Osiris, and Horus, with a large amount of snake, crocodile, and lizard mixed in, suggested by Grafton Elliot Smith. From the Nile, this Dragon prototype spread north, east, and west, where it was transformed and assimilated by other cultures, symbolizing the personification of evil.

(5) Based on dinosaur or pterodactyl fossils.

(6) In the seventeenth century, the Olm (*Proteus anguineus*), a cave-dwelling, aquatic salamander of Yugoslavia and northern Italy, was thought to be the offspring of a Dragon. It has an eel-like body, white skin, three pairs of external gills, four tiny legs, and vestigial eyes. It grows to about 12 inches long. When washed out of their caves by heavy rainfall, Olms gather in deep pools, but they will not voluntarily leave the water.

*Sources:* "Hymn to Apollo," in *Hesiod, the Homeric Hymns, and the Homerica,* trans. Hugh G. Evelyn-White (Cambridge, Mass.: Harvard University Press, 1982), pp. 339–351; Ulisse Aldrovandi, *Serpentum, et draconu historiae libri duo* (Bologna, Italy: C. Ferronium, 1640); Athanasius Kircher, *Mundus subterraneus* (Amsterdam: Joannem Janssonium, 1668); Johann Jakob Scheuchzer, *Helvetica* (Leiden, the Netherlands: Petri Vander Aa, 1723); Johann Heinrich Zedler, *Grosses Vollständiges Universal-Lexikon aller Wissenschafften und Kunste* (Halle, Germany: J. H. Zedler, 1732–1750), vol. 34, pp. 1793–1796; Grafton Elliot Smith, *The Evolution of the Dragon* (New York: Longmans, Green, 1919); Ernest Ingersoll, *Dragons and Dragon Lore* (New York: Payson and Clarke, 1928); Wilhelm Bölsche, *Drachen: Sage und Naturwissenschaft* (Stuttgart, Germany: Kosmos, 1929); Othenio Abel, *Das Reich der Tiere: Tiere der Vorzeit in ihrem Lebensraum* (Berlin: Deutscher Verlag, 1939), pp. 82–83; Ludwig Bechstein, *Märchen und Sagen* (Berlin: T. Knaur, 1940), p. 209; Joseph Fontenrose, *Python: A Study of the Delphic Myth and Its Origins* (Berkeley: University of California Press, 1959); Sidney Bernard, "Swiss Terrors of the Past," *Contemporary Review* 209 (1966): 293–295; Erich Thenius, *Fossils and the Life of the Past* (New York: Springer-Verlag, 1973), pp. 37–38; Julio Carlo Baroja, *Ritos y mitos equívocos* (Madrid: Ediciones Istmo, 1974), pp. 167, 205; Paul Norman, *The Hill of the Dragon* (Totowa, N.J.: Rowman and Littlefield, 1980); Michel Meurger, *Histoire naturelle des dragons* (Rennes, France: Terre de Brume, 2001).

## Dragon Bird

Legendary BIRD of East Asia.

*Variant names:* Hai riyo (Japanese), O-gon-cho, Schachi hoko, Tobi tatsu.

*Physical description:* Golden feathers on wings, body, and tail. Dragonlike head. Bearded. Clawed feet.

*Behavior:* Call is a blood-curdling howl.

*Distribution:* Japan.

*Significant sighting:* Every fifty years, in a lake near Kyoto, a white Dragon called Ukisima is said to take the form of a golden songbird called O-gon-cho. Last seen in April 1834, its appearance presaged disease and starvation.

*Possible explanation:* Said to be the Japanese equivalent of the winged stage of the Chinese DRAGON (Ying lung).

*Sources:* Charles Gould, *Mythical Monsters* (London: W. H. Allen, 1886), pp. 249–255; Karl Shuker, *Dragons: A Natural History* (New York: Simon and Schuster, 1995), pp. 92–93.

## Dre-Mo

Mystery PRIMATE or BEAR of Central Asia, often confused with the YETI.

*Etymology:* Tibetan (Sino-Tibetan) word, apparently with various meanings, among them: a female demon, a person who has gone astray from a religious life, a she-bear, and the red and blue varieties of the brown bear.

*Variant names:* Chemo ("big"), Chemong, Dredmo ("brown bear"), Dremo.

*Physical description:* Looks like a bear or large monkey. Taller than a human. Shaggy reddish, black, or dark-gray hair. Sometimes white head-hair. Small eyes. Pointed mouth.

*Behavior:* Nocturnal. Walks on all fours as well as bipedally. Growls and whistles. Omnivorous. Looks for food under large rocks. Throws rocks. Kills with its hands (or paws).

*Distribution:* Eastern Tibet; Bhutan.

*Significant sighting:* Somewhere southwest of Alamdo, Tibet, in July 1986, Reinhold Messner encountered a large, dark-haired animal that emerged from rhododendron bushes onto the path about 30 feet ahead of him. It rose on its hind legs, turned, and ran away on all fours. Local Tibetans told him it was a Chemo.

*Possible explanations:*

(1) The Brown bear (*Ursus arctos*), especially the isabelline or red variety found in the eastern and central Himalayas, is known in the Karakoram Range of Baltistan, Pakistan, as the *dreng mo;* to the Ladakhs in Jammu and Kashmir as *drin mor;* and in Tibet as the *dred mong.* Considered by some a subspecies (*U. a. isabellinus*), the red bear is generally 5 feet 6 inches–8 feet long, with a reddish, grizzled coat. It eats grasses, roots, and scavenged kills such as ibex.

(2) The blue or horse variety of brown bear, sometimes considered a subspecies (*U. a. pruinosus*), is found in eastern Tibet and Sichuan Province, China. Its blue-tinted brown hairs are tipped with gold or slate-gray. A yellowish-brown or whitish cape forms a saddle mark over its shoulders, hence the name "horse bear."

(3) The Chemo may refer to the YETI or DZU-TEH, while the Dre-mo is a bear.

*Sources:* Edmund Hillary and Desmond Doig, *High in the Cold Thin Air* (Garden City, N.Y.: Doubleday, 1962), pp. 100–101, 119–123; Odette Tchernine, *The Yeti* (London: Neville Spearman, 1970), p. 175; Terry Domico, *Bears of the World* (New York: Facts on File, 1988); Reinhold Messner, *My Quest for the Yeti* (New York: St. Martin's, 2000).


## Dsonoqua

CANNIBAL GIANT of western Canada.

*Etymology:* Kwakiutl (Wakashan), "wild woman of the woods."

*Variant name:* Tsonoqua.

*Physical description:* Covered with long, black hair. Long arms. Hairy hands. Sharp claws. Short hind legs.

*Behavior:* Antisocial. Upright gait. Most often described as a cannibal woman.

*Distribution:* Southwestern British Columbia.

*Significant sighting:* Represented on carved, wooden masks used for ritual purposes. Its face also appears on totem-pole carvings.

*Sources:* Franz Boas and George Hunt, "Kwakiutl Texts," *Memoirs of the American Museum of Natural History* 5 (1902): 431–436; Franz Boas, "Kwakiutl Tales, New Series," *Contributions to Anthropology, Columbia University* 26 (1935): 147–156; Joseph H. Wherry, *Indian Masks and Myths of the West* (New York: Funk and Wagnalls, 1969), pp. 114–121; Grant R. Keddie, "On Creating Un-humans," in Vladimir Markotic and Grover Krantz, eds., *The Sasquatch and Other Unknown Hominoids* (Calgary, Alta., Canada: Western Publishers, 1984), pp. 22–29.


## Du

Mystery BIRD of Oceania.

*Etymology:* Ajië (Austronesian) word.

*Physical description:* Red plumage. Star-shaped, bony structure on its head.

*Behavior:* Flightless. Can run swiftly with wings outstretched. Aggressive. Its single egg hatches in four months. Said to put its egg in a banyan tree to lure a giant lizard into incubating it.

*Distribution:* Isle of Pines, New Caledonia, in the South Pacific.

*Significant sighting:* On both New Caledonia and the Isle of Pines, there are large, moundlike structures, as much as 50 feet in diameter and 5 feet high. François Poplin suggests that they are not burial mounds but piles of earth constructed by the Du to incubate its eggs.

*Present status:* At first treated as sacred by local people, the bird may have become a handy food source that was eventually exterminated, perhaps by A.D. 300. However, Lars Thomas reported that it was still thought to be alive in 1991.

*Possible explanation:* Surviving Giant megapode (*Sylviornis neocaledoniae*), a large galliform bird whose subfossil bones were first discovered on the Isle of Pines in 1974. They were radiocarbon-dated to about 1500 B.C., after the Melanesians settled the island. The bird was 5–6 feet tall and flightless.

*Sources:* Paul Griscelli, "Deux oiseaux fossiles de Nouvelle-Calédonie," *Bulletin de la Société d'Études Historiques de Nouvelle-Calédonie* 29 (1976): 3–6; François Poplin and Cécile Mourer-Chauviré, "*Sylviornis neocaledoniae* (Aves, Galliformes, Megapodiidae), oiseau géant éteint de l'Île des Pins (Nouvelle-Calédonie)," *Géobios* 18 (February 1985): 73–97; Cécile Mourer-Chauviré and François Poplin, "Le mystère des tumulus de Nouvelle-Calédonie," *La Recherche* 16 (September 1985): 1094; Lars Thomas, *Mysteriet om Havuhyrerne* (Copenhagen: Gyldendal Boghandel, 1992).


## Duende

LITTLE PEOPLE of Central and South America.

*Etymology:* Spanish, "goblin" or "dwarf." From *dueno de casa* ("lord of the house"), referring to a Spanish household spirit. Used as early as 1653 for a bandit in Peru; since then, the term has expanded to include ghosts and other supernatural creatures.

*Variant names:* Alar (Cabécar/Chibchan), Dominguito (in Honduras), el Duendi, Duenos del monte ("mountain lords"), Dwendi, Mauh (Chortí/Mayan, "not good"), Pombero, el Silborcito (in Brazil, "little whistler"), el Sombrero'n ("big hat"), Tata (Mayan for "grandfather") duende.

*Physical description:* Height, 1 foot 3 inches–4 feet 6 inches. Covered in thick brown or black hair. Red fur or hair (in Honduras, Peru, and Venezuela). Blond, gray, or red hair (in Panama). Flat, yellowish-brown, wrinkled face. Blue eyes (in Panama). Pointed ears (in Costa Rica). Large teeth. Long, white beard (in Guatemala). Heavy shoulders. Hair especially thick and coarse down the neck and back. Long arms. Chubby (in Colombia, Peru, and Argentina). Thick calves (in Belize). Chickenlike feet (in Argentina and Costa Rica). Reversed feet. Pointed heels. Female Duendes are rare.

*Behavior:* Mostly nocturnal. Inquisitive. Makes cries like a baby as well as loud roars and also chatters, squeaks, or cackles. Eats fruit, molasses, livestock, and fishes. Attacks dogs and carries them off. Plaits the manes of horses (in Colombia). Said to wear skins, rags, red or green clothes, and especially a big straw hat. Sometimes rescues humans lost in the forest. Folklore credits the Duende with a facility for language, making music, hypnotic powers, invisibility, and shape-shifting.

*Tracks:* Small and deep, with pointed heels.

*Habitat:* Caves, mines, mountainous forests, deep canyons and valleys, rivers, abandoned houses, plantations, vineyards.

*Distribution:* Throughout Central and South America.

*Possible explanation:* A fairy-tale creature with no objective reality, possibly a mix of European folktales and Indian trickster myths.

*Sources:* Alberto Uribe Holguín, *La leyenda de los duendes* (Bogotá, Colombia: Editorial Marconi, 1927); Aimé F. Tschiffely, *Tschiffely's Ride: Ten Thousand Miles in the Saddle from Southern Cross to Pole Star* (New York: Simon and Schuster, 1933), p. 182; Charles Wisdom, *The Chortí Indians of Guatemala* (Chicago: University of Chicago Press, 1940), p. 408; Carlos López Narvaez, "Presentación folklórica del duende," *Revista de Folklore* (Bogotá) 2 (1947): 1–5; Ivan T. Sanderson, *Abominable Snowmen: Legend Come to Life* (Philadelphia: Chilton, 1961), pp. 164–166; Virginia Rodriguez Rivera, "Los duendes en Mexico (el alux)," *Folklore Americano* 10, no. 10 (1962): 68–85; Nicholas M. Fintzelberg, "The Form,

Meaning, and Function of a Duende Legend in the Santa Elena Peninsula, Ecuador," Ph.D. diss., University of California, Davis, 1975; Luis Millones, "Las duendes del casma: Religion popular en un valle de la Costa Norte," *Folklore Americano* 23 (1975): 81–92; Alan Rabinowitz, *Jaguar: Struggle and Triumph in the Jungles of Belize* (New York: Arbor House, 1986); Meg Craig, *Characters and Caricatures in Belizean Folklore* (Belize City: Belize UNESCO Commission, 1991); Mark Sanborne, "An Investigation of the Duende and Sisimite of Belize: Hominoids or Myth?" *Cryptozoology* 11 (1992): 90–97; Mark Sanborne, "On the Trail of the Duende and Sisimite of Belize," *Strange Magazine*, no. 11 (Spring-Summer 1993): 10–13, 54–57; John E. Roth, *American Elves* (Jefferson, N.C.: McFarland, 1997), pp. 34–36, 54–62, 97–104, 156–160.

## Dulugal

Alternate name for the YOWIE of Australia.

*Etymology:* From the Dhurga or Thurawal (Australian) *duligaal* ("wild blackfellow").

*Variant names:* Dhuligal, Doolagard, Doolagarl, Dooligal, Douligah, Dulagarl, Thoolagal.

*Physical description:* Covered with hair.

*Behavior:* Nocturnal. Raids Aboriginal camps.

*Habitat:* Mountain ranges.

*Distribution:* Southern coast of New South Wales.

*Significant sighting:* In June 1970, near Geehi, New South Wales, mountaineers Ron Bartlett and Frank Sinclair spotted a 7-foot-tall human-like figure after finding odd tracks in the snow.

*Sources:* Te Whare [Henry V. Edwards], *A Bush Cinema Made in Australia* (Sydney, Australia: Te Whare, 1922), p. 8; Roland Robertson, *Black-Feller White-Feller* (Sydney, Australia: Angus and Robertson, 1958); Graham Joyner, *The Hairy Man of South Eastern Australia* (Kingston, A.C.T., Australia: Graham Joyner, 1977).

## Dwayyo

HAIRY BIPED of Maryland.

*Etymology:* Letters forming this word originated with a police teletype message accompanying a November 1965 report. Coined by reporter George May of the *Frederick (Md.) News Post.*

*Variant names:* Dwayo, Wago.

*Physical description:* Height, 6 feet. Black hair. Bushy tail.

*Behavior:* Runs on four legs. Screams like a puma.

*Distribution:* Frederick County, Maryland.

*Significant sighting:* John Becker claimed to have fought a hairy, black creature in his backyard on Fern Rock Road, 10 miles out of Frederick, Maryland, in late November 1965.

*Possible explanations:* Probable hoax.

*Sources: Frederick (Md.) News Post,* November 30, December 2–3, 6, 8, 15, 1965; Mark Chorvinsky and Mark Opsasnick, "Notes on the Dwayyo," *Strange Magazine,* no. 2 (1988): 28–29.

## Dzoavits

CANNIBAL GIANT of the western United States.

*Etymology:* Shoshoni (Uto-Aztecan), "stone giant."

*Distribution:* Wyoming; Idaho; Nevada.

*Source:* Kyle Mizokami, Bigfoot-Like Figures in North American Folklore and Tradition, http://www.rain.org/campinternet/bigfoot/bigfoot-folklore.html.

## Dzu-Teh

GIANT HOMINID or unknown BEAR of Central Asia, often confused with the smaller YETI.

*Etymology:* Lepcha (Sino-Tibetan) word. Said to be pronounced "chu-tay." Meaning and origin not established, though one derivation is *dzu* ("livestock") + *teh* ("animal"). Another is that *teh* is the same as *dred* ("bear"). In modern Tibetan, *te* is a particle attached to a verb and means "when," "after," "thus," or "although" and sometimes forms a gerund ("-ing").

*Variant names:* Chhudi (in Sikkim), Churails, Chu-teh, Chutey.

*Physical description:* Bearlike but bigger. Height, 6–9 feet. Shaggy reddish, black, or dark-gray hair. Flat head. Pronounced browridge. Long, powerful arms. Huge hands.

*Behavior:* Walks on all fours as well as bipedally. Kills and eats yaks and cattle by catching them by their horns and twisting their necks. Said by the Sherpas to be seen at altitudes of 13,000–15,000 feet.

*Tracks:* Huge and human-looking.

*Distribution:* Sikkim State, India; Bhutan; Tibet.

*Possible explanations:*

(1) The red or isabelline variety of the Brown bear (*Ursus arctos isabellinus*) has a pale, reddish-brown coat and stands around 6 feet 6 inches tall. It is found in Alpine meadows between the tree line and the snow line. A rarer blue variety (*U. a. pruinosus,* with bluish-brown hairs frosted with gold or slate-gray) is also known, especially in Tibet; skins of this bear obtained in Nepal in 1959 and 1960 by journalist Desmond Doig were touted as YETI skins, but there is considerable doubt that the locals made any such claim.

(2) An evolved *Gigantopithecus blacki.* This huge-jawed Pleistocene ape lived as recently as 500,000 years ago in southern China and Vietnam, while a smaller species, *G. giganteus,* dates to 9–6 million years ago in the Siwalik Hills of India and Pakistan. Both species are known only from jaw fragments and isolated teeth.

*Sources:* Ralph Izzard, *The Abominable Snowman* (Garden City, N.Y.: Doubleday, 1955), p. 100; Ivan T. Sanderson, *Abominable Snowmen: Legend Come to Life* (Philadelphia: Chilton, 1961), pp. 268, 325; Edmund Hillary and Desmond Doig, *High in the Cold Thin Air* (Garden City, N.Y.: Doubleday, 1962), pp. 31, 101, 117; Odette Tchernine, *The Yeti* (London: Neville Spearman, 1970), p. 176; Loren Coleman, *Tom Slick and the Search for the Yeti* (Boston: Faber and Faber, 1989), pp. 97–98.

# E

## Ea

The earliest known named MERBEING, from the Middle East.

*Etymology:* Akkadian (Semitic), "he who does good to men."

*Variant names:* Dagon (Hebrew/Semitic), Enki (Sumerian), Oannes (Greek).

*Physical description:* Human or goatlike above the waist, fish tail below. Sometimes shown as completely human with a fishlike cloak or with waves springing from its shoulders.

*Behavior:* Stays in the water at night, comes on land in the day. Speaks and acts like a human being.

*Distribution:* Red Sea.

*Significant sighting:* The fish-tailed sea god of the Akkadians, Ea is said to have provided them with technological and agricultural skills and the beginnings of their culture. According to the Babylonian writer Berosus, Ea came out of the Red Sea and taught things by day, retiring to the sea at night.

*Sources:* "Fragments of Chaldaean History," in Isaac Preston Cory, ed., *Ancient Fragments* (London: W. Pickering, 1832), pp. 18–19, 27–29; Gwen Benwell and Arthur Waugh, *Sea Enchantress: The Tale of the Mermaid and Her Kin* (New York: Citadel, 1961), pp. 23–28.

## Earth Hound

Mystery RODENT of Scotland.

*Variant names:* Yard dog, Yird swine.

*Physical description:* Ratlike. The size of a ferret. Brown. Long, doglike head. Prominent snout like a pig's. Large incisors. Molelike feet. Short, bushy tail.

*Behavior:* Burrows in graves. Eats corpses.

*Habitat:* Graveyards and alluvial plains.

The EARTH HOUND, a mystery mammal of northeastern Scotland, said to be found in graveyards. (William M. Rebsamen/Fortean Picture Library)

*Distribution:* Aberdeenshire.

*Significant sightings:* About 1867, a Scottish gardener plowed up an Earth hound, killing it after it bit and cut his boot. He took the carcass home.

In 1915, an Earth hound was turned up by a plow and killed in the parish churchyard of Mastrick, Aberdeenshire. It was about the size of a rat but had molelike feet.

*Sources:* Walter Gregor, *Notes on the Folk-Lore of the North-East of Scotland* (London: Folk-Lore Society, 1881), p. 130; Alexander Fenton and David Heppell, "The Earth Hound: A Living Banffshire Belief," *Scottish Studies* 31 (1992–1993): 145–146; Karl Shuker, *Mysteries of Planet Earth* (London: Carlton, 1999), pp. 28–29.

## Eastern Puma

The puma is indeed returning to its former range in the eastern United States and Canada, though this fact has only been reluctantly accepted by naturalists in recent years. Part of the reluctance has been based on imprecise witness descriptions, which can sometimes venture into the bizarre or supernatural. (Reports from southern states technically constitute sightings of the Southern puma, which still exists in very small numbers in southern Florida.)

Although a case could be made for separating Eastern puma reports into two categories—those conforming to a traditional puma that is expanding its range and those involving a melanistic or other aberrant animal turning up in areas that probably could not support a large cat—the task is too daunting. Black animals are reported from likely Eastern puma habitats, and perfectly reasonable puma reports come from areas where the animal has never been seen before or since. Misidentifications are rampant, and ALIEN BIG CATS (whatever they may be) seem to have a presence in North America. Therefore, all the reports are lumped together in this category except for the MANED AMERICAN LIONS, which at least has an easily identifiable nonpuma characteristic.

*Scientific names:* *Puma concolor couguar,* since 1993; formerly, *Felis concolor,* a name given by Carl von Linné in 1771.

*Variant names:* BEAST OF BLADENBORO, BOOGER, Catamount, Catawampus, Critter, Devil cat, Eastern panther, Gallywampus, Ghost cat, GLAWACKUS, Indian devil, Mansfield mystery cat (in Massachusetts), Michichibi, Montie the Monster, NELLIE THE LION, OZARK HOWLER, Phantom panther, SANTER, VARMINT, Wampus cat, Whirling whimpus, Whistling wampus, Woofin nanny, Wooleneag, Wowzer, Yati wasagi (Mikasuki/Muskogean, "separated man"), Zoominzacker (in North Carolina).

*Physical description:* Powerfully built large cat. Length, 7–8 feet including tail. Shoulder height, 2–3 feet. Weight, more than 200 pounds. Most reports are of tan pumas, though about 16–30 percent describe a gray or black pelt. Head described as both large and small. Short, pricked ears. Glowing greenish, yellow, or red eyes. Long, slender tail.

*Behavior:* Nocturnal. A female in heat calls with loud screams. When its natural prey (deer) has been reduced, it will attack livestock, especially chickens and rabbits, though goats, sheep, pigs, dogs, cats, and cows are also vulnerable. Sometimes, only parts of animals are favored, such as pigs' ears. Pumas tend to avoid people, although attacks have increased since 1990; consequently, it is difficult to reconcile stories of fearless, intelligently aggressive behavior toward persons that are occasionally reported, from midwestern states especially. At least two or three reports involve an animal that can stand on its hind legs, a feat that no puma can perform.

*Tracks:* Front feet are larger than hind feet and are ahead of or partially overlapped by the rear feet. Length, 3.5–6 inches wide, 3–5 inches long. Heel pads have squared-off fronts and three lobes at the rear. Toes are small, teardrop-shaped, and widely spaced compared to a dog's. Rear feet are asymmetrical. Prints are 25–30 inches apart. Claw marks are sometimes reported; though this is more characteristic of a dog, big cat tracks will show claws in certain terrain or when the animal is sprinting or leaping.

*Habitat:* Mountains, forests, swamps.

*Distribution:* Southeastern Canada and the United States east of the Mississippi River. Black pumas have also been reported in Washington, Texas, and California. In some places in the United States and Canada, puma sightings correlate closely with BIGFOOT "hot spots."

A partial list of places where Eastern pumas have been reported follows:

*Alabama*—South Mobile County, Nauvoo, Tuscaloosa.

*Arkansas*—Logan County, Mena, Russellville.

*Connecticut*—Chaplin, Glastonbury, and places in the northeastern portion of the state.

*Delaware*—Concord, Harrington, Wilmington.

*Georgia*—Bulloch County, Savannah, Stockbridge.

*Illinois*—Alexander County, Centralia, Champaign County, Clarksdale, Decatur, East Carondelet, Edwardsville, Forest Park, Hampton, Itasca, Jasper County, Kaskaskia, Mahomet, Momence, Olive Branch, Oquawka, Pana, Peoria, Plainfield, and many places in the northeastern portion of the state.

*Indiana*—Hancock County, Knox County,

*The Puma (*Puma concolor*). (© 2002 ArtToday.com, Inc., an IMSI Company)*

Lebanon, Monument City, Paradise, Perry County, Richmond, Rising Sun, South Bend.

*Kentucky*—Floyd County, Russellville.

*Louisiana*—St. Mary Parish, Vidalia.

*Maine*—Baxter State Park, Blue Hill Mountain, Cape Elizabeth, Fryeburg, Hartland, Little St. John Lake, Waldo County, Westport Island.

*Maryland*—Clinton, Frostburg, Garrett County, Harford County, Street.

*Massachusetts*—Hockomock Swamp, Mansfield, Shutesbury, Truro.

*Michigan*—Canton Township, Cass County, Clare County, Oakland County, Perronville, Seul Choix Point, Sturgis, the Upper Peninsula from Watersmeet to Drummond Island.

*Minnesota*—Bemidji, Hopkins, Hugo, Plymouth, St. Louis County, Watonwan County.

*Mississippi*—Bay Springs.

*Missouri*—Lamar, Mound Creek, Maries County, Phelps County, Pulaski County, Wellsville.

*New Brunswick, Canada*—Albert County, Fredericton, Juniper, Mundleville, St. John County, Queens County, Waasis.

*New Hampshire*—Benton, Stewartstown.

*New Jersey*—Cumberland County, Maurice River, Salem County, Sussex County.

*New York*—Brookhaven, Eden, Elmira, Ronkonkoma, Spencer, Van Etten.

*North Carolina*—Bladenboro, Concord, Fontana Dam, Greensboro, Rowan County, Sampson County, and places in the northwestern portion of the state.

*Ohio*—Allen County, Bluffton, Cincinnati (forested areas), Coshocton County, Kirkwood, Minerva, Oak Harbor, Richard Township, Springfield, Urbana, Wellston, Westerville.

*Oklahoma*—Arkoma, Verdigris.

*Ontario, Canada*—Algoma, Bruce Peninsula, Marathon, Orient Bay, Saugeen River.

*Pennsylvania*—Allegheny County, Armstrong County, Cameron County, Clarion County, Clearfield County, Clinton County, Crawford County, Erie County, Forest County,

Lycoming County, Pottstown, Schuylkill County, Sullivan County, Tarentum.

*South Carolina*—Charleston area, Georgetown County, Santee River, White Oak Swamp.

*Tennessee*—Carthage, Crossville, Indian Mound.

*Vermont*—Berlin, Bethel, Bridport, Craftsbury, Orwell, Rutland.

*Virginia*—Abingdon, Bedford County, Prince William County, Purgatory Mountain, Wise County.

*West Virginia*—Hardy County, Pocahontas County, Randolph County, Wyoming County.

*Wisconsin*—Lincoln County, Manitowoc, Oneida County, Rhinelander, Sauk County.

Black pumas in western states and Central America:

*California*—East Bay area, Las Trampas Regional Park, Marin County, Ventura County.

*Honduras*—Puerto Castillo.

*Mexico*—Sinaloa State.

*Texas*—Fort Worth.

*Washington*—Port Angeles.

*Significant sightings:* Reports of mystery felines prior to the 1950s were not taken seriously by zoologists. Stories of the GLAWACKUS, NELLIE THE LION, and Wampus cats in the South were collected primarily by folklorists and Forteans.

Marian Harpan Peduzzi saw a glossy, black panther in 1946 near Berlin, Vermont. It was 4 feet in length, with an elegant, curved tail.

On March 29, 1947, Bruce S. Wright discovered three sets of unmistakable puma tracks (two adults and one cub) on the border between Albert and St. John Counties, New Brunswick. These were the first puma tracks recorded in eastern Canada in more than 100 years.

Game warden Paul G. Myers shot and wounded a black cat near Decatur, Illinois, on October 25, 1955.

Walter Bigelow and his wife saw a strange animal cross the road near Mound Creek, Missouri, in the path of their car headlights in mid-July 1957. It was black, 3 feet long, tailless, and "rather stubby." Later, whatever the animal was, it scared some hunting dogs that tried to flush it.

On June 2, 1963, Bill Chambers watched a jet-black puma for fifteen minutes from his pickup truck near Mahomet, Illinois. It was hunting in a clover patch 190 yards away. He estimated its shoulder height as 14–15 inches and its total length as 4 feet 6 inches to 5 feet.

In the summer of 1966, a hairy, catlike animal locals called the Woofin nanny killed a number of animals and pets near Greensboro, North Carolina, bleeding the carcasses dry through puncture wounds.

Bruce S. Wright and his wife were driving west from Fredericton, New Brunswick, on September 28, 1966, when they saw a puma cross the road in front of them in broad daylight.

On April 10, 1970, Mike Busby was stopped by the side of the road south of Olive Branch, Illinois, when an animal with glowing, greenish eyes, 6 feet tall, black, and standing upright, attacked him. Tumbling him about, the creature tore his shirt and inflicted some scratches on his arm, chest, and abdomen. It was scared by a passing truck and loped away.

In September 1975, citizens of Stockbridge, Georgia, reported a black panther that screamed at night. After the newspaper stories broke, James Rutledge revealed that he had shot and killed a black cat the previous spring, but he declined to reveal where he had buried it.

In April 1976, a large male cougar was shot and killed in Pocahontas County, West Virginia, after it had killed a farmer's sheep. Two days later, an apparently pregnant female was captured alive. The state's Department of Natural Resources no longer has the paperwork on the case.

From April to June 1977, Sampson County, North Carolina, was plagued by a mystery animal that mangled pet cats and dogs, damaged trailer homes and porch screens, and left numerous clawless, four-toed tracks in the vicinity.

Charles and Helen Marks found more than 200 prints, some with claws, around their trailer court in Westerville, Ohio, on June 10, 1979. There were sightings of panthers in the area in May and June.

In late September 1981, William and Marsha Medeiros got within 50 feet of a puma along a trail in the Cape Cod National Seashore near Truro, Massachusetts.

On June 1, 1982, a Pittsburgh television crew

filmed a thirty-second videotape of a tan puma near Tarentum, Pennsylvania, on Ruth O'Brien's property, where a series of sightings and puma screams had been reported since July 1979.

On April 24, 1989, Hubert Graham watched a tawny, juvenile puma sunning itself for twenty minutes in a clearing below his fire-watch tower on Blue Hill Mountain, south of Bangor, Maine.

A videotape of a puma was taken near Waasis, New Brunswick, in the spring of 1990 by Roger Noble.

In August 1992, a couple in Street, Maryland, watched for twenty minutes and took photographs of a light-brown puma the size of a German shepherd dog. The cat was seen by others as it wandered east through Harford County.

A 3-foot, white puma was seen in the winter of 1992–1993 around Stewardstown, New Hampshire.

In December 1993, Wayne Perri of Hartland, Maine, was walking his dogs when he encountered a puma near Decker Pond. He took a photo of the animal, which shows it accompanied by two of his hounds.

Near Craftsbury, Vermont, in the winter of 1994–1995, game wardens found tracks, scat, and other physical evidence that produced a DNA match with a puma.

In June 1997, a small female puma was hit by a truck in western Floyd County, Kentucky. The witness said the animal was following a larger cat with another small cat. He picked up the carcass and turned it over to the Kentucky Department of Fish and Wildlife Resources, which kept the carcass in a freezer. It was determined to be an 8-pound puma kitten with all its claws intact and no tags or collars on it.

On July 15, 2000, on railroad tracks near Fort Kaskaskia State Historic Site, Illinois, a 110-pound, male puma carcass was discovered, killed by a train. Necropsy results showed that the animal had all its claws and had been feeding on white-tailed deer, indicating that it was wild and not an escapee.

*Present status:* The eastern subspecies (*Puma concolor couguar*) once ranged from New Brunswick south to South Carolina and west to

Illinois. Because of persistent yet unconfirmed reports, the U.S. Fish and Wildlife Service (USFWS) added the Eastern puma to the endangered species list in 1973. Puma sign found in 1981 convinced Robert Downing of the USFWS that the animal survived in Virginia and West Virginia. However, federal and state agencies have avoided expending scarce conservation resources on an animal still presumed extinct. In March 1993, based on tracks, hair, and fecal samples collected near Juniper in November 1992, the New Brunswick Department of Natural Resources acknowledged the presence of a puma population in the province.

The southern subspecies (*P. c. coryi*) formerly ranged from Georgia and Florida west to Arkansas and Louisiana; it is now estimated that only thirty to fifty adults exist in small pockets of southern Florida. A plan for genetic restoration of the remaining animals began in 1995 with the release of eight female Texas pumas (*P. c. stanleyana*) into south Florida.

In eastern and southern states, both pumas and the deer they fed on were greatly reduced in numbers in the early nineteenth century as white settlements advanced into the Appalachians. Deer did not become extensively stocked and protected again until the establishment of state and national parks from the 1930s to the 1950s. There is evidence that a few pumas survived the critical period between 1900 and 1930 and thus might be responsible for increasing populations in the East.

Many recent witnesses have reported seeing Eastern pumas at close range, but few have produced supporting evidence. The few specimens reported killed have not been preserved.

Pumas in the western United States are thriving. Arizona, California, Colorado, Idaho, Montana, Nevada, New Mexico, Oregon, Texas, Utah, Washington, and Wyoming have healthy populations, as do British Columbia and Alberta in Canada. Reports in adjacent states and provinces occur less frequently and may represent transients.

*Black pumas*—The leopard and jaguar are the two cats with the most frequently reported instances of melanism (black coloration). Melanism is virtually unknown in pumas, with only three questionable specimens killed in

Brazil (in 1843), Gunnison, Colorado (in 1912), and Costa Rica (in 1959) and unconfirmed rumors from Nicaragua, Panama, and Argentina. Normal puma coloration is either tawny or silver-gray, with no gradations in between. Its rain forest coloration is sometimes a dark red-brown. If a recessive gene for melanism were present in the North American puma population, it would likely have turned up more frequently in the wild and in captive breeding populations. Frederick Boyle, *A Ride across a Continent: A Personal Narrative of Wanderings through Nicaragua and Costa Rica* (London: R. Bentley, 1868); William Thompson, *Great Cats I Have Met* (Boston: Alpha, 1896); Angel Cabrera and José Yepes, *Historia natural ediar* (Buenos Aires: Compañía Argentina de Editores, 1940); Jim Bob Tinsley, *The Puma: Legendary Lion of the Americas* (El Paso: University of Texas at El Paso, 1987).

*Possible explanations:*

(1) A black Labrador retriever may sometimes be mistaken for a melanistic cat. Other large dog breeds, seen at night from a distance, might also be taken for a large cat.

(2) A black feral Domestic cat (*Felis silvestris catus*) has been mistaken for a mystery felid when the witness has unintentionially exaggerated its size.

(3) The Bobcat (*Lynx rufus*) is found within the Eastern puma's range but is considerably smaller (25–30 inches in length) and has a short tail. Bobcat melanism is rare but known in Florida.

(4) A number of other animals can make a noise like a puma's scream, among them a feral domestic cat, bobcat, Gray fox (*Urocyon cinereoargenteus*), and Eastern screech-owl (*Otus asio*).

(5) Tracks most commonly mistaken for pumas are made by a Domestic dog (*Canis familiaris*), bobcat, and Black bear (*Ursus americanus*).

(6) Deer kills in the wild are mostly inconclusive as evidence, since both released pet pumas and wild bobcats can kill, drag, and cover adult deer in the same way that a puma does.

(7) Pumas mark their paths frequently by scraping up a patch of dirt with their hind feet and urinating or spraying on it. Other animals make similar scratch hills, among them the Wild boar (*Sus scrofa*), Collared peccary (*Tayassu tajacu*), bobcat, Jaguar (*Panthera onca*), Ruffed grouse (*Bonasa umbellus*), Wild turkey (*Meleagris gallopavo*), black bear, dog, Squirrel (*Sciurus* spp.), Skunk (*Mephitis* spp.), and fox.

(8) An escaped circus animal is often offered as an explanation, but few correlations between puma sightings and escape incidents have been documented.

(9) An escaped or released exotic pet, especially a melanistic Leopard (*Panthera pardus*), is a possibility, although how a single, large, probably declawed animal that has lost its hunting skills could persist in the wild for very long without getting caught poses a problem.

(10) Puma pelts that have been intentionally dyed black as a hoax are not unknown.

(11) The return of pumas to the East may be the result of the persistence of the original *Puma concolor couguar* subspecies, migration of western or Florida subspecies, or individuals or groups released into the wild at different times and places.

As early as 1959, Canadian researcher Bruce S. Wright came to believe that pumas were still present in New Brunswick and in almost every eastern state from the Canadian border to Florida, but the animals had become scarce, cautious, and primarily nocturnal. By 1972, he had documented 304 solid reports from eastern Canada and 44 scattered sightings from Maine to Alabama.

Naturalist Helen Gerson studied 318 reports from Ontario received by the provincial Ministry of Natural Resources between 1935 and 1983. More than half were logged as "probable." Only 9 percent involved black specimens.

John and Linda Lutz, in *Eastern Puma Network News,* based in Baltimore, Maryland, recorded 615 reports in the United States from 1983 to 1989, with the greatest numbers by far in Maryland (135),

Pennsylvania (131), and West Virginia (113). Of these reports, 44 percent involved multiple witnesses, 27 percent were observations by hunters, and 37 percent involved black specimens. The Eastern Puma Research Network logged 567 sightings in 1991, 435 sightings in 1993 (over half of them in Pennsylvania), 245 sightings in 1994, 510 sightings in 1995, and 397 sightings in 1999. The percentage of melanistic individuals seems to be falling, from 31 percent in 1990 to 16 percent in 1999.

Todd Lester's Eastern Cougar Foundation, based in North Spring, West Virginia, logged 673 sightings from 1995 to 1999. The greatest number of melanistic pumas were in West Virginia (122), North Carolina (17), and Virginia (16).

The Lutzes ranked the top ten states with the most number of reports as of January 2001 as follows: *Tan pumas*—Pennsylvania (920), New York (442), Maryland (361), West Virginia (330), Virginia (180), Michigan (158), New Jersey (128), Maine (124), Illinois (121), and Ohio (118). *Black pumas*—Pennsylvania (282), New York (146), Wisconsin (98), Maryland (86), West Virginia (76), New Jersey (61), Illinois (47), Michigan (46), Virginia (35), and Tennessee (30).

(12) Bruce Wright has suggested that melanism may have evolved in isolated Eastern puma populations to increase elusiveness and ensure survival.

(13) Chad Arment points out that the evidence for black pumas seems extremely sparse before the 1940s, and he suggests that prior to that time, an unknown group of sport hunters introduced a group of melanistic Leopards (*Panthera pardus*) from European zoos into the Appalachians or Ozarks.

(14) Loren Coleman believes that returning Eastern pumas cannot account for all American mystery felids, especially those with black coloration and/or aggressive behavior patterns. He suggests that the black cats are surviving female American lions (*Panthera atrox*), a Pleistocene lion that died out 9,000 years ago, while the males are reported as MANED AMERICAN LIONS.

(15) Errant or escaped mustelids such as the Fisher (*Martes pennanti*) or Wolverine (*Gulo gulo*) might account for observations of smaller black felids such as the Woofin nanny. The fisher is particularly catlike, 2 feet long, dark brown to black in color, with a 15-inch, bushy tail. Its normal range is northern New England, Canada, and portions of the Rockies and Coast Ranges in the West.

(16) The Jaguarundi (*Herpailurus yaguarondi*) is found from south Texas to Paraguay and has been introduced in Florida. A medium-sized cat (3 feet–4 feet 6 inches long) with short legs and a long tail, the jaguarundi tends to a dark brown or black color in tropical rain forests. An escapee might be mistaken for a larger cat.

(17) Jaguars (*Panthera onca*) also have a melanistic morph but are no longer found north of Mexico, except possibly in Arizona. *See* ARIZONA JAGUAR.

*Sources:* "A Tiger in Kentucky," *Lexington (Ky.) Gazette,* July 17, 1823; Stanley P. Young and Edward A. Goldman, *The Puma: Mysterious American Cat* (Washington, D.C.: American Wildlife Institute, 1946); G. H. Pipes, *Strange Customs of the Ozark Hillbilly* (New York: Hobson, 1947); John Harden, *The Devil's Tramping Ground* (Chapel Hill: University of North Carolina Press, 1949), pp. 147–154; Vance Randolph, *We Always Lie to Strangers* (New York: Columbia University Press, 1951); Gerald T. Bue and Milton H. Stenlund, "Are There Mountain Lions in Minnesota?" *Conservation Volunteer* 15 (September 1952): 32–37; Herbert Ravenel Sass, "The Panther Prowls the East Again!" *Saturday Evening Post* 226 (March 13, 1954): 31, 133–136; Dunbar Robb, "Cougar in Missouri," *Missouri Conservationist* 16 (July 1955): 14; "Dogs Routed by 'Panther,'" *Kansas City (Mo.) Times,* July 23, 1957, business sec., p. 5; Bruce S. Wright, *The Ghost of North America: The Story of the Eastern Panther* (New

York: Vantage, 1959); Farnum Gray, "'Woofin Nanny' Has Mamas on Edge," *Winston-Salem (N.C.) Journal and Sentinel,* July 9, 1966, p. 1; Farnum Gray, "'Armed Men Staked Out to Await 'Woofin Nanny,'" *Winston-Salem (N.C.) Journal and Sentinel,* July 12, 1966, p. 1; R. E. Buehler, "Looking through the Archives: The Big Cat," *Journal of the Ohio Folklore Society* 1 (Winter 1966): 75–78; Loren Coleman, "Mystery Animals in Illinois," *Fate* 24 (March 1971): 48–54; Jerome Clark and Loren Coleman, "On the Trail of Pumas, Panthers and ULAs (Unidentified Leaping Animals)," *Fate* 25 (June 1972): 72–82, and (July 1972): 92–102; Bruce S. Wright, *The Eastern Panther: A Question of Survival* (Toronto, Canada: Clark, Irwin, 1972); Loren Coleman, "Phantom Panther on the Prowl," *Fate* 30 (November 1977): 62–67; Susan Power Bratton, "Is the Panther Making a Comeback?" *National Parks and Conservation Magazine* 52 (July 1978): 10–13; Loren Coleman, "Black 'Mountain Lions' in California?" *Pursuit,* no. 46 (Spring 1979): 61–62; Paul B. Thompson, "The Sampson County Mystery Animal," *Pursuit,* no. 56 (1981): 149–151; E. J. Kahn Jr., "Stalking the Cape Cod Cougar," *Boston Magazine,* July 1982; John Brinkley, "American Sues over 'Invasion,'" *USA Today,* August 8, 1983, p. 9A; Robert L. Downing, "The Search for Cougars in the Eastern United States," *Cryptozoology* 3 (1984): 31–49; Jim Bob Tinsley, *The Puma: Legendary Lion of the Americas* (El Paso: University of Texas at El Paso, 1987); Helen Gerson, "Cougar, *Felis concolor,* Sightings in Ontario," *Canadian Field Naturalist* 102 (1988): 419–424; "The Eastern Puma: Evidence Continues to Build," *ISC Newsletter* 8, no. 3 (Autumn 1989): 1–8; E. Randall Floyd, *Great Southern Mysteries* (Little Rock, Ark.: August House, 1989); Karl Shuker, *Mystery Cats of the World* (London: Robert Hale, 1989), pp. 151–166; Jay W. Tischendorf, "The Eastern Panther on Film? Results of an Investigation," *Cryptozoology* 10 (1990): 74–78; "Eastern Puma Officially Acknowledged in Canada," *ISC Newsletter* 12, no. 2 (1993–1996): 9–11; Gene Letourneau, "Sportsmen Say," *Portland Maine Sunday Telegram,* February 6 and 13 and March 20, 1994; Charles R. Humphreys, *Panthers of the Coastal Plain* (Wilmington, N.C.: Fig Leaf Press, 1994); Mark A. Hall, "The Eastern Catamount (Felis concolor)," *Wonders* 3, no. 1 (March 1994): 21–29; Chad Arment, "The Eastern Cougar in Harford County, Maryland," *INFO Journal,* no. 71 (Autumn 1994): 21–23; Joseph A. Citro, *Green Mountain Ghosts, Ghouls and Unsolved Mysteries* (Montpelier: Vermont Life, 1994), pp. 88–93; Chris Bolgiano, *Mountain Lion: An Unnatural History of Pumas and People* (Mechanicsburg, Pa.: Stackpole, 1995); Jay W. Tischendorf and Steven J. Ropski, eds., *Proceedings of the Eastern Cougar Conference, 1994* (Fort Collins, Colo.: American Ecological Research Institute, 1996); John A. Lutz, *All You Need to Know about the Eastern Cougar* (Baltimore, Md.: Eastern Puma Research Network, 1997); Gerry R. Parker, *The Eastern Panther: Mystery Cat of the Appalachians* (Halifax, N.S., Canada: Nimbus, 1998); E. Randall Floyd, "Tales of 'Cat Creature' Abound in Swamps," *Augusta (Ga.) Chronicle,* May 17, 1998; Chad Arment, "Black Panthers in North America: Examining the Published Explanations," *North American BioFortean Review* 2, no. 1 (2000): 38–56, http://www. strangeark.com/nabr/NABR3.pdf; Chad Arment, "Devil Monkeys or Wampus Cats?" *North American BioFortean Review* 2, no. 2 (2000): 45–48, http://www.strangeark. com/nabr/NABR4.pdf; Brad LaGrange, "Black Panthers in Perry County, Indiana," *North American BioFortean Review* 2, no. 3 (December 2000): 4, http://www.strangeark. com/nabr/NABR5.pdf; Loren Coleman, *Mysterious America,* rev. ed. (New York: Paraview, 2001), pp. 105–126; Paul Eno, *Footsteps in the Attic* (Woonsocket, R.I.: New River Press, 2001); Todd Lester, "Search for Cougars in the East," *North American BioFortean Review* 3, no. 2 (October 2001): 15–17, http://www.strangeark.com/ nabr/NABR7.pdf; Kelvin McNeil, "Some Little Known Cougar Sightings in New Hampshire," *North American BioFortean Review* 3, no. 2 (October 2001): 20–23, http://www.strangeark. com/nabr/NABR7.pdf; John A. Lutz and Linda

A. Lutz, "Century-Old Mystery Rises from the Shadows," *North American BioFortean Review* 3, no. 2 (October 2001): 30–50, http://www.strangeark.com/ nabr/NABR7.pdf; Robert Prevo, "Arkansas' Black Panthers," *North American BioFortean Review* 3, no. 2 (October 2001): 51–53, http://www.strangeark.com/nabr/NABR7.pdf; Chad Arment, "Possible Cougar Photographed in Maryland," *North American BioFortean Review* 3, no. 2 (October 2001): 54–55, http://www.strangeark.com/nabr/NABR7.pdf; Florida Panther Net, http://www.panther.state.fl.us; Eastern Cougar Foundation, http://www.geocities.com/rainforest/vines/1318/; Patrick Rusz, *The Cougar in Michigan: Sightings and Related Information* (Bath: Michigan Wildlife Habitat Foundation, 2001), available at http://www.mwhf.org/pdffiles/cougar.pdf; Chester Moore Jr., "Are U.S. 'Black Panthers' Actually Jaguarundi?" The Anomalist Online, 2002, at http://www.anomalist.com/features/jag.html.

## Ecuadorean Giant

GIANT HOMINID of South America.

*Physical description:* Long head-hair. Beard. Large eyes.

*Behavior:* Bloodthirsty. Lives in villages. Has knowledge of wells and masonry. Wears animal skins. Rapes women and kills men. Openly practices sodomy.

*Distribution:* Santa Elena, Ecuador.

*Significant sighting:* A group of giant men is said to have landed their seagoing rafts near Santa Elena, Ecuador, in the remote past. Large fossil ribs, skulls, and teeth were discovered by Capt. Juan de Olmos of Trujillo in the area in 1543 and attributed to the myth.

*Possible explanations:*

(1) Memory of a pre-Columbian visit by Polynesians.

(2) Legends surrounding the Las Vegas preceramic culture, which lived in the area 8000–4700 B.C. In 1977, a burial site yielded some 200 interments from the period, including the double burial of a man and woman entwined in each others' arms; they became affectionately known as the "Lovers of Sumpa," as the area is called.

*Sources:* Garcilaso de la Vega, *The Incas: The Royal Commentaries of the Inca* [1617], ed. Alain Gheerbrant (New York: Avon, 1964), pp. 327–328 (bk. IX); Adolph F. Bandelier, "Traditions of Precolumbian Landings on the Western Coast of South America," *American Anthropologist* 7 (1905): 250–270.

## Ecuadorean Ground Sloth

SLOTH-like mammal of South America.

*Physical description:* Length, 10 feet. Long hair. Long, horselike snout.

*Behavior:* Can stand on hind legs but walks on all fours.

*Habitat:* Caves. Browses on vegetation.

*Distribution:* Ecuador.

*Present status:* Only one report from Ecuador in the 1980s.

*Possible explanation:* A surviving Giant ground sloth (*Megatherium*), which could walk either bipedally or quadrupedally. This sloth was a large-bodied browser that lived in South America from the Late Pliocene to the Pleistocene, 1.9 million–8,000 years ago.

*Source:* "Giant Ground Sloth Survival Proposed Anew," *ISC Newsletter* 12, no. 1 (1993– 1996): 1–5.

## Eelpoot

FRESHWATER MONSTER of Maryland.

*Variant names:* Haneturtle, Hoopinflinder, Lun.

*Behavior:* Unpleasant odor.

*Distribution:* Zekiah Swamp, Charles County, Maryland.

*Possible explanation:* Tall tale invented by an old storyteller who used to cross the swamp once a week to get supplies at the store.

*Source:* Amy Gibson Compton, "Tales of the Zekiah Swamp," *Maryland Magazine* 7, no. 3 (Spring 1975): 14–17.

## Elbst

FRESHWATER MONSTER of Switzerland.

*Etymology:* From the Old German *albiz* ("swan").

*Physical description:* Serpentine. Sometimes looks like a drifting log or floating island. Reddish color. Head the size of a pig's. Scales. Clawed feet.

*Behavior:* Favors stormy weather. Creates a big wake. Travels on land at night. Eats cattle.

*Distribution:* Selisbergsee, Canton Uri, Switzerland.

*Significant sightings:* First reported in 1585 and last seen in 1926 by workers building a new road.

*Sources:* Renward Cysat, *Collectanea chronica und denkwürdige Sachen pro chronica Luchernensi et Helvetiae* [1614], vol. 4 (Lucerne, Switzerland: Diebold Schilling Verlag, 1961–1972); C. Kohlrusch, ed., *Schweizerisches Sagenbuch* (Leipzig, Germany: R. Hoffmann, 1854); Josef Müller, *Sagen aus Uri aus dem Volksmunde gesammelt,* vol. 1 (Basel, Switzerland: Gesellschaft für Folkskunde, 1926).

## Elephant-Dung Bat

Small, unknown BAT of East Africa.

*Physical description:* Silver, brownish-gray fur. Paler underparts. Very small wingspan, possibly only 5 inches.

*Behavior:* Roosts on the ground in piles of dried elephant dung.

*Distribution:* Marsabit Forest and Mount Kulal, Kenya.

*Significant sighting:* Terence Adamson briefly ran across this bat in the 1950s in two different locations in Kenya.

*Possible explanation:* The small Horn-skinned bat (*Eptesicus floweri*), suggested by Karl Shuker, has a habit of roosting in acacia roots, which are possibly comparable in texture to dried dung. It is known in Mali and southern Sudan.

*Sources:* John G. Williams, "An Unsolved Mystery," *Animals* 10 (June 1967): 73–75; Karl Shuker, "A Belfry of Crypto-Bats," *Fortean Studies* 1 (1994): 235–245.

## ELEPHANTS (Unknown)

There are three species of living Elephants (Order Proboscidea): the African bush elephant (*Loxodonta africana*), the African forest elephant (*L. cyclotis*), and the Asian elephant (*Elephas maximus*). DNA tests conducted in 2001 confirmed that the two African species are genetically distinct and probably diverged about 2.6 million years ago. *Elephas* evolved in Africa but migrated to Eurasia around the same time. Asian elephants are smaller, with humped or rounded backs, smaller ears, and one finger instead of two on the tip of the trunk.

The African bush elephant is the largest known living terrestrial animal. The average adult male stands 9 feet 10 inches–12 feet 2 inches at the shoulder and weighs 4.4–7.7 tons. The largest specimen on record had a shoulder height of 13 feet and an estimated weight of 13.5 tons; it was shot in Angola on November 7, 1974.

Though elephants are best known for their elongated trunks, their earliest ancestors completely lacked them. The hippo-sized *Moeritherium* of the Late Eocene (35 million years ago) had nasal bones placed far forward on its face, indicating a lack of large muscles necessary for a trunk. The identifying characteristics of proboscideans are much less obvious and involve particular skull and shoulder-blade features, teeth with unique cusps, hind feet with a specific ankle formation, and wrists with serial bone arrangement. The earliest was *Phosphatherium,* which weighed about 33 pounds and stood 2 feet at the shoulder. Proboscideans first evolved, probably in North Africa, near the end of the Paleocene, 55 million years ago, from primitive hoofed mammals called condylarths.

The best-known extinct proboscideans are mastodons and mammoths, which were contemporaneous in North America for about 4 million years in the Pliocene and Pleistocene. American Mastodons (Family Mammutidae) were browsers that split off from the elephant family tree in the Oligocene, nearly 30 million years ago, while the Mammoths (Family Elephantidae) were grazers with a slender build; a taller skull; inwardly-curving tusks that projected well below the horizontal; and flat, ridged teeth.

Mammoths died out relatively recently at the end of the Pleistocene in both Eurasia and North America. They are featured in about 400 cave paintings in Europe; in Ukraine, archaeologists have discovered dwellings constructed partially from mammoth bones and tusks. In North America, there is considerable evidence that the Paleo-Indians hunted or scavenged mammoths as recently as 10,000 years ago. Folk traditions of these interactions may be preserved in myths of the MAMANTU in Siberia and China and the STIFF-LEGGED BEAR in North America. Another group of proboscideans, the gomphotheres, may have lingered in Southeast Asia and provided inspiration for the MAKARA.

The BEAST OF BARDIA, the PYGMY ELEPHANT, and the THAI MAMMOTH could represent new species or distinct variations in known forms.

## Mystery Elephants

BEAST OF BARDIA; MAKARA; MAMANTU; PINK-TUSKED ELEPHANT; PYGMY ELEPHANT; STIFF-LEGGED BEAR; THAI MAMMOTH

## El-Ish-Kas

CANNIBAL GIANT of the northwestern United States.

*Etymology:* Makah (Wakashan) word.

*Variant name:* Kakawat.

*Distribution:* Olympic Peninsula, Washington.

*Source:* Alice Henson Ernst, *The Wolf Ritual of the Northwest Coast* (Eugene: University of Oregon Press, 1952), p. 74.

## Ellengassen

Unknown SLOTH-like mammal of South America.

*Etymology:* Tehuelche (Chon) word.

*Variant names:* Lobo-toro (Spanish equivalent of Araucanian word meaning "wolf bull"), Lofo-toro.

*Physical description:* The size of a bull. Long hair.

*Behavior:* Roars or howls like a wolf. Herbivore. Makes its den in a cave.

*Tracks:* Like a wooden shoe with two cleats across the sole, according to a lone report from 1898.

*Distribution:* Patagonia, especially in Lago Buenos Aires area of Santa Cruz Province, Argentina; southern Mendoza Province, Argentina.

*Present status:* Probably extinct.

*Possible explanations:*

(1) May represent a recently surviving Patagonian cave-dwelling sloth (*Mylodon darwinii*), subfossil remains of which are known from the Cueva del Milodón in southern Chile. Manuel Palacios told Bruce Chatwin there was a rock painting of a *Mylodon* in the Monumento Natural los Bosques Petrificados, Santa Cruz Province, Patagonia.

(2) Muddled Indian legends of Jaguars (*Panthera onca*) and feral oxen.

*Sources:* Francisco P. Moreno, *Viaje á la Patagonia austral, emprendido bajo los auspicios del gobierno nacional, 1876–1877* (Buenos Aires: La Nación, 1879), p. 395; Santiago Roth, "Descripción de los restos encontrados en la Caverna de Ultima Esperanza," in "El mamífero misterioso de la Patagonia, II," *Revista del Museo de La Plata* 9 (1899): 421–453; H. Hesketh Prichard, *Through the Heart of Patagonia* (New York: D. Appleton, 1902); Robert and Katharine Barrett, *A Yankee in Patagonia: Edward Chace* (Boston: Houghton Mifflin, 1931), p. 30; Carlos Rusconi, "La supuesta existencia de Milodontes en la Patagonia Austral (*Milodon listai*)," *Revista del Museo de Historia Natural de Mendoza* 3 (1949): 252–264; Bruce Chatwin, *In Patagonia* (New York: Summit, 1977), p. 72.

## Emela-Ntouka

Unknown DINOSAUR-like reptile or HOOFED MAMMAL of Central Africa.

*Etymology:* Bomitaba (Bantu), "killer of elephants" or "eater of the tops of the palms."

*Variant names:* Aseka-moke, CHIPEKWE, Emeula natuka, Emia-ntouka (in the Congo), Forest rhinoceros, Ngamba-namae, Ngoulou (Baka/Ubangi), NSANGA, NYAMA.

*Physical description:* As large as an elephant or larger. Reddish-brown to gray. Hairless. Single, large, curved, ivory horn on its nose. Beaked mouth. Short, frilled neck. Massive legs. Heavy tail like a crocodile's.

*Behavior:* Amphibious. Foul-tempered.

*The EMELA-NTOUKA, an elephant-killing, dinosaur-like animal of Central Africa. (William M. Rebsamen)*

Snorts, howls, and roars. Feeds on a wide variety of leaves, including the Malombo liana (like the MOKELE-MBEMBE). Disembowels elephants, buffalos, and hippopotamuses with its horn.

*Tracks:* Like a rhinoceros.

*Habitat:* Dense rain forest.

*Distribution:* Liberia; Boumba and Ngoko Rivers, eastern Cameroon; Gabon; Loubomo, Kellé, Ouesso, Impfondo, Dongou, and Epéna in the Republic of the Congo; Central African Republic; Zambia.

*Significant sightings:* In 1913, Hans Schomburgk heard stories from the Klao tribe about a small rhinoceros that lived in the mountains of Liberia.

In 1950, a French official named Millet, stationed at Kellé in the Republic of the Congo, heard of a rhinoceros that lived in the forests. Inhabitants of the district drew sketches of its footprint, which resembled that of a rhinoceros.

In August or September 1966, Atelier Yvan Ridel photographed some 10-inch-wide, three-toed footprints along a riverbank northeast of Loubomo, Republic of the Congo.

Roy Mackal collected information on the Emela-ntouka during his expeditions to the Congo in 1980 and 1981, noting that lore about the animal is often confused with that of the MOKELE-MBEMBE.

*Possible explanations:*
(1) A semiaquatic rhinoceros that inhabits the rain forest, suggested by Lucien Blancou, though the large tail argues against it. A semiaquatic fossil rhino named *Teleoceras* is known from 17–5 million years ago in Late Miocene river and lake sediments of North America. It was hippolike, with short limbs, a massive body, and high-crowned teeth.
(2) Roy Mackal has proposed a surviving ceratopsian dinosaur like *Monoclonius,* a quadrupedal herbivore about 18 feet long with a backwardly curved nose horn and a bony neck frill. *Monoclonius* fossils have been found in Montana and Alberta and date from the Late Cretaceous, about 70 million years ago. However, no ceratopsians are known from Africa. Also, they were egg-laying dinosaurs, and no reports of the Emela-ntouka describe it as oviparous.
(3) The elephant-sized *Elasmotherium* was a Pleistocene rhino with a 7-foot horn in the center of its forehead. It is known from grasslands in Europe, Siberia, and China.

*Sources:* Lucien Blancou, "Notes sur les mammifères de l'Equateur Africain Français: Un rhinocéros de fôret?" *Mammalia* 18 (December 1954): 358–363; Georges Trial, *Dix ans de chasse au Gabon* (Paris: Crépin-Leblond, 1955); Herman A. Regusters, "Mokele-Mbembe: An Investigation into Rumors Concerning a Strange Animal in the Republic of the Congo, 1981," *Munger Africana Library Notes,* no. 64 (1981): 1–27; Roy P. Mackal, *A Living Dinosaur? In Search of Mokele-Mbembe* (Leiden, the Netherlands: E. J. Brill, 1987), pp. 44, 235–249, 316–321.

# Engbé

SMALL HOMINID of West Africa.

*Etymology:* Dida (Kru) word. In the Central African Republic, this is the Banda-Yangere (Ubangi) term for the Moustached monkey (*Cercopithecus cephus*).

*Variant name:* Egbéré (in Sierra Leone).

*Physical description:* Long hair.

*Behavior:* Said to kidnap people and bring them to their villages upon occasion.

*Habitat:* Villages in the deep forest.

*Distribution:* Southern Côte d'Ivoire; Sierra Leone.

*Source:* Gaston Joseph, "Notes sur les Avikams de la lagune de Lahou et les Didas de la région du Bas-Bandama," *Bulletins et Mémoires de la Société Anthropologique de Paris,* ser. 6, 1 (1910): 234–247.

# Engôt

GIANT HOMINID of Central Africa.

*Etymology:* Seki (Bantu), "ogre."

*Variant names:* Éngunguré (Fang/Bantu), En-zinzi, Ézôzôme, Ntyii.

*Physical description:* Feet are turned the wrong way around.

*Behavior:* Eats humans.

*Distribution:* Gabon.

*Possible explanation:* Muddled folk memory of encounters with Gorillas (*Gorilla gorilla*) in the remote past.

*Sources:* Richard Lynch Garner, *Gorillas and Chimpanzees* (London: Osgood, McIlvaine, 1896), pp. 208–211; Bernard Heuvelmans, *Les bêtes humaines d'Afrique* (Paris: Plon, 1980), pp. 553–554.


## Enkidu

The original mythical image of the WILDMAN in the Middle East, from the Babylonian *Epic of Gilgamesh.* Enkidu was a wild warrior and companion to Gilgamesh, the king of the city-state of Uruk (Erech in modern Iraq) who lived around 2800 B.C.

*Etymology:* Sumerian word, possibly meaning "created by EA," "Lord of the Good Place," or "wild one."

*Variant name:* Ea-bani.

*Physical description:* Covered with hair. Long head-hair.

*Behavior:* Swift runner. Eats wild plants. Drinks from a water hole. Becomes civilized when he is seduced by a sacred temple girl.

*Possible explanation:* Ancient tradition of WILDMEN in the mountains of West Asia.

*Sources: The Epic of Gilgamesh,* trans. N. K. Sandars (Baltimore, Md.: Penguin, 1970); *The Epic of Gilgamesh,* trans. Danny P. Jackson (Wauconda, Ill.: Bolchazy-Carducci, 1993).


## ENTITIES

Despite the desire of most cryptozoologists to explain all observations of unknown animals in zoological terms, there nonetheless exist experiences that seem to belong to the psychic, the paranormal, or the spiritual world rather than the physical. BLACK DOGS disappear into thin air, FLYING HUMANOIDS with batlike wings appear in an area of concentrated unidentified flying object (UFO) reports, and HAIRY BIPEDS are impervious to bullets.

These high-strangeness cases often, but not always, intertwine the metaphysical or cultural belief systems of the observer with whatever stimulus happens to be physically present. A furtive shadow on the wall in a New Delhi suburb becomes a menacing MONKEY MAN; an out-of-place moose is transformed into an extraterrestrial-like DOVER DEMON; a distant house cat provides the inspiration for an ALIEN BIG CAT report; an odd combination of unrelated sights, sounds, and smells creates the illusion of a mythical CHUPACABRAS with fiery red eyes.

Sometimes, the stimulus is a cryptid whose appearance is magnified through a filter of fear and preconception. Undoubtedly, the BEAST OF GÉVAUDAN (a hyena) contributed to WEREWOLF lore in eighteenth-century France. The sighting of an errant BIGFOOT reinforces the CANNIBAL GIANT folklore espoused by a startled Native American or muddles the sensory input of a midwestern couple unfamiliar with the physical characteristics of a Pacific Northwest hominid.

Of course, it is possible that genuine apparitions of a psychic nature are responsible for some of these phenomena. Whether or not that's true, cryptozoologists still need to study Entity cases in order to discern the differences between SMALL HOMINIDS and LITTLE PEOPLE or between BLACK DOGS and German shepherds on the loose. Are fiery red eyes always a characteristic of a paranormal entity? Or are there circumstances under which the eyes of real animals can appear red and luminous? We need to know the precise mechanisms by which observers perceive or misperceive unusual, unexpected, or inexplicable events and how belief and veridical experience interact.

Eleven of the fifteen Entities in this section have a roughly human or humanoid shape.


### Mystery Entities

ALIEN BIG CAT; BIG GREY MAN; BLACK DOG; BRENIN LLWYD; CANNIBAL GIANT; CHUPACABRAS; DAISY DOG; DOVER DEMON; FLYING HUMANOID; HAIRY BIPED; LITTLE PEOPLE; LIZARD MAN; MONKEY MAN; PHANTOM WOLF; WEREWOLF


## Esakar-Paki

Mystery piglike HOOFED MAMMAL of South America.

*Etymology:* Shuar (Jivaroan) word.

*Physical description:* Small peccary. Reddish-brown fur.

*Behavior:* Aggressive. Lives in troops of fifty to sixty individuals. Attacks humans.

*Distribution:* Sangay National Park, Ecuador, east to the Peruvian border.

*Significant sighting:* Caver Marcelo Churuwia was chased by a troop of these peccaries on the Ecuador-Peru border.

*Source:* Angel Morant Forés, "An Investigation into Some Unidentified Ecuadorian Mammals," October 1999, http://perso.wanadoo.fr/cryptozoo/expeditions/ecuador_eng.htm.

## Esti Capcaki

CANNIBAL GIANT of the southeastern United States.

*Etymology:* Seminole (Muskogean), "tall man."

*Distribution:* Florida.

*Source:* Kyle Mizokami, Bigfoot-Like Figures in North American Folklore and Tradition, http://www.rain.org/campinternet/bigfoot/bigfoot-folklore.html.

## Ethiopian Deer

Unidentified HOOFED MAMMAL of East Africa.

*Distribution:* Southern Ethiopia.

*Significant sighting:* Apparently known to the ancient Egyptians.

*Possible explanations:*

(1) Surviving fossil giraffid (*Climacoceras*) that lived in the Miocene (18–6 million years ago) in East Africa. It had branched, antlerlike cranial appendages.

(2) Ethiopian subspecies of the Fallow deer (*Dama dama*) found in Ethiopia and Egypt from the Late Pliocene to the Late Pleistocene (2 million–10,000 years ago).

*Source:* Christine Janis, "A Reevaluation of Some Cryptozoological Animals," *Cryptozoology* 6 (1987): 115–118.

## Ethiopian Hyrax

Unknown HYRAX of East Africa.

*Distribution:* Southern Ethiopia.

*Significant sighting:* There is a vague tradition of a giant herbivorous mammal, 4 feet long and 2 feet high at the shoulder, in the Ethiopian desert.

*Possible explanation:* Unknown species of Hyrax (*Procavia*) or Bush hyrax (*Heterohyrax*).

*Source:* Bernard Heuvelmans, "Annotated Checklist of Apparently Unknown Animals with which Cryptozoology Is Concerned," *Cryptozoology* 5 (1986): 1, 20.

## Ethiopian Vampire Bat

Unknown BAT of East Africa.

*Variant name:* Death bird.

*Physical description:* Wingspan, 12–18 inches.

*Behavior:* Said to feed on the blood of animals and humans, causing puncture wounds and debilitating sickness.

*Distribution:* Devil's Cave, somewhere near Nek'emte, in the Welega division of Ethiopia.

*Significant sighting:* In the 1930s, Byron de Prorok explored a cave said by the locals to be haunted by hyena-men and a death bird. The hyenas proved real enough, and so did the death birds, in the form of a huge swarm of bats. De Prorok noted that goatherds in the area looked very debilitated, and they blamed their condition on bites from these bats.

*Possible explanations:*

(1) The only known sanguinivorous bats are found in Mexico, Central, and South America. Infected bites from parasites carried by the bats might be mistaken for bat bites.

(2) Fungal spores from guano or *Leptospira* bacteria causing Weil's syndrome, which produces liver and kidney problems, meningitis, and vomiting, could be mistakenly blamed on bat bites.

(3) African vampire legends might also exaggerate a normal bat's activities.

*Sources:* Byron Khun de Prorok, *Dead Men Do Tell Tales* (New York: Creative Age Press, 1942); Karl Shuker, "A Belfry of Crypto-Bats," *Fortean Studies* 1 (1994): 235–245.

## European Flying Snake

FLYING REPTILE of Southern Europe.

*Physical description:* Length, 3–6 feet. Green, black, gray, or white. Sometimes said to have wings.

*Behavior:* Moves along the ground in a straight line. Emits a peculiar vocalization.

*Distribution:* Alpes-Maritimes Department, France; near Sarajevo, Bosnia; Bulgaria.

*Significant sightings:* In 1930 or 1931, a green snake with wings frightened the mother of André Mellira in a forest near La Bollène-Vésubie, Alpes-Maritimes Department, France.

In Bulgaria in the summer of 1947, Hazel Göksu surprised a group of snakes 3–6 feet long on a footpath. With a peculiar cry, they flew into the air 6–9 feet above the ground toward a spring and vanished behind some trees.

*Sources:* François de Sarre, "Are There Still Dragons in Southern France?" *INFO Journal,* no. 71 (Autumn 1994): 44–45; Izzet Göksu (letter), "Flying Snakes of Bulgaria," *Fortean Times,* no. 78 (December 1994-January 1995): 57; Karl Shuker, "Flying Snakes," *Strange Magazine,* no. 17 (Summer 1996): 26–27.

# F

## Fairy

LITTLE PEOPLE of Western Europe with magical powers.

*Etymology:* From the Old French *fae* or *fée* ("fairy"), deriving from the Latin *fatum* ("destiny").

*Variant names:* Brownie, Fary (in Northumberland), Fay, Fayry, Fenoderee (Manx), Ferier (in Suffolk), Ferrish (Manx), Frairy (in East Anglia), Gentle folk, The Gentry, Good people, Gwyllion (Wales), Huldre (Norwegian), Huldufolk (Icelandic), Klippe (in Forfarshire), Korrigan (Breton), Leprechaun (Irish), Lutin (French), Mound folk, Nis, Nisse (Norwegian), Piskie (in Cornwall), Pixy (in Somerset), Poldie (in Cheshire), Sídhe (Irish), Sìth (Gaelic), Sleagh Maith (Irish, "good people"), Spyris (Cornish), S'thich (Gaelic), The Strangers (in Lincolnshire), Tomte (Swedish), Tylwyth teg (Welsh).

*Physical description:* Height, 2–5 feet, or smaller. Generally good-looking but usually with some deformity that is difficult to hide. Red hair. Hairy face. Long arms. Large feet.

*Behavior:* Clever and mischievous. Eats barley meal and oatmeal. Lives in megalithic structures. Wears clothes, often red or green. Said to be vengeful, especially when cheated or when its home or environment is destroyed. Has supernatural powers and can become invisible or alter its form at will. Associated with buried treasure. Said to be fond of braiding horse's manes. Appears to children more often than adults. Steals human children and replaces them with their own (changelings). Carries people away to Fairyland or detains them there if they enter a Fairy hill and can be tricked into tasting Fairy food or drink. Causes paralytic seizures. Social Fairies engage in such complex social structures as government, art, music, marriage, labor, funerals, and war.

*Distribution:* Worldwide but especially known in Ireland, Scotland, Iceland, Isle of Man, and Norway.

*Significant sightings:* In 1188, a Welsh cleric named Elidyr told Gerald of Wales that when he was twelve years old, he had encountered two tiny men who led him through a dark tunnel and into a fantastic realm of little people ruled by a king. He returned to visit several times until he tried stealing a golden ball. The little men pursued and took it back from him, after which he could no longer find the tunnel.

In 1757, when British cleric Edward Williams was seven years old, he and some other children playing in a field in Wales saw a group of tiny couples dressed in red and carrying white kerchiefs. One of the little men, who had an "ancient, swarthy, grim complexion," chased the children. The incident puzzled Williams all his life.

In the early twentieth century, W. Y. Evans-Wentz traveled throughout Ireland, Scotland, Wales, Cornwall, and Brittany, gathering many oral traditions of Fairies from all social classes. One informant, named Neil Colton, told him about Fairies he had seen in the mid-nineteenth century at Lough Derg, County Donegal, Ireland. He and some other children were gathering berries when they heard music and saw six to eight of them dancing a few hundred feet away. A little woman came running toward them and hit a girl on the face with a green rush. The girl fainted after they all ran home and was revived only with the help of a priest.

The notorious Cottingley Fairy photographs, taken by Frances Griffiths and Elsie Wright, somehow fooled many people over the years. It

*Artist's conception of a group of* FAIRIES. *Drawn by Arthur Rackham for J. M. Barrie's* Peter Pan in Kensington Gardens. *(New York: Charles Scribner's Sons, 1906). (Fortean Picture Library)*

was only in 1983 that the women finally admitted to using cutouts from *Princess Mary's Gift Book* (London: Hodder and Stoughton, 1914), by Princess Mary, Countess of Harewood, on two of the photos taken in 1917. Three other photos taken in August 1920 with a different camera were probably double exposures. However, they never denied seeing real Fairies in the beck near Cottingley, Yorkshire, and claimed the hoax was done to demonstrate their reality. The photos and related documents sold for £22,000 at an auction in July 1998.

On April 30, 1973, Mary Treadgold was traveling by bus on the Island of Mull in Scotland when she looked out the window and saw a small figure, about 18 inches high, who appeared to be digging peat with a spade. It was dressed in bright-blue pants and suspenders and a white shirt with rolled-up sleeves, and it remained completely still as the bus passed.

Unexpected mishaps during construction of a new road at Akureyri, Iceland, in 1984 were blamed on the local fairies. Helgi Hallgrimsson, director of the Akureyri Natural History Museum, has collected many eyewitness reports from the district around Eyjafjörður, where a Fairy town is said to be located.

Brian Collins, age fifteen, was vacationing on Aran Island, County Donegal, Ireland, around 1992 when he saw two men about 3 feet 6 inches tall, talking in Irish and dressed in green with brown boots. They were sitting on a bank, fishing in the ocean, but suddenly they jumped away and disappeared. Collins retrieved a pipe one had been smoking, but it later disappeared from a locked drawer.

*Present status:* Not a traditional cryptozoological puzzle in that these diminutive entities do not seem to belong to the purely physical realm. However, Fairies serve as a reminder that even Western cultures can have difficulty separating the real world from the paraphysical.

*Possible explanations:*

(1) Folk memories of a race of small-statured people said to have existed in Europe in antiquity, suggested by Elizabeth Andrews and others. The survival of megalithic monuments that were apparently built by shorter people contributed to this belief.

(2) Folk memories of Celtic or other pagan gods, dimly remembered from pre-Christian times.

(3) Folk memories of an ancient cult of the dead or actual spirits of the dead. Various types might be classed as the evil dead, the recently dead, the heathen dead, and the ancient dead.

(4) Nature spirits of gardens and glens; elemental personifications of trees, plants, earth, and water.

(5) Hallucinations of some kind, perhaps by fantasy-prone individuals.

(6) Paranormal or supernatural apparitions, fallen angels, or a race of beings halfway between the material and the spiritual.

(7) A premodern manifestation of entities related to the unidentified flying object (UFO) phenomenon. Individuals who are "taken by the fairies" have been compared to those who claim abduction by UFO aliens.

*Sources:* Gerald of Wales, *The Journey through Wales* [1188], trans. Lewis Thorpe (New York: Penguin, 1978), pp. 133–136 (I.8); Robert Kirk, *The Secret Common-Wealth of Elves, Fauns and Fairies* [1691] (London: D. Nutt, 1893); Thomas Keightley, *The Fairy Mythology* (London: H. G. Bohn, 1850); James Bowker, *Goblin Tales of Lancashire* (London: W. Swan Sonnenschein, 1878); Wirt Sikes, *British Goblins* (London: S. Low, Marston, Searle and Rivington, 1880); David MacRitchie, *Fians, Fairies and Picts* (London: Kegan Paul, Trench, Trubner, 1893); W. Y. Evans-Wentz, *The Fairy-Faith in Celtic Countries* (London: H. Frowde, 1911); Elizabeth Andrews, *Ulster Folklore* (London: Elliot Stock, 1913); Arthur Conan Doyle, "Fairies Photographed," *Strand Magazine* 60 (December 1920): 462–467; Arthur Conan Doyle, *The Coming of the Fairies* (London: Hodder and Stoughton, 1922); Geoffrey Hodson, *Fairies at Work and at Play* (London: Theosophical Publishing House, 1925); Edward L. Gardner, *Fairies: The Cottingley Photographs and Their Sequel* (London: Theosophical Publishing House, 1945); Lewis Spence, *British Fairy Origins* (London: Watts,

1946); Diarmuid A. MacManus, *The Middle Kingdom* (London: Max Parrish, 1959); Tor Åge Bringsværd, *Phantoms and Fairies from Norwegian Folklore* (Oslo: Johan Grundt Tanum Forlag, 1970), pp. 95–102; Keith Thomas, *Religion and the Decline of Magic* (London: Weidenfeld and Nicolson, 1971); Mary Treadgold (letter), *Journal of the Society for Psychical Research* 48 (September 1975): 186–187; Katharine M. Briggs, *A Dictionary of Fairies* (London: Allen Lane, 1976); Alan Boucher, ed., *Elves, Trolls and Elemental Beings: Icelandic Folktales II* (Reykjavik: Iceland Review Library, 1977); Nancy Arrowsmith and George Moorse, *A Field Guide to the Little People* (New York: Hill and Wang, 1977); Katharine M. Briggs, *The Vanishing People: A Study of Traditional Fairy Beliefs* (London: Batsford, 1978); Geoffrey Crawley, "That Astonishing Affair of the Cottingley Fairies," *British Journal of Photography,* in 10 parts, December 24, 1982, to April 8, 1983; "Icelandic Fairies," *Fortean Times,* no. 43 (Spring 1985): 45–46; Ulrich Magin, "Yeats and the 'Little People,'" *Strange Magazine,* no. 4 (1989): 10–13, 55–58; Joe Cooper, *The Case of the Cottingley Fairies* (London: Robert Hale, 1990); Ulrich Magin, "The Akureyri Fairies Revisited," *INFO Journal,* no. 66 (June 1992): 18–19; Jerome Clark, *Encyclopedia of Strange and Unexplained Physical Phenomena* (Detroit, Mich.: Gale Research, 1993), pp. 59–61, 95–101; "More Fairies Seen," *Fate* 46 (April 1993): 14–15; David Lazell, "Modern Fairy Tales," *Fortean Times,* no. 71 (October-November 1993): 39–41; Peter Narváez, ed., *The Good People: New Fairylore Essays* (Lexington: University Press of Kentucky, 1997); Janet Bord, *Fairies: Real Encounters with Little People* (New York: Carroll and Graf, 1997); John E. Roth, *American Elves* (Jefferson, N.C.: McFarland, 1997), pp. 43–50; Carole G. Silver, *Strange and Secret Peoples: Fairies and Victorian Consciousness* (New York: Oxford University Press, 1998); Bob Curran, *The Truth about the Leprechaun* (Dublin: Wolfhound Press, 2000); Diane Purkiss, *Troublesome Things: A History of Fairies and Fairy Stories* (London: Penguin, 2001).

## Fangalabolo

Giant BAT of Madagascar.

*Etymology:* Betsileo Malagasy (Austronesian), "that which seizes the hair."

*Physical description:* Giant bat.

*Behavior:* Dives on humans and tears their hair.

*Distribution:* Madagascar.

*Possible explanation:* Unknown species of fruit bat larger than the Madagascan flying fox (*Pteropus rufus*), which has a wingspan of 5 feet.

*Sources:* Raymond Decary, *La faune malgache, son rôle dans les croyances et les usages indigènes* (Paris: Payot, 1950), p. 206; Bernard Heuvelmans, *On the Track of Unknown Animals* (New York: Hill and Wang, 1958), p. 516.

## Fantasma de los Riscos

GIANT HOMINID of South America.

*Etymology:* Spanish, "ghost of the badlands."

*Physical description:* Naked, hairy man.

*Behavior:* Howls.

*Distribution:* Pie de Palo area, San Juan Province, Argentina.

*Source:* Fabio Picasso, "South American Monsters and Mystery Animals," *Strange Magazine,* no. 20 (December 1998): 28–35.

## Farishta

WILDMAN of Central Asia.

*Etymology:* Arabic (Semitic), an Islamic "angel."

*Distribution:* Pamir Mountains, Tajikistan.

*Source:* Bernard Heuvelmans and Boris F. Porshnev, *L'homme de Néanderthal est toujours vivant* (Paris: Plon, 1974), p. 109.

## Father-of-All-the-Turtles

A giant TURTLE, one category of SEA MONSTER identified by Bernard Heuvelmans.

*Etymology:* From a Sumatran legend.

*Variant name:* Aspidochelone (Greek, "snake-turtle").

*Physical description:* Tortoiselike head. Large, prominent eyes. Wide mouth. No teeth. Medium-length, slender neck. Rounded carapace with a saw-toothed ridge. Large scales on the back. Two pairs of flippers.

*Behavior:* Breathes through its mouth, making a whistling noise.

*Distribution:* North Atlantic Ocean and the Caribbean Sea.

*Significant sightings:* As his three caravels sailed east along the southern coast of the Dominican Republic in early September 1494, Christopher Columbus and his crew saw a whale-sized turtle that kept its head out of the water. It had a long tail with a fin on either side.

On March 30, 1883, the schooner *Annie L. Hall* sighted what looked like a capsized ship on the Grand Banks (or off the Azores if you go by the longitude provided) in the North Atlantic Ocean. However, it turned out to be a turtle "40 feet long, 30 feet wide, and 30 feet from the apex of the back to the bottom of the under shell." The flippers were 20 feet long. This report was apparently confused in contemporary newspapers with a more conventional, 100-foot-long SEA MONSTER sighting that took place in November 1883 by Capt. W. L. Green and some fishermen off Long Branch, New Jersey.

On March 8, 1955, L. Alejandro Velasco was stranded on a raft off the Gulf of Urabá, Colombia, when he saw a yellow turtle about 14 feet long.

In June 1956, the cargo steamer *Rhapsody* reported a 45-foot turtle with a white carapace south of Nova Scotia. It had flippers 15 feet long and could raise its head 8 feet out of the water.

On September 13, 1959, Tex Geddes and James Gavin were fishing off Soay in the Inner Hebrides, Scotland, when they saw the head and back of a huge animal approach them until it was only 20 yards away. Its body was 4–8 feet broad at the water line, and the back was 2 feet–2 feet 6 inches high. They watched it for five minutes, after which it dived and swam further out to sea.

In August 1971, NESSIE-hunter Tim Dinsdale discovered a huge, dead turtle in a storage shed in Mallaig on the coast of Scotland not far from Soay. He estimated its weight at 1,500 pounds.

*Possible explanations:*

(1) The Atlantic leatherback turtle (*Dermochelys coriacea coriacea*) ranges throughout the North Atlantic from Newfoundland to the British Isles and can reach a length of 7–8 feet. It has longitudinal ridges on its back but no jagged ridge. It rarely exceeds 800 pounds.

(2) A surviving *Archelon ischyros,* the largest known turtle, measured up to 16 feet long and 12 feet wide, and may have weighed as much as 11,000 pounds. It lived some 70 million years ago in marine seas of the Late Cretaceous. Fossils have been found in South Dakota, Kansas, and Colorado.

*Sources:* T. H. White, *The Bestiary: A Book of Beasts* (New York: G. P. Putnam's Sons, 1960), pp. 197–198; Hernando Colon, "The Life of the Admiral by His Son, Hernando Colon," in J. M. Cohen, ed., *The Four Voyages of Christopher Columbus* (New York: Penguin, 1969), p. 184; "A Large Turtle," *Scientific American* 48 (1883): 292; "Ship Reports Giant Sea Turtle," *New York Herald Tribune,* June 8, 1956; Tex Geddes, *Hebridean Sharker* (London: H. Jenkins, 1960); Maurice Burton, "The Soay Beast," *Illustrated London News* 236 (1960): 972–973; Maurice Burton, "Was the Soay Beast a Tourist?" *Illustrated London News* 239 (1961): 632; Bernard Heuvelmans, *In the Wake of the Sea-Serpents* (New York: Hill and Wang, 1968), pp. 271, 564–565; Tim Dinsdale, *Project Water Horse* (London: Routledge and Kegan Paul, 1975), p. 167; Ulrich Magin, "In the Wake of Columbus' Sea Serpent: The Giant Turtle of the Gulf Stream," *Pursuit,* no. 78 (1987): 55–56; X, "The Gigantic Turtle of 1883," *INFO Journal,* no. 70 (January 1994): 14–15.

## Fating'ho

SMALL HOMINID of West Africa.

*Etymology:* Mandinka (Mande) word.

*Variant name:* Kudeni.

*Physical description:* Large head. Long head-hair. Red eyes. Shaggy, black body-hair.

*Behavior:* Walks upright. Grunts.

*Habitat:* Dense forests.

*Distribution:* Senegal; northern Guinea.

*Significant sightings:* Long ago, a hunter from around Diaroumé, Senegal, captured a female

Fating'ho and succeeded in taming it and having children by it. A daughter, named Na Fancani, was said to be of great beauty, and her descendants still lived in the area in 1945.

In November 1992, entomologist Malang Mane was collecting specimens at an altitude of 3,600 feet in a forested area of northern Guinea when he saw a man-sized, black-haired creature walk to within a few feet of him before it ran away.

*Sources:* Coly Dembo, "Étrange métissage (Casamance)," *Notes Africaines,* no. 27 (July 1945): 18–19; Karl Shuker, "The Secret Animals of Senegambia," *Fate* 51 (November 1998): 46–50.

## Faun

Rural deity of Southern Europe, the Roman equivalent of the Greek SATYR.

*Etymology:* From the Latin fertility god Faunus or Fanus, based on the Greek PAN; plural, *Fauni. See also* SILVANUS.

*Physical description:* Less bestial than the satyr. Small horns on forehead. Pointed ears. Beard. Tail.

*Behavior:* Associated with drunken debauchery.

*Source: New Larousse Encyclopedia of Mythology* (New York: Putnam, 1968), pp. 207–208, 215.

## Fei-Fei

Mystery PRIMATE of East Asia.

*Etymology:* Chinese (Sino-Tibetan), "baboon."

*Physical description:* Height, 10 feet. Human face. Long lips. Long hair down the neck. Feet point backward.

*Behavior:* Runs swiftly. Call sounds like human laughter. Its paws are eaten by the local people.

*Distribution:* Western Sichuan Province, China; Lishui, Zhejiang Province, China.

*Sources:* Bernard E. Read, *Chinese Materia Medica: From the Pen ts'ao kang mu Li Shih-chen, A.D. 1597* (Beijing: Peking Natural History Bulletin, 1931), pt. 51; Robert Hans

van Gulik, *The Gibbon in China: An Essay in Chinese Animal Lore* (Leiden, the Netherlands: E. J. Brill, 1967); John Napier, *Bigfoot: The Yeti and Sasquatch in Myth and Reality* (New York: E. P. Dutton, 1973), pp. 29–30.

## Filipino Secretary Bird

Unknown BIRD of prey of Southeast Asia.

*Physical description:* Feathers longer on the lower portion of the crest. Outermost tail feathers longer than the inner feathers.

*Distribution:* Philippine Islands.

*Significant sighting:* French naturalist Pierre Sonnerat visited the Philippines from 1771 to 1772 and reported the existence of a large bird of prey similar to the secretary bird of sub-Saharan Africa.

*Present status:* No known species of secretary bird exists outside Africa.

*Possible explanations:*

(1) Specimens of the African Secretary bird (*Sagittarius serpentarius*), a long-legged, crested, eaglelike bird, may have been bought to the Philippines by traders.

(2) Sonnerat might have heard secondhand accounts of the Philippine eagle (*Pithecophaga jefferyi*), which was unknown in the eighteenth century.

*Sources:* Pierre Sonnerat, *Voyage dans la Nouvelle Guinée* (Paris: Ruault, 1776); Karl Shuker, "All New Talon Show," *Fortean Times,* no. 105 (December 1997): 49.

## FISHES (Unknown)

In general, a fish is a streamlined animal with a backbone that swims by undulations and breathes through gills. It has a cranium and a muscular tail with a tail fin. There are several different groups, the most important three being the sharks, rays, and chimaeras with cartilaginous skeletons (Class Chondrichthyes); the familiar ray-finned fishes with bony skeletons (Actinopterygii); and the lobefins and lungfishes that belong to the Sarcopterygii, which have pairs of fleshy fins or limbs with a series of internal bones, only one of which is attached to the shoulder girdle or pelvis. This last group is the one from which humans—indeed, all tetrapods (four-legged animals)—evolved.

The earliest recognizable bony fishes arose in the Ordovician period, more than 450 million years ago. Called ostracoderms, they lacked movable jaws but had a distinctive brain encased in cranial bones and often were armored with bony plates and scales. Other jawless fish, the still-existing lampreys and hagfishes, are about 300 million years old, from the late Carboniferous; they probably arose from a different group of ostracoderms than the jawed fishes.

The earliest fish with jaws probably appeared as early as the late Ordovician, 450 million years ago. Microscopic scraps of sharklike skin denticles have been found in the United States dating from that time and in Mongolia from the Silurian, about 420 million years ago. The most anatomically primitive shark fossils date from the Late Devonian, 360 million years ago. Even at this early stage, this class of animal had an elongate body, large triangular fins, an upturned tail, and a mouth filled with rows of teeth.

The placoderms were a formidable group of jawed fishes from the Late Devonian and were the largest vertebrates of their time, some of them reaching 26–33 feet in length. Nearly 200 fossil placoderm genera are known.

The earliest fossil lobefins are from China, Spitsbergen, Norway, and Canada and date from the Early Devonian, 400 million years ago. For a time, before bony fishes got into their stride, they were the dominant fishes of the Devonian. Lungfishes and coelacanths are the only finned relatives of land animals to have survived the major extinction at the end of the Permian, 251 million years ago. The COELACANTHS include only two living saltwater species (*Latimeria chalumnae* and *L. menadoensis*). Some of the more primitive coelacanths had a sharklike, asymmetrical tail with special muscles that allowed it to twitch. Coelacanths of the Early Cretaceous, 120 million years ago, such as *Mawsonia* (10 feet long), lived in brackish waters in Brazil and Africa.

Isolated scales from relatives of bony fishes have been obtained from the late Silurian of Russia and China, and fragmentary bones and teeth have been found in similar strata in Estonia. One of the most primitive bony species known is *Cheirolepis,* from the Middle Devonian, 380 million years ago. Its anatomy suggests it was a swift swimmer, and the pointed teeth in its large mouth indicate it was an efficient predator.

Almost half of all known species of vertebrates now alive are ray-finned bony fishes; the 23,681 species cataloged in 1994 by Joe Nelson are probably a vast underestimate, since isolated pools and streams in tropical forests can, over time, evolve new varieties. Sturgeons (Family Acipenseridae) belong to a primitive group called chondrosteans that are separate from the more advanced teleost fishes. Often listed as candidates for FRESHWATER MONSTERS, sturgeons have largely cartilaginous skeletons and live in the sea but travel a long distance up rivers and into lakes to breed. Teleost fishes fall into four major groupings: the Bonytongues (Osteoglossomorpha); Eels and Tarpons (Elopomorpha); Herrings and their relatives (Clupeomorpha); and everything else, from salmon to minnows (Euteleostei).

The largest living fish is the Whale shark (*Rhincodon typus*), found in tropical oceans. In early 1919, an unverified 55-foot whale shark became wedged in a fish trap off Ban Ko Chik, Thailand. A huge individual known as Sapodilla Tom frequented the waters off Honduras for fifty years and was said to measure 60–70 feet in length. The largest official specimen measured 41 feet 6 inches and was caught in the Indian Ocean off Karachi, Pakistan, on November 11, 1949.

The largest carnivorous fish is the Great white shark (*Carcharodon carcharias*), which averages 14–15 feet in length. Outsize specimens grow to at least 20 feet.

The largest bony fish is the Ocean sunfish (*Mola mola*), which averages 6 feet from snout to tail and 8 feet in vertical length; an outsize specimen caught in 1908 off Sydney, Australia, measured 10 by 14 feet. The largest freshwater fish is the Giant catfish or Pa beuk (*Pangasianodon gigas*), found in the Mekong River and its tributaries in Southeast Asia. A specimen 9 feet 10 inches long and weighing 533 pounds was caught in the Ban Mee Noi River, Thailand.

Of the twenty-three fishes in this list, seven appear to be unknown ray-finned fishes, eight appear to be sharks or rays, one might be a surviving placoderm, three are lungfishes or lobefins, two are of ambiguous provenance, and two others are probably not fishes at all.

Some SEA MONSTERS, FRESHWATER MONSTERS, and MERBEINGS may also involve known or unknown species of fishes.

**Mystery Fishes**

BEEBE'S ABYSSAL FISHES; BEEBE'S MANTA; BLACK FISH (VENOMOUS); CAPTAIN HANNA'S FISH; CHALLENGER DEEP FLATFISH; CLEAR LAKE CATFISH; COELACANTH (Unrecorded Populations); DAKUWAQA; GIANT COOKIECUTTER SHARK; GIANT LUNGFISH; GIANT RAT-TAIL; GIANT SALMON; GLOWING MUDSKIPPER; GROUND SHARK; GUARAÇAÍ AIR-BREATHER; JAPANESE HAIRY FISH; LAKE SENTANI SHARK; LORD OF THE DEEP; MALPELO MONSTER; MANGURUYÚ; MOHA-MOHA; MONGITORE'S MONSTROUS FISH; SEA MONK

## Five-Lined Constellation Fish

One of BEEBE'S ABYSSAL FISHES of the North Atlantic Ocean.

*Scientific name:* *Bathysidus pentagrammus*, given by William Beebe.

*Physical description:* Roundish body. Five lines of purple and yellow photophores on the sides. Large eyes. Small pectoral fins.

*Distribution:* North Atlantic Ocean.

*Significant sighting:* Observed only once at 1,900 feet by William Beebe in a bathysphere off Bermuda in the early 1930s.

*Possible explanation:* Carl Hubbs thought Beebe had seen a mass of jellyfish distorted by the mist of his breath on the bathysphere's porthole.

*Source:* William Beebe, *Half Mile Down* (New York: Harcourt, Brace, 1934).

## Five-Toed Llama

Unknown HOOFED MAMMAL of South America.

*Significant sighting:* Pottery fragments showing animals that look like llamas with five toes were discovered in the 1920s in the pre-Incan Paracas culture area near Pisco, Ica Department, Peru, by Julio C. Tello. Estimates of the culture's age are now put between 600 B.C. and A.D. 200. Tello also apparently came across bones that seemingly belonged to such an animal. Llamas normally have only two toes, and most fossil camelids are similarly cloven-hoofed.

*Possible explanations:*
(1) A Llama (*Lama glama*) with polydactyly, a condition producing more than the normal number of toes, might have been given special treatment by Paracan artists and shamans. Llamas have been domesticated for thousands of years; their ancestors are represented in cave paintings in the Cueva de las Manos, Río Pinturas, Santa Cruz Province, Argentina.
(2) An unknown variety of litoptern, an order of odd-toed South American ungulates that ultimately died out in the Pleistocene. Some were medium-sized and horselike (*Thoatherium*), while others were long-necked with an elongated tapirlike snout (*Macrauchenia*). Most had three toes, but some were reduced to one toe.

*Sources:* Julio César Tello, "Andean Civilization: Some Problems of Peruvian Archaeology," in *Proceedings of the Twenty-third International Congress of Americanists,* September 1928 (New York, 1930), pp. 259–290; Karl Shuker, "Hoofed Mystery Animals and Other Crypto-Ungulates, Part III," *Strange Magazine,* no. 11 (Spring-Summer 1993): 25–27, 48–50.

## Flying Humanoid

Paranormal winged ENTITY found in many traditions worldwide.

*Variant names:* ALAN, BATSQUATCH, CHICK-CHARNEY, Ch'uan-t'ou (in China), GARUDA, GWRACH-Y-RHIBYN, HARPY, Houston bat man, JERSEY DEVIL, MOTHMAN, ORANG BATI, OWLMAN, POPOBAWA, SASABONSAM, SPRINGHEEL JACK, TENGU, Vietnamese flying lady.

*Physical description:* A human being, either naked or clothed, flying through the air by means of wings most often described as batlike.

*Tracks:* Human.

*Distribution:* United States; Bahamas; Brazil; Cornwall, Wales, and London, England; Turkey; Tanzania; India; Philippines; Indonesia; Vietnam; China; Japan; Siberia.

*Significant sightings:* William H. Smith saw a winged human form moving over Brooklyn, New York, on September 18, 1877.

In September 1880, a black man with bat's wings was seen flying 1,000 feet over Coney Island toward New Jersey. The figure had a "cruel and determined expression."

On April 14, 1897, B. C. Wells, the mayor of Mount Vernon, Illinois, and other citizens saw something that resembled the "body of a huge man swimming through the air with an electric light on his back." The incident occurred during a wave of unidentified flying object (UFO) sightings throughout the midwestern United States.

On July 11, 1908, writer Vladimir K. Arsen'ev encountered what he thought was a flying man in the Sikhote Alin Mountains, Siberia. Local hunters had apparently seen it on other occasions.

On January 6, 1948, Bernice Zaikowski and a group of children in Chehalis, Washington, claimed to have seen a man flying by means of long, mechanical wings that he manipulated with instruments on his chest.

In 1952, U.S. Air Force private Sinclair Taylor saw what he thought was an enormous bird while on guard duty at Camp Okubo near Kyoto, Japan. As it hovered not far away, he saw it had the body of a man about 7 feet tall and a 7-foot wingspan. He emptied his carbine into it as it landed but couldn't find a trace of it.

Hilda Walker and two other residents of Houston, Texas, on June 18, 1953, watched a bat-winged figure wearing a cape and bathed in a dim, gray light glide into a pecan tree at 118 East Third Street. The light faded, and the figure disappeared as a torpedo-shaped object swooped overhead.

U.S. Marine Earl Morrison was on guard duty near Da Nang, Vietnam, in the summer of 1969 when he and two other soldiers saw a naked woman with black, furry skin fly about 6–7 feet over their heads just after 1:00 A.M. It was about 5 feet high; flew by flapping black wings attached to its arms, hands, and fingers; and gave off an eerie greenish glow.

*Possible explanations:*
(1) Test flights of personal aeronautical devices.
(2) UFO phenomena or paranormal apparitions.
(3) Misidentifications of various birds or bats.

*Sources: New York Times,* September 12, 1880, p. 6; "Sees Man Fishing from Air Ship," *Chicago Tribune,* April 16, 1897, p. 4; Vladimir K. Arsen'ev, *V gorakh Sikhote-Alinia* (Moscow, 1937); E. D. Edwards, *The Dragon Book* (London: William Hodge, 1938); William C. Thompson, "Houston Bat Man," *Fate* 6 (October 1953): 26–27; Sinclair Taylor, "The Bird Thing," *Fate* 13 (December 1960): 53–54; Samuel Kamakau, *Ka Po'e Kahiko: The People of Old* (Honolulu, Hawaii: Bernice P. Bishop Museum, 1964), pp. 47–53; Joseph Mitchell Johnson, *The Story of a County Pastor* (New York: Vantage, 1967), pp. 245–247; Beulah M. D'Olive Price, "Angels over Milan, Tennessee: A Legend?" *Mississippi Folklore Register* 5 (1971): 122–123; Jerome Clark and Loren Coleman, "Winged Weirdies," *Fate* 25 (March 1972): 80–89; Don Worley, "The Winged Lady in Black," *Flying Saucer Review Case Histories,* no. 10 (June 1972): 14–16; Yuri B. Petrenko, "Forerunner of the Flying 'Lady' of Vietnam?" *Flying Saucer Review* 19, no. 2 (March-April 1973): 29–30; "Homens alados em Pelotas," *SBEDV Boletin,* no. 112–115 (September 1976–April 1977).

# FLYING REPTILES

In the animal world, there are three types of flight: gliding, soaring, and powered. *Gliding* is passive, involving extended body surfaces that transform a vertical fall into a gradual transverse descent. A diapsid reptile named *Coelurosauravus,* found in Western Europe and Madagascar, was the first vertebrate to develop this method in the Late Permian, 260 million years ago; it was 1–2 feet long and had large, retractable wings supported by twenty-two rodlike bones that were not attached to the skeleton and probably were evolved specifically for flight. Another gliding reptile, *Icarosaurus siefkeri,* lived during the

Triassic, 175–200 million years ago, and developed a gliding flight from membranes attached to hinged wing struts extended from its ribs; it was 7 inches long and had a 10-inch wingspan.

The Common flying lizard (*Draco volans*) of Malaysia is similarly outfitted with wings supported by elongated ribs; these are spread when it jumps from trees and glides as far as 30 feet to the ground. Other gliders include the Chinese gliding tree frog (*Polypedates dennysi*), which has suction cups on its toes that enable it to stick to the bark and leaves of the tropical trees and webbed feet that act as parachutes as it falls from tree to tree; the 7-inch Flying gecko (*Ptychozoon lionotum*), which has a loose fold of skin along its legs that allows it to glide; and the 3- to 4-foot Golden tree snake (*Chrysopelea ornata*) of Malaysia and Indonesia that jumps from tree to tree by contracting its body into a concave surface to slow its fall and by undulating in the air to change course (broad, keeled scales on its underside allow it to grip the tree bark).

*Soaring* is accomplished by using extended body surfaces to navigate air currents or rising columns of air, as hawks and eagles do. *Powered flight* is generated by flapping paired aerodynamic wings up and down, creating lift.

The flying reptiles of the Mesozoic were capable of both soaring and true flight. Pterosaur wing membranes were attached to a tremendously elongated fourth finger, supported by three other digits that allowed the animal to crawl on the ground or clamber through trees after the wing was folded. The early, crow-sized, short-tailed pterodactyls were powered, flapping fliers, slightly unstable yet agile. Long-tailed, narrow-winged pterosaurs such as *Rhamphorhynchus* were probably accomplished soarers. *Pteranodon*, with a wingspan of 23 feet, was capable of short, powered flights but was better at long, soaring flights over the sea. The large, short-tailed Cretaceous pterosaurs, such as *Quetzalcoatlus* with a wingspan of 38 feet, were limited to continuous gliding and soaring under mild weather conditions.

Of the thirteen flying reptiles in this section, five are clearly identified as flying (or gliding) snakes. The other eight have been compared to pterosaurs by observers or commentators. The evidence for any one of these cryptids is not particularly strong, but collectively, these creatures embody an intriguing tradition.

Additional types of animals or legends may be involved, such as unknown types of BATS, BIG BIRDS, or various DRAGON traditions. A Mayan relief sculpture of a bird with reptilian characteristics was discovered in the 1960s by archaeologist José Diaz-Bolio at the ruins of El Tajín in Veracruz State, Mexico, which flourished from A.D. 600 to 1200. Such representations are considered by Mayan scholars to depict either stylized birds such as Macaws (parrots of the genera *Ara* and *Anodorhynchus*) or myths such as the celestial bird Itzam-yeh, which represents nature tamed by the Mayans. Some have compared the El Tajín sculpture to a primitive bird such as an *Archaeopteryx* or a pterosaur. "Serpent-Bird of the Mayans," *Science Digest* 64 (November 1968): 1.

### Mystery Flying Reptiles

ARABIAN FLYING SNAKE; CRETAN PTEROSAUR; EUROPEAN FLYING SNAKE; KONGAMATO; MANAUS PTEROSAUR; NAHUEL HUAPÍ PTEROSAUR; NAMIBIAN FLYING SNAKE; OLITIAU; ROPEN; SNALLYGASTER; T'ANG FLYING SNAKE; TRAPPE PTEROSAUR; WELSH WINGED SNAKE

## Fontoynont's Tenrec

Unusual INSECTIVORE of Madagascar.

*Scientific name:* *Dasogale fontoynonti,* given by Guillaume Grandidier in 1929.

*Physical description:* Length, 3.75 inches. Brownish-yellow. Spines on the back and flanks but none on the head or belly. Spines are white at the base, with a black band above and a white tip. Solid claws.

*Distribution:* Eastern Madagascar.

*Significant sighting:* Known only from two poorly preserved specimens, one in the Paris Museum.

*Possible explanations:*

(1) A misidentification of a juvenile Greater hedgehog tenrec (*Setifer setosus*), a common spiny species.

(2) Ross MacPhee acknowledges it as a new species in the genus *Setifer*.

(3) A new species, intermediate between the Spiny tenrecs (Subfamily Tenrecinae) and the Furred tenrecs (Subfamily Oryzorictinae).

*Sources:* Guillaume Grandidier, "Un nouveau type de mammifère insectivore de Madagascar, le *Dasogale fontoynonti* G. Grand.," *Bulletin de l'Academie Malgache,* new ser. 11 (1929): 85–90; Ross D. E. MacPhee, "Systematic Status of *Dasogale fontoynonti* (Tenrecidae, Insectivora)," *Journal of Mammalogy* 68 (1987): 133–135.

## Fotsiaondré

Mystery PRIMATE of Madagascar.

*Etymology:* Betsileo Malagasy (Austronesian), "white sheep."

*Variant name:* Habéby.

*Physical description:* Size of a sheep. White coat, spotted with buff or black. Long, furry ears. Large, staring eyes. Long muzzle. Said to have cloven hooves.

*Behavior:* Nocturnal. Herbivorous.

*Distribution:* Isalo Massif, Madagascar.

*Possible explanation:* An unknown species of lemur, one of a group of three prosimian families (Lemuridae, Cheirogaleidae, and Indriidae) endemic to Madagascar.

*Source:* Raymond Decary, *La faune malgache, son rôle dans les croyances et les usages indigènes* (Paris: Payot, 1950), pp. 203–204.

## FRESHWATER MONSTERS

A wide variety of unidentified animals have been reported in freshwater lakes and rivers around the world. These animals often go under the generic name of lake monsters or river monsters. Most also are known by the name of the lake or river plus the word "monster," as in the "Payette Lake monster."

A comprehensive, country-by-country list is found in the "Lake and River Monster" section on pages 655–690. More specific information can be found under the following names:

AFANC; AIDAKHAR; ALTAMAHA-HA; AMHÚLUK; ANFISH; ANGEOA; ANGONT; ARCHIE; ASHUAPS; ATÚNKAI; BALONG BIDAI; BIG WALLY; BOZHO; BROSNIE; BUNYIP; CANAVAR; CARABUNCLE; CECIL; CHAMP; CHAN; CHAOUSAROU; CHRISTINA; CHUNUCKLAS; CRESSIE; CUERO; DAKWA; EELPOOT; ELBST; GAUARGE; GJEVSTROLL; GROOT SLANG; GRYTTIE; GUÀI WÙ; GUIRIVILU; HAMLET; HAPYXELOR; HIPPOGRIFF; HIPPOTURTLEOX; HORSE'S HEAD; HUILLA; IGOPOGO; ILLIE; INKANYAMBA; ISSIE; KA-IS-TO-WAH-EA; KLATO; KOLOWISI; KTCHI PITCHKAYAM; KUDDIMUDRA; KURREA; KUSHII; LAGARFLJÓTSORMURINN; LAU; LENAPÍZHA; LINDORM; LIZZIE; LLAMHIGYN Y DWR; LUKWATA; MAASIE; MAMLAMBO; MANETÚWI MSÍ-PISSÍ; MANIPOGO; MEMPHRÉ; MESSIE; MI-NI-WA-TU; MISHIPIZHIW; MISIGANEBIC; MJOSSIE; MONTANA NESSIE; MORAG; MOSQUETO; NAHUELITO; NAITAKA; NAMPÈSHIU; NESSIE; NHANG; NINKI NANKA; NYCKER; NYKKJEN; OGOPOGO; OGUA; OLD NED; ON NIONT; ONYARE; OOGLE BOOGLE; PADDLER; PAMBA; PATAGONIAN PLESIOSAUR; PÉIST; PICTISH BEAST; PINK EYE; PINKY; PIRANU; PONIK; PUFF; RASSIC; ROCKY; RØMMIE; SEILEAG; SELMA; SHARLIE; SHUSWAGGI; SINT-HOLO; SKRIMSL; SLAL'I'KUM; SLIMY CASPAR; SOUTH BAY BESSIE; STORSJÖODJURET; TAG; TAHOE TESSIE; TANIWHA; TCINTO-SAKTCO; TEGGIE; THREE-TOES; TIRICHUK; TLANÚSI; TRELQUEHUECUVE; TSINQUAW; UKTENA; ULAR TEDONG; UNKTEHI; VASSTROLLET; WAKANDAGI; WATER BULL; WATER HORSE; WEE OICHY; WEWIWILEMITÁ MANETÚ; WHITEY; WINNIPOGO; WIWILIÁMECQ'; WURRUM; YERO; ZEMO'HGÚ-ANI

Some of the most common descriptions are of animals that resemble either a telephone pole, a big fish, an overturned boat (especially larger individuals), a head and long neck, a floating log, a serpent, or a string of humps. The shape of the head is most often described as either horselike, cowlike, serpentine, crocodilian, or fishlike. Ridges, spines, or horns are sometimes reported. Appendages include fins, flippers, or feet. The skin is as often smooth as scaly. Whiskers and a mane are frequent adornments.

It is difficult and probably ill advised to place specific Freshwater monsters into distinct categories. Eyewitness descriptions are often imprecise, and in many cases, the animals themselves appear to lack any distinguishing features other than unusual size. Several different descriptions may be given for animals seen in one specific lake. Though this could mean that several unknown species live in the same body of water, it is more likely that (1) descriptions of an unexpected, exotic animal viewed at a distance under unfavorable atmospheric or optical conditions by untrained observers may often be less than accurate, and/or (2) cultural preconceptions of what a Freshwater monster should look like may override an observer's objectivity.

However, in general, at least four categories of unknown animals seem to be involved:

(1) A long-necked animal, perhaps a freshwater equivalent of Bernard Heuvelmans's marine LONGNECK. Some of the most notable are CHAMP, NESSIE, and STORSJÖODJURET. In general, this type has a small, flat head on a long neck; small eyes; small horns, which may actually be either ears or some form of breathing apparatus; a mane; two or three dorsal humps; little or no tail; and four webbed feet or paddles. The animal is most often seen in warm weather when the water surface is flat calm. It can swim and dive swiftly.

(2) A serpentine freshwater animal similar to Bernard Heuvelmans's MULTIHUMPED SEA MONSTER, often perceived only as a string of humps or coils. The Canadian MANIPOGO and OGOPOGO could be placed in this category. This type has a serpentine body and a long tail, and it usually swims by vertical undulations.

(3) The European WATER HORSE appears to be a freshwater form of Bernard Heuvelmans's MERHORSE. Its distinctive, horselike head gives credence to the legend that the animal can pass as a domestic horse.

(4) Different types of large FISHES, especially those that have the distinctive, razorback scutes of a sturgeon along the spine.

*Possible explanations:*

(1) A surviving basilosaurid, a family of archaic CETACEANS that lived 42–33 million years ago in the Middle to Late Eocene, could account for the serpentine type of Freshwater monster. These animals possessed torpedo-shaped bodies with flexible vertebrae, elongated skulls, limber necks, twin flippers derived from forelegs, small dorsal fins, and long, fluked tails. The hind limbs, about 2 feet long in *Basilosaurus isis,* were reduced but functional. Nostrils were placed at the top of the snout. Unlike modern whales, their teeth were differentiated into incisors, canines, and molars. Basilosaurids came in a wide range of sizes, 6–82 feet long. They are known to have inhabited shallow coastal waters and swamps as well as open oceans. Fossils have been found in Egypt (especially the Zeuglodon Valley), India, and North America.

(2) A surviving plesiosaur matches the morphology of many lake monsters. This group of marine reptiles lived 238–65 million years ago and swam with paddlelike limbs. Some, but not all, species had long necks, and body length varied from 6 to 46 feet. The continuing evolution of a surviving species from the Mesozoic might account for such characteristics as cold-weather adaptation, a freshwater habitat, and the ability to crawl on land.

(3) An unknown, large, long-necked SEAL (Suborder Pinnipedia) could account for animals existing in cold, northern lakes. This theory was first advanced for NESSIE by the Dutch zoologist Anthonie Cornelis Oudemans in 1935, based on his idea in the 1890s that most SEA MONSTER sightings involved such an animal. However, the fossil evidence is lacking. A seal would likely be seen at the surface more often than most lake monsters.

(4) Various species of Sturgeon (Family Acipenseridae), large marine fishes with several rows of dorsal plates, swim up rivers into lakes for spawning and are likely candidates for certain lake monsters.

(5) An unknown species of giant Freshwater eel (Family Anguillidae) would account for scarce surface sightings of serpentine animals but is more problematic for the humped variety.

(6) A surviving duck-billed DINOSAUR (Family Hadrosauridae), one of the most

abundant types in North America in the Late Cretaceous, 80 million years ago, was suggested by Loren Coleman. These browsing vegetarians often had bony head crests that probably served as visual and audible signaling devices. However, because of their stiff tails, short toes, and small hands, it's unlikely they spent much of the time in the water.

(7) Boat wakes, logs, swimming deer, and other misidentifications.

(8) Mats of rotting vegetation.

(9) The Northern river otter (*Lontra canadensis*) and European otter (*Lutra lutra*) can appear mysterious to the untrained eye. The otter's streamlined body is almost serpentine, and several animals swimming in a row could simulate a monster from a distance.

(10) Wayward Sharks (Subclass Elasmobranchii) often travel many miles upriver. In 1978, a shark was found in the water intake of Edison's Trenton Channel Power Plant off the Detroit River in Michigan.

(11) Wayward seals also migrate into freshwater environments occasionally.

(12) A surviving *Phobosuchus*, a 50-foot marine CROCODILIAN that lived in the Cretaceous, 70 million years ago, has been proposed by Mark A. Hall.

(13) A surviving mosasaur, such as *Clidastes propython*, a marine reptile that lived 88–74 million years ago, also was suggested by Mark A. Hall. It was about 12 feet long, ate fishes and squid, and swam by horizontal undulations. Fossils have been found in Kansas and Alabama. *Plotosaurus* was nearly three times as large (33 feet) and was also found in Kansas. The end of its tail extended into a vertical fin. *Platecarpus*, another Kansas mosasaur, was about 21 feet long and died out at the end of the Cretaceous.

For an overview of Freshwater monsters, see Peter Costello, *In Search of Lake Monsters* (New York: Coward, McCann and Geoghegan, 1974); Michel Meurger and Claude Gagnon, *Lake Monster Traditions: A*

*Cross-Cultural Analysis* (London: Fortean Tomes, 1988); Karl Shuker, *In Search of Prehistoric Survivors* (London: Blandford, 1995), pp. 78–113; John Kirk, *In the Domain of the Lake Monsters* (Toronto, Canada: Key Porter, 1998).

## Freshwater Octopus

Medium-sized CEPHALOPOD occasionally found in rivers of North America.

*Physical description:* Length, 2–3 feet.

*Distribution:* Licking River, Kentucky; Kanawha and Blackwater Rivers, West Virginia; Ohio River, at Louisville and Cincinnati.

*Significant sightings:* On December 24, 1933, Robert Trice and R. M. Saunders were fishing on the Kanawha River near Charleston, West Virginia, when they hauled in a 3-foot octopus. Recent research by Mark Hall has proven this incident a hoax.

On January 30, 1959, a gray octopus was seen surfacing and moving onto the bank of the Licking River near Covington, Kentucky.

On November 19, 1999, a dead octopus was found on the bank of the Ohio River at the Falls of the Ohio State Park, Jeffersonville, Indiana, on some fossil beds. It was identified as either a Caribbean armstripe octopus (*Octopus burryi*) or a Bumblebee two-stripe octopus (*O. filosus*), both Atlantic species, and was not in a state of decomposition. Both are available through aquariums.

*Present status:* All known cephalopod species are exclusively marine. Octopuses, even more than squid, require high salinity levels.

*Possible explanations:*

(1) Discarded aquarium pets.

(2) Wandering, senescent individuals at the end of their life cycle, though the Ohio River is a bit far to stray in essentially toxic water.

(3) An unknown species of octopus adapted to a low-saline environment.

*Sources:* "Octopus Caught by Two Boatmen on Kanawha River," *Charleston (W. Va.) Gazette,* December 25, 1933; "Octopus Story Just a Hoax," *Charleston (W. Va.) Gazette,* December 29, 1933; "More Details Needed,"

*Doubt,* no. 16 (1946): 242; "Displaced Critters," *Doubt,* no. 48 (1955): 341; Chad Arment and Brad LaGrange, "A Freshwater Octopus?" *North American BioFortean Review* 2, no. 3 (December 2000): 47–51, http://www.strangeark.com/nabr/NABR5.pdf; Mark A. Hall, "Mysteries of West Virginia," *Wonders* 6, no. 4 (December 2001): 113–126.

## Furred Sea Monster

SEA MONSTER of the Indian and North Pacific Oceans.

*Physical description:* Length, 25–47 feet. Covered in thick, white fur. Tapering head like an elephant's, 5 feet long. Trunklike appendage, 5 feet long and 14 inches in diameter. Tail, 14 feet long, beginning at the rib section.

*Behavior:* Fights with whales.

*Distribution:* The coasts of South Africa and Alaska.

*Significant sightings:* On November 1, 1922, Hugh Ballance saw two whales fighting with an unusual animal some 1,300 yards off the shore near Margate, KwaZulu-Natal Province, South Africa. Through binoculars, it looked like a huge polar bear with a tail with which it struck the whales repeatedly. Crowds of people watched the battle for three hours until the monster was killed. The next night, the carcass washed ashore and lay on the beach for ten days. It had an elephant's trunk and was covered in snow-white hair.

In November 1930, the carcass of a 25-foot animal with a long and tapering head washed up on Glacier Island, Alaska. W. J. McDonald, supervisor of the Chugach National Forest, and six others examined the body, which had very little flesh left on it. The widest part of the skeleton was 3 feet 2 inches. Its weight was estimated at 1,000 pounds.

*Possible explanations:*

(1) A decomposing shark or whale, where the dried-out, fibrous connective tissue looks like white fur.

(2) An unknown marine mammal, completely unlike anything else in the fossil record.

*Sources: Daily Mail* (London), December 27, 1924; "Ice Bares Strange Animal," *New York Times,* November 26, 1930; "Monster in Ice Has Long Snout," *New York Sun,* November 28, 1930; "Confirm Finding of Pre-Historic Monster in Ice," *New York Evening World,* November 28, 1930; "Furry Beast 20 Feet Long Is Washed Ashore Lifeless," *New York Times,* October 3, 1944; Thomas Victor Bulpin, *Your Undiscovered Country* (Durban, South Africa: Total Oil Products, 1965).

# G

## Gabon Orangutan

Mystery PRIMATE of Central Africa.

*Variant name:* AIZ 6624 (specimen catalog number).

*Physical description:* An unusually small variety of chimpanzee. Weight, 6 pounds. Dark, gray-brown skin on face, ears, back, and lateral portion of limbs. High, hairless forehead. Small face. Protuberant eyes. Narrow nose. Hands and feet small in relation to the body. Lacks thumbs and big toes.

*Behavior:* Said to travel in a group of 100.

*Distribution:* Gabon.

*Significant sighting:* A specimen was obtained in August 1957 by Phillip J. Carroll and sent to the Anthropological Institute of the University of Zürich. It had fallen from a tree, injured itself, and died three weeks later. Carroll claimed it was with a group of 100 other chimps of the same size.

*Probable explanation:* The individual was a young Chimpanzee (*Pan troglodytes*) with severe deformities of the skull and skeleton.

*Sources:* Adolph H. Schultz, "Acrocephalo-Oligodactylism in a Wild Chimpanzee," *Journal of Anatomy* 92 (1958): 568–579; Ivan T. Sanderson, *Abominable Snowmen: Legend Come to Life* (Philadelphia: Chilton, 1961), p. 186; Michael K. Diamond, "Setting the Record Straight on the 'Gabun Orangutan,'" *Pursuit,* no. 48 (Fall 1979): 142–145.

## Gabriel Feather

A single BIRD feather housed in El Escorial Palace near Madrid, Spain.

*Etymology:* Said to have come from a wing of the Archangel Gabriel.

*Physical description:* A rose-colored feather of extraordinary beauty.

*Present status:* Acquired by El Escorial sometime after the palace was built between 1563 and 1584 by King Philip II. Seen in 1787 by William Beckford. Apparently, the Monastery of San Lorenzo at the palace no longer owns this relic.

*Possible explanations:*

(1) A feather of the Resplendent quetzal (*Pharomachrus mocinno*) of Central America, a bird of the Trogon family sacred to the Aztec and Maya Indians and famous for its plumage. But the prized wing and tail feathers of the male are green, not rose.

(2) A plume from one of New Guinea's Birds of paradise (Family Paradisaeidae). Survivors of Ferdinand Magellan's voyage around the world took back to Portugal skins of these birds obtained from the island's inhabitants in 1522. Count Raggi's bird of paradise (*Paradisea raggiana*) has rose-colored plumes, which Karl Shuker suggests might account for the Gabriel feather.

*Sources:* William Beckford, *Italy: With Sketches of Spain and Portugal,* vol. 2 (London: R. Bently, 1834); Karl Shuker, "Angel Feathers and Feathered Snakes," *Strange Magazine,* no. 19 (Spring 1998): 24–25; Karl Shuker, *Mysteries of Planet Earth* (London: Carlton, 1999), pp. 150–153.

## Gabriel Hound

BLACK DOG of northern England.

*Etymology:* First recorded around 1665. Originally, a spectral dog in a pack led by the pre-Christian spirits Herne or Gwyn that escorted souls to the underworld; in Christian folklore, the pack was transferred to the care of the

Archangel Gabriel. An alternate explanation is that the word is derived from an ancient word, *gabbara* ("dead body").

*Variant names:* Gabble ratchet, Gabriel ratchet, Sky yelper.

*Physical description:* Huge dog with a human head. Sometimes described as a spectral bird with glowing eyes.

*Behavior:* Makes eerie howls. Said to travel high in the air and hover over a house when misfortune is about to occur.

*Distribution:* Lancashire, Derbyshire, and Cleveland, England.

*Possible explanation:* It is said that the sound of migrating Bean geese (*Anser fabalis*) flapping their wings can be mistaken for the baying of a pack of these hounds. The howling or "gabbling" sounds might also be produced by a Curlew (*Numenius arquata*), Eurasian wigeon (*Anas penelope*), or Eurasian teal (*Anas crecca*).

*Sources:* Lewis Spence, *The Fairy Tradition in Britain* (London: Rider, 1948); Katharine M. Briggs, *A Dictionary of Fairies* (London: Allen Lane, 1976), p. 183; Jacqueline Simpson and Steve Roud, *A Dictionary of English Folklore* (New York: Oxford University Press, 2000), p. 139.


## Gally-Trot

BLACK DOG of southern England.

*Etymology:* Possibly from the French *gardez le tresor* ("guard the treasure"); from *gally* ("frighten") + the German *Trötsch* ("spirit"); or from the Frisian *glay* or *gley* ("shining") + *Trötsch*.

*Variant names:* Galley trot, Hound of the hill, White hound of Cator.

*Physical description:* Size of a bullock. White, shaggy coat. Red ears.

*Behavior:* Chases people who try to run away from it.

*Habitat:* Lives in hollow hills.

*Distribution:* Norfolk and Suffolk; Leek Brook, Staffordshire; Pluckley, Kent; Wellington, Somerset; Bunbury, Cheshire; Dartmoor, Devon.

*Sources:* Alasdair Alpin MacGregor, *The Ghost Book* (London: Robert Hale, 1955), pp. 55–81; Ruth L. Tongue, "Traces of Fairy

Hounds in Somerset," *Folklore* 67 (1956): 233–234; Ruth E. Saint Leger-Gordon, *The Witchcraft and Folklore of Dartmoor* (London: Robert Hale, 1965), p. 188; Katharine M. Briggs, *A Dictionary of Fairies* (London: Allen Lane, 1976), pp. 183, 225–226; Karl Shuker, "White Dogs and Fairy Hounds," *Strange Magazine*, no. 19 (Spring 1998): 12–13.


## Gambo

SEA MONSTER of West Africa.

*Etymology:* Coined by Karl Shuker after the name of the country, The Gambia.

*Variant name:* Kunthum belein (Mandinka word for dolphin, literally "cutting jaws").

*Physical description:* Smooth, scaleless skin. Length, 15 feet. Width, 5 feet. Dark brown on top, white below. Dolphinlike head. Small, brown eyes. Jaws, 18 inches in length, with eighty sharp, conical, uniform teeth. No blowhole. Nostrils are at the tip of the jaws. Short neck. No dorsal fin. Four paddle-shaped flippers, each 18 inches long. Pointed tail, 5 feet long. No flukes.

*Distribution:* Kotu, The Gambia.

*Significant sighting:* On June 12, 1983, Owen Burnham discovered the carcass of an odd sea creature washed up on the beach near the Bungalow Beach Hotel at Kotu. Local people were in the process of cutting off the head to sell when he found it.

*Possible explanations:*

(1) The combination of four paddles, eighty teeth, lack of scales and blowhole, and long tail rules out seals, known cetaceans, sirenians, modern reptiles, and fishes.

(2) Fossil archaic basilosaurid whales only had forty teeth.

(3) Shepherd's beaked whale (*Tasmacetus shepherdi*) matches somewhat in coloration, but it has a blowhole, tail flukes, a dorsal fin, a much shorter beak, no nostrils, and no pelvic flippers. In addition, this rare cetacean prefers the cold water of New Zealand and the South Atlantic.

(4) A surviving pliosaur, a member of a group of short-necked plesiosaurs with large heads, elongated jaws with massive teeth,

GAMBO, *an odd sea monster that washed up on the beach in Gambia in 1983. (William M. Rebsamen)*

two sets of flippers, and pointed tails. In some larger species such as *Kronosaurus queenslandicus* (over 40 feet), the skull was as much as 10 feet long. These marine reptiles lived 200–65 million years ago (from the Early Jurassic to the end of the Cretaceous), swam underwater aerodynamically like penguins, and were probably pursuit predators.

(5) A surviving mosasaur, a group of twenty genera that included some of the largest marine reptiles ever, frequently exceeding 33 feet in length. They lived in the Late Cretaceous, 95–65 million years ago, and had large, conical teeth, each set in a deep socket. The plioplatecarpines and tylosaurines had short bodies and long, narrow tails.

(6) A surviving metriorhynchid archosaur, a member of a group of thalattosuchians with flippers, no dermal armor, and an expansion at the end of the tail. These reptiles lived 200–95 million years ago, from the Early Jurassic to the Late Cretaceous. Like mosasaurs, they moved through the water by undulating trunk and tail.

(7) A surviving ichthyosaur, a group of dolphinlike reptiles with narrow, pointed snouts and spindle-shaped bodies. They lived 245–65 million years ago, from the Early Triassic to the end of the Cretaceous, reaching their greatest size (about 48 feet) in the Late Triassic. The ichthyosaur had big eyes, nostrils placed well back from the tip of the snout, a dorsal fin, and a fishlike tail that did all the work of moving the animal through the water.

(8) A surviving champsosaur, a freshwater, crocodile-like animal with a flat skull and slender snout that lived from the Late Cretaceous to the Oligocene, 70–30 million years ago. It had well-ossified limbs and could probably walk on land.

*Sources:* Karl Shuker, "Gambo: The Beaked

Beast of Bungalow Beach," *Fortean Times,* no. 67 (February-March 1993): 35–37; Karl Shuker, *In Search of Prehistoric Survivors* (London: Blandford, 1995), pp. 116–118.

## Ganba

Mythical giant SNAKE of Australia.

*Etymology:* Mirning (Australian) word.

*Variant name:* Jeedarra.

*Physical description:* Huge size.

*Behavior:* The gurgling sound of underground streams is said to be caused by its breath. Seizes and eats people.

*Habitat:* Underground caves; also the ocean.

*Distribution:* Nullarbor Plain, South Australia.

*Possible explanations:*

(1) The Amethystine python (*Morelia amethystina*) of Cape York, Queensland, is Australia's largest snake. Average specimens are 15 feet long, though outsize individuals attain nearly 24 feet. Its scales have an iridescent sheen. However, this snake strictly lives in the forest.

(2) The Taipan (*Oxyuranus scutellatus*), one of Australia's most dangerous snakes, is found along the coast of Queensland and the Kimberley region.

*Sources:* Daisy Bates, *The Passing of the Aborigines* (London: John Murray, 1938), p. 132; Charles Barrett, *The Bunyip and Other Mythical Monsters and Legends* (Melbourne, Australia: Reed and Harris, 1946), pp. 47–48.

## Gargouille

Legendary DRAGON of France.

*Etymology:* French, "gargler."

*Variant name:* Gargoyle.

*Physical description:* Serpentine. Scaly head. Slender snout. Eyes that gleam like moonstones. Long neck. Four membranous flippers.

*Behavior:* Shoots jets of water from its mouth. Lives in a cave on the riverbank. Capsizes boats. Eats people.

*Distribution:* Seine River in Normandy, France.

*Significant sighting:* A scaly monster emerged from the Seine River near Rouen, France, in the early seventh century and caused flooding by emitting jets of water from its mouth. It was subdued by St. Romain (Romanus), archbishop of Rouen (from A.D. 626 to 640), who led it back to town, where it was burned to death.

*Present status:* Served as the inspiration for the architectural gargoyles that began to adorn French churches in the thirteenth century as waterspouts.

*Sources:* *Histoire véritable de la Gargouille: Complainte en 32 couplets* (Caen, France: Chez Renardini, 1826), http://users.skynet.be/dhs/gargouilles/legende.htm; Karl Shuker, *Dragons: A Natural History* (New York: Simon and Schuster, 1995), pp. 18–19; *La fête de la Gargouille à Rouen,* http://www.france-pittoresque.com/traditions/24.htm.

## Garuda

FLYING HUMANOID or bird of the Indian subcontinent. In Hindu mythology, it is the king of the birds and is identified with fire and the sun.

*Etymology:* Sanskrit (Indo-Aryan) word.

*Variant names:* Garutmat, Gerda (in Malaysia), Kruth (in Cambodia), Nagantaka ("destroyer of snakes"), Sitanana ("white face"), Taraswin, Vinayaka ("destroyer of obstacles").

*Physical description:* Large bird with human arms and legs. Said to be as bright as the sun. Its white face is half human, half bird. Golden or green body feathers. Scarlet wings.

*Behavior:* Enemy of snakes (NAGA).

*Distribution:* India; Southeast Asia.

*Present status:* In Hindu mythology, this entity is the vehicle of the god Vishnu. Earlier depictions show the Garuda as an eaglelike bird; later artwork makes it more human. In Indonesia, the Garuda has survived modernization to become the national emblem.

*Possible explanation:* The Brahminy kite (*Haliastur indus*) is a common hawk found from India east to the Solomon Islands. It has a distinctive, deep-chestnut color and a white head and neck. Length, 20 inches. Hindus consider it a sacred bird associated with Garuda.

*Sources:* *The Mahārbhārata,* ed. J. A. B. van Buitenen (Chicago: University of Chicago

Press, 1973–1978), vol. 1, pp. 78–92, 419, and vol. 3, pp. 389–395 (Āstēka, I. V. 19–31; Gālava, V. LIV. 99–103); *The Garuda Purānam*, trans. Manmatha Nath Dutt Shastrī (Varanasi, India: Chowkhamba Sanskrit Series Office, 1968); *Hindu Myths: A Sourcebook Translated from the Sanskrit*, ed. Wendy Doniger O'Flaherty (New York: Penguin, 1975), pp. 221–228; Joe Nigg, *A Guide to the Imaginary Birds of the World* (Cambridge, Mass.: Apple-Wood, 1984), pp. 73–75; Shanti Lal Nagar, *Garuda, the Celestial Bird* (New Delhi: Book India, 1992); Garuda, 2001, http://www.khandro.net/mysterious_garuda.htm.

## Gassingrâm

TIGRE DE MONTAGNE of Central Africa.

*Etymology:* Yulu (Nilo-Saharan) word, possibly incorrectly transcribed.

*Variant name:* Vassoko.

*Physical description:* Larger than a lion. Reddish-brown. Eyes glow at night like headlights. Small ears like a dog's. Long fangs that extend beyond its lips.

*Behavior:* Primarily nocturnal. Bellows like an elephant. Carnivorous. Carries off its prey to the mountains.

*Tracks:* Larger than a lion's.

*Habitat:* Caves in the mountains.

*Distribution:* Massif des Bongos, near Ouanda Djallé, Central African Republic.

*Sources:* Bernard Heuvelmans, *On the Track of Unknown Animals* (New York: Hill and Wang, 1958), p. 465; Bernard Heuvelmans, *Les derniers dragons d'Afrique* (Paris: Plon, 1978), pp. 263, 266, 383, 392, 395.

## Gauarge

Mythical FRESHWATER MONSTER of Australia.

*Etymology:* Australian word.

*Physical description:* Like a featherless emu.

*Behavior:* Drags bathers down into a whirlpool.

*Habitat:* Water holes.

*Possible explanation:* Folk memory or extrapolation based on fossils of an Australian theropod dinosaur such as *Kakuru*, which lived in the Early Cretaceous, 110 million years ago, in South Australia.

*Sources:* Gilbert Whitley, "Mystery Animals of Australia," *Australian Museum Magazine* 7 (1940): 132–139; Bernard Heuvelmans, *On the Track of Unknown Animals* (New York: Hill and Wang, 1958), pp. 193–194.

## Genaprugwirion

Mystery LIZARD of Wales.

*Etymology:* From Welsh *genau* ("mouth") + *pryf* ("insect") + *gwirion* ("silly") = "silly insect-eater" (?).

*Variant name:* Cenaprugwirion.

*Physical description:* Length, 12 inches. Muddy-brown color. Head is the size of an orange. Pronounced dewlap. Long tongue.

*Behavior:* Rolls its eyes continually. Lives in a burrow, poking its head out to catch flies or insects.

*Distribution:* Aber Sôch, Lleyn Peninsula, Gwynedd, Wales.

*Present status:* Now rare but said to be common long ago.

*Possible explanations:*

(1) Naturalized population of a nonnative lizard, such as an Iguana (Family Iguanidae), Agama (Family Agamidae), Skink (Family Scincidae), or Chameleon (Family Chamaeleonidae). However, the Welsh climate is not suitable for a sustained population of these tropical lizards.

(2) Karl Shuker has suggested a naturalized population of Tuataras (*Sphenodon punctatus* or *S. guntheri*) of New Zealand, lizardlike reptiles that were often kept as exotic pets in the nineteenth century. Adults measure 16–26 inches long and have such a low metabolic rate that they can go an hour without breathing and subsist indefinitely on two earthworms a week. Their maximum life expectancy in the wild could be 100 years or more. Able to withstand a temperate climate, the Tuatara is the last living representative of the Order Sphenodontida and is now confined to about twenty small islands off the northeast coast of New Zealand and in Cook Strait.

Sphenodonts were once widespread, and fossils from the Late Triassic through the Jurassic, 210–140 million years ago, have been found in England and continental Europe.

*Sources:* Richard Wallis (letter), *British Herpetological Society Bulletin,* Autumn-Winter 1987, p. 65; Karl Shuker, "Land of the Lizard King," *Fortean Times,* no. 95 (February 1997): 42–43.


## Ge-No'sgwa

CANNIBAL GIANT of the northeastern United States and Canada.

*Etymology:* Seneca (Iroquoian), "stone giant."

*Variant names:* Ot-ne-yar-hed (Onondaga /Iroquoian), Stone giant.

*Behavior:* Rubs its body with tree resin and sand.

*Distribution:* New York; Ontario, Canada.

*Sources:* Hartley Burr Alexander, *North American Mythology* (Boston: Marshall Jones, 1916), p. 29; William Martin Beauchamp, *Iroquois Folk Lore, Gathered from the Six Nations of New York* (Syracuse, N.Y.: Dehler, 1922); Marvin A. Rapp, "Legend of the Stone Giants," *New York Folklore Quarterly* 12 (1956): 280–282; Joseph Bruchac, *Stone Giants and Flying Heads* (Trumansburg, N.Y.: Crossing, 1979); Marianne Mithun and Myrtle Peterson, "Ge:no:sgwa' (The Stonecoat)," in Marianne Mithun and Hanni Woodbury, eds., *Northern Iroquoian Texts* (Chicago: University of Chicago Press, 1980), pp. 110–122.


## Gérésun Bamburshé

WILDMAN of Central Asia.

*Etymology:* Tibetan (Sino-Tibetan), "wild man."

*Physical description:* Covered in long hair.

*Behavior:* Stands erect. Wears clothing made of skins. Throws stones at travelers.

*Distribution:* Central Tibet.

*Sources:* William Woodville Rockhill, *The Land of the Lamas* (New York: Century, 1891), pp. 116–117, 150–151; William Woodville Rockhill, "Explorations in Mongolia and Tibet," *Annual Report of the Smithsonian Institution,* 1892, pp. 669–670; William Montgomery McGovern, *To Lhasa in Disguise* (New York: Century, 1924), pp. 118–121.


## Gerit

GIANT HOMINID of East Africa.

*Etymology:* Kalenjin (Nilo-Saharan) word.

*Variant names:* Gereet, Gereit, Kereet, N'gugu (Masai/Nilo-Saharan), Tiondo (Kalenjin/Nilo-Saharan).

*Physical description:* Slightly larger than a human. Dull reddish-yellow.

*Behavior:* Bipedal. Steals honey from beehives.

*Habitat:* Caves. Does not frequent the thick forests.

*Distribution:* Western Kenya.

*Present status:* The name is easily confused with that of the hyena-like GETEIT.

*Source:* Bernard Heuvelmans, *Les bêtes humaines d'Afrique* (Paris: Plon, 1980), p. 544.


## Geteit

Local name for the NANDI BEAR of East Africa.

*Etymology:* Kalenjin (Nilo-Saharan), "brain-eater."

*Variant names:* Gadett (Masai/Nilo-Saharan), Keteit, Ketit.

*Behavior:* Said to break into native huts at night, kill the occupants, and eat their brains. Kills goats and sheep in the same way. Rises on hind legs to attack.

*Distribution:* Western Kenya.

*Significant sighting:* In the 1920s, a Gadett was said to have eaten the brains of fifty-seven goats and sheep over a period of ten days, leaving thirteen victims alive. In this instance, the animal turned out to be an unusually large Spotted hyena (*Crocuta crocuta*).

*Source:* Charles R. S. Pitman, *A Game Warden among His Charges* (London: Nisbet, 1931), pp. 287–302.


## Get'qun

CANNIBAL GIANT of Alaska.

*Etymology:* Na-Dené word.

GIANT ANACONDA *reportedly killed by the Brazilian Boundary Commission in 1932. (William M. Rebsamen/Fortean Picture Library)*

*Distribution:* Lake Iliamna, Alaska.

*Source:* Kyle Mizokami, Bigfoot-Like Figures in North American Folklore and Tradition, http://www.rain.org/campinternet/bigfoot/bigfoot-folklore.html.

## Giant Anaconda

Individual SNAKE or a separate species of anaconda of South America that exceeds the accepted length of 30 feet.

*Variant names:* Boiúba, Boiúna, Camoodi, Cobra-grande, Controller, Ibibaboka, Lampalagua (in Argentina), Matatora (Spanish, "bull-killer"), SACHAMAMA, Sucuriju gigante (Portuguese, "giant anaconda"), Yaurinka.

*Physical description:* Length, 40–150 feet. Dark-chestnut color. Diameter, more than 2 feet 6 inches. Weight, up to 5 tons. Triangular head. Two horns above the eyes. Large eyes glow phosphorescent blue at night. Off-white spots on belly.

*Behavior:* Semiaquatic. A swift swimmer. Creates a huge wake.

*Tracks:* Wide furrow through the swamp, with trees pushed up.

*Distribution:* Amazon River basin, Brazil; less credible reports are from Argentina, Venezuela, and Guyana.

*Significant sightings:* Explorer Percy H. Fawcett shot a 62-foot anaconda on the Rio Abuna, Acre State, Brazil, near the Peruvian border, in January 1907. However, its largest diameter was 12 inches, which seems small for the length.

Fr. Victor Heinz saw a giant water snake on the Amazon near Alenquer, Brazil, on October 29, 1929. Its blue-green, phosphorescent eyes were at first mistaken for a riverboat's lights.

A photo made into a postcard shows a 105-foot snake with shining eyes that was reportedly killed by the Brazilian Boundary Commission on the frontier with Venezuela in 1932. It was said to be 4 feet thick. No hint of the snake's

*Large, but not giant, anaconda* (Eunectes murinus) *at the Bronx Zoo. From the National Geographic Society's* Scenes from Every Land. *(© 2002 ArtToday.com, Inc., an IMSI Company)*

size is provided, though three out-of-focus humans can be seen in the background.

Another photo taken in 1948 or 1949 by Joaquim Alencar shows a huge snake, variously said to be 115 or 147 feet long, floating on the Rio Abuna, Acre State, Brazil.

In 1977, Amarilho Vincente de Oliveira saw a giant snake with horns and greenish eyes on a tributary of the Rio Purus, Brazil.

*Present status:* The record length for an anaconda of 37 feet 6 inches, reported in 1939 or 1940 by Robert Lamon, is not universally accepted. John Murphy and Robert Henderson point out that enormous snakes have fewer places in which to hide from predators, and their great size would cause problems with maintaining blood pressure in the tail. After surveying the ratio of minimum adult length to the record length of many snake species, Peter Pritchard has concluded that the maximum length of a snake is 1.5–2.5 times its shortest adult length; small anacondas are 10–12 feet long, making the largest theoretical length 30 feet. Aaron Bauer estimates that Fawcett's 62-

foot snake had to have been at least 30 inches in diameter and spend virtually all its time in the water.

*Possible explanations:*
(1) Skins of normal-sized Anacondas (*Eunectes murinus*) are often dried and stretched, resulting in a much greater length. Even a skin that is not stretched can be 10–20 percent longer than the live snake.
(2) Field estimates are often unreliable, especially for snakes that are partially submerged.
(3) An unknown species of anaconda that normally attains such lengths.
(4) Reports of horns might be caused by protruding eyes, fleshy outgrowths caused by an injury, or barlike markings on the head.

*Sources:* Algot Lange, *In the Amazon Jungle* (New York: G. P. Putnam's Sons, 1912), pp. 229–238; Afranio do Amaral, "Serpentes gigantes," *Boletim do Museu Paraense Emílio Goeldi* 10 (1948): 211–237; Percy H. Fawcett, *Exploration Fawcett* (London: Hutchinson,

1953), pp. 92–93; Bernard Heuvelmans, *On the Track of Unknown Animals* (New York: Hill and Wang, 1958), pp. 284–298; Paul Gregor, *Amazon Fortune Hunter* (London: Souvenir, 1962), pp. 58–65, 85–90; Tim Dinsdale, *The Leviathans* (London: Routledge and Kegan Paul, 1966), pp. 120–130; "Giant Snakes," *Pursuit*, no. 6 (April 1969): 36–37; Richard Perry, *The World of the Jaguar* (New York: Taplinger, 1970), pp. 97–100; Gerald L. Wood, *The Guinness Book of Animal Facts and Feats* (Enfield, England: Guinness Superlatives, 1982), pp. 107–108; J. Richard Greenwell, "Colonel Fawcett and the Giant Anaconda," *ISC Newsletter* 11, no. 2 (1992): 8–11; Bob Rickard and John Blashford-Snell, "The Expeditionist," *Fortean Times,* no. 70 (August-September 1993): 30–34; Peter C. H. Pritchard (letter), "The Tympanum," *Bulletin of the Chicago Herpetological Society* 29, no. 2 (1994): 37–39; John C. Murphy and Robert W. Henderson, *Tales of Giant Snakes: A Historical Natural History of Anacondas and Pythons* (Malabar, Fla.: Krieger, 1997), pp. 23–45; Jeremy Wade, "Snakes Alive!" *Fortean Times,* no. 97 (May 1997): 34–37; Gary S. Mangiacopra, Michel M. Raynal, Dwight G. Smith, and David F. Avery, "Snake Bounty on Giant Boas," *Fortean Studies* 5 (1998): 202–207.

## Giant Ape

Unknown PRIMATE of South Africa.

*Physical description:* Size and color of a gorilla.

*Behavior:* Walks on its hind legs but also drops on all fours and uses the edges of its front feet.

*Tracks:* Gorilla-like. Five toes. Length, 5.5 inches.

*Distribution:* Magaliesberg Range, near Pretoria, South Africa; area around Outjo, Namibia.

*Significant sightings:* Rock art near Goedgegeven, Free State Province, South Africa, apparently shows a battle between Khoisan people wielding spears and apelike creatures using stones as weapons.

A gorilla-like ape was seen several times near Outjo, Namibia, in November 1959.

*Possible explanations:*

(1) An out-of-place Gorilla (*Gorilla gorilla*); the nearest of these animals live 1,000 miles away in the Democratic Republic of the Congo. However, it's unlikely that a zoo escapee would have gone unnoticed. The arid environment of Namibia would be inhospitable for a forest ape.

(2) Surviving australopiths, whose fossils have been found in the region around Johannesburg. Australopiths were a family of Pliocene fossil hominids that persisted into the Early Pleistocene, 4.4–1.4 million years ago. They had apelike skulls, hominid teeth and projecting jaws, and an upright, bipedal gait, although they apparently also climbed trees.

*Sources:* John Sanderson, "Memoranda of a Trading Trip into the Orange River (Sovereignty) Free State, and the Country of the Transvaal Boers, 1851–52," *Journal of the Royal Geographical Society* 30 (1860): 253; "Ape in SWA May Be Gorilla," *Salisbury (Zimbabwe) Evening Standard,* November 18, 1959; D. Neil Lee and Herbert C. Woodhouse, *Art on the Rocks of Southern Africa* (Cape Town, South Africa: Purnell, 1970), p. 148; Bernard Heuvelmans, *Les bêtes humaines d'Afrique* (Paris: Plon, 1980), pp. 548–552.

## Giant Aye-Aye

Mystery PRIMATE of Madagascar.

*Distribution:* Northwestern Madagascar.

*Significant sighting:* An exceptionally large aye-aye skin was found around 1930 by a government official named Hourcq at a native's home near Andranomavo, Soalala District, Madagascar.

*Possible explanation:* A surviving fossil lemur, *Daubentonia robusta,* that was probably contemporaneous with humans. It was three to five times heavier than the living Aye-aye (*D. madagascariensis*). No skull has yet been found, but postcranial bones are larger and much more robust than those of the living form. Teeth perforated for stringing offer direct evidence that the animal was hunted by humans, and it is virtually certain that this species was driven to extinction by human action in the past 2,000 years.

*Source:* W. C. Osman Hill, *Primates: Comparative Anatomy and Taxonomy* (Edinburgh: Edinburgh University Press, 1953), vol. 1, *Strepsirhini.*

## Giant Beaver

Legend of a large, beaverlike RODENT in eastern Canada.

*Physical description:* Not much information is available from the Native American legends, even on size. However, the animal must have been larger than the American beaver (*Castor canadensis*), which is about 2 feet–2 feet 6 inches long, with a 10-inch tail.

*Distribution:* Labrador, New Brunswick, and Nova Scotia. Mishtamishku-shipu (Giant Beaver River), Labrador, 54°40′ N, 62°25′ W, was named for the animals that were killed there, according to the local Montagnais-Naskapi people.

*Significant sighting:* The Giant beaver created much damage with its huge dams. The Indian folk hero Gluskap set out to destroy it but wound up chasing it further west.

*Possible explanation:* Surviving Giant beaver (*Castoroides ohioensis*), a large North American beaver that ranged from the Yukon south to Florida and from New York west to Nebraska and apparently died out 10,000 years ago. It was 7 feet 6 inches long and may have weighed 440 pounds. Its cutting teeth were 6 inches long, with strong enamel ridges. The tail was apparently round, not flat.

*Sources:* Jane C. Beck, "The Giant Beaver: A Prehistoric Memory?" *Ethnohistory* 19 (Spring 1972): 109–122; *Mishtamishku-shipu:* Giant Beaver River, http://www.innu.ca/beaver1.html.

## Giant British Octopus

Large CEPHALOPOD of the North Atlantic Ocean.

*Physical description:* Octopus with arms 6 feet or more in length.

*Distribution:* Western coast of Scotland and Cornwall.

*Significant sighting:* On January 12, 1952, Constable John Morrison came across a cephalopod lying half out of the water near Broadford, Isle of Skye, Scotland. He gave it a hefty kick, but it gripped him by the left ankle with a 6-foot tentacle. Morrison slipped out of his boot and eventually killed the animal with rocks and garden shears. It was later identified as a Red flying squid (*Ommastrephes bartrami*), apparently an occasional visitor to Britain.

*Possible explanation:* The Common octopus (*Octopus vulgaris*) has an average radial size of 1–2 feet, but outsize specimens with a radial spread of just over 6 feet have been reported in British waters.

*Sources:* Gerald L. Wood, *The Guinness Book of Animal Facts and Feats* (London: Guinness Superlatives, 1982), p. 194; Ulrich Magin, "Is There a British Monster Octopus?" *INFO Journal*, no. 51 (February 1987): 5–7.

## Giant Bushbaby

Unknown PRIMATE of Central and West Africa.

*Physical description:* Large galago (also known as a bushbaby) the size of a cat. Pale-gray fur.

*Distribution:* Cameroon; Côte d'Ivoire; Senegal.

*Significant sightings:* Owen Burnham saw a Giant bushbaby accompanied by some young ones in Casamance Forest, Senegal, in June 1985.

A specimen was photographed in Cameroon in 1994 by an assistant of zoologist Simon K. Bearder.

*Present status:* Greater galagos (*Otolemur* spp.) are not known in Cameroon. The largest, almost cat-sized at 15 inches in length not counting the tail, is the Silver galago (*O. argentatus*), found only in Kenya. Smaller galagos, especially the Senegal galago (*Galago senegalensis*), which has a body length of only 6.5 inches, do occur in Cameroon.

*Sources:* Karl Shuker, "A Supplement to Dr. Bernard Heuvelmans' Checklist of Cryptozoological Animals," *Fortean Studies* 5 (1998): 208, 216; Karl Shuker, "The Secret Animals of Senegambia," *Fate* 51 (November 1998): 46–51.

## Giant Centipede

Oversized segmented INVERTEBRATE of Missouri and Arkansas.

*Physical description:* Length, 7–18 inches. Arthropod with multiple legs.

*Behavior:* Female wraps itself around newly hatched young.

*Distribution:* Ozark Mountains near Gainesville, Bradleyville, Stone County, and Taney County, Missouri; Marion County, Arkansas.

*Significant sighting:* S. C. Turnbo collected stories of large centipedes in the Ozarks in the mid-nineteenth century. An 18-inch centipede was said to have been captured alive by Bent Music on Jimmie's Creek in Marion County, Arkansas, in 1860. It was placed in a jar of alcohol in a drugstore in Yellville, but people lost track of it during the Civil War.

*Possible explanation:* The largest known species of centipede in North America is the Giant desert centipede (*Scolopendra heros*), a black-and-orange banded animal with yellow legs that grows to more than 8 inches. It is found in Mexico and the southern United States. Females guard their hatchlings closely for a few days after birth. A related species, the Galapagos centipede (*S. galapagensis*), is the largest in the world, growing to 17 inches.

*Sources:* Desmond Walls Allen, ed., *Turnbo's Tales of the Ozarks: Snakes, Birds and Insect Stories* (Conway: Arkansas Research, 1989); Silas Claiborn Turnbo, *The White River Chronicles of S. C. Turnbo: Man and Wildlife on the Ozarks Frontier* (Fayetteville: University of Arkansas Press, 1994); Chad Arment, "Giant Centipedes in the Ozarks," *North American BioFortean Review* 1, no. 2 (June 1999): 5–6, http://www.strangeark.com/nabr/NABR2.pdf.

## Giant Cookiecutter Shark

Unknown FISH of the North Pacific Ocean.

*Behavior:* Takes large circular bites out of whales and dolphins.

*Distribution:* Arctic waters off Alaska.

*Significant sighting:* While working in Alaska, a colleague of Eugenie Clark reported that a dead Narwhal (*Monodon monoceros*) was pulled up alongside their research boat for examination one night. In the morning, as the scientists began to examine it, they found round bites on the animal that strongly resembled those left by the Cookiecutter sharks (*Isistius* spp.). However, they were much bigger than bites made by known cookiecutters.

*Possible explanations:*

(1) The Cookiecutter shark (*Isistius brasiliensis*) and the Largetooth cookiecutter shark (*I. plutodus*) are both subtropical species. Neither grows much longer than 18–20 inches. The cookiecutter clamps onto the flesh of its much larger prey with its jaws and bites down with the sharp teeth on its lower jaw to extract circular chunks of flesh. Tunas, elephant seals, dolphins, whales, swordfish, and other large marine animals have been found with large gouges that were probably inflicted by this shark. A giant species of cookiecutter would theoretically take much larger bites.

(2) The Pacific sleeper shark (*Somniosus pacificus*) frequently scavenges whale carcasses.

(3) The Greenland shark (*Somniosus microcephalus*) also leaves razor-edged, circular bites on narwhals and seals.

*Source:* Ben S. Roesch, "Do *Giant* Cookiecutter Sharks Exist?" http://www.ncf.carleton.ca/~bz050/HomePage.giantcookiecutter.html.

## Giant Ethiopian Lizard

Mystery LIZARD of East Africa.

*Physical description:* Length, 10–12 feet. Loose, gray skin. Dorsal crest. Huge, clawed, three-toed feet.

*Distribution:* Ethiopia.

*Significant sighting:* Adrian Conan Doyle interviewed a big-game hunter who said he had come across a large, lizardlike animal, about 10–12 feet long, on the border of Ethiopia and Sudan.

*Sources:* Adrian Conan Doyle, *Heaven Has Claws* (London: John Murray, 1952), pp. 29–31; Bernard Heuvelmans, *Les derniers dragons d'Afrique* (Paris: Plon, 1978), pp. 148–151.

# GIANT HOMINIDS

In this category are humanlike, hairy creatures that are described as 6 feet 6 inches tall or greater. Their bipedal gait, appearance, and behavior indicate a closer relationship to humans (HOMINIDS) than to the apes (PRIMATES).

The only known fossil that comes close to giant status is *Gigantopithecus,* a huge ape first recognized by Dutch anthropologist G. H. R. von Koenigswald from a single molar he purchased in a Hong Kong pharmacy in 1935. Since then, more than 1,000 other teeth and a few mandibles have been recovered. There are two known species, *G. blacki* of China and Vietnam and *G. giganteus* of India. *G. giganteus* is older and smaller, dating from the Late Miocene, 9–6 million years ago. By the time *G. blacki* roamed East Asia in the Early and Middle Pleistocene, 1 million–400,000 years ago, *Gigantopithecus* had become extremely robust. One estimate puts its height at 9–10 feet tall and its weight at 900–1,200 pounds. However, no weight-bearing bones have been recovered, and it is possible that the animal's teeth and jaws were disproportionate to its body size.

Another von Koenigswald discovery involved two large fossil mandibles recovered in Java in 1939 and 1941. Designated *Meganthropus palaeojavanicus,* the specimens were described by German anatomist Franz Weidenreich only from casts that von Koenigswald made and sent to Beijing before he was captured and interred by the Japanese during World War II. A handful of other fragmentary finds have been included in this taxon, but there is no consensus on the creature's status. Many regard it as belonging to *Homo erectus,* though some consider it pathologically oversized.

Legends of a race of giants are found in many cultures, making it all the more strange that the known hominid fossil record is so sparse. The CYCLOPS of Greece, GRENDEL of the Anglo-Saxons, and the NEFILIM in the Book of Genesis are examples of this rich tradition of ancient giants. However, many legends were based on the discovery of fossil elephants and other extinct megafauna; the ancient Greeks and Romans tended to identify huge skulls and femurs with various mythical heroes, even if they were not particularly human-looking. Still, persistent reports of the discovery of GIANT HUMAN SKELETONS have come from both Europe and North America.

According to the *Guinness Book of Records,* the tallest man in medical history was Robert Pershing Wadlow (1918–1940) of Alton, Illinois, who measured 8 feet 11.1 inches shortly before his death. He weighed 491 pounds on his twenty-first birthday.

Modern Giant hominids have been reported from every continent except Antarctica. BIGFOOT of the Pacific Northwest is probably the most familiar variety. Interestingly, a diverse tradition of hairy giants also occurs in Siberia, the logical origination for hominids migrating into North America.

TRUE GIANTS constitute a subcategory of Giant hominids 10 feet or more tall.

## Mystery Giant Hominids
### Africa
ENGÔT; GERIT; MULAHU; NANAUNER; NDESU; NGOLOKO; WA̓AB

### Asia, Central
DZU-TEH; JEZ-TYRMAK; NYALMO; RIMI

### Asia, East
SHĀN GUI

### Asia, Southeast
KAPRE; KUNG-LU; ORANG DALAM; ORANG GADANG; TOK

### Asia, West
NART; NEFILIM; TORCH

### Australasia and Oceania
JOGUNG; QUINKIN; SPINIFEX MAN

### Central and South America
CURINQUÉAN; DIENTUDO; ECUADOREAN GIANT; FANTASMA DE LOS RISCOS; PATAGONIAN GIANT; UCUMAR

### Europe
AFONYA; CYCLOPS; GRENDEL; GYONA PEL; TROLL

North America
BIGFOOT; CANNIBAL GIANTS; GIANT HUMAN SKELETONS; GILYUK; PITT LAKE GIANT; TALLEGWI

Siberia
CHUCHUNAA; KHEYAK; KILTANYA; KUL; MECHENY; MIRYGDY; PIKELIAN; TUNGU; ZEMLEMER

## Giant Human Skeletons

Subfossil bones, skulls, and skeletons of humans or GIANT HOMINIDS of North America and Europe.

*Physical description:* Height, 8 feet or more.

*Variant name:* TALLEGWI.

*Distribution:* Paleo-Indian mounds in the eastern United States and other sites in North America; scattered sites in Europe.

A partial list of places where Giant human skeletons have been reported follows:

*Arizona*—Fort Crittenden, Winslow.

*British Columbia, Canada*—Neskain Island.

*California*—Cascade Mountains, Lompoc, Minarets Wilderness, Santa Rosa Island.

*England*—Gateshead, Durham; Repton, Derbyshire; Rotherhithe, Greater London; St. Bees, Cumbria.

*France*—Angers, Rouen, Soyons.

*Greece*—Aléa.

*Indiana*—Brewersville, Potato Creek, Walkerton.

*Ireland*—Donadea, Leixlip.

*Italy*—Mazzarino.

*Kentucky*—Allen County, Carroll County, Christian County, Holly Creek.

*Mexico*—Río Baluarte in Sinaloa State.

*Minnesota*—Chatfield, Clearwater, Dresbach, Koronis Lake, LaCrescent, McKinstry Mounds, Moose Island, Pine City, Rainy River, Sauk Rapids, Warren.

*Montana*—Fish Creek.

*Nevada*—Spring Valley.

*New York*—East Randolph, Tug Hill Plateau.

*Ohio*—Zanesville.

*Pennsylvania*—Bedford County, Bradford, Ellisburg, Gastonville, Greensburg, Hanover, Sayre, Sterling Run, West Hickory.

*Switzerland*—Luzern.

*Tennessee*—White County.

*West Virginia*—Salem.

*Significant sightings:* The earliest known discovery of giant's bones was near Aléa, Arcadia, Greece, about 560 B.C., when a blacksmith uncovered a 10-foot-long coffin containing a huge skeleton. Hailed as the bones of the Spartan hero Orestes, they were reburied in that city with great honor. Some scholars think they were the fossil remains of large animals, discovered and interred in a coffin at a much earlier time. Adrienne Major calls Herodotus's account of this event the earliest fossil measurement ever recorded.

In 1509, some workers digging ditches near Rouen, France, uncovered a stone tomb that contained the skeleton of a man of enormous size. The skull was large enough to hold a bushel of corn, and the shinbone measured 4 feet in length; from this, the full height was estimated at 17 feet. On the tomb was a copper plate that identified the body as Chevalier Ricon de Vallemont.

From the 1860s to the 1880s, settlers in Minnesota digging into Indian mounds excavated human skeletons 7–8 feet tall. In December 1868, quarry workers at Sauk Rapids found a petrified skeleton 10 feet 9.5 inches tall in a grave chamber capped by a limestone slab about 7 feet below the surface. The skull was flat on the top and measured 31.5 inches in circumference. The femur was 26.25 inches long, and the fibula was 25.5 inches.

A skeleton measuring 8 feet in length was discovered by George B. Dresbach Jr. while leveling an earthwork near Dresbach, Minnesota, in the nineteenth century.

Four skeletons of men 7–9 feet tall were unearthed in two mounds near Salem, West Virginia, in 1930. However, by the time anthropologist D. T. Stewart reached the site, most of the bones had disintegrated or become lost. The few remaining bones were considered to be only average size.

In 1965, Kenneth White dug up a perfectly preserved skeleton under a rock ledge near Holly Creek, Kentucky, that was 8 feet 9 inches tall when reassembled. Its arms were relatively long, its hands large, and its feet relatively small. The skull was 30 inches in circumference. The eye and nose sockets were slits instead of cavities. The jawbone was solidly fused to the skull. Folklorist Michael Henson was able to examine the skeleton before it was reburied by White.

In August 1965, physician Robert W. Denton discovered the top portion of an unusual skull in a boggy area in the Minarets Wilderness in northern California. It was examined by Gerald K. Ridge, a pathologist at the Ventura County General Hospital, and by University of California, Los Angeles (UCLA) archaeologists Herman Bleibtreul and Jack Prost. The nuchal ridge was unusally developed, leading Ridge to think it was not human. The specimen has since been misplaced, although, according to Matt Moneymaker, it may be languishing unidentified in UCLA's off-campus museum annex in Chatsworth.

*Present status:* Most reports are old, unconfirmed, and of dubious provenance, but they are nonetheless intriguing as possible physical evidence for BIGFOOT or other GIANT HOMINIDS.

*Possible explanations:*

(1) Two diseases caused by oversecretion of the human growth hormone (HGH) by the pituitary gland can result in enlarged body size. Acromegaly, a disorder usually caused by a benign pituitary tumor, is marked by gradual and permanent enlargement of the jaw, hands, feet, internal organs, nose, lips, and tongue. In most cases, onset occurs between the ages of thirty and fifty after normal bone growth has stopped, resulting in bones that become deformed rather than elongated. Overgrowth in the jaw causes it to protrude, and the ribs thicken, creating a barrel chest. Much rarer is gigantism, which begins abruptly in childhood before the end plates of the long bones have closed. The condition leads to exaggerated bone growth and abnormal height (with a growth rate of as much as 6 inches per year). Afflicted adults may reach a height of more than 6 feet 8 inches.

(2) The unearthing of mastodon bones and other megafaunal remains undoubtedly contributed to many of these accounts.

*Sources:* Herodotus, *The Histories,* trans. Aubrey de Sélincourt (New York: Penguin,

1996), pp. 26–28 (I. 67–68); Philostratus of Lemnos, *On Heroes,* VII. 9, VIII. 3–14; Phlegon of Tralles, *Phlegon of Tralles' Book of Marvels,* trans. William Hansen (Exeter, England: University of Exeter Press, 1996); Lewis Collins, *Historical Sketches of Kentucky* (Maysville, Ky.: Lewis Collins, 1848), pp. 168, 229; Edward J. Wood, *Giants and Dwarfs* (London: R. Bentley, 1868); "Giant Skeleton," *New York Times,* December 25, 1868; "Ancient American Giants," *Scientific American* 43 (1880): 106; *History of Bedford, Somerset, and Fulton Counties, Pennsylvania* (Chicago: Waterman, Watkins, 1884); *St. Paul Pioneer-Press,* June 29 and July 1, 1888; George M. Gould and Walter L. Pyle, *Anomalies and Curiosities of Medicine* (Philadelphia: W. B. Saunders, 1897), chap. 7; Newton H. Winchell, *The Aborigines of Minnesota* (St. Paul: Minnesota Historical Society, 1911), pp. 80, 89–90, 301, 341, 372–373; "A Nine-Foot Skeleton," *Scientific American* 124 (1921): 203; "Archaeological No-Man's-Land," *Science News-Letter* 18 (1930): 6; Jesse James Benton, *Cow by the Tail* (Boston: Houghton Mifflin, 1943), p. 170; Phyla Phillips, "Giants in Ancient America," *Fate* 1 (Spring 1948): 126–127; Jack Clayton, "The Giants of Minnesota," *Doubt,* no. 35 (1952): 120–122; Henry Winfred Splitter, "The Impossible Fossils," *Fate* 7 (January 1954): 65–66; Leland Lovelace, *Lost Mines and Hidden Treasure* (New York: Ace, 1956), pp. 57–67; Ivan T. Sanderson, *Abominable Snowmen: Legend Come to Life* (Philadelphia: Chilton, 1961), pp. 36–37; *Hanover (Pa.) Sun,* June 22, 1963; H. E. Krueger, "The Lesser Wilderness: Tug Hill," *The Conservationist* 21 (December 1966-January 1967): 12–16, 38; Robert R. Lyman, *Amazing Indeed!* (Coudersport, Pa.: Potter Enterprise, 1971), pp. 6–10; "Giant Skeletons," *Pursuit,* no. 23 (July 1973): 69–70; Dorothy P. Dansie, "John T. Reid's Case for the Redheaded Giants," *Nevada Historical Society Quarterly* 18 (1975): 152–167; B. Ann Slate and Alan Berry, *Bigfoot* (New York: Bantam, 1976), pp. 160–165; Michael Paul Henson, *Tragedy at Devil's Hollow, and Other Haunting Tales from Kentucky* (Bowling Green,

Ky.: Cockrel, 1984); Mark A. Hall, "Giant Bones," *Wonders* 2, no. 1 (March 1993): 3–13; William R. Corliss, ed., *Biological Anomalies: Humans III* (Glen Arm, Md.: Sourcebook Project, 1994), pp. 43–46; Charles DeLoach, *Giants: A Reference Guide from History, the Bible, and Recorded Legend* (Metuchen, N.J.: Scarecrow, 1995); Adrienne Mayor, *The First Fossil Hunters: Paleontology in Greek and Roman Times* (Princeton, N.J.: Princeton University Press, 2000), pp. 104–156; Ross Hamilton and Patricia Mason, "A Tradition of Giants and Ancient North American Warfare," *Ancient American,* no. 36 (December 2000): 6–13; Matt Moneymaker, Buried Treasure: The Minaret Skull, http://www.bfro.net/ref/theories/mjm/minaret.htm.

## Giant Hyrax of Shaanxi

Unknown HYRAX of East Asia.

*Physical description:* A stocky quadruped with a head like a hyrax's, single hooves, and a short, hippopotamus-like tail.

*Distribution:* Shaanxi Province, China.

*Significant sighting:* Chinese bronze statuettes, at least one of which is from Shaanxi Province and dates from the Period of Warring States (403–221 B.C.), depict this animal.

*Present status:* Presumably extinct.

*Possible explanation: Pliohyrax,* a hyrax about the size of a large pig that lived in China 2 million years ago, in the Late Pliocene. It is only known from fossil skulls, but the eye placement indicates a semiaquatic adaptation similar to that of the hippopotamus. However, modern hyraxes have hooflike nails, not hooves, and *Pliohyrax* was probably no different—unless it evolved a padded foot structure that was imperfectly conveyed by the sculptor.

*Sources:* F. Martin Duncan, "A Chinese Noah's Ark," *The Field* 166 (November 30, 1935): 1286–1287; Karl Shuker, *In Search of Prehistoric Survivors* (London: Blandford, 1995), pp. 158–159; Karl Shuker, "A Giant Owl and a Giant Hyrax . . . ?" *Strange Magazine,* no. 21 (Fall 2000), on line at http://www. strangemag.com.

## Giant Jellyfish

Unknown marine INVERTEBRATE.

*Distribution:* North Atlantic and South Pacific Oceans.

*Significant sightings:* In 1953, an Australian deep-sea diver watched a shapeless, brown mass engulf a shark.

Divers Richard Winer and Pat Boatwright encountered a huge jellyfish, 50–100 feet in diameter, when they were diving 14 miles southwest of Bermuda in November 1969. It was deep purple with a pinkish outer rim.

In January 1973, in the South Pacific between Australia and Fiji, the Australian ship *Kuranda* collided with a colossal jellyfish that draped itself over the forecastle head. One crew member came too close to one of the flailing tentacles and died from the sting. Capt. Langley Smith estimated that some of the tentacles were 200 feet long and that the deck was covered in a slimy mass 2 feet deep. An SOS eventually brought a deep-sea salvage tugboat, the *Hercules,* to the rescue, and the animal was dislodged with the aid of high-pressure hoses. Samples of the remaining substance on the deck were analyzed in Sydney and tentatively identified as a lion's mane jelly.

*Possible explanation:* The largest known jellyfish is the Lion's mane jelly (*Cyanea capillata*) found in the North Atlantic and North Pacific Oceans, most often in shallow coastal waters. Large individuals are deep-red or purple in color, while smaller ones are more yellow or brown. The nematocysts produce painful stings but are not usually fatal. One specimen examined in 1865 by Alexander Agassiz in Massachusetts Bay had a bell measuring 7 feet 6 inches across and tentacles stretching 120 feet long.

*Sources:* Eric Frank Russell, *Great World Mysteries* (New York: Roy, 1957); Gary Mangiacopra, "A Monstrous Jellyfish?" *Of Sea and Shore* 7, no. 3 (Summer 1976): 169; James B. Sweeney, *Sea Monsters: A Collection of Eyewitness Accounts* (New York: David McKay, 1977); "Jellyfish Ate My Wife," *Fortean Times,* no. 55 (Autumn 1990): 26.

## Giant Kangaroo

Mystery MARSUPIAL of Australia.

*Variant name:* Kuperree.

*Distribution:* Eyre Peninsula, South Australia.

*Significant sighting:* The Aborigines of Port Lincoln had a legend of the extinction of a Giant kangaroo named Kuperree.

*Possible explanation:* The Eastern gray kangaroo (*Macropus giganteus*) was much larger during the Late Pleistocene, perhaps twice as massive. Specimens now grow no greater than a body length of 5 feet, with a tail at least 3 feet long. Buried in a swamp near Lancefield, Victoria, are the remains of thousands of giant *Macropus* fossils that died about 26,000 years ago.

*Sources:* William Anderson Cawthorne, *Legend of Kuperree, or the Red Kangaroo: An Aboriginal Tradition from the Port Lincoln Tribe* (Adelaide, Australia: Alfred N. Cawthorne, 1858); Charles Barrett, *The Bunyip and Other Mythical Monsters and Legends* (Melbourne, Australia: Reed and Harris, 1946), pp. 79–80.

## Giant Lungfish

Unknown FISH of Southeast Asia.

*Physical description:* Eel-like fish with both lungs and gills. Length, 6 feet.

*Distribution:* Vietnam.

*Present status:* The only extant genera of Lungfishes (Class Dipnoi) are found in South America, Australia, and Africa.

*Source:* Karl Shuker, *Extraordinary Animals Worldwide* (London: Robert Hale, 1991), p. 61.

## Giant Malagasy Tortoise

Mystery TURTLE of Madagascar.

*Variant name:* Ranta (Malagasy/Austronesian).

*Physical description:* Large tortoise.

*Habitat:* Caves.

*Distribution:* Southwest coast of Madagascar.

*Possible explanation:* The Giant Malagasy tortoise (*Geochelone grandidieri*), an extinct turtle with a carapace 4 feet long, may have persisted until the 1940s.

*Source:* Raymond Decary, *La faune malgache,*

*son rôle dans les croyances et les usages indigènes* (Paris: Payot, 1950), pp. 205–206.

## Giant Mediterranean Octopus

Large CEPHALOPOD of the Mediterranean Sea.

*Variant names:* Hydra, Scylla.

*Physical description:* Octopus with arms longer than 10 feet.

*Distribution:* Mediterranean Sea off France and Greece.

*Significant sightings:* A large octopus might have inspired the Greek myths of the six-necked, twelve-footed, cave-dwelling sea creature Scylla that attacked Odysseus's crew in the Strait of Messina, Italy, or the nine-headed, serpentine Hydra that Herakles killed in the marshes of Lerna south of Árgos in the Peloponnesus, Greece.

An octopus with arms 13 feet long was reported in 1912 off Toulon, France.

Sometime in the 1950s, a diver encountered a huge octopus either in the Gulf of Corinth or off Piraiéus in the Aegean Sea.

*Possible explanation:* The Common octopus (*Octopus vulgaris*) has an average radial size of 1–2 feet, but outsize specimens with spans of 8–9 feet are known. Reports of sizes greater than this are sparse and vague.

*Source:* Ulrich Magin, "Danger under the Waves: The Giant Octopus of the Mediterranean," *Pursuit,* no. 71 (1985): 128–129.

## Giant North American Lizard

Large, unknown LIZARDS of North America.

*Variant names:* Canip monster lizard, Crosswick monster, Giant pink lizard, GOWROW, Mini-rex, Mountain boomer, River dino, River lizard.

*Physical description:* Various sizes and descriptions.

*Behavior:* Some are bipedal, others quadrupedal.

*Tracks:* Three- or four-toed.

*Distribution:* British Columbia, Canada; Colorado; Texas; South Dakota; Ohio; Kentucky; Pennsylvania.

*Significant sightings:* Prior to 1820, when a

drought exterminated them, pink lizards 3–8 feet long were said to inhabit "Catlick Creek Valley," which Mark Hall has identified as Scippo Creek in Pickaway County, Ohio. The animals were said to have horns like a cow's.

In the late nineteenth century, two young boys fishing in a stream near Crosswick, Ohio, were attacked by a lizard that stood 12–16 feet tall. Three men rescued the boy, but the lizard escaped into a huge hollow tree. Later in the day, townsfolk came to cut the tree down, but the animal ran away on its two hind legs.

Myrtle Snow claimed to have seen five "baby dinosaurs" near Chromo, Colorado, in May 1935 when she was three years old. John Martinez had shot one a few months earlier after it killed some sheep. It was 7 feet tall, gray, had a head like a snake's, short front legs with claws, large hind legs, and a long tail. Snow saw similar animals near a cave in 1937 and October 1978.

Several reports of smallish, bipedal lizards have come from Vancouver and Texada Islands, British Columbia. In one instance, railroad workers came across a nest of 12-inch-tall lizards that scampered away on two legs.

In July 1975, there were several sightings of a large, black-and-white-striped lizard with a red, forked tongue near Canip Creek in Trimble County, Kentucky. It left clawed tracks that were 5 inches long by 4.5 inches wide. Clarence and Garrett Cable saw it on three occasions in a junkyard near Milton. It appeared to be about 15 feet long.

In 1981, a 2-foot, green, crested lizard was chased by some boys along a railroad track in New Kensington, Pennsylvania.

In the early 1990s, Jimmy Ward investigated rumors of a green or brown, bipedal lizard with a booming voice in west Texas near the Big Bend National Park. It was called the Mountain boomer and stood 5–6 feet tall on its hind legs.

In 2000, Ron Schaffner obtained some photos showing small, dinosaur-like lizards allegedly taken in the Fountain Creek, Colorado, area, but the animals might well be rubber models.

*Possible explanations:*

(1) Unknown monitor lizards (Family Varanidae), though existing species are known only from Africa, Asia, and Australasia.

(2) Surviving *Matthewichnus caudifer,* a fossil amphibian whose tracks are known from the Carboniferous period, 300 million years ago, in Tennessee, suggested by Mark Hall.

(3) A neotenic Mole salamander (*Ambystoma* spp.), also suggested by Hall. However, this overgrown, underdeveloped larva (axolotl) does not leave the water.

(4) Escaped pet Colombian black-and-white tegu (*Tupinambis teguixin*), which looks somewhat like a monitor lizard and grows to 4 feet long, suggested by Chad Arment for the Canip Creek animal.

(5) Escaped pet Green basilisk (*Basiliscus plumifrons*), a bright-green, arboreal lizard from Central America that grows to 3 feet and has a banded tail and dorsal crest, suggested by Chad Arment for the New Kensington lizard.

(6) The Eastern collared lizard (*Crotaphytus collaris collaris*) is, for unknown reasons, also called the Mountain boomer, though it has no vocal cords. A Western subspecies (*C. c. baileyi*) is found in the Big Bend area and grows to about 2 feet in length. It runs on its hind legs.

*Sources:* "More Monsters," *Doubt,* no. 16 (1946): 236–237; Erasmus Foster Darby [David Knowlton Webb], *A True Account of the Giant Pink Lizard of Catlick Creek Valley, Being a Tale of South Central Ohio Pioneer Days* (Chillicothe, Ohio: [Ross County Historical Society], 1954); Hazel Spencer Phillips, *Crosswick Monster: Folklore Series,* no. 11 (Lebanon, Ohio: Warren County Historical Society, 1978); Myrtle Snow (letter), Empire Magazine, *Denver Post,* August 22, 1982; Mark A. Hall, *Natural Mysteries,* 2d ed. (Minneapolis, Minn.: Mark A. Hall, 1991), pp. 27–42; Jimmy Ward, "The Mountain Boomer," *Far Out* 1, no. 4 (1993): 45–46; Chad Arment, "Dinos in the USA: A Summary of North American Bipedal 'Lizard' Reports," *North American BioFortean Review* 2, no. 2 (2000): 32–39, http://www.strangeark.

com/nabr/NABR4.pdf; K. Strong, "Reports of Unknown Reptiles on Vancouver Island," *BCSCC Quarterly,* no. 39 (January 2000): 5.

# Giant North American Snake
Unknown SNAKE of the United States.

*Variant names:* Big Jim, GIANT PENNSYLVANIA SNAKE, PENINSULA PYTHON, Pete the Python, Salem serpent.

*Physical description:* Length, 8–30 feet. As big around as a stovepipe.

*Behavior:* Eats chickens. Can raise its neck and head in the air.

*Tracks:* As wide as an automobile tire track and 4 inches deep.

*Habitat:* Wetlands.

*Distribution:* A partial list of places where Giant North American snakes have been reported follows:

*Alabama*—Clanton, Kilpatrick.

*Arkansas*—Foreman.

*Georgia*—Seney.

*Indiana*—Adams County, Dubois County, Fort Wayne, Indianapolis, Knox County, Orange County, Pike County, Ripley County, Shelby County.

*Kansas*—Fredonia.

*Kentucky*—Hazel.

*Maryland*—Hall's Springs, Harford County.

*Massachusetts*—Bridgewater.

*Michigan*—Hastings, Salem.

*Missouri*—Lock Springs.

*Montana*—Cascade.

*Nebraska*—Holdrege.

*New York*—Dresden.

*Ohio*—Doylestown, Kenton, Loudonville, Peninsula, Rogues Hollow.

*Oklahoma*—Wewoka.

*Pennsylvania*—Allentown, Broad Top Mountains, Gettysburg, Jenners, Morgantown, Pocono Mountains, Somerset County, York County.

*South Dakota*—Moccasin Creek.

*Tennessee*—Nashville.

*Texas*—Ames.

*Significant sightings:* In January or February 1871, a snake 38 feet 9 inches long and 43 inches in circumference was killed near Fredo-nia, Kansas. However, in the nineteenth century, Kansas was widely regarded as an area for exaggeration and tall tales.

A dead snake 13 feet 6 inches long was found behind the Clyde Myers home near Doylestown, Ohio, on May 1, 1944. It was 6 inches in diameter and had bent and broken the tall grass in an area at least 30 feet in diameter with its thrashings. It was on display at a service station in Barberton for a week before health officials ordered it buried.

An 8-foot snake with a diamond shape on its flat head struck at Orland Packer's horse as he was riding near Kenton, Ohio, on June 9, 1946.

The D. A. Crance family was driving next to Spy Run Creek in Fort Wayne, Indiana, on June 13, 1952, when they saw an 18-foot, grayish-blue snake with a head as big as a bulldog cross the road. The *Fort Wayne Journal-Gazette* nicknamed it "Pete the Python" after a hunt organized by Sheriff Harold Zeis had gone on for three days without finding anything. Additional sightings ended with a hoax story on June 18.

Eileen Blackburn was driving on I-15 south of Cascade, Montana, in October 1978 when she struck a snake 20–30 feet long that was lying in the road with its head and neck 2–3 feet in the air. It was gray-white with a tan stripe and had a flat head.

Clifton Louviere shot a 25-foot snake on his pig farm near Ames, Liberty County, Texas, on April 10, 1982. However, the carcass disappeared the next day, and Louviere supposed the snake had only been stunned.

*Possible explanations:*
(1) The Black rat snake (*Elaphe obsoleta obsoleta*) typically grows no longer than 7 feet, although an 8-footer has been recognized. It is a uniform black with faint spotting and is found in the east from Kansas to Connecticut.
(2) The Northern black racer (*Coluber constrictor constrictor*) does not grow much longer than 6 feet. It is black, with dark, middorsal blotches, and is found from southern Maine to northern Alabama. The Southern black racer (*C. c. priapus*) is similar and ranges from southern Indiana to Florida.
(3) The Eastern coachwhip (*Masticophis*

*flagellum flagellum*) is typically 4–5 feet long, with oversize individuals reported up to 8 feet 6 inches. The head and neck are dark brown or black, gradating to a lighter color ventrally. Found in the South from North Carolina to Florida and west to Texas.

(4) The Eastern cottonmouth (*Agkistrodon piscivorus piscivorus*) is a brown, black, or olive semiaquatic snake normally only 3–4 feet long, with a maximum length of 6 feet. Its range is from southeastern Virginia to central Georgia. The Western cottonmouth (*A. p. leucostoma*) maxes out at 5 feet and is found from southern Illinois to Alabama and eastern Texas.

(5) An escaped Indian python (*Python molurus*), an Asian snake that has an average length of 13 feet and an outsize length of 20 feet.

*Sources:* "More Monsters," *Doubt,* no. 15 (Summer 1946): 228; Gus Larson, "Python Posse," *Nebraskaland,* October 1970, pp. 8–9; Howard Coffin, "Lopsided Legend 'Circles' Hillsides of Vermont," *Christian Science Monitor,* August 12, 1975; "Monster Snake," *Fate* 32 (May 1979): 20–22; Cindy Horswell, "Welder Reports 25-Foot Snake," *Houston Chronicle,* April 22, 1982, sec. 4, p. 5; Mark A. Hall, "Giant Snakes and Mystery Mounds in North America," *Wonders* 3, no. 4 (December 1994): 93–116; Mark A. Hall, "Giant Snakes in the Twentieth Century," *Wonders* 4, no. 1 (March 1995): 11–29; Mark A. Hall, "More Giant Snakes Alive!" *Wonders* 4, no. 3 (September 1995): 80–89; Brad LaGrange, "Cryptoherps of Indiana," *North American BioFortean Review* 1, no. 1 (April 1999): 27, http://www.strangeark.com/nabr/NABR1.pdf; Loren Coleman, *Mysterious America,* rev. ed. (New York: Paraview Press, 2001), pp. 76–82.

## Giant Owl

Mystery BIRD of the central and eastern United States.

*Variant names:* Bighoot, Booger owl, Flying head, In-da-dhin-ga (Omaha-Ponca/Siouan), MOTHMAN, WOOO-WOOO.

*Physical description:* Length, 4 feet. White. Wingspan, 10–12 feet.

*Behavior:* Nocturnal. Said to be able to carry off lambs, calves, dogs, and small children.

*Distribution:* South Texas; the Ozark Mountains, Arkansas; southern Ohio; northern New Jersey; West Virginia.

*Significant sightings:* Iroquoian legends of "flying heads" may be related to large owls.

A woman saw a large bird at Rocky Fork Lake, southern Ohio, at sundown in August 1982. It looked just like a 10-foot tree until it moved into a clearing and unfolded its wings. One year later, near the same spot on the same lake, she saw it again, this time noticing its yellowish legs and feet.

*Possible explanations:*

(1) The Snowy owl (*Nyctea scandiaca*) is the closest in plumage but is only half as large, with a wingspan of 4 feet 4 inches. It breeds in the Arctic but winters as far south as Minnesota, Michigan, and New York. Strays are occasionally found much farther south, often in the daytime and usually sick or hungry.

(2) The largest living owl is the Eurasian eagle-owl (*Bubo bubo*), which reaches 30 inches in length and is only found in Europe and Asia. Its feet are the size of a man's hand.

(3) A giant flightless owl (*Ornimegalonyx oteroi*) that exceeded 3 feet in length is known from the Pleistocene of Cuba, but there is no evidence of its persistence into modern times.

*Sources:* James Owen Dorsey, "Siouan Folk-Lore and Mythologic Notes," *American Antiquarian* 7 (1885): 107; William Elsey Connelley, *Wyandot Folk-Lore* (Topeka, Kans.: Crane, 1899), pp. 85–86; Vance Randolph, *We Always Lie to Strangers* (New York: Columbia University Press, 1951), pp. 63–66; Virginia M. Miller (letter), "The 'Mothman' Visits," *Fate* 29 (March 1976): 127–129; Joseph Bruchac, *Stone Giants and Flying Heads* (Trumansburg, N.Y.: Crossing, 1979); Mark A. Hall, *Thunderbirds! The Living Legend of Giant Birds* (Minneapolis, Minn.: Mark A. Hall, 1988), pp. 48–49, 84; Mark A. Hall, "Bighoot: The Giant Owl,"

Wonders 5, no. 3 (September 1998): 67–79;
Karl Shuker, "A Giant Owl and a Giant
Hyrax . . . ?" *Strange Magazine,* no. 21 (Fall
2000), on line at http://www.strangemag.com.

## Giant Pennsylvania Snake

Large SNAKE of south-central Pennsylvania and
northern Maryland.

*Variant names:* Big snake, Boss snake, The
Devil, Devil snake, Heap big snake, Log snake.

*Physical description:* Length, 15–20 feet. Diameter, 8–10 inches, or as thick as a stovepipe.
Black with some gray, dark gray with yellow
markings, or dirty tan with variegated markings.
Huge mouth.

*Behavior:* Sometimes blocks rural roads. Coils
its tail around a tree branch and swings its head
to and fro. Said to be able to move with its head
and neck erect. Hisses or groans. Eats roosters
and cats.

*Habitat:* Forests, mountains, rocky areas.

*Distribution:* Southern Pennsylvania; northern Maryland.

*Significant sightings:* Emanuel Bushman's
brother and six others saw a Devil snake on Big
Round Top, south of Gettysburg, Pennsylvania,
in April 1833. Other reports place the snake in
Devil's Den. It was probably gone by the Battle
of Gettysburg in July 1863, but the name
"Devil's Den" may have originated with this
creature rather than Confederate sniper fire during the battle.

A black snake 25–35 feet long was seen in the
vicinity of Allentown, Pennsylvania, in 1870
and 1871, catching and eating roosters and cats.

A 15-foot "anaconda" was reported around
Hall's Springs, Maryland, in the summer of
1875. Its track was measured at 11.5–15 inches
wide. It swallowed pigs, a turkey, and a chicken
in a trap set for it, but it eluded capture.

*Present status:* Possibly the same as other GIANT
NORTH AMERICAN SNAKES reported elsewhere.

*Possible explanations:*

(1) The Black rat snake (*Elaphe obsoleta
obsoleta*) is the largest snake in Pennsylvania,
growing to 7 feet in length. It is solid black
with faint traces of a spotted pattern.

(2) The Northern black racer (*Coluber
constrictor constrictor*) is the second-largest
snake in Pennsylvania but does not grow
much longer than 6 feet and is more slender
than the rat snake. It is bluish-gray to black
on top, with some white on the chin.

(3) The Bullsnake (*Pituophis catenifer sayi*),
a yellowish colubrid snake with dark
blotches, grows to over 8 feet long but is
only found in isolated pockets in the East.

(4) A large, unknown subspecies of
bullsnake, suggested by Chad Arment.

*Sources:* Emanuel Bushman, "Big Snake,"
*Gettysburg (Pa.) Compiler,* August 12, 1875;
Thomas Turner Wysong, *The Rocks of Deer
Creek, Harford County, Maryland: Their Legends
and History* (Baltimore, Md.: Sherwood, 1879),
p. 38; Salome Myers Stewart, "Reminiscences
of Gettysburg," *Chattanooga (Tenn.) News,*
October 30, 1913; Annie Weston Whitney and
Caroline Caulfield Bullock, "Folk-Lore from
Maryland," *Memoirs of the American Folklore
Society* 18 (1925): 193; Jon Baughman and
Ron Morgan, *Tales of the Broad Top* (Saxton,
Pa.: Jon Baughman and Ron Morgan, 1977);
Jon Baughman, *Strange and Amazing Stories of
Raystown Country* (Saxton, Pa.: Broad Top
Bulletin, 1987); Chad Arment, "Giant Snake
Stories in Maryland," *INFO Journal,* no. 73
(Summer 1995): 15–16; Garry E. Adelman and
Timothy H. Smith, *Devil's Den: A History and
Guide* (Gettysburg, Pa.: Thomas Publications,
1997), pp. 11, 141; Jeffrey R. Frazier, *The
Black Ghost of Scotia and More Pennsylvania
Fireside Tales,* vol. 2 (Lancaster, Pa.: Egg Hill,
1997); Patty A. Wilson, *Haunted Pennsylvania*
(Laceyville, Pa.: Belfry Books, 1998), pp.
37–41; Chad Arment, "Giant Snakes in
Pennsylvania," *North American BioFortean
Review* 2, no. 3 (December 2000): 36–43,
http://www.strangeark.com/nabr/NABR5.pdf.

## Giant Python

Individual SNAKE or a separate species of reticulated python of East Asia that exceeds the accepted length of 33 feet.

*Physical description:* Length, 33–70 feet.

*Distribution:* India; Bangladesh; Thailand;
Indonesia; the Philippines.

*Significant sightings:* On May 21, 1877, the crew of the barque *Georgina* saw a large, gray-and-yellow snake, 40–50 feet long, swimming in the Indian Ocean west of Sumatra, Indonesia.

Third Officer S. Clayton, of the China Navigation Company's *Taiyuan,* observed what looked to be a 70-foot, cane-colored python swimming with horizontal undulations in the Celebes Sea in the summer of 1907.

*Present status:* The greatest official length for a Reticulated python (*Python reticulatus*) was 32 feet 9.75 inches, recorded in 1912 on the north coast of Sulawesi, Indonesia. It is the only species that regularly exceeds 20 feet in length.

*Sources:* Spenser St. John, *Life in the Forests of the Far East* (London: Smith, Elder, 1862), pp. 256–261; Bernard Heuvelmans, *In the Wake of the Sea-Serpents* (New York: Hill and Wang, 1968), pp. 272, 382–383; Gerald L. Wood, *The Guinness Book of Animal Facts and Feats* (Enfield, England: Guinness Superlatives, 1982), pp. 106–109; Bernard Heuvelmans, "Annotated Checklist of Apparently Unknown Animals with Which Cryptozoology Is Concerned," *Cryptozoology* 5 (1986): 1–26; John C. Murphy and Robert W. Henderson, *Tales of Giant Snakes: A Historical Natural History of Anacondas and Pythons* (Malabar, Fla.: Krieger, 1997), pp. 27, 47–50.

## Giant Rabbit

Mystery MARSUPIAL of Australia.

*Habitat:* Desert.

*Distribution:* Central Australia.

*Significant sighting:* Gold prospectors are said to have reported rabbits 9 feet long, according to Bernard Heuvelmans.

*Possible explanations:*

(1) Naturalist Ambrose Pratt suggested a surviving *Diprotodon optatum,* the largest marsupial that ever lived, which was probably contemporaneous with earlier generations of Aborigines and died out 18,000–6,000 years ago. This large-snouted browser was nearly 10 feet long and almost 8 feet high at the shoulder.

(2) *Palorchestes azael,* a Late Pleistocene herbivore, was a 1-ton marsupial the size of a horse that could balance on its powerful tail and hind limbs like a kangaroo while reaching up with huge, curved claws on its forelimbs to pull trees and branches into the reach of its short, elephant-like trunk.

(3) Christine Janis pointed out that the Sthenurinae, an extinct subfamily of kangaroos, may be better candidates than the lumbering *Diprotodon optatum.* The sthenurine kangaroos had shorter tails, bigger forearms, and possibly longer ears, making them look more rabbitlike.

*Sources:* Bernard Heuvelmans, *On the Track of Unknown Animals* (New York: Hill and Wang, 1958), pp. 206–209; Christine Janis, "A Reevaluation of Some Cryptozoological Animals," *Cryptozoology* 6 (1987): 115–118. Unfortunately, the original source for the prospectors' report is unknown.

## Giant Rat-Tail

Unknown marine FISH.

*Physical description:* Length, 6–10 feet.

*Distribution:* North Atlantic Ocean.

*Significant sightings:* Two observations—one in deep water near Bermuda in the 1930s, another above the seafloor in the Gulf of Mexico in the 1960s.

*Possible explanation:* Unknown species of Grenadier or Rat-tail (Family Macrouridae) of exceptional size. Macrourids are deepwater fishes with sharply narrowing tails. There are 285 species, the largest of which is the Giant grenadier (*Albatrossia pectoralis*), which grows to 5 feet long.

*Source:* Gardner Soule, *Mystery Monsters of the Deep* (New York: Franklin Watts, 1981).

## Giant Salmon

Mystery FISH of northern China.

*Physical description:* Red fish similar to the taimen but five times as large. Length, up to 33 feet. Head, 3 feet across. Spiny dorsal rays. Tail fins.

*Behavior:* Causes huge waves. Said to feed on cattle and sheep.

*Distribution:* Lake Hanas, Xinjiang Uygur Autonomous Region, China.

*Significant sightings:* Biologist Xiang Lihao and his students visited the lake in July 1985 and photographed a school of some sixty of these fishes. However, the photos have apparently never been published.

Three 13-foot specimens were reported by fishermen in July 1988.

*Possible explanation:* Enormous species of Taimen (*Hucho taimen*), a large, freshwater salmon that resembles a pike or muskellunge and normally only grows to 6 feet long.

*Sources:* Wen Jiao, "Does China Have a Loch Ness Monster?" *China Reconstructs* 35 (April 1986): 28–29; "Giant Fish Reported in China," *ISC Newsletter* 5, no. 3 (Autumn 1986): 7–8; *Detroit (Mich.) News,* August 10, 1988.

## Giant Spider

Unknown arthropod INVERTEBRATE of Central Africa and Australasia.

*Physical description:* Huge spider.

*Distribution:* Democratic Republic of the Congo; Papua New Guinea.

*Significant sightings:* R. K. Lloyd and his wife were motoring in the Democratic Republic of the Congo in 1938 when they saw a large object crossing the trail in front of them. At first, they thought it was a cat or a monkey, but they soon realized it was a spider with legs nearly 3 feet long.

An Australian soldier, on patrol along the Kokoda Trail in Papua New Guinea in the fall of 1942, claims to have run across a spider the size of a puppy dog that had spun a 10- to 15-foot web.

*Present status:* The largest known living spider is the Goliath birdeater (*Theraphosa blondi*) of northern South America, a tarantula with a 13-inch legspan and a body length of 3.5 inches. Other big species are the Brazilian salmon tarantula (*Lasiodora parahybana*) and the Brazilian tawnyred (*Grammostola mollicoma*), both with legspans of 10 inches. The largest known fossil spider was *Megarachne servinei*, a 16-inch-long giant with a legspan greater than 20 inches that lived in San Luis Province, Argentina, 300 million years ago, during the Carboniferous period.

*Sources:* Karl Shuker, "From Dodos to Dimetrodons," *Strange Magazine,* no. 19 (Spring 1998): 22–23; Chad Arment, "CZ Conversations: Giant Spiders," *North American BioFortean Review* 3, no. 2 (October 2001): 28–29, http://www.strangeark.com/nabr/NABR7.pdf.

## Giant Tongan Skink

An unknown LIZARD of Oceania.

*Physical description:* Total length, about 18 inches long. Diameter, 1–1.5 inches. Dull green, with blackish markings.

*Behavior:* Runs away swiftly.

*Habitat:* Wooded areas.

*Distribution:* Tongatapu Island, Tonga.

*Significant sightings:* Lannon Oldenburg saw a lizard about 18 inches long running across the ground in a plantation near Tupou College. It was dull green with black markings and lacked a middorsal ridge.

Peter Chignell observed a large lizard in an isolated stand of forest near an ancient burial mound during a burn-off of some scrubland. It was about 1.4 inches in diameter.

*Present status:* John R. H. Gibbons conducted an extensive search for skinks in the remaining stand of forest on Tongatapu in 1985. No unknown species were found. Deforestation and the introduction of cats and rats have made it difficult for any large lizards to survive on Tongatapu. Neighboring Eua' Island holds more promise.

*Possible explanations:*

(1) A Giant skink (*Tachygyia microlepis*), known only from two specimens probably collected from Tongatapu Island in 1827 by J. R. C. Quoy and J. P. Guimard, was over 12 inches long. However, its coloration was dark brown, with no markings. Also, this species probably crawled rather than ran.

(2) An unnamed, olive-brown skink (*Emoia trossula*) found in Tonga and Fiji is more likely, though it rarely exceeds 6 inches in length.

*Source:* Ivan Ineich and George R. Zug, "*Tachygyia,* the Giant Tongan Skink: Extinct or Extant?" *Cryptozoology* 12 (1996): 30–35.

## Giant Vampire Bat

Unknown BAT of South America.

*Physical description:* Large size.

*Behavior:* Said to attack cattle and horses.

*Distribution:* Rio Ribeira valley, São Paulo State, Brazil.

*Possible explanation:* Surviving population of the extinct Giant vampire bat (*Desmodus draculae*), a large bat that lived 500,000 years ago, in the Pleistocene.

*Source:* E. Trajano and M. de Vivo, "*Desmodus draculae* Morgan, Linares, and Ray, 1988, Reported for Southeastern Brazil, with Paleoecological Comments (Phyllostomidae, Desmodontidae)," *Mammalia* 55, no. 3 (1991): 456–459.


## Gigantic Octopus

A huge CEPHALOPOD of the North Atlantic Ocean.

*Scientific names: Octopus giganteus,* given by Addison E. Verrill in 1897; *Otoctopus giganteus,* proposed by Michel Raynal in 1986.

*Variant names:* Bermuda blob, LUSCA.

*Distribution:* From the east coast of Florida to Bermuda, Belize, and south Texas.

*Significant sightings:* On November 30, 1896, a huge carcass was found washed up on Anastasia Beach, near St. Augustine, Florida. DeWitt Webb, a local medical doctor, examined and took several photographs of it. The specimen was 20 feet long, 4 feet high, and 5 feet wide. Its estimated weight was 5 tons. It had a pear-shaped, pinkish body with a silvery cast and was covered with 3–6 inches of extremely tough connective tissue. The stumps of five arms were evident, and some of the detached arms, one of which was 28 feet long and 8 inches thick, were found lying several feet away. Some of the internal organs were still present. A storm carried the mass out to sea in early January 1897, but it reappeared 2 miles farther south, and Webb managed to haul it up to higher ground using horses, tackle, and windlass. He identified the remains as an octopus and sent descriptions, photos, and tissue samples to Yale cephalopod expert Addison E. Verrill. Verrill first identified the mass as a Giant squid (*Architeuthis*) but

*DeWitt Webb shown in front of the remains of what might have been a GIANT OCTOPUS washed ashore near St. Augustine, Florida, in 1896. (Fortean Picture Library)*

changed his designation to Gigantic octopus long enough to give it a scientific name; he then retracted that statement after looking more closely at Webb's tissue samples and suggested the mass may have come from the nose of a Sperm whale (*Physeter catodon*).

A blob of similar matter with five arms washed into Mangrove Bay on Bermuda in May 1988. Discovered by Teddy Tucker, it was a mass of tough, white, fibrous substance 8 feet long and about 3 feet thick.

*Tissue analysis:* Fortunately, a sample from the 1896 stranding sent to William H. Dall at the Smithsonian Institution had been retained, though the bulk of it is now lost. Three analyses have been performed on this material: histological tests in 1963 by Joseph Gennaro, amino-acid analysis in 1986 by Roy Mackal, and electron-microscope and biochemical procedures in 1994 by Sidney Pierce. The first two analyses indicated the substance was connective tissue similar to that found in an octopus; the last suggested that both it and the 1988 Bermuda sample consisted of collagen—whale blubber in the first instance and the thick skin of a fish in the second. Probably only a sophisticated collagen electrophoresis test or amino-acid sequence analysis will resolve this discrepancy. Unfortunately, the Florida specimen may be too contaminated now to be tested successfully.

*Possible explanations:*

(1) A gigantic North Atlantic variety of octopus. The largest known species is the Giant Pacific octopus (*Enteroctopus dofleini*), which can exceed a radial spread of 20 feet and a weight of 100 pounds. A smaller relative, *E. megalocyathus,* is found in the eastern South Pacific and South Atlantic in Chilean and Argentinan waters. Michel Raynal has suggested that a giant form of cirrate octopus, such as *Cirroteuthis,* might be involved.

(2) The spermaceti tank from a sperm whale's head, which has a baglike shape, weighs several tons and is rich in collagen.

(3) A decomposed Ocean sunfish (*Mola mola*) because of its unusual shape. The heaviest of all bony fishes, with a maximum weight of 4,400 pounds, the sunfish looks like a big head with long dorsal and anal fins. The scaleless body is covered with thick, elastic skin. It grows to a maximum length of nearly 11 feet and is common in warm and temperate waters of the Atlantic. However, a sunfish does not come close to matching the description of the original specimen.

*Sources:* Addison E. Verrill, "A Gigantic Cephalopod on the Florida Coast," *American Journal of Science,* ser. 4, 3 (January 1897): 79; Addison E. Verrill, "Additional Information Concerning the Giant Cephalopod of Florida," *American Journal of Science,* ser. 4, 3 (February 1897): 162–163; Addison E. Verrill, "The Florida Sea-Monster," *American Naturalist* 31 (April 1897): 304–307; Addison E. Verrill, "The Supposed Giant Octopus of Florida: Certainly Not a Cephalopod," *American Journal of Science,* ser. 4, 3 (April 1897): 355–356; Forrest G. Wood and Joseph G. Gennaro, "An Octopus Trilogy," *Natural History* 80 (March 1971): 15–24, 84–87; Gary S. Mangiacopra, "*Octopus giganteus* Verrill: A New Species of Cephalopod," *Of Sea and Shore* 6, no. 1 (Spring 1975): 3–10, 51–52; Gary S. Mangiacopra, "More on *Octopus giganteus,*" *Of Sea and Shore* 8, no. 3 (Fall 1977): 174, 178; "Giant Octopus Blamed for Deep Sea Fishing Disruptions,"

*ISC Newsletter* 4, no. 3 (Autumn 1985): 1–6; Roy P. Mackal, "Biochemical Analyses of Preserved *Octopus giganteus* Tissue," *Cryptozoology* 5 (1986): 55–62; "Bermuda Blob Remains Unidentified," *ISC Newsletter* 7, no. 3 (Autumn 1988): 1–6; Richard Ellis, *Monsters of the Sea* (New York: Alfred A. Knopf, 1994), pp. 303– 322; Michel Raynal, "The Case for the Giant Octopus," *Fortean Studies* 1 (1994): 210–234; Sidney K. Pierce, Gerald N. Smith Jr., Timothy K. Maugel, and Eugenie Clark, "On the Giant Octopus (*Octopus giganteus*) and the Bermuda Blob: Homage to A. E. Verrill," *Biological Bulletin* 188 (1995): 219–230; Gary S. Mangiacopra et al., "An Open Forum on the *Biological Bulletin*'s Article on the *Octopus giganteus* Tissue Analysis," *Of Sea and Shore* 19, no. 1 (Spring 1996): 45–50; John Moore, "What Are the Globsters?" *Cryptozoology Review* 1, no. 1 (Summer 1996): 20–29; Michel Raynal, "Debunking the Debunkers of the Giant Octopus," *INFO Journal,* no. 74 (Winter 1996): 24–27; Gary S. Mangiacopra, Michel P. R. Raynal, Dwight G. Smith, and David F. Avery, "'Him of the Hairy Hands': *Octopus giganteus,* Speculation on the Eared Octopus," *Cryptozoology Review* 1, no. 3 (Winter-Spring 1997): 13–18; Gary S. Mangiacopra, "Another *Octopus giganteus* Rebuttal—Again!" *Of Sea and Shore* 21, no. 4 (Winter 1999): 233–234; Michel Raynal, "Le 'Monstre de Floride' de 1896: Cachalot ou pieuvre géante?" May 2000, http://perso. wanadoo.fr/cryptozoo/floride/intro.htm.

## Gigantic Pacific Octopus

A huge CEPHALOPOD of the Pacific Ocean.

*Variant name:* GLOBSTER.

*Physical description:* Grayish-brown. Arms, 8–75 feet long.

*Distribution:* Hawaiian Islands; the Philippines.

*Significant sightings:* In 1928, when stationed at Pearl Harbor, Robert Todd Aiken discovered a group of six large octopuses, 40 feet from tip to tip, off the shore of Oahu. In July 1936, he brought film director Robert Hale to the spot to film a documentary. Whether it was ever filmed or not is unknown.

In 1950, Madison Rigdon saw an octopus the size of a car surrounded by sharks off Lahilahi Point, near Makaha, Oahu, Hawaii. It defended itself with a 30-foot tentacle that had suckers the size of dinner plates.

Also in 1950, off the Kona Coast, Hawaii, fisherman Val Ako saw a monster octopus, with tentacles 75 feet long and suckers as big as auto tires, resting underwater on a reef. His *kupuna* (family adviser) told him later that the octopus came to the island every year for a month with a female.

On December 24, 1989, a group of fourteen people on an 18-foot motorized canoe in Iligan Bay in the Philippines watched as a huge octopus with 8-foot tentacles seized the boat and started rocking it. After ten minutes, the boat capsized. The passengers were either saved by fishermen or swam to shore.

*Possible explanation:* A gigantic Pacific variety of octopus. The largest known octopus is the Giant Pacific octopus (*Enteroctopus dofleini*), which can exceed a radial spread of 20 feet and a weight of 100 pounds. One individual captured near Victoria, British Columbia, in 1967 weighed 156 pounds and was almost 23 feet from arm tip to arm tip. There are unofficial records of a few individuals greater than 300 pounds and one that was more than 400 pounds. This creature ranges from the coast of southern California north to Alaska and into Asia south to Japan. A smaller relative, *E. megalocyathus,* is found in the eastern South Pacific and South Atlantic in Chilean and Argentinan waters.

*Sources:* "Terrors from the Deep," *Fortean Times,* no. 56 (Winter 1990): 14; Nick Sucik, "Just When You Thought It Was Safe to Go Snorkeling: Hawaii's Giant Octopuses," *North American BioFortean Review* 2, no. 3 (December 2000): 11–17, http://www.strangeark.com/nabr/NABR5.pdf.

## Gilyuk

TRUE GIANT hominid or CANNIBAL GIANT of Alaska.

*Etymology:* Probably Na-Dené word, "big man with the little hat."

*Variant name:* Nandna (Tanaina/Na-Dené).

*Physical description:* Covered with shaggy hair.

*Behavior:* Whistles. Wears a little hat. Twists birch saplings. Said to abduct and eat humans.

*Distribution:* Southern Alaska.

*Sources:* Cornelius B. Osgood, "The Ethnography of the Tanaina," *Publications in Anthropology, Yale University* 16 (1937): 171–173; Russell Annabel, "Long Hunter— Alaskan Style," *Sports Afield* 150 (July 1963): 34–36, 102–108.

## Gjevstroll

FRESHWATER MONSTER of Norway.

*Distribution:* Fyresvatn, Telemark County, Norway.

*Significant sighting:* Two men saw a large animal in the lake in 1918, and one of them fired three rounds at it with his Krag Jørgensen rifle.

*Source:* Erik Knatterud, "Sea Serpents in Norwegian Lakes," March 2001, http://www.mjoesormen.no.

## Glaucous Macaw

Colorful BIRD of the Parrot family (Psittacidae) in South America, presumed extinct since the 1930s.

*Scientific name: Anodorhynchus glaucus,* given by Louis Pierre Vieillot in 1818.

*Physical description:* Dull-gray turquoise parrot with brownish-gray throat and sooty face. Total length including tail is about 29 inches. Black bill.

*Habitat:* Swampy riverlands.

*Distribution:* Recent reports have been from Corrientes and Misiones Provinces, Argentina; Artigas department, Uruguay; Rio Grande do Sul and Santa Caterina States, Brazil.

*Significant sightings:* Unconfirmed reports of this species were made in 1951, 1970, and 1988. In 1992, one of a pair of Lear's macaws was identified as a female Glaucous macaw by several, but not all, macaw experts.

*Present status:* Most likely extinct since the 1930s, if not earlier. The last known specimen died at the London Zoo in 1912. Another was housed in the Buenos Aires Zoo as late as 1936, but it may have been a Lear's macaw. Destruction of its primary food source, the Yatay palm (*Butia yatay*), has been the major factor in its disappearance.

*Possible explanations:*

(1) The Hyacinth macaw (*Anodorhynchus hyacinthinus*) is larger and bluer.

(2) The Lear's macaw (*A. leari*) is also slightly larger and brighter.

*Sources:* Rosemary Low, *The Complete Book of Macaws* (New York: Barron's, 1990); Tony Pittman, "The Glaucous Macaw: Does It Still Exist?" *Parrot Society Magazine* 26 (1992), at http://www.bluemacaws.org/glau2.htm; Nigel J. Collar, et al., *Threatened Birds of the Americas* (Washington, D.C.: Smithsonian Institution, 1992); Tony Pittman, "Some New Information on the Glaucous Macaw," *Parrot Society Magazine* 31 (1997), at http://www.bluemacaws.org/glau1.htm; Errol Fuller, *Extinct Birds* (Ithaca, N.Y.: Cornell University Press, 2001), pp. 236–239.

## Glawackus

EASTERN PUMA of Connecticut.

*Etymology:* From town name—Glastonbury, Connecticut—plus "wacky."

*Variant names:* Granby panther, Injun devil.

*Physical description:* Looks variously like a large cat or dog. Length, 4 feet. Height, 2 feet–2 feet 6 inches. Black or tawny in color. Long tail, sometimes described as bushy.

*Behavior:* Emits blood-curdling screams.

*Tracks:* Like a puma's.

*Distribution:* North-central Connecticut.

*Significant sightings:* After weeks of hearing odd animal cries, finding large cat tracks, and seeing glimpses of a large black cat or dog, residents of Glastonbury, Connecticut, organized an all-day hunt into the surrounding hills on January 14, 1939. Nothing was found, but sightings continued until February 24, when Harold Roberts found a 2-mile trail of pawprints 4 miles east of town.

Another cluster of sightings took place in the mid-1950s, with a climax in the summer of 1959 around Granby, Connecticut, where farm

animals were attacked by something pumalike. Black cats continued to be reported through 1967. Since then, reports of tawny animals have been more frequent.

*Possible explanations:*

(1) The EASTERN PUMA returning to its former range. The last killing of a Puma (*Puma concolor couguar*) in New England was in 1881 at Barnard, Vermont.

(2) Accidental or deliberate reintroduction of pet or show pumas.

(3) Escaped melanistic Jaguars (*Panthera onca*) or Leopards (*Panthera pardus*).

(4) Loren Coleman suggests that the black cats are surviving maneless, female American lions (*Panthera atrox*), a Pleistocene lion that died out 9,000 years ago, while the males are reported as MANED AMERICAN LIONS.

*Source:* Gary S. Mangiacopra and Dwight G. Smith, "Connecticut's Mystery Felines: The Glastonbury Glawackus, 1939–1967," *The Anomalist*, no. 3 (Winter 1995–1996): 90–123.

## Globster

Beached specimens of GIGANTIC PACIFIC OCTOPUS or other organic masses of the Pacific Ocean.

*Etymology:* Coined by Ivan T. Sanderson to describe the 1960 carcass.

*Variant names:* Jughead, The Thing.

*Physical description:* No apparent bone structure. Ivory-colored, rubbery, stringy, extremely tough skin. Covered with fine hair or fiber, like greasy sheep's wool. No defined head. No visible eyes. Five gill-like, hairless slits on each side. Smooth, gullet-like orifice.

*Distribution:* Pacific Ocean.

*Significant sightings:* A headless carcass about 20–22 feet long and 4 feet wide was found on the Pacific coast near Delake, Oregon, in March 1950. Local people nicknamed it "Jughead" and cut off pieces of it as souvenirs.

A huge mass of organic material was found on the beach north of the Interview River, Tasmania, Australia, in August 1960 by Ben Fenton, Jack Boote, and Ray Anthony. It measured 20 feet long by 18 feet wide by 4 feet 6 inches thick, and it weighed 5–10 tons. It appeared to be made up of "tendon-like threads welded together with a fatty substance." Over eighteen months, it showed no signs of decomposition. An on-site analysis of the material by Bruce Mollison of the Commonwealth Scientific and Industrial Research Organization (CSIRO) on March 7, 1962, was unable to provide an identification. A second CSIRO analysis on March 17, 1962, indicated protein and collagen as primary components and suggested the material was "not inconsistent with blubber."

Another fibrous Globster was found in March 1965 on Muriwai Beach, North Island, New Zealand. It was 30 feet long, 8 feet high, and covered with hair 4–6 inches long.

Ben Fenton found a third Australasian Globster south of Sandy Cape, Tasmania, in November 1970. This one was 8 feet long.

An 8- to 10-foot specimen that washed ashore near Wanganui, New Zealand, in early October 1997 was dismissed as a partly decomposed sperm whale.

A 4-ton, 20-foot fibrous mass with six tentacles washed up on Four Mile Beach, northwest of Zeehan, Tasmania, in late December 1997 and was similarly diagnosed as whale blubber.

*Possible explanations:*

(1) A Pacific manta ray (*Manta hamiltoni*), which can weigh up to 1.5 tons, was suggested by A. M. Clark.

(2) Whale blubber, especially the 1965 New Zealand carcass, though no one noticed any characteristic oil, odor, bones, or internal organs.

(3) Remains of a GIGANTIC PACIFIC OCTOPUS.

(4) The Basking shark (*Cetorhinus maximus*) attains a length of 40 feet. Decomposition causes its muscle fibers to appear stringy.

*Sources: Victoria (B.C.) Daily Times,* March 7, 1950; Ivan T. Sanderson, "Monster on the Beach," *Fate* 15 (August 1962): 24–35; Tim Dinsdale, *Monster Hunt* (Washington, D.C.: Acropolis, 1972), pp. 159–160; John Michell and Robert J. M. Rickard, *Living Wonders: Mysteries and Curiosities of the Animal World* (London: Thames and Hudson, 1982), pp. 27–31; "Bermuda Blob Remains Unidentified," *ISC Newsletter* 7, no. 3 (Autumn 1988): 1–6; Richard Ellis, *Monsters of the Sea* (New York: Alfred A. Knopf, 1994), pp. 303–322; John Moore, "What Are the Globsters?" *Cryptozoology Review* 1, no. 1 (Summer 1996): 20–29; "Mystery Blobby Found in Tasmania," *Fortean Times,* no. 109 (April 1998): 21.

## Glowing Mudskipper

Mystery FISH of Southeast Asia.

*Physical description:* Looks like a Mudskipper (Family Gobiidae). Glows pulsating red at night.

*Distribution:* Ceram, Indonesia.

*Significant sighting:* Agriculturalist Tyson Hughes observed this fish in a Ceram river in 1986. He attempted to catch a specimen but failed.

*Source:* Karl Shuker, "A Supplement to Dr. Bernard Heuvelmans' Checklist of Cryptozoological Animals," *Fortean Studies* 5 (1998): 208–229.

## Gnéna

LITTLE PEOPLE of West Africa that share some attributes with the legendary Islamic DJINN (intermediaries between humans and angels) and SMALL HOMINIDS.

*Etymology:* Bambara (Mande) word.

*Variant names:* Gnéna or Guinné is the generic name for all of these entities, including some larger ones. The smaller ones include: Artakourma (Zarma/Songhay), Asamanukpa (in Ghana), Attakourma, Bâri (Susu/Mande), Bésonroubé (Manza/Ubangi), Dato (Senoufo/Gur), Datobou, Déguédégué (Songhay), Dioudiou (Fulfulde/Fulani), Doudo (Baka/Ubangi), Gotteré (Fulfulde/Fulani), Kélékongbo (Banda/Ubangi), Kélékumba (Banda/Ubangi), Kinpélili (Gbaya/Ubangi), Kitikpili (Bokoto/Ubangi), Konkimbu (Lobi/Gur), Konkoma (Malinke/Mande), Kontimbié (Lobi/Gur), Kontoma (Dagaari/Gur), Korokombo (Banda/Ubangi), Mokala (Manza/Ubangi), Nyama (Bambara/Mande), Ouokolo, Pori (Gourmantché/Gur), Sonkala (Manza/Ubangi), Tikirga (Mõõre/Gur), Wokolo (Bambara/Mande), Wouoklo, Yamana (Bambara/Mande).

*Physical description:* Height, 2–4 feet. Covered in long, black or dirty-gray hair. Sometimes said to be covered with sharp spines. Large head. Yellow eyes. Pointed beard. Long arms. Knock-kneed. Feet are turned toward the rear or webbed.

*Behavior:* Walks on the outside of its feet. Very strong. Malicious. Sleeps in the trees. Collects sticks and bundles them together. Shoots tiny arrows at people. Said to switch its infants with human babies like the European FAIRY does.

*Distribution:* Côte d'Ivoire; Senegal; Guinea; Burkina Faso; Mali; Niger; Ghana; Cameroon; Central African Republic.

*Possible explanation:* Legends about the ancestors of the Mbuti, Twa, and Mbenga peoples—short-statured, forest-dwelling Pygmies.

*Sources:* Abbé Joseph Henry, *L'âme d'un peuple africain,* Anthropos Ethnologische Bibliothek, Band 1, Heft 2 (Munich, Germany: Aschendorffsche Verlag, 1910); Victor François Equilbecq, *Essai sur la littérature merveilleuse des noirs, suivi de contes indigènes de l'Ouest-Africain français* (Paris: E. Leroux, 1913–1916), vol. 1, pp. 106–135; Margaret J. Field, "Gold Coast, Ethnography: The Asamanukpai of the Gold Coast," *Man* 34 (December 1934): 186–187; Eugène-René Viard, *Les Guérés, peuple de la forêt* (Paris: Société d'Éditions Geographiques, Maritimes et Coloniales, 1934), pp. 11–13; Antonin Marius Vergiat, *Les rites secrets des primitifs de l'Oubangi* (Paris: Payot, 1936), pp. 60–64; Mamby Sidibé, "Légendes autour des génies nains en Afrique Noir," *Notes Africaines,* no. 47 (1950): 100; Bernard Heuvelmans, *Les bêtes humaines d'Afrique* (Paris: Plon, 1980), pp. 483–487, 496–498.

## Goatman

HAIRY BIPED of Maryland.

*Variant name:* Abominable phantom.

*Physical description:* Height, 6 feet. Covered with hair. Blazing red eyes. Lower body resembles a goat's.

*Behavior:* Bipedal. Makes a high-pitched squeal. Said to be responsible for mutilated pets.

*Habitat:* Country roads and forests.

*Distribution:* Prince George's and Calvert Counties, Maryland.

*Significant sightings:* Reverty Garner and his wife ran into a hairy wildman as they pulled into their driveway in Upper Marlboro, Maryland, on August 1, 1957.

On November 3, 1971, April Edwards saw a large creature in her backyard along Fletcher-town Road in Bowie, Maryland; her dog, Ginger, disappeared shortly afterward. Willie Gheen and John Hayden discovered the dog's head the next morning. Kathy Edwards and a group of girls saw a humanlike form climb off a pickup truck and walk into the woods on November 17.

*Possible explanations:*

(1) An elderly human hermit.

(2) Urban legend.

*Sources:* *Washington (D.C.) Evening Star,* August 5, 1957; *Washington (D.C.) Daily News,* August 7, 1957; *Prince George's County (Md.) News,* October 27, November 10, and November 24, 1971; Bob Weller, ed., *Prince George's Community Collage: An Oral History Collection* (Largo, Md.: Prince George's Community College, 1986); Mark Opsasnick, "On the Trail of the Goatman," *Strange Magazine,* no. 14 (Fall 1994): 18–21; Sean Daly, "The Legend of Goatman," *Washington (D.C.) City Paper,* September 18, 1998, on line at http://www.washingtoncitypaper.com/archives/cover/1998/cover0918.html; John Lawson, The Goatman Legend of Prince George's County, http://azaz.essortment.com/goatmanlegend_rhcn.htm.

## Goazi

SMALL HOMINID of South America.

*Etymology:* Tupinambá (Tupí) word.

*Variant name:* Guayazi.

*Distribution:* Brazil; Colombia.

*Habitat:* Forests.

*Possible explanation:* Probably short-statured Indians.

*Sources:* Simão de Vasconcellos, *Noticias curiosas, y necessarias das cousas do Brasil* (Lisbon: I. da Costa, 1668); Robert, marquis de Wavrin, *Chez les indiens de Colombie* (Paris: Plon, 1953).

## Golden Ant

Legendary INVERTEBRATE or RODENT of Central Asia.

*Physical description:* Size of a fox. Skin like a leopard's.

*Behavior:* Moves swiftly. Digs holes in the winter.

*Distribution:* Highlands around the Indus River area in Pakistan; Jammu and Kashmir State, India.

*Significant sighting:* The ancient Greek historian Herodotus wrote of an area in northern India where large, vicious ants dug burrows that turned up a large quantity of gold-bearing sand. The Persians went to the region in the morning to bag the sand and take it back while the ants were still underground.

*Present status:* Not taken very seriously for nearly 2,500 years, until Michel Peissel visited the region in 1996.

*Possible explanations:*

(1) The Long-tailed marmot (*Marmota caudata*) of Baltistan in northern Pakistan, Michel Peissel notes, burrows in gold-bearing soil. The ancient Persian word for these animals translates as "mountain ant," which resulted in Herodotus's misidentification.

(2) Other sources mention giant insects in northern China. Berthold Laufer has suggested a confusion of the Mongolian word for ant, *shorgoolj,* with the Shirongol Mongols, now more commonly known as the Dongxiang, who currently live in Gansu Province, China, east of Linxia.

*Sources:* Herodotus, *The Histories,* ed. John Marincola (New York: Penguin, 1996), pp. 194–195 (III. 102–105); *The Classic of Mountains and Seas,* trans. Anne Birrell (New York: Penguin, 1999), pp. 146, 200, 211 (bk. 12); Strabo, *The Geography,* trans. Horace Leonard Jones (Cambridge, Mass.: Harvard University Press, 1969), vol. 7, pp. 75–77 (XV. 1.44); David Hawkes, ed., *Ch'u tz'u: The Songs of the South* (Oxford: Clarendon, 1959), p. 104; Berthold Laufer, "Die Sage von dem goldgrabenden Ameisen," *T'ung Pao,* ser. 2, vol. 11 (1908); Peter Costello, *The Magic Zoo* (New York: St. Martin's, 1979), pp. 90–93; Peter Humi, "Solving the Mystery of the 'Golden Ants,'" CNN News report, December 2, 1996.

## Golden Ram

Sheep of West Asia; *see* SEMIMYTHICAL BEASTS.

*Physical description:* Like a normal ram but with wings and golden wool.

*Behavior:* Flies easily through the air. Has the ability to reason and speak.

*Distribution:* The Black Sea coast of Georgia.

*Significant sighting:* Chrysomallus was a winged ram with golden wool that was sent to earth by Hermes. The animal was the object of Jason's voyage in the ship *Argo* to the shore of Colchis on the Black Sea in 1400 B.C. or earlier.

*Possible explanations:*

(1) Jason's expedition might really have been in search of gold said to be found in the rivers of Colchis (modern Georgia). In the first century B.C., Strabo described an ancient method of extracting gold from alluvial sands by sifting water over ramskins that retained the grains of gold in their fleece. The Egyptians used this method, as documented on a wall painting in the Temple of Ramses III at Medinet-Habu. The myth of a Golden ram derived from this process.

(2) The fleece of the Golden ram could have possessed extremely fine fibers like those from the wool of the Spanish or American merino breeds, making it considerably valuable for weaving. Scythian fine-wooled sheep apparently existed around the Black Sea as early as the fifth century B.C.

(3) In 1932, Claude Rimington, of the Wool Industries Research Association in Leeds, investigated a golden-brown pigment found in varying intensities within the suint secreted by the sweat glands of certain sheep. Later research showed it was caused by bilirubin from the sheep's liver that had passed into the sweat, producing an abnormal golden coloration. The condition was stimulated when the sheep ate the leaves of certain plants, including Shrub verbena (*Lantana* spp.) and the Olive tree (*Olea*

*europaea*), that prevent the liver from excreting bilirubin effectively.

*Sources:* Strabo, *The Geography,* XI. 2.19; Claude Rimington and A. M. Stewart, "A Pigment Present in the Sweat and Urine of Certain Sheep," *Proceedings of the Royal Society,* ser. B, 110 (1932): 75–91; J. M. M. Brown, Barbara Sawyer, et al., "Studies on Biliary Excretion in the Rabbit II," *Proceedings of the Royal Society,* ser. B, 157 (1963): 473–491; M. L. Ryder and J. W. Hedges, "Ancient Scythian Wool from the Crimea," *Nature* 242 (1973): 480; G. J. Smith, "Jason's Golden Fleece Explained?" *Nature* 327 (1987): 561; Patrick Moyna and Horacio Heinzen, "Why Was the Fleece Golden?" *Nature* 330 (1987): 28; Karl Shuker, "The Search for the Real Golden Fleece," *Fate* 42 (September 1989): 46–52; Maria Rosario Belgiorno, "Il vero significato del mito del vello d'oro e del viaggio degli Argonauti," *Studi Micenei,* 2000, on line at http://www.area.fi.cnr.it/r&f/n17/belgiorno1.htm.

## Golub-Yavan

WILDMAN of Central Asia.

*Etymology:* Tajik (Persian), "wild man."

*Variant names:* Galub-yavan, Ghool-biaban, Gul-biavan, Gul'bi-yavan, Guli-b'yabon, Gulibyavan, Khaivan-akvan, Voita. *See also* GUL.

*Physical description:* Height, 5 feet–6 feet 6 inches. Covered in reddish-gray or black hair. Older individuals are grayer. Head-hair is thick and matted. Slanting forehead. Brows and cheekbones are prominent. Face is bare. Glowing eyes. Wide, flat nose. Ears stick out. Lower jaw is massive. Large teeth. Short neck. Thick hair on chest and hips, close-cropped and thick on the belly. Long arms. Buttocks are relatively hairless. Knees are calloused. Feet and palms are hairless. Feet are wider and shorter than a man's.

*Behavior:* Call is a mewing or whistling sound. Strong odor. Food includes berries. Searches for marmots under rocks, making piles of rocks in the process. Uses caves as shelters. Uses sticks as weapons. Said to attack humans.

*Tracks:* Humanlike but shorter and broader. The four smaller toes are wider than a human's.

*Distribution:* Pamir and Tian Shan Mountains, Kyrgyzstan; eastern Tajikistan. Possibly extends into the Kunlun Mountains of western China south of Taxkorgan, the Karakoram Range in northern India, and the Hindu Kush of eastern Afghanistan.

*Significant sightings:* Maj. Gen. Mikhail Topilski, head of a scouting party in the fall of 1925, ran across a group of Golub-yavan during a skirmish with White Russian guerrillas in the Vanch District, Tajikistan; the guerrillas had taken refuge in an ice cave that the creatures apparently used as a shelter. One wildman was shot and inspected by the party's physician. The dead creature was 5 feet 6 inches tall and looked much more human than apelike, though it was covered with dense hair except for its face, palms, soles, knees, and buttocks. It had heavy browridges, a flat nose, and a massive lower jaw. The foot was noticeably wider than a human's. The soldiers could not take the body with them, so they buried it under a heap of stones.

A resident of Imeni Kalinina, Tajikistan, was attacked by a Gul-biavan while hunting in 1939. He wrestled it to the ground but lost consciousness. Villagers found him later, along with evidence of a struggle.

Hunters in the mountains around Vanch, Tajikistan, call the wildman a Voita and say it is taller than a man and covered with short, black hair.

Alexander G. Pronin saw a Golub-yavan on a cliff in the Balyandkiik Valley, Tajikistan, on August 12, 1957. It walked out of a cave and was visible for several minutes before it disappeared from view.

*Possible explanation:* A surviving early hominid. Artifacts and *Homo erectus*–like remains, dated at 125,000 years ago, have been found at Selungur Cave, Kyrgyzstan. Early Paleolithic stone flakes and cores about 850,000 years old were discovered at Kuldara, Tajikistan. A Neanderthal burial and grave goods have been found in the Teshik-Tash Cave in Uzbekistan.

*Sources:* Kirill V. Staniukovich, "Golub-Javan: Svendeniia o 'snezhnom cheloveke' na Pamire," *Geograficheskoe Obshchestvo SSSR Izvestia,* 1957, no. 4, pp. 343–345, and 1957, no. 5, p. 89; "Vstrecha so 'Snezhnom

Chelovekom,'" *Komsomol'skaia Pravda,* January 15, 1958, p. 4; Boris F. Porshnev, *Sovremennoe sostoianie voprosa o reliktovykh hominoidakh* (Moscow: Viniti, 1963); Kirill V. Staniukovich, *Po sledam udivitel'noi zagadki* (Moscow: Molodaia Gvardiia, 1965); Bernard Heuvelmans and Boris F. Porshnev, *L'homme de Néanderthal est toujours vivant* (Paris: Plon, 1974), pp. 83–88, 98; Myra Shackley, *Still Living? Yeti, Sasquatch and the Neanderthal Enigma* (New York: Thames and Hudson, 1983), pp. 117–126; Valentin B. Sapunov, "Results of Chimpanzee Pheromone Use in Snowman (Wildman) Field Investigations," *Cryptozoology* 8 (1989): 64–66; Dmitri Bayanov, *In the Footsteps of the Russian Snowman* (Moscow: Crypto-Logos, 1996), pp. 71–75, 81–84.

## Goodenough Island Bird

Mystery BIRD of Australasia.

*Physical description:* Black plumage. Size of a small crow. Long tail.

*Behavior:* Call is a short, explosive rattle.

*Distribution:* Goodenough Island, Papua New Guinea.

*Significant sightings:* One member of the 1953 Fourth Archbold Expedition saw a crow-sized, black bird high in the treetops on Goodenough.

Zoologist James Menzies saw a group of these birds in the forest canopy of Mount Oiamadawa on December 28, 1975.

*Possible explanations:*

(1) An unknown species of Bird of paradise (*Astrapia* sp.), Honeyeater (*Meliphaga* sp.), or Drongo (Family Dicruridae).

(2) The Paradise crow (*Lycocorax pyrrhopterus*), suggested by Karl Shuker, although this is thought to be endemic to Halmahera, Bacan, Obi, and adjacent islands in Indonesia.

*Source:* Bruce M. Beehler, *A Naturalist in New Guinea* (Austin: University of Texas Press, 1991), pp. 81–83, 104–105; Karl Shuker, *Mysteries of Planet Earth* (London: Carlton, 1999), pp. 54–56.

## Gorillaï

Either a WILDMAN of West Africa or an ancient encounter with an anthropoid ape.

*Etymology:* Unknown; probably derived from an African word that Hanno the Carthaginian heard. One hypothesis is that it represents the Kongo (Bantu) word *ngò dìida* ("powerful animal that beats its chest"), which assumes Hanno made it south of the equator; other hypotheses include the Wolof (Atlantic) word *goloh* ("ape") in Senegal and the Benga (Bantu) word *ngiya,* used for the gorilla in Gabon.

*Physical description:* Hairy and wild.

*Behavior:* The males run away into the mountains and let the females be slaughtered.

*Significant sighting:* Hanno the Carthaginian discovered Gorillaï on his voyage along the African coast in the early fifth century B.C. and is said to have taken two skins back to Carthage, where they were displayed in the Temple of Juno.

*Distribution:* An island on the West African coast. Bernard Heuvelmans thought the location was Morocco; others think it was Senegal, Sierra Leone, Cameroon, or Gabon.

*Possible explanations:*

(1) Neanderthals (*Homo neanderthalensis*) surviving into historical times in North Africa, suggested by Heuvelmans. However, so far there is no unambiguous evidence that Neanderthals migrated into Africa from either Europe or West Asia. Also, there is no nautical reason to suppose that Hanno did not actually travel past Cap Vert to the Gulf of Guinea.

(2) An early account of the Lowland gorilla (*Gorilla gorilla*), if Hanno made it as far as Cameroon. The name of the animal itself derives from Hanno's account. Gorillas can't swim, however, so if Hanno actually found them on an island, this explanation is problematic.

*Sources:* Hanno, *The Periplus of Hanno* (Philadelphia: Commercial Museum, 1912); Jona Lendering, Hanno, http://www.livius.org/ha-hd/hanno/hanno02.html#Translation; Pomponius Mela, *De chorographia,* III. 9; Pliny the Elder, *Historia naturalis,* in John F. Healy, ed., *Natural History: A Selection* (New York: Penguin Classics, 1991), p. 71 (VI. 200); Bernard Heuvelmans, *Les bêtes humaines d'Afrique* (Paris: Plon, 1980), pp. 168–201.

## Gougou

CANNIBAL GIANT of eastern Canada.

*Etymology:* Micmac (Algonquian) word.

*Variant names:* Gugu, GUGWÉ, Kuhkw.

*Physical description:* Female monster taller than a ship.

*Behavior:* Carries a pouch in which it puts humans to be eaten later. Whistles shrilly.

*Distribution:* Bonaventure Island, Québec; Miscou Island, New Brunswick.

*Sources:* Samuel de Champlain, *Des Sauvages* [1603], in Henry Percival Biggar, ed., *The Works of Samuel de Champlain* (Toronto, Canada: Champlain Society, 1922), vol. 1, p. 186–187; Sidney W. Dean and Marguerite Mooers Marshall, *We Fell in Love with Quebec* (Philadelphia: Macrae Smith, 1950), pp. 221–222; Richard S. Lambert, *Exploring the Supernatural* (Toronto, Canada: McClelland and Stewart, 1955), p. 181; Bruce S. Wright, "The Gougou: The Bigfoot of the East," *Bigfoot Bulletin,* no. 25 (1971).

## Gowrow

Giant LIZARD of Arkansas.

*Etymology:* From the sound the lizards make.

*Physical description:* Length, up to 20 feet. Tusklike teeth.

*Behavior:* Makes an assortment of groans and hisses.

*Habitat:* Caves.

*Distribution:* Boone and Searcy Counties, northern Arkansas.

*Significant sighting:* Sometime before 1935, E. J. Rhodes heard a commotion in a deep cavern called Devil's Hole, 3 miles northwest of Myrtle, Arkansas. He crawled down 200 feet to investigate, but couldn't see anything. Later, when he lowered a flatiron on a rope into the cavern, something bit through the rope.

*Present status:* Only insubstantial rumors and folktales exist.

*Possible explanations:*

(1) Classic example of Ozark folk humor.

(2) A legend based on the Alligator (*Crocodylus acutus*), which lives in the southern two-thirds of Arkansas and grows to 12 feet long.

*Sources:* Vance Randolph, *We Always Lie to Strangers* (New York: Columbia University Press, 1951), pp. 43–44; Brad LaGrange, "The Gowrow vs. Occam's Razor: An Exercise in Folklore," *North American BioFortean Review* 2, no. 2 (2000): 4–5, http://www.strangeark.com/nabr/NABR4.pdf.

## Great Auk

Flightless sea BIRD of the Auk family (Alcidae) of the North Atlantic Ocean, extinct since 1844.

*Scientific name: Alca impennis,* given by Carl von Linné in 1758.

*Variant name:* Garefowl.

*Physical description:* Length, 2 feet 6 inches. Upper parts black, white below. Oval, white patch in front of the eye. Large, black beak with white grooves.

*Behavior:* Flightless.

*Distribution:* Canada; Greenland; Iceland; Scotland; Norway.

*Significant sightings:* In 1867, a native Greenlander is said to have captured a Great auk on an island in Qeqertarsuaq Tunua, Greenland, and eaten it.

In the 1920s and 1930s, supposed Great auks reported in the Lofoten Islands, Nordland County, Norway, turned out to be penguins brought from Australia and released by whalers.

*Present status:* The last known breeding pair of Great auks were killed by three fishermen on the island of Eldey, Iceland, on June 3, 1844. The body parts of these two are kept in specimen jars at the University Zoological Museum in Copenhagen, Denmark. Errol Fuller thinks the skins are at the Los Angeles County Natural History Museum and the Royal Institute of Natural Sciences in Brussels. Other scattered specimens may have lingered after 1844 but not for long.

*Sources:* Isaac J. Hayes, *The Land of*

The GREAT AUK (Alca impennis), a flightless bird of the North Atlantic Ocean, extinct since 1844. (© 2002 Art-Today.com, Inc., an IMSI Company)

*Desolation, Being a Personal Narrative of Adventure in Greenland* (London: Sampson Low, Marston, Low, and Searle, 1871); "Raiders of the Lost Auk," *ISC Newsletter* 6 (Spring 1987): 5–7; Errol Fuller, *The Great Auk* (New York: Harry N. Abrams, 1999); Christopher Cokinos, *Hope Is the Thing with Feathers* (New York: Jeremy P. Tarcher, 2000), pp. 305–336; Nick Warren, "The End of the Auk," *Fortean Times*, no. 145 (May 2001): 48; Jeremy Gaskell, *Who Killed the Great Auk?* (New York: Oxford University Press, 2001).

## Greek Dolphin

Unknown CETACEAN of the Mediterranean Sea.

*Physical description:* Similar to the striped dolphin but without the diagnostic dark stripe running from the underside of the tail to the eye or the pale-gray, finger-shaped marking below the dorsal fin.

*Significant sighting:* Seen several times in the Mediterranean by Willem Mörzer Bruyns.

*Possible explanation:* Significant variations in the markings of the Striped dolphin (*Stenella coeruleoalba*) are known to occur both individually and geographically.

*Source:* W. F. J. Mörzer Bruyns, *Field Guide of Whales and Dolphins* (Amsterdam: Tor, 1971).

## Grendel

Legendary GIANT HOMINID of Northern Europe, as portrayed in the oldest narrative epic poem in English or Teutonic literature, *Beowulf.* The text in its known form dates from a copy made around A.D. 1000, but it represents a tradition that dates from a much earlier time.

*Etymology:* Old English, "grinder" or "destroyer."

*Variant names:* Eoten, Feond, Thyrs.

*Physical description:* Large. Gorilla-like. Covered with hair.

*Behavior:* Nocturnal. Eats humans. Able to change shape.

*Habitat:* Marshes.

*Significant sighting:* The sixth-century Scandinavian hero Beowulf traveled from Geatland in southern Sweden to aid the Danish king Hrothgar, whose great hall, Heorot (possibly located on the site of modern Lejre near Copenhagen in Denmark), was under attack by the giant Grendel. Beowulf killed both Grendel and its mother in two separate battles.

*Possible explanations:*

(1) Folk memory of Neanderthals (*Homo neanderthalensis*) or other early hominids who coexisted with Europeans in ancient or medieval times.

(2) Folk tradition of Scandinavian corpse-eating ghosts (Draugr).

*Sources:* Seamus Heaney, ed., *Beowulf: A New Verse Translation* (New York: Farrar Straus Giroux, 2000); Nicolas K. Kiessling, "Grendel: A New Aspect," *Modern Philology* 65 (1968): 191–201; Christie Ward, The Walking Dead: Draugr and Aptrgangr in Old Norse Literature, 1996, http://www.vikinganswerlady.org/ghosts.htm.

# Griffin

SEMIMYTHICAL BEAST of Central Asia.

*Etymology:* From the Latin *gryphus,* a misspelling of *grypus,* derived from the Greek *gryps* ("hooked") and possibly related to the Persian *giriften* (to "grip" or "seize").

*Variant names:* BRENTFORD GRIFFIN, Griffon, Gryphon, Gryps.

*Physical description:* Size of a wolf. Has scales or feathers. Crest (in Asia) or mane (in Greece). Fiery red eyes. Knob or short horn on head might represent long, upright ears. Strong beak like an eagle's. Neck is variegated, with blue feathers. Black feathers on back, red feathers on breast. Wing feathers are white. Four legs. Large claws. Further symbolic embellishments were added to this profile in the Middle Ages.

*Behavior:* Flightless despite its wings. Lays eggs in burrows in auriferous deposits. Said to guard gold and be protective of its young. Attacks horses, mountain lions, elk, geese, deer, and humans. Can be captured with a baited trap.

*Distribution:* The Altai and Tien Shan Mountains and the Gobi Desert of China, Mongolia, and adjacent regions.

*Significant sightings:* The Griffin appears on Scythian gold, bronze, wood, and leather artifacts from 3000 to 100 B.C. It became a popular theme in Greek art around 600 B.C. and in Roman art until A.D. 300.

The Griffin was first described in literature by Aristeas in his lost epic the *Arimaspea,* written about 675 B.C., as an animal known to Scythian nomads who traded with the Greeks and traveled as far west as the Altai Mountains of Mongolia and China. Intriguingly, the Scythian word *arimaspu,* which refers to the CYCLOPS (Arimaspeans) who try to steal gold from the Griffins, is linguistically related to the Mongolian ALMAS. Aristeas's story was repeated by the Greek playwright Aeschylus in his tragedy *Prometheus Bound* (lines 790–805) and by other classical authors.

*Possible explanations:*

(1) The Tibetan mastiff, a large guard dog bred for centuries in the Himalayas to protect monasteries, villages, nomadic camps, and livestock herds, was suggested by Valentine Ball.

*The GRIFFIN has appeared in artistic representations since 3000 B.C. (© 2002 ArtToday.com, Inc., an IMSI Company)*

(2) A local Jerboa (Family Allactaginae) or Squirrel (Family Sciuridae) because of its burrowing activities.

(3) The Lammergeier (*Gypaetus barbatus*), a large, carrion-eating bird of Central Asia, has been suggested by Peter Costello; also an eagle or another bird of prey. However, as early as Aeschylus, the flightless *gryps* was distinguished from the winged eagle *aetos.*

(4) A literary invention symbolizing vigilance, the difficulty of mining gold on the Asian steppes, swiftness, the sun, the sky, death, or loyalty.

(5) A speculative re-creation based on the fossil remains of the Woolly mammoth (*Mammuthus primigenius*), discovered in antiquity near auriferous sands in Siberia, proposed by Adolph Erman.

(6) As suggested by Adrienne Mayor, the Griffin may be based on the fossil remains of ceratopsian dinosaurs, especially

*Protoceratops,* a Late Cretaceous herbivore that averaged 7–8 feet in length and whose bones are commonly found in the Gobi, Turpan, and Junggar Deserts along the caravan route between the Tien Shan Mountains of Kyrgyzstan and China and the Altai Mountains of Mongolia. These mountains and their alluvial basins were the source of the gold mined by the Scythians and other ancient peoples, and the proximity of the desert fossils accounts for the ancient association of gold and Gryps. *Protoceratops* had a powerful beak and a dorsal shield like a rearward-projecting horn. Its bones are common even in modern times, and the area is a rich source of fossil eggs and clutches of young dinosaurs.

*Sources:* Herodotus, *The Histories,* ed. John Marincola (New York: Penguin, 1996), p. 221 (IV. 13); Ctesias, *Indika,* in J. W. McCrindle, ed., *Ancient India* (Calcutta, India: Thacker, Spink, 1882), pp. 17, 44–46; Thomas Browne, *Pseudodoxia Epidemica* [1672] (Oxford: Clarendon, 1981), pp. 199–201, 822–823; Adolph Erman, *Travels in Siberia, Including Excursions Northwards* (London: Longman, Brown, Green, and Longmans, 1848), vol. 2, pp. 87–89, 377–382; Valentine Ball, "The Identification of the Pygmies, the Martikhora, the Griffin, and the Dikarion of Ktesias," *The Academy* 23 (1883): 277; Edward Peacock, "The Griffin," *The Antiquary* 10 (September 1884): 89–92; George Jennison, *Animals for Show and Pleasure in Ancient Rome* (Manchester, England: Manchester University Press, 1937), p. 115; Sergei I. Rudenko, *Sibirskaia kollektsiia Petra Pervogo* (St. Petersburg, Russia: Akademiia Nauk SSSR, 1962); Anna Maria Bisi, *Il grifone: Storia di un motivo iconografico nell'antico Oriente mediterraneo* (Rome: Centro di Studi Semitici, Istituto di Studi del Vicino Oriente, Universita di Roma, 1965); Engeborg Flagge, *Untersuchungen zur Bedeutung des Greifen* (Sankt Augustin, Germany: Hans Richarz, 1975); Peter Costello, *The Magic Zoo* (New York: St. Martin's, 1979), pp. 71–82; Joe Nigg, *The Book of Gryphons* (Cambridge, Mass.: Apple-wood Books, 1982); Laskarina Bouras, *The Griffin through the Ages* (Athens: Midland Bank, 1983);

Adrienne Mayor, "Griffin Bones: Ancient Folklore and Paleontology," *Cryptozoology* 10 (1991): 16–41; Adrienne Mayor and Michael Heaney, "Griffins and Arimaspeans," *Folklore* 104 (1993): 40–66; Adrienne Mayor, "Guardians of the Gold," *Archaeology* 47 (November-December 1994): 53–58; Kenneth Carpenter, *Eggs, Nests, and Baby Dinosaurs* (Bloomington: Indiana University Press, 1999); Adrienne Mayor, *The First Fossil Hunters: Paleontology in Greek and Roman Times* (Princeton, N.J.: Princeton University Press, 2000), pp. 15–53.

## Groot Slang

FRESHWATER MONSTER of South Africa.

*Etymology:* Afrikaans, "great serpent."

*Variant names:* Kayman, Ki-man (Nama/Khoisan), !Koo-be-eng (Nama/Khoisan), !Kouteign !koo-rou ("master of the water," Nama/Khoisan).

*Physical description:* Length, 20–39 feet. Larger than a hippo. Black skin. Head, 7–8 inches wide. Neck, 8–10 feet long.

*Tracks:* Width, 18 inches.

*Habitat:* Rivers, lakes, and swamps.

*Distribution:* Orange and Vaal Rivers, South Africa.

*Significant sightings:* A Nama rock painting on Cathedral Peak, KwaZulu-Natal Province, South Africa, depicts a great horned serpent called !Koo-be-eng. Others appear in Brakfontein Cave near Koesberg; in the cave near Klein Aasvogelkop; and in the cave of the Great Black Serpent in Rockwood Glen, near the Upper Orange River.

About 1867, Hans Sauer saw a large, black snake in the Orange River near Aliwal North, Eastern Cape Province.

In 1899, merchant G. A. Kinnear was crossing the Orange River near Upington, Northern Cape Province, when he saw the head of a monstrous serpent emerge from the water. About 8–10 feet of head and neck were visible.

In 1910, Frederick C. Cornell was camping about 20 miles from Augrabiesvalle, Northern Cape Province, with two companions, one an American named Kammerer, who was bathing in a pool nearby. Suddenly, Kammerer came back shouting and said that a great wave had come up behind him and that a head with massive jaws belonging to a giant snake had risen 12 feet in the air.

In May 1920, at the confluence of the Great Fish and Orange Rivers, Frederick C. Cornell and others in his party saw the head and neck of a large snake swimming in the water.

John Clift saw a 20-foot crocodilian emerge from the Big Hole, an abandoned mine crater near Kimberley, Northern Cape Province, in November 1947.

In November 1963, newspapers started reporting various encounters with a water monster in the Vaal Dam, Free State Province. Most of the reports were vague. Stanley Jacob and his father, David, watched a monster surface 110 yards from their boat, near Oranjeville on February 16, 1964. At first, it looked like a swimming horse. They went to fetch a gun, then returned. The animal had grayish-brown skin, smoother than a hippo's.

*Possible explanations:*

(1) A large variety of African rock python (*Python sebae*), which often grows to 30–33 feet.

(2) The Water monitor (*Varanus niloticus*) is Africa's largest lizard, reaching more than 5 feet.

(3) An unknown species of monitor lizard, suggested by naturalist Mike Meyring.

(4) Bernard Heuvelmans equated this animal with his LONGNECK variety of seal, which he thought might be responsible for NESSIE and other lake monsters.

*Sources:* James Edward Alexander, *An Expedition of Discovery into the Interior of Africa* (London: H. Colburn, 1838), pp. 114–115; George William Stow, *The Native Races of South Africa* (London: Swan Sonnenschein, 1905); Frederick C. Cornell, *The Glamour of Prospecting* (New York: Frederick A. Stokes, 1920), pp. 142, 181; "River Monster with a 10 Ft. Neck," *Daily Mail* (London), February 8, 1921, p. 1; George William Stow, *Rock-Paintings in South Africa* (London: Methuen, 1930); Hans Sauer, *Ex Africa* (London: Geoffrey Bles, 1937), pp. 102–103; "Monster Lurking in 'Big Hole' at Kimberley," *Johannesburg Sunday Times,* November 30, 1947; Lawrence G. Green, *To the River's End* (Cape Town, South Africa: Howard B. Timmins, 1948), pp. 126–129; Frank Day, "Police Fire on Mysterious Vaal 'Monster,'" *Rand Daily Mail,* November 11, 1963; Harald L. Pager, *Stone Age Myth and Magic as Documented in the Rock Paintings of South Africa* (Graz, Austria: Akademische Druck-und Verlagsanstalt, 1975), p. 47; Bernard Heuvelmans, *Les derniers dragons d'Afrique* (Paris: Plon, 1978), pp. 74–109.

## Grotte Cosquer Animal

Paleolithic cave art depicting a SEA MONSTER in France.

*Physical description:* Fat, bulky body. Small head on a relatively long neck. Two flexible front flippers and two pointed rear flippers.

*Distribution:* Grotte Cosquer, Cap Morgiou, near Marseille, France.

*Present status:* This underwater cave was discovered in 1985 by Henri Cosquer, who also found the artwork six years later. The entrance, 120 feet below water level, would have been above water during the Ice Age. The charcoal drawings of animals in the cave were confidently dated by Jean Courtin and Jacques Collina-Girard in 1994 as 18,000–19,000 years old. Most of the images are of land animals, especially horses, but fully 11 percent depict marine life, including auks, fishes, seals, and jellyfish.

*Possible explanations:*

(1) Commonly accepted by archaeologists as depicting a Penguin (Family Spheniscidae), though it looks nothing like this Antarctic bird. However, during the Pleistocene, the colder European climate would have been favorable to penguins.

(2) A Fur seal or Sea lion (Family Otariidae), though probably an unknown species.

(3) The image is remarkably close to Bernard Heuvelmans's LONGNECK variety of Sea monster.

*Sources:* Jean Clottes and Jean Courtin, *The Cave Beneath the Sea* (New York: Harry N.

Abrams, 1996); La grotte Cosquer: Les animaux marins, http://www.culture.fr/culture/archeosm/fr/fr-cosqu4.htm.

## Ground Shark

Unknown FISH of the Indian Ocean.

*Physical description:* Larger than a great white shark, which has an average length of 14 feet. No prominent dorsal fin.

*Behavior:* Lies in wait for other fishes on the ocean floor. Said to be a man-eater.

*Distribution:* Timor Sea.

*Possible explanation:* A giant form of Wobbegong shark (Family Orectolobidae), suggested by Karl Shuker. The Spotted wobbegong (*Orectolobus maculatus*) inhabits Australasian waters and grows to 10 feet 6 inches. It feeds on the bottom but attacks waders and fishers in tidal pools.

*Sources:* Willy Ley, *The Lungfish and the Unicorn* (New York: Modern Age, 1941); Karl Shuker, "The Search for Monster Sharks," *Fate* 44 (March 1991): 41–49.

## Gryttie

FRESHWATER MONSTER of Sweden.

*Etymology:* After the lake.

*Physical description:* Serpentine. Length, 100 feet.

*Distribution:* Gryttjen lake, Gävleborgs County, Sweden.

*Source:* Gryttie the Lake Monster of Hälsingland, http://hem.passagen.se/gryttie/about.html.

## Guài Wù

FRESHWATER MONSTER of China.

*Etymology:* Mandarin Chinese (Sino-Tibetan), "strange beast."

*Physical description:* Size of an ox. Black with white underparts. Large, seal-like head. Long neck.

*Distribution:* Chon-Ji Lake (also called Tianchi, Changbai, or Dragon Lake), Jilin Province, China.

*Significant sightings:* Reports date back to the nineteenth century, but Chinese researchers claim to have collected 100 reports between 1962 and 1994.

In August 1980, a party of meteorologists saw a large animal with a 3-foot neck, a cow-shaped head, and a duck-shaped beak.

In early January 1987, a group of fifty tourists was surprised when a lake monster surfaced near the eastern shore. One witness, Shen Ruder, said it roared like a locomotive and sprayed water out of its nose.

Photos and a video of a dragonlike animal were taken on September 2, 1994. The creature swam for ten minutes on the surface, raising waves 6 feet high.

Four black animals were seen frolicking in the lake by more than 200 people in 1996 and were allegedly captured on film by photographer Wang Ling.

*Present status:* The volcano where this lake is located erupted in 1702, so presumably, anything in it has been imported since that date.

*Sources:* Steve Moore, "Water Dragons," *Fortean Times*, no. 36 (Winter 1982): 47; "China Bits and Pieces Not a Crock," *INFO Journal*, no. 51 (February 1987): 27–28; "Another Chinese Lake Monster," *Fortean Times*, no. 48 (Spring 1987): 11; "Lake Monsters Ahoy!" *Fortean Times*, no. 77 (October-November 1994): 16; Karl Shuker, *In Search of Prehistoric Survivors* (London: Blandford, 1995), p. 35; "Chinese Lake Monster," *INFO Journal*, no. 77 (Spring 1997): 43; Karl Shuker, "Freshwater Monsters: The Next Generation," *Fate* 51 (February 1998): 19.

## Guaraçaí Air-Breather

Mystery FISH of South America.

*Physical description:* Length, 5 inches. Barbels. Only one gill. Two small, articulated limbs, with a membrane between the toes.

*Behavior:* Can stay alive out of the water. Surfaces every five minutes to breathe.

*Distribution:* Southern Brazil.

*Significant sighting:* The only known specimen was caught in September 1995 by Paulinho Clemente in a lake near Guaraçaí, São Paulo State, Brazil.

*Sources:* *O Estado* (São Paulo), September 22, 1995; "Fish Caught Walking Underwater," *Fortean Times*, no. 86 (May 1996): 40.

## Gugwé

CANNIBAL GIANT of eastern Canada.

*Etymology:* Micmac (Algonquian) word.

*Variant names:* Chenoo, Djenu, Kookwe.

*Physical description:* Tall. Face like a bear's. Big hands.

*Behavior:* Whistles like a Gray partridge (*Perdix perdix*), which emits a repeated "kishrrr," "ksheeerik," or "keeeah."

*Distribution:* New Brunswick and Nova Scotia.

*Sources:* Elsie Clews Parsons, "Micmac Folklore," *Journal of American Folklore* 38 (1925): 55–133; Wilson D. Wallis and Ruth Sawtell Wallis, *The Micmac Indians of Eastern Canada* (Minneapolis: University of Minnesota Press, 1955), pp. 348, 417.

## Guiafairo

BAT-like animal of West Africa.

*Etymology:* Said to mean the "fear that flies by night."

*Physical description:* Gray. Human face. Bat-like wings. Clawed feet.

*Behavior:* Nocturnal. Has a nauseating odor that engenders fear. Said to be able to appear behind locked doors.

*Habitat:* Rocky outcrops.

*Distribution:* Senegal.

*Source:* Karl Shuker, "The Secret Animals of Senegambia," *Fate* 51 (November 1998): 46–51.

## Güije

LITTLE PEOPLE of the West Indies.

*Etymology:* Possibly of Arawakan origin.

*Variant name:* Jigüe.

*Physical description:* Half monkey, half human. Height, 3 feet. Black skin. Flattened, oversized head. Long beard. Big belly. Breastlike navel. Powerful claws. Short feet.

*Behavior:* Nocturnal. Extraordinarily strong.

*Habitat:* Ponds.

*Distribution:* Eastern Cuba.

*Sources:* Antonio Bachiller y Morales, "Jigues: Tradición Cubana," *Archivos del Folklore Cubano* 2, no. 2 (1926): 169–173; Gertrudis Gómez de Avellaneda y Arteaga, "Supersticiones," in *La enciclopedia de Cuba* (San Juan, Puerto Rico: Enciclopedia y Clásicos Cubanos, 1975–1977), vol. 8, pp. 217–222; Scott Corrales, "Aluxoob: Little People of the Maya," *Fate* 54 (June 2001): 30–34.

## Guirivilu

FRESHWATER MONSTER of South America.

*Etymology:* Araucanian, "fox serpent."

*Variant names:* Glyryvilu, Neguruvilu.

*Physical description:* Serpentine. Foxlike head. Long tail with a double row of pointed nails and a claw at the tip.

*Behavior:* Said to kill animals and people by enveloping them.

*Habitat:* Lakes.

*Distribution:* Southern Chile; Neuquén Province, Argentina.

*Sources:* Giovanni Ignazio Molina, *The Geographical, Natural, and Civil History of Chili* (London: Longman, Hurst, Rees, Orme, 1809); Robert Lehmann-Nitsche, "La pretendida existencia actual del Grypotherium," *Revista del Museo de La Plata* 10 (1902): 277–279; Julio Vicuña-Cifuentes, *Mitos y supersticiones recogidos de la tradicion oral Chilena* (Santiago de Chile: Universitaria, 1915), pp. 65–66; Hartley Burr Alexander, *Latin American Mythology* [1920] (New York: Cooper Square, 1964), p. 328.

## Gul

WILDMAN of Central Asia.

*Etymology:* Tajik (Persian), from the Arabic *ghul.*

*Variant name:* Adzhina.

*Physical description:* Height, 5 feet–6 feet 6 inches. Silvery-gray or black body-hair. Short neck.

*Behavior:* Feeds on mice and gophers. Uses a forked stick to catch mice. Said to have a hypnotic power.

*Tracks:* Length, 14 inches. Width, 6 inches at the toes. Big toe considerably larger than the others. Toes are slightly spread. Foot is flat. Prints are 4 feet apart.

*Distribution:* Pamir Mountains, western Tajikistan.

*Significant sightings:* Western Tajikistan has been the traditional origin of a curative drug said to be made from the skin of wildmen. Called *mu-gö* or *mu-miyo* (possibly from the Farsi *mum,* "wax," though *mu* also means "hair"), the preparation was carried by pilgrims to Mecca and was at one time said to be one of the sources of wealth for the emir of Bukhara. The village of Khakimi in the Karatag Valley was once a production center.

Igor Tatsl and Igor Bourtsev found Gul tracks near Khakimi, Tajikistan, on August 15 and 21, 1979.

Ukrainian library-school student Nina Grinyova came close to a Gul nicknamed "Gosha" in the Varzob River gorge, Tajikistan, on August 20, 1980, during an expedition to search for the creatures. Grinyova offered to stay alone in the woods one night in order to encourage a close encounter with a Gul that had been leaving tracks in the area. The Gul approached, but Grinyova inadvertently scared it away by offering it a squeaky rubber toy. She experienced a fugue walking back to camp and believes that the creature had a psychic effect on her.

Vadim Makarov discovered a four-toed, 19.25-inch print on the banks of the Varzob River on September 29, 1981.

*Sources:* Bernard Heuvelmans and Boris F. Porshnev, *L'homme de Néanderthal est toujours vivant* (Paris: Plon, 1974), pp. 109, 155–161; Myra Shackley, *Still Living? Yeti, Sasquatch and the Neanderthal Enigma* (New York: Thames and Hudson, 1983), pp. 117–126; Dmitri Bayanov, "A Field Investigation into the Relict Hominoid Situation in Tajikistan, USSR," *Cryptozoology* 3 (1984): 74–79; Dmitri Bayanov, *In the Footsteps of the Russian Snowman* (Moscow: Crypto-Logos, 1996), pp. 85–103, 114–120; Ioann Gornenskii, *Legendy Pamira i Gindukusha* (Moscow: Aleteia, 2000), pp. 10–11, 29–30, 136, 157, 159, 161–164.

## Gulebaney
WILDMAN of West Asia.

*Etymology:* Possibly Azerbaijani (Turkic), "wild man."

*Variant names:* Biaban-guli, Kulieybani, Vol'-moshin' (for the female).

*Behavior:* In the summer, catches fishes, crustaceans, and frogs in the rivers. Approaches villages in the autumn to raid vegetable gardens.

*Distribution:* Talysh Mountains, Azerbaijan.

*Significant sightings:* One evening in the 1890s, noted zoologist K. A. Satunin watched a female Biaban-guli cross a clearing in the Talysh Mountains, Azerbaijan.

In the summer of 1947, a soldier in the Azerbaijani militia named Ramazan was walking home at night when a shaggy wildman attacked him and dragged him to the foot of a nearby tree, where a female was waiting. The two creatures examined his face and clothes, then seemed to get into a gutteral argument and shoving match. Near dawn, they left him alone.

*Sources:* Konstantin A. Satunin, "Biabanguli," *Priroda i Okhota,* no. 7 (1899): 28–35; Odette Tchernine, *The Yeti* (London: Neville Spearman, 1970), pp. 22, 179; Bernard Heuvelmans and Boris F. Porshnev, *L'homme de Néanderthal est toujours vivant* (Paris: Plon, 1974), pp. 162–164.

## Gwrach-y-Rhibyn
Mythical FLYING HUMANOID of Wales.

*Etymology:* Welsh, "hag of the warning."

*Physical description:* Thin female figure. Swarthy skin. Long, black hair. Sunken, piercing, black or gray eyes. Long, batlike wings. Crooked back.

*Behavior:* Flies low over rivers and streams. Wears long, black robes. Flapping wings can be heard against windowpanes. Prefigures a death.

*Distribution:* Wales, especially Ceredigion.

*Significant sighting:* Said to have been seen often in the latter half of the eighteenth century inhabiting Caerphilly Swamp.

*Source:* Marie Trevelyan, *Folklore and Folk-Stories of Wales* (London: Elliot Stock, 1909), p. 65.

## Gwyllgi
BLACK DOG of Wales.

*Etymology:* Welsh, "dog of darkness."

*Variant names:* Black dog of Hergest, Cŵn annwfn ("dog of the otherworld"), Cŵn annwn, Cŵn bendith y mamau ("fairy dog"), Cŵn cyrff ("corpse dog"), Cŵn toili, Cŵn wybr ("sky dog").

*Physical description:* As large as a calf. Color said to be black, red-gray, or snow-white. Glowing red eyes.

*Behavior:* Often runs in a pack. Screams and howls. Walks behind people, snarling. Dogs are terrified of it.

*Tracks:* Doglike.

*Distribution:* Powys, South Wales.

*Significant sightings:* Sir Arthur Conan Doyle heard of the Black dog of Hergest while staying near Clyro, Powys, and was inspired to write his Sherlock Holmes story "The Hound of the Baskervilles." He agreed with the Welsh Baskerville family to set the scene in Dartmoor rather than Wales.

Dozens of sheep near Clyro were found with their throats ripped out in August 1989. At least two people saw the predator, which they thought was a large, dark-colored dog.

*Sources:* Edmund Jones, *A Relation of Apparitions of Spirits, in the County of Monmouth and the Principality of Wales* (Newport, Wales: E. Lewis, 1813); Marie Trevelyan, *Folklore and Folk-Stories of Wales* (London: Elliot Stock, 1909), p. 52; *The Independent,* September 2, 1989; James MacKillop, *Oxford Dictionary of Celtic Mythology* (New York: Oxford University Press, 1998), pp. 122, 263.

## Gyedarra

Mystery MARSUPIAL of Australia.

*Etymology:* Australian word.

*Physical description:* Size of a horse.

*Behavior:* Semiaquatic. Eats grass.

*Habitat:* Creek beds, where it excavates large holes in the banks.

*Distribution:* Near Gowrie Station, Queensland, Australia.

*Significant sighting:* Aborigines claimed that the fossil bones of extinct diprotodonts belonged to large animals that were alive several generations earlier.

*Present status:* Extinct but known as living animals to the ancestors of the Aborigines.

*Possible explanation:* Surviving *Diprotodon optatum,* a fossil wombatlike marsupial, the largest known, that lived from 2.5 million to as recently as 6,000 years ago. It was the size of a modern rhinoceros, about 10 feet long, and had a heavy skull nearly 3 feet long. It had massive jaws and a large lower incisor.

*Source:* George Bennett, "A Trip to Queensland in Search of Fossils," *Annals and Magazine of Natural History,* ser. 4, 9 (1872): 315.

## Gyedm Gylilix

CANNIBAL GIANT of western Canada.

*Etymology:* Nass-Gitksian (Penutian), "man of the woods."

*Variant names:* Gyedm gyilhawli (Tsimshian/ Penutian), Gyedm lakhs sgyinist ("man of the jackpines").

*Distribution:* West-central British Columbia.

*Source:* Bruce Rigsby, "Some Pacific Northwest Native Language Names for the Sasquatch Phenomenon," *Northwest Anthropological Research Notes* 5 (1971): 153–156.

## Gyona Pel

GIANT HOMINID of northern Russia.

*Etymology:* Komi (Uralic), "hairy eared."

*Distribution:* Komi Republic, European Russia.

*Source:* Dmitri Bayanov, *In the Footsteps of the Russian Snowman* (Moscow: Crypto-Logos, 1996), p. 141.

# H

## Hadjel

TIGRE DE MONTAGNE of West Africa.

*Etymology:* Daju (Nilo-Saharan) word.

*Variant name:* Biscoro (Tupuri/Ubangi).

*Physical description:* Larger than a lion. Long teeth. Maned. Short tail like a hyena's.

*Behavior:* Said to be painful for it to open its mouth because of its teeth. Only eats small prey.

*Habitat:* Mountains.

*Distribution:* Near Temki, Chad.

*Possible explanation:* A surviving saber-toothed cat, possibly a *Megantereon,* which lived in South Africa 3 million years ago.

*Source:* Jeanne-Françoise Vincent, *Le pouvoir et le sacré chez les Hadjeray du Tchad* (Paris: Éditions Anthropos, 1975), pp. 100–101.

## Haietluk

SEA MONSTER of coastal British Columbia, Canada, apparently corresponding to CADDY.

*Etymology:* Nootka (Wakashan), "lightning snake" or "wriggler."

*Variant names:* Heitlik, Hiaschuckaluck (Chinook Jargon/Pidgin, *hayash* ["big"] + *olêk* ["snake"]), Hiyitl'iik, NUMKSE LEE KWALA (Comox/Salishan), T'chain-ko (Sechelt/Salishan).

*Physical description:* Serpentine. Length, 7–8 feet. Horselike head. Head and back covered with hair or a mane. Prominant teeth. Four legs.

*Behavior:* Can move on land by wriggling like a snake.

*Distribution:* Strait of Georgia, British Columbia.

*Sources:* Charles F. Newcombe, "Petroglyphs in British Columbia," *Victoria (B.C.) Times,* September 10, 1907; Beth and Ray Hill, *Indian Petroglyphs of the Pacific Northwest* (Saanichton, B.C., Canada: Hancock House, 1974); David W. Ellis and Luke Swan, *Teachings of the Tides: Uses of Marine Invertebrates by the Manhousat People* (Nanaimo, B.C., Canada: Theytus, 1981).

## Hairy Biped

A humanlike or apelike ENTITY of North America, possessing some of the characteristics of GIANT HOMINIDS or NORTH AMERICAN APES.

*Etymology:* Coined by Jerome Clark as a catchall term for humanoids reported in the midwestern and eastern United States and Canada.

*Variant names:* Big hairy monster (BHM), Billiwack monster (in southern California), BOOGER, Buenafoot (in southern California), CANNIBAL GIANT, DWAYYO, Eastern bigfoot, Fluorescent Freddie, GOATMAN, Goonyak (in Vermont), Grassman (in Ohio), LAKE WORTH MONSTER, Manbeast, Manimal, MOMO, Old slipperyskin (in Vermont), Old yellow top (in Ontario), Ole woolly, Orange eyes (in Ohio), Precambrian Shield man, Taku he (Dakota/Siouan, "what's that?"), Wejuk (in Vermont), Wood devil, Wookie, Woolly booger, Yeahoh (in Kentucky).

*Physical description:* Not as uniform as the BIGFOOT of the Pacific Northwest, though always covered with hair and walking on two legs (hence its name). It's difficult to generalize traits from reports that might have multiple causes, but some of the following features are usually present. Height, 4–9 feet, though sizes up to 12 feet are mentioned. Hair or fur is reddish-brown to black, often described as 6–8 inches long. Often distinctly lacking in facial features, but a catlike face is occasionally reported. Red, or-

*Artist's conception of a* HAIRY BIPED. *(William M. Rebsamen)*

ange, yellow, or green glowing eyes. Flat, broad nose. Pointed ears. Werewolflike fangs. Mane. Long arms. Hands are sometimes clawed. Long legs.

*Behavior:* Primarily nocturnal. Usually has an awkward, bipedal gait but sometimes runs on all fours. Said to be able to swim. Occasionally seen with young. Reported calls are moans, grunts, howls, high-pitched shrieks. Strong, putrid odor like decaying flesh or rotten eggs. These creatures are sometimes ascribed such paranormal features as invulnerability, transparency, insubstantiality, invisibility, and the ability to disappear instantaneously. Appears to show interest in and have no fear of human dwellings. Dislikes cars and dogs, which often react with great fright. Sometimes associated with unidentified flying object (UFO) sightings.

*Tracks:* Anywhere from two- to six-toed. Three-toed are perhaps commonest and have been reported from the South, the Midwest, Pennsylvania, Maryland, and southern California. Length, up to 14 inches. Stride, up to 5 feet. Hair samples have been found.

All primates have five toes. Any Hairy biped that leaves clear imprints showing anything less than five toes constitutes an extreme evolutionary anomaly. Pentadactyly (having five fingers or toes) is a common and primitive feature of reptiles and mammals. However, it is not an essential requirement, and many animals have modified the plan: frogs only have four digits, cows have two, horses have dropped all but one, and snakes have gotten rid of legs altogether. Most birds get by walking on only four (three in front and one behind), while the Ostrich (*Struthio camelus*) only has two. If three-toed, humanlike bipeds really exist as flesh-and-blood creatures and are not paranormal apparitions, it would be most interesting to find out more about their foot structure. Perhaps three toes is better than five when you've chosen a swamp or wetland as your habitat.

*Habitat:* Secluded areas, often forested wetlands or mountainous regions.

*Distribution:* Nearly every U.S. state and Canadian province. Most sightings represent only transient individuals.

A partial list of places where Hairy bipeds have been reported follows:

*Alabama*—Choccolocco Valley, Town Creek.

*Arkansas*—Center Ridge, Greene County, Jonesboro, Leachville, Poinsett County, St. Francis County, South Crossett, Springdale.

*California*—Antelope Valley, Borrego Sink, Lytle Creek, Pearblossom, San Gorgonio Mountains, Santa Paula.

*Colorado*—Green Mountain Falls.

*Connecticut*—Bristol, Crystal Lake Reservoir, Winsted.

*Delaware*—Selbyville.

*Georgia*—Edison.

*Illinois*—Big Muddy River, Cairo, Centerville, Chittyville, Creve Coeur, East Peoria, Effingham, Farmer City, Kickapoo Creek, Murphysboro.

*Indiana*—Attica, French Lick, Galveston, Hoosier National Forest, Knox County, Pike County, Richmond, Rising Sun, Roachdale, Sharpsville, Winslow.

*Iowa*—Clinton.

*Kentucky*—Albany, Leslie County, Trimble County.

*Labrador, Canada*—Goose Bay.

*Louisiana*—Cotton Island, Honey Island Swamp.

*Maine*—Durham.

*Manitoba, Canada*—Gypsumville, Steinbach, Whiteshell Provincial Park.

*Maryland*—Calvert County, Churchville, Dickerson, Harford County, Kingsville, Prince George's County, Sykesville.

*Massachusetts*—Bridgewater, Raynham Center.

*Michigan*—Byron, Charlotte, Dowagiac Swamp, Fenton, Houghton Lake State Forest, Lake City, Marshall, Mason, Mio, Monroe, Oscoda County, Port Huron, Saginaw, Shiawassee River, Sister Lakes, Tuscola County, Yale.

*Minnesota*—northern part of state.

*Mississippi*—Meridian, Winona.

*Missouri*—Louisiana, Pacific, Troy.

*Montana*—Monarch, Vaughn.

*Nebraska*—south of Lincoln.

*Nevada*—Nevada Test Site.

*New Hampshire*—Hollis, Salisbury.

*New Jersey*—Great Bear Swamp, High Point, Middletown, Vineland.

*New York*—Burlington County, Ellisburg, Morristown, Mount Misery, Richmondtown, Sherman, Watertown, Whitehall.

*Newfoundland, Canada*—Trinity Bay.

*North Carolina*—Dismal Swamp, Tabor City.

*Ohio*—Alliance, Brookside Park, Carlisle, Coshocton County, Defiance, Eaton, Huron, Kenmore, Kimbolton, Mansfield, Minerva, Monroeville, Muskingum County, Newcomerstown, Point Isabel, Rome.

*Oklahoma*—Canton, Kiamichi Mountains, Mountain Fork River, Nowata, Noxie, Tahlequah, Wann.

*Ontario, Canada*—Cobalt, Webequie, Weenusk Indian Reservation.

*Oregon*—Conser Lake, Roseburg.

*Pennsylvania*—Allegheny County, Allison, Beaver County, Bradford County, Buffalo Mills, Chester County, Chestnut Ridge, Derry Township, East Pennsboro Township, Edinboro, Fayette County, Gray Station, Indiana County, Jeannette, Lancaster, Latrobe, Lock Haven, Somerset County, Uniontown, Westmoreland County, Whitney.

*Saskatchewan, Canada*—Grand Rapids.

*South Dakota*—Standing Rock Indian Reservation.

*Tennessee*—Charlotte, Flintville, Lascassas, Knox County, Monteagle Mountain.

*Texas*—Bells, Caddo, Denton, Haskell, Lamar County, Lake Worth, Newton County, Paris, Peerless, Polk County.

*Vermont*—Chittenden, Hartland, Rutland County, Williamstown.

*Virginia*—Colonial Beach, Middletown.

*West Virginia*—Cacapon Bridge, Davis, Hickory Flats, Marlinton, Parsons.

*Wisconsin*—Benton, Cashton, Deltox Swamp, Grafton, Granton, Jefferson, Medford.

*Significant sightings:* Riley W. Smith saw a naked hairy man, about 6 feet tall, while picking berries near Winsted, Connecticut, on August 17, 1895. The incident was the first of about twenty that allegedly took place in western Connecticut and the Catskill Mountains of New York over the next few weeks. Widely and possibly erroneously regarded as a hoax by newspaperman Louis T. Stone, the original incident may have involved a bear.

An apelike, bipedal creature with a yellow head and mane was seen by workers near the Violet Mine east of Cobalt, Ontario, in September 1906. In 1923, two prospectors saw a similar yellow-headed, black-haired animal eating blueberries; they thought it was a bear until they threw a rock at it, prompting it to get up and walk away on two legs. Later sightings earned it the nickname "Old yellow top." The last sighting was in August 1970 when Aimée Latreille, the driver of a bus carrying twenty-seven miners, was forced to swerve after he saw an apelike creature with a light mane cross the road; the bus nearly had a fatal crash down a nearby rock cut.

In August 1963, Harlan E. Ford and a friend encountered a huge humanoid in Honey Island Swamp near Slidell, Louisiana. It glared menacingly at them and ran away on two legs.

In May 1964, near Sister Lakes, Michigan, Gordon Brown and his brother saw a hairy man about 9 feet tall who made a whimpering sound. Shortly afterward, three teenagers saw a 7-foot creature with a black face running through the underbrush in Silver Creek Township. Many other witnesses came forth and were named in extensive newspaper coverage.

A green, 10-foot-tall monster with glowing red eyes was seen in March 1965 by teenagers in the woods south of French Lick, Indiana. They called it "Fluorescent Freddie."

In 1965, two teenagers were chased from their campfire by a 9- to 10-foot hairy creature on the north slope of the San Gorgonio Mountains, California.

On August 13, 1965, Christine Van Acker and her mother were driving near Monroe, Michigan, when a hairy, 7-foot giant stepped in front of their car. Van Acker hit the brakes, stalling the car, and the creature reached through the open window and grabbed the top of her head. The women's screams and horn honking apparently made it retreat.

On May 19, 1969, George Kaiser saw a man-sized creature covered in black fur on his farm near Rising Sun, Indiana. It made a strange grunting sound, jumped over a ditch, and swiftly ran down the road. Later, footprints with three small toes and a big toe were found. A greenish-white UFO was seen by a neighbor the next night.

Odd, froglike noises woke up teenagers Wayne Hall and Dave Chapman early on July 24, 1972, at the latter's home near Crystal Lake Reservoir in northwestern Connecticut. Looking outside, they saw an 8-foot hairy creature. It crossed a road and moved around in the shadows near a horse barn. After forty-five minutes, it crossed the road again and disappeared in the woods by the lake.

On the night of April 22, 1973, William Roemermann, Brian Goldojarb, and Richard Engels saw a BIGFOOT-like creature near the Sycamore Flats campground in Big Rock Canyon, Los Angeles County, California. It chased their truck for about twenty seconds, its long arms swinging in front of its chest. On returning, they found many huge, three-toed tracks.

In May and June 1973, an apelike creature terrorized the area around Sykesville in Carroll County, Maryland. Five-toed, 13-inch footprints were found, separated by a stride of 6 feet.

On June 25, 1973, Randy Needham and Judy Johnson were parked near a boat ramp on the Big Muddy River near Murphysboro, Illinois, when they heard a piercing cry that came from the nearby woods. They looked up and saw the sound came from a huge shape lumbering toward them. The creature was about 7 feet tall and covered with a matted, whitish hair. Others saw and heard the same creature over the next two weeks, and it reappeared in the summers of 1974, 1975, 1988, and 1989.

At 4:30 A.M. on September 2, 1973, Chester Yothers woke up and saw a BIGFOOT-like creature only 5 feet away outside his trailer near Whitney, Pennsylvania, apparently looking at the house next door. He woke his wife and called the police, who arrived shortly afterward. The monster was gone, but they found wet footprints on the concrete and in the flower bed.

Dennis Smith and Jimmy Slate heard pounding and shrieking noises in the woods next to Overlook Drive, near Watertown, New York, in the early morning of August 10, 1976. As the sun was rising, they saw an erect, black hominid walking down the road about two city blocks away. When Smith yelled, the creature turned around and ran in the opposite direction. Later, two 15-inch-long tracks, trampled grass, and some long hairs were found.

On May 18, 1977, two thirteen-year-old boys were walking their dog near the historic Roberts Covered Bridge south of Eaton, Ohio, when the dog got frightened and they smelled a rotten-meat odor. Turning around, they saw a 9-foot, apelike creature with dirty brown hair, white eyes, and long arms; it chased them toward the road. Both boys were terrified for weeks after the incident. Two 14-inch, humanlike prints were found near Seven Mile Creek on a nearby farm.

Some twenty-eight sightings of BIGFOOT-like creatures 6–9 feet tall were reported in wooded areas around Little Eagle in the Standing Rock Reservation in South Dakota from September to November 1977. Numerous large footprints were found, and high-pitched shrieks were heard repeatedly. Cecelia Thunder Shield said the being was tall with gray, shining hair and a black face.

In January 1980, an employee of Reynolds Electrical and Engineering Company saw a 6- to 7-foot hairy creature while driving along a highway at the northern end of the Nevada Test Site. It disappeared in the sagebrush.

James Guyette saw a huge hairy humanoid walking and swinging its arms along an interstate highway near Hartland Dam, Vermont, in April 1984. It moved down the embankment and headed west.

A woodsman of Gray Station, Pennsylvania, was walking at the forest edge at dusk on December 13, 1986, when something threw a large piece of wood at him. He looked up and saw a hairy creature, standing 8–9 feet tall with wide shoulders and long arms, blocking the path.

After a moment, it turned, stooped, and ran into the woods.

Gary Lee Hayes was hunting near a tract of the Houghton Lake State Forest, Michigan, on November 25, 1990, when he saw a tall, upright creature moving on the crest of a nearby hill. It had black hair all over its body and was 7 feet tall. The creature walked down to a large beaver dam, squatted down, stood up, then went back uphill.

Robert Toal found huge, human-shaped tracks in the snow on his property in Kingsville, Maryland, on the night of February 4–5, 1995. Field investigators from the Baltimore-area Enigma Project arrived a few days later and photographed the tracks, which were 20 inches long, 11 inches wide, but only 1 inch deep in the powdery snow. The tracks had an average stride of 4 feet 10 inches in a straight line and apparently passed through a 4-foot-high wire fence. Since even humans weighing less than 200 pounds made deeper impressions in the snow, the Enigma group thought these were the full-body impressions of a much lighter animal, possibly a jumping rabbit.

Early in the morning of March 28, 2000, James Hughes was driving his newspaper route near Grafton, Wisconsin, when he saw an 8-foot hairy humanoid standing by the side of the road. The creature was carrying something that looked like a dead goat.

Human tracks 14 inches long and 5 inches wide were found in early June 2001 on the Weenusk Indian Reservation at the mouth of the Winisk River on Hudson Bay, Ontario. The stride measured 6 feet.

*Present status:* Distinctions between NORTH AMERICAN APES, DEVIL MONKEYS, Hairy bipeds, and BIGFOOT are nebulous and possibly arbitrary. In general, NORTH AMERICAN APES are tailless and primarily quadrupedal, and they resemble chimpanzees; DEVIL MONKEYS are tailed and resemble baboons; Hairy bipeds cover a wide range of descriptions, from apes to WILDMEN and even paranormal ENTITIES; BIGFOOT is a robust, tall hominid with a range that seems restricted to the Pacific Northwest.

*Possible explanations:*
(1) Many hoaxes, such as pranksters wearing masks or suits. The Selbyville, Delaware,

swamp monster of 1964 was admittedly a hoax perpetrated by a man in a monster suit.

(2) Mentally unstable or homeless humans living in the woods. This explanation may have been especially true for nineteenth-century reports.

(3) Misidentified American black bears (*Ursus americanus*).

(4) Monkeys or apes escaped from zoos or circuses.

(5) ENTITIES associated with UFOs, suggested by Stan Gordon and Don Worley.

(6) Occurrences of BIGFOOT outside its traditional range in the Pacific Northwest. The only comparative analysis of Hairy biped data in eastern North America has been done by Craig Heinselman, who looked at 654 reports from fifteen eastern and northeastern states between 1838 and 2001 and found few differences in height or other narrowly selected physical characteristics from the Pacific Northwest BIGFOOT. He arrived at a tentative population estimate of 210–420 adult individuals for all fifteen states.

*Sources:* Leonard Roberts, *South from Hell-fer-Sartin: Kentucky Mountain Folk Tales* (Lexington: University of Kentucky Press, 1955), p. 162; *Indianapolis News,* March 15–17, 1965; "Monster Season," *Newsweek,* August 30, 1965, p. 22; Gene Caesar, "The Hellzapoppin' Hunt for the Michigan Monster," *True,* June 1966, pp. 59–60, 84–85; Ivan T. Sanderson, "Wisconsin's Abominable Snowman," *Argosy,* April 1969, pp. 27–29, 70; Warren Smith, "America's Terrifying Woodland Monster-Men," *Saga,* July 1969, pp. 34–37, 92–94; John A. Keel, *Strange Creatures from Time and Space* (Greenwich, Conn.: Fawcett, 1970); Jerome Clark, "On the Trail of Unidentified Furry Objects," *Fate* 26 (August 1973): 56–64; Allen V. Noe, "ABSMal Affairs in Pennsylvania and Elsewhere," *Pursuit,* no. 24 (October 1973): 84–89; Stan Gordon, "UFOs in Relation to Creature Sightings in Pennsylvania," in Walter H. Andrus Jr., ed., *MUFON 1974 UFO Symposium Proceedings* (Seguin, Tex.: Mutual

UFO Network, 1974), pp. 132–154; Jerome Clark and Loren Coleman, "Swamp Slobs Invade Illinois," *Fate* 27 (July 1974): 84–88; Berthold Eric Schwarz, "Berserk: A UFO-Creature Encounter," *Flying Saucer Review* 20, no. 1 (July 1974): 3–11; Milton LaSalle, "Bigfoot Sighting," *Pursuit,* no. 40 (Fall 1977): 120–123; Mark A. Hall, "Contemporary Stories of 'Taku He' or 'Bigfoot' in South Dakota as Drawn from Newspaper Accounts," *Minnesota Archeologist* 37 (1978): 63–78; Jerome Clark and Loren Coleman, *Creatures of the Outer Edge* (New York: Warner, 1978); Mark A. Hall, "Stories of 'Bigfoot' in Iowa during 1978 as Drawn from Newspaper Sources," *Minnesota Archeologist* 38 (1979): 2–17; S. Stover, "Does Maryland Have a Sasquatch?" *INFO Journal,* no. 34 (March-April 1979): 2–6; Dennis Pilichis, *Night Siege: The Northern Ohio UFO-Creature Invasion* (Rome, Ohio: Dennis Pilichis, 1982); Bruce G. Hallenbeck, Bob Bartholomew, and Paul Bartholomew, "Bigfoot in the Adirondacks," *Adirondack Bits 'n Pieces* 1, no. 3 (Spring-Summer 1984): 21–26, 49–50, 58–61; Mark Opsasnick, *The Maryland Bigfoot Reference Guide* (Greenbelt, Md.: Mark Opsasnick, 1987); Mike Marinacci, *Mysterious California* (Los Angeles, Calif.: Panpipes, 1988), pp. 84–86, 93–94; Mark Chorvinsky and Mark Opsasnick, "The Selbyville Swamp Monster Exposed," *Strange Magazine,* no. 4 (1989): 6–8; Michael T. Shoemaker, "Searching for the Historical Bigfoot," *Strange Magazine,* no. 5 (1990): 18–23, 57–62; David E. Philips, *Legendary Connecticut* (Willimantic, Conn.: Curbstone, 1992), pp. 175–177; Michael T. Shoemaker, "The Winsted Wild Man Revisited," *Strange Magazine,* no. 11 (Spring-Summer 1993): 30–31, 59; Joseph A. Citro, *Green Mountain Ghosts, Ghouls and Unsolved Mysteries* (Montpelier: Vermont Life, 1994), pp. 93–101; Michael A. Frizzell, "The Kingsville Tracks," *INFO Journal,* no. 74 (Winter 1996): 17–21; Loren Coleman, "Three Toes Are Better than Five," *Fortean Times,* no. 98 (June 1997): 44; Christopher L. Murphy, *Bigfoot in Ohio: Encounters with the Grassman* (New Westminster, B.C., Canada: Pyramid, 1997); Christopher Kiernan Coleman, *Strange Tales of* the *Dark and Bloody Ground* (Nashville, Tenn.: Rutledge Hill, 1998), pp. 53–55; Don Keating, "Active Sasquatch in Coshocton County, Ohio," *North American BioFortean Review* 1, no. 1 (April 1999): 5, 41, http://www.strangeark.com/nabr/NABR1.pdf; Dana Holyfield, *Encounters with the Honey Island Swamp Monster* (Pearl River, La.: Honey Island Swamp Books, 1999); Keith Edwards, "Wisconsin a New Home for Bigfoot?" *Milwaukee Journal Sentinel,* April 4, 2000; Tim Swartz, "The Hairy Ones," *Strange Magazine,* no. 21 (Fall 2000), on line at http://www.strangemag.com; Ron Schaffner, "Retrospective: Preble County, Ohio Incident," *Crypto Hominology Special,* no. 1 (April 7, 2001), pp. 50–58, at http://www.strangeark.com/crypto/Cryptohominids.pdf; Francine Dubé, "Big Footprints Stir Sasquatch Speculation," *National Post* (Canada), June 25, 2001; Joe Nickell, "Tracking the Swamp Monsters," *Skeptical Inquirer* 25, no. 4 (July 2001): 15; Craig Heinselman, "Eastern Sasquatch Analysis: Potential Patterns or Dubious Data?" paper presented at the Third East Coast Bigfoot Researchers Meeting, September 22, 2001, Delmont, Pennsylvania; Chester Moore Jr., "Monsterous Sounds: A Field Investigation of Texas Bigfoot Vocalizations," *The Anomalist,* no. 10 (2002): 13–19.

## Hairy Jack
BLACK DOG of central England.

*Physical description:* Black, shaggy dog. Long tail.

*Behavior:* Sometimes becomes invisible and is only felt. Accompanies people walking by themselves.

*Distribution:* Lincolnshire, especially near Willoughton.

*Significant sighting:* In 1933, a Willoughton man was pushed up against his gatepost by what seemed to be a large but unseen dog that placed its paws on his shoulders.

*Sources:* Ethel H. Rudkin, "The Black Dog," *Folklore* 49 (1938): 111–131; Katharine M. Briggs, *A Dictionary of Fairies* (London: Allen Lane, 1976), p. 216; K. Miller, "The Black Dog and Other Canine Apparitions in Lincolnshire," in N. Field and A. White, eds., *A Prospect of Lincolnshire: Collected Articles in Honour of E. H. Rudkin* (Lincoln, England, 1984).

## Hairy Lizard
Mystery LIZARD of Australasia.

*Physical description:* Lizard with hairy or furred skin.

*Habitat:* Caves.

*Distribution:* Mount Albert Edward, Papua New Guinea.

*Significant sighting:* Gold miners on the Aikora River told Charles Monckton on April 17, 1906, that they had encountered reptiles with hair.

*Present status:* Only one report.

*Possible explanation:* A small mammal, rather than a lizard, which does not have hair.

*Source:* Charles A. W. Monckton, *Last Days in New Guinea* (London: John Lane, 1922).

## Haitló Laux
CANNIBAL GIANT of western Canada.

*Etymology:* Lillooet (Salishan) word.

*Physical description:* Covered in black, brown, or red hair. Bearlike. Height, up to 10 feet.

*Behavior:* Nocturnal.

*Distribution:* Southern British Columbia.

*Source:* James A. Teit, "Traditions of the Lillooet Indians of British Columbia," *Journal of American Folklore* 25 (1912): 287, 346–347.

## Hamlet
FRESHWATER MONSTER of California.

*Etymology:* From Shakespeare's *Hamlet,* the chief resident of Elsinore Castle in Denmark.

*Variant name:* Elsie.

*Physical description:* Serpentine. Length, 12 feet. Diameter, 3 feet.

*Behavior:* Swims by vertical undulations.

*Distribution:* Lake Elsinore, California.

*Significant sightings:* First reported in 1884. The lake dried up in both 1951 and 1955, but sightings persisted. Bonnie Pray saw the monster twice in the winter of 1970.

*Sources: Los Angeles Daily Illustrated News,* May 6, 1942; "The Endless Search," *Fate* 23

(November 1970): 32–36; John Kirk, *In the Domain of Lake Monsters* (Toronto, Canada: Key Porter Books, 1998), pp. 171–172.

## Hantu Sakai

Unknown PRIMATE of Southeast Asia.

*Etymology:* Malay (Austronesian), "demon Sakai"; *Sakai* is a generic derogatory term for the Senoi, nomadic hunters and gatherers of Malaysia.

*Variant names:* Hantu raya, MAWAS.

*Physical description:* Height, 5 feet 10 inches. Thick body-hair. White or pinkish skin. Long, black head-hair. Sad-looking face. Receding forehead. Projecting brow. Bushy eyebrows. Red eyes. Long mustache. Long canine teeth. The back of the forearm is said to have a sharp bone.

*Behavior:* At ease in water. Moves easily through the trees. Hops on the ground on its heels. Croaks like a bird. Keen sense of smell. Strong animal odor. Uses forearm to cut foliage. Timid. Recognizes guns and is afraid of them. Said to kill and eat humans, especially thin ones. Wears a bark loincloth.

*Distribution:* Peninsular Malaysia.

*Significant sightings:* A. D. Frederickson was visiting the maharajah of Johor, Malaysia, in the 1870s when he observed a captive wildman that had been found in the interior. It was allegedly being taken to a learned society in Calcutta. He drew a sketch of it for his notebook.

Two males and one female were seen by sixteen-year-old Wong Yee Moi at a rubber plantation at Terolak, Perak State, Malaysia, as she was tapping a rubber tree on December 25, 1953. The two males stood behind her as the female approached and got her attention by touching her shoulder, offering a fang-filled smile, and croaking like a bird. The girl screamed and ran. Over the next few days, the creatures were seen by five or six others, including Corporal Wahab of the Malayan Home Guard.

*Possible explanations:*

(1) A surviving *Homo erectus,* fossils of which have been found in Java. The projecting brow is a feature of *erectus* fossils but not orangutans.

*A captive wildman, or HANTU SAKAI, found in the interior of Malaysia in the 1870s. From Aug Daniel Frederickson,* Ad Orientem *(London: W. H. Allen, 1889). (From the original in the Northwestern University Library)*

(2) A surviving mainland population of the Orangutan (*Pongo pygmaeus*), which is now limited to the islands of Borneo and Sumatra. Orangutan fossils from around 2 million years ago have been found in Laos, Vietnam, and southern China, as well as the islands of Sumatra, Java, and Borneo. These apes are more distantly related to humans than are Chimpanzees (*Pan troglodytes*) and Gorillas (*Gorilla gorilla*). The lineage is unclear, but the likeliest theory is that they derived from *Sivapithecus,* an extinct ape that lived in India and Pakistan in the Late Miocene 12–8 million years ago. The arboreal abilities of Hantu Sakai favor this theory.

*Sources:* Aug Daniel Frederickson, *Ad Orientem* (London: W. H. Allen, 1889), pp. 276–277; Walter William Skeat and Charles Otto Blagden, *Pagan Races of the Malay*

*Peninsula* (London: Macmillan, 1906), vol. 2, pp. 282–283; Bernard Heuvelmans, *On the Track of Unknown Animals* (New York: Hill and Wang, 1958), pp. 104–105; Ivan T. Sanderson, *Abominable Snowmen: Legend Come to Life* (Philadelphia: Chilton, 1961), pp. 227–232; Ronald McKie, *The Company of Animals* (New York: Harcourt, Brace, World, 1966), pp. 30, 196–197; "Abominable Jungle-Men," *Pursuit,* no. 10 (April 1970): 36–37.

## Hapyxelor

FRESHWATER MONSTER of Ontario, Canada.

*Etymology:* Name given by Donald Humphreys because it "popped into his head" when he saw the creature in 1968.

*Variant names:* Hapaxelor, Mussie.

*Physical description:* Length, 14–24 feet. Silvery-green or dark-brown color. Head like an alligator's. Crest or mane. Three bright eyes. Three ears. One big tooth. Slender neck. Two humps. One big fin. Two flippers.

*Behavior:* Eats fishes.

*Distribution:* Muskrat Lake, near Cobden, Ontario.

*Significant sightings:* A. W. Peever saw an animal the size of a horse crossing the lake in 1941.

In the spring of 1968, Donald Humphreys saw a silver-green animal, 24 feet long (later revised to 14–16 feet long), at the southern end of the lake. It had a large head with one tooth and a pair of front flippers.

Sonar tracings on an Eagle Z-5000 "fish-finder" portable device, taken by Michael Bradley on October 5, 1988, showed two 8- to 10-foot objects swimming side by side and heading toward the surface from a depth of 54 feet.

*Possible explanations:*

(1) A stray seal.

(2) A Lake sturgeon (*Acipenser fulvescens*), though none are officially known to live in Muskrat Lake.

*Sources: Pembroke (Ont.) Observer and Upper Ottawa Advertiser,* September 10, 1880; *Philadelphia Evening Bulletin,* July 8, 1969; Michael Bradley, *More than a Myth: The Search for the Monster of Muskrat Lake* (Willowdale, Ont., Canada: Hounslow Press, 1989).

## Harimau Jalur

Mystery CAT of Southeast Asia.

*Etymology:* Malay (Austronesian), "striped tiger."

*Physical description:* Large tiger with stripes that run from head to tail rather than downward.

*Distribution:* Terengganu State, Malaysia.

*Possible explanation:* An observational trick of the light, suggested by Arthur Locke.

*Source:* Arthur Locke, *The Tigers of Trengganu* (London: Museum Press, 1954).

## Harpy

Mythical FLYING HUMANOID of Southern Europe.

*Etymology:* From the Greek *hárpyia* ("snatchers").

*Physical description:* Body, wings, and claws like an eagle's. Ears like a bear's. Head and breasts like a woman's.

*Behavior:* Flies as swiftly as the wind. Has a foul stench. Swoops down and snatches food from tables. Spreads disease with its excrement.

*Distribution:* European Turkey; southern Greece.

*Significant sighting:* Phineus, the seer and blind king of Thrace, was tormented by Harpies sent by the gods to steal his food and make him starve. When the Argonauts visited the area (modern Kiyiköy on the Black Sea in European Turkey), they drove the birds away to the Strophades Islands in the Ionian Sea. In return, Phineus gave them some Black Sea navigation tips.

*Possible explanations:*

(1) The Hoopoe (*Upupa epops*) has a distinctive crest and a call that is a swiftly repeated "hoop." It is found throughout Europe, sub-Saharan Africa, and Asia. It is notoriously foul and unhygienic.

(2) Wood hoopoes (Family Phoeniculidae) have a preen gland that produces a bad odor. They are found in Central, East, and South Africa.

(3) The Hoatzin (*Opisthocomus hoazin*) was suggested by Raymond Manners, who thinks Jason and the Argonauts may have made it to South America.

*Sources:* Apollonius Rhodius, *Argonautica,* trans. R. C. Seaton (Cambridge, Mass.: Harvard University Press, 1912), II; Raymond D. Manners, "The Geography of the *Argonautica,*" *INFO Journal,* no. 60 (June 1990): 4–12.

## Harrum-Mo
WILDMAN of Central Asia.

*Etymology:* Lepcha (Sino-Tibetan) word.

*Behavior:* Avoids human dwellings. Speaks an unknown language. Eats snakes and vermin. Uses bows and arrows.

*Distribution:* Lunak Valley, Nepal.

*Source:* Joseph Dalton Hooker, *Himalayan Journals* (London: Ward, Lock, Bowden, 1891), p. 298.

## Havhest
WATER HORSE of Northern Europe.

*Etymology:* Norwegian, "sea horse." In Norway, this is also a common name for the Northern fulmar (*Fulmarus glacialis*), a stocky, thick-necked seabird that breeds along the North Atlantic coast.

*Physical description:* Horse's head. Small, yellow eyes. Double row of teeth. Long canines. Scaly body. Long mane. Front flippers or hooves. Long, curved fish tail.

*Behavior:* Stinking breath. Lashes water with tail.

*Distribution:* Norway.

*Significant sighting:* Represented in traditional Scandinavian folk art as a horse-fish hybrid with a long, scaly, curved tail.

*Possible explanation:* The Walrus (*Odebenus rosmarus*) is found in the Norwegian dependency of Svalbard and may have occasionally strayed to the Norwegian coast in earlier eras.

*Sources:* Kristian Bugge, *Folkeminneoptegnelser* (Oslo: Norsk Folkeminnelag, 1934), pp. 103–105; Halvor J. Sandsdalen, *Ormen i Seljordsvatnet* (Oslo:

Noregs Boklag, 1976); Michel Meurger and Claude Gagnon, *Lake Monster Traditions: A Cross-Cultural Analysis* (London: Fortean Tomes, 1988), pp. 28, 223–225.

## Havmand
Fish-tailed MERBEING of Northern Europe.

*Etymology:* Danish and Norwegian, "sea man."

*Variant names:* Havfrue (for the female), Havmaður (Icelandic), Maremind ("mermaid"), Marmaele ("sea children"), Marmennill (Icelandic), Meerfrau (German), Meerminnen (Dutch), Meerweib (German), Merminne (German), Merriminni (German).

*Physical description:* Green or black hair. Bearded. Handsome. The female is beautiful, with long brown hair.

*Behavior:* Males are friendly. Females are often friendly but sometimes predatory and seductive. Likes to comb its hair. Sits on submerged rocks with its baby but jumps into the sea when approached. Presages stormy weather. Said to gather the souls of the dead.

*Habitat:* The sea or on rocky cliffs along the shore.

*Distribution:* Scandinavia; Germany.

*Significant sightings:* Many of these creatures are said to have appeared once near Assens, Fyn County, Denmark.

*Sources:* Erik Pontoppidan, *The Natural History of Norway* (London: A. Linde, 1755), pp. 186–195; Benjamin Thorpe, *Northern Mythology* (London: Edward Lumley, 1851), vol. 2, pp. 27–28, 76–77, 170–174; W. A. Craigie, *Scandinavian Folklore* (Paisley, Scotland: Alexander Gardner, 1896), pp. 220–231; Nelson Annandale, *The Faroes and Iceland* (Oxford: Clarendon, 1905); Gwen Benwell and Arthur Waugh, *Sea Enchantress* (London: Hutchinson, 1961), pp. 180–182.

## Hecaitomixw
CANNIBAL GIANT of the northwestern United States.

*Etymology:* Quinault (Salishan), "devil of the forest."

*Distribution:* Olympic Peninsula, Washington.

*Source:* Kyle Mizokami, Bigfoot-Like Figures in North American Folklore and Tradition, http://www.rain.org/campinternet/bigfoot/bigfoot-folklore.html.

## Hessie

SEA MONSTER of Northern Europe.

*Etymology:* After Hessafjorden, off Ålesund, Norway.

*Physical description:* Length, 80–100 feet. Width, 5 feet. Brown color. Square head like an anaconda's. Squarish dorsal fin is about 15 inches high.

*Behavior:* Seems to move by both horizontal and vertical undulations. Eats carrion.

*Distribution:* Norwegian Sea off central Norway.

*Significant sighting:* On June 2, 1999, Arnt Helge Molvær watched a Hessie for ten minutes through binoculars. It was feeding on the carcass of a whale off the shore near Ålesund, Møre og Romsdal County, Norway. He ran home to get a video camera and returned fifty minutes later to shoot some footage.

Two fisherman on the fishing vessel *Klaring* saw an animal with two humps swimming at high speed about 300 feet away from their boat in the Storfjorden south of Sula Island on March 18, 2001.

*Sources:* Erik Knatterud, The Hessa Serpent, http://www.mjoesormen.no/thehessaseaserpentI.htm, http://www.mjoesormen.no/thehessaserpentII.htm; Erik Knatterud, The Sula Sea Serpent, http://www.mjoesormen.no/thesulaseaserpent.htm.

## Hibagon

Unknown PRIMATE of Japan.

*Etymology:* After the mountain.

*Physical description:* Apelike. Height, 5 feet. Covered in dark hair. Dark-brown, triangular face. Large, glaring eyes. Snub nose.

*Behavior:* Smells like rotten flesh.

*Tracks:* Length, 10 inches. Width, 6 inches.

*Habitat:* Foothills.

*Distribution:* Mount Hiba, Hiba-Dogo-Taishaku-Quasi National Park, Hiroshima Prefecture, Japan.

*Significant sighting:* In the fall of 1972, Reiko Harada and her small son saw a gorilla in the underbrush near Hiwa.

*Possible explanation:* The Japanese macaque (*Macaca fuscata*) has gray to light-brown fur, is 3–4 feet long, and has a hairless, red face. Its coat grows thicker in the winter. These monkeys are known to wash sweet potatoes in water and roll snowballs.

*Source:* Janet and Colin Bord, *Alien Animals* (Harrisburg, Pa.: Stackpole, 1981), pp. 179–180.

## High-Finned Sperm Whale

Unknown CETACEAN of the North Atlantic Ocean.

*Scientific name: Physeter tursio,* given by Carl von Linné in 1758.

*Variant name:* High-finned cachalot.

*Physical description:* Like a sperm whale. Length, 60 feet. Teeth are only in the lower jaw. Large dorsal fin looks like a ship's mast.

*Distribution:* North Atlantic Ocean, off the Shetland Islands and Nova Scotia.

*Significant sightings:* Two stranded specimens were reported in the seventeenth century.

On either August 27 or September 27, 1946, a black whale with a high dorsal fin was seen to enter Annapolis Basin, Nova Scotia, Canada, and was apparently trapped there for two days. Its length was variously estimated between 10 and 100 feet.

*Sources:* Robert Sibbald, *Phalainologia nova* (Edinburgh: Joannis Redi, 1692), pp. 13–19; Carl von Linné, *Systema naturae per regna tria naturae,* 10th ed. (Stockholm: Laurentii Salvii, 1758–1759), vol. 1, p. 77; "No Such Animal," *Doubt,* no. 16 (1946): 237.

## Hippogriff

FRESHWATER MONSTER of New York.

*Etymology:* "Winged horse-griffin," from the Greek *hippos* ("horse") + the Latin *gryphus* ("griffin").

*Distribution:* Lake George, New York.

*Significant sightings:* Around 1904, reports of a lake monster were generated by a 10-foot cedar log manipulated from the shore by acclaimed local artist Harry W. Watrous. The device is now housed at the Lake George Historical Association.

*Sources:* Curtis MacDougall, *Hoaxes* (New York: Dover, 1958), p. 14; Harry Henck, "The Lake George Monster," *Adirondack Life,* March-April 1980, pp. 37–41; Joseph W. Zarzynski, "The Lake George Monster Hoax of 1904," *Pursuit,* no. 51 (Summer 1980): 99–100; Ginger Henry, The Lake George Monster, March 3, 1998, http://tracylee.com/haguechronicle/monster.shtml.

## Hippoturtleox
FRESHWATER MONSTER of Tibet.

*Etymology:* Coined by J. Richard Greenwell in 1986 after the composite nature of the animal.

*Physical description:* Oxlike body. Skin like a hippopotamus's. Short, curled horns. Legs like a turtle's.

*Distribution:* Lake Duobuzhe, Tibet.

*Significant sighting:* In 1972, Chinese soldiers reportedly killed an animal fitting this description.

*Source:* J. Richard Greenwell, "Hippoturtleox," *ISC Newsletter* 5, no. 1 (Spring 1986): 10.

## Hoàn Kiem Turtle
Giant freshwater TURTLE of Southeast Asia.

*Etymology:* Vietnamese (Austroasiatic), "returned sword," after the legend by which the lake got its name.

*Scientific name: Rafetus hoankiemensis,* given by Ha Dinh Duc.

*Physical description:* Possibly the world's largest freshwater turtle. Length, 5 feet–6 feet 6 inches. Width, 3 feet. Weight, 440 pounds. Gray, mottled upper shell. Pinkish belly.

*Distribution:* Hoàn Kiem Lake in Hanoi, Vietnam.

*Significant sightings:* The turtle's first appearance was around 1428. When King Le Thai To was boating on the lake in celebration of his successful martial exploits against the Chinese, a gigantic tortoise rose from the depths and pulled the king's sword from his hands. Le Thai To re-

named the lake Hoàn Kiem ("returned sword") because he believed that in this way, the blade had been restored to his DRAGON protector.

A stuffed specimen of this turtle exists in Ngoc Son Temple on an island in the middle of the lake. An amateur cameraman took a video of three turtles when they surfaced March 24, 1998. Still photos were taken of specimens on the surface in November 1993 and on March 14, 2000, by Ha Dinh Duc.

A 1-inch × 2-inch egg thought to be from one of the turtles was found April 7, 2000; an unsuccessful attempt was made to incubate it.

*Present status:* The number of individuals remaining in the lake is unknown.

*Possible explanations:*

(1) An outsize specimen of Swinhoe's softshell turtle (*Rafetus swinhoei*), which grows to more than 300 pounds. The only known captive specimen is in the Shanghai Zoo.

(2) The Asian giant softshell turtle (*Pelochelys bibroni*) has a carapace nearly 4 feet long and is found elsewhere in Vietnam, China, the Philippines, and Papua New Guinea.

(3) A surviving *Stupendemys geographicus,* the largest fossil freshwater turtle, known from the Early Pliocene 6 million years ago in Venezuela and Brazil. Its carapace was nearly 7 feet in diameter.

*Sources:* CNN, "Giant Turtle Sightings Set Vietnam Capital Abuzz," April 13, 1998, http://www.cnn.com/EARTH/9804/13/vietnam.turtles.ap/; Karl Shuker, "Turning Turtle?" *Fortean Times,* no. 113 (August 1998): 18; Viet Nam News, "A Hoàn Kieám Turtle Pops Up to Say Hello," March 16, 2000, http://vietnamnews.vnagency.com.vn/2000–03/15/Miscellany.htm, accessed in 2001; Craig Heinselman, "Hoan Kiem Turtle: A Tale of the Sword," *Crypto* 3, no. 3 (May 2000): 15–18, http://www.strangeark.com/crypto/Crypto7.pdf.

## Hominids
In the late 1950s and early 1960s, before the methodology of cladistics was defined and the genetic analysis of relationships in the human

family tree became possible, it was the established practice to classify humanlike primates (Hominidae) separately from apelike primates (Pongidae). Two classic works of cryptozoology dealing with sightings of hairy, primitive-looking creatures were written in this transitional period by Bernard Heuvelmans (*On the Track of Unknown Animals,* 1958) and Ivan T. Sanderson (*Abominable Snowmen: Legend Come to Life,* 1961). Traditionally, human beings (genus *Homo*) and their ancestors all the way back to *Australopithecus* were placed in the Hominidae, and the great apes (except the gibbons) were classed in the Pongidae. Some of the writers who have consulted these works have not updated the terminology and the concepts that underlie the new taxonomy, which may be confusing.

Molecular studies have shown that modern Humans (*Homo sapiens*) shared a common ancestor with Chimpanzees (*Pan troglodytes*) and Gorillas (*Gorilla gorilla*) only about 7–5 million years ago. Because 98 percent of the DNA of modern humans and chimps is more or less the same and that of gorillas is nearly so, many anthropologists have placed the African apes and humans in the same subfamily (Homininae) of the Hominidae. At the present writing, all primates are divided into three suborders—the Prosimians (lemurs, lorises, and bushbabies), the Tarsiiformes (tarsiers), and the Anthropoidea (monkeys, gibbons, orangutans, African apes, and humans).

There is still no complete consensus on the various branches of the human family tree, but anthropology is resilient and can accommodate multiple hypotheses until clear evidence is uncovered. However, this can be confusing at first to those who haven't kept up to date with taxonomic theory.

For example, bipedalism was at one time stressed as a hominid characteristic, but the apelike nature of upright australopiths has called this into question. The discovery of tool use among chimps and other animals has also led to downgrading primitive toolkits as exclusively human. A complex brain and the capacity for structured speech remain two of the major characteristics that humans have over apes. After further molecular work is done, one of the following scenarios will most likely be adopted: (1) humans and australopiths will stay in the Hominidae, and everything not quite so bipedal will go to the Pongidae; (2) chimps and gorillas will join humans in the Hominidae, while *Sivapithecus* and the Orangutans (*Pongo pygmaeus*) will be reserved for the Pongidae; or (3) the DNA linkage will become overwhelming, all the apes will stay with their cousins the humans in the Family Hominidae, and the Pongidae will be abandoned.

For cryptozoology, which does not have the luxury of examining crania, mandibles, and femurs (not to mention the DNA that orders their construction), it seems more practical to call anything walking on its hind legs a hominid. Anything that seems uncomfortable with bipedalism (no matter how brainy it seems or what toolkit it's using) will get conservatively lumped into an unknown PRIMATE category, deferring for the time being its proper classification.

*Mystery hominids and hominid-like creatures:* CANNIBAL GIANT, FLYING HUMANOID, GIANT HOMINID, HAIRY BIPED, LEAST HOMINID, LITTLE PEOPLE, LIZARD MAN, MARKED HOMINID, MONKEY MAN, NEO-GIANT, PROTO-PIGMY, SHORTER HOMINID, SMALL HOMINID, SUB-HOMINID, SUB-HUMAN, TALLER HOMINID, TRUE GIANT, WILDMAN.

## *Homo ferus*

WILDMAN of Europe.

*Etymology:* Latin, "wild man."

*Present status: Homo ferus* was the Swedish taxonomist Carl von Linné's classification of what were probably feral children, allegedly raised in the wilderness by animals. In the tenth edition of his *Systema naturae* in 1759, Linné divided the genus *Homo* into seven racial types, more or less based on skin color: *europaeus* (white), *asiaticus* (yellow), *americanus* (red), *afer* (black), *troglodytes* (orangutan), *monstrosus* (giants and mutants), and *ferus* (hairy, mute, and walking on all fours). His successor Johann Friedrich Blumenbach dropped the *ferus* and *troglodytes* categories in 1775 and added a Malayan race (brown) in 1795. These racial types became the basic anthropological designations used until the mid-twentieth century.

*Sources:* Carl von Linné, *Systema naturae per regna tria naturae,* 10th ed. (Stockholm: Laurentii Salvii, 1758–1759); Johann Friedrich Blumenbach, *De generis humani varietate nativa* (Göttingen, Germany: F. A. Rosenbuschii, 1775); Franck Tinland, *L'homme sauvage* (Paris: Payot, 1968); Lucien Malson, *Wolf Children and the Problem of Human Nature* (New York: Monthly Review Press, 1972).

### Homo nocturnus

PRIMATE of Southeast Asia.

*Etymology:* Latin, "night man."

*Present status: Homo nocturnus* was the Swedish taxonomist Carl von Linné's original classification of the Orangutan (*Pongo pygmaeus*) in the first edition of his *Systema naturae* in 1735. Linné used as a type specimen the young orangutan described by Jakob de Bondt as *Homo sylvestris* ("forest man") in a work published posthumously in 1658.

*Source:* Carl von Linné, *Systema naturae, 1735* (facsimile of the first edition), M. S. J. Engel-Ledeboer and H. Engel, eds. (Nieuwkoop, the Netherlands: B. de Graaf, 1964).

### Homo troglodytes

PRIMATE of Southeast Asia.

*Etymology:* From the Latin *homo* ("man") + the Greek *troglodytaï* ("cave"). *See also* TROGLODYTE.

*Present status: Homo troglodytes* was the Swedish taxonomist Carl von Linné's reclassification of the Orangutan (*Pongo pygmaeus*) in the tenth edition of his *Systema naturae* in 1759. The type description was also loosely based on stories of an albino tribe on the island of Ternate in the Moluccas in Indonesia and a white tribe in Central Africa whose members had membranes over their eyes. The specific name *troglodytes* has been perpetuated in the scientific name for the Chimpanzee (*Pan troglodytes*).

*Sources:* Carl von Linné, *Systema naturae per regna tria naturae,* 10th ed. (Stockholm: Laurentii Salvii, 1758–1759), p. 24; Carl von Linné, *Dissertatio academica, in qua Anthropomorpha, respondent C. E. Hoppius* (Uppsala, Sweden: Carl von Linné, 1760); Charles Wardell Stiles and Mabelle B. Orleman, *The Nomenclature for Man, the Chimpanzee, the Orang-utan, and the Barbary Ape* (Washington, D.C.: Government Printing Office, 1927), p. 9; Bernard Heuvelmans, *Les bêtes humaines d'Afrique* (Paris: Plon, 1980), pp. 36–44.

## HOOFED MAMMALS (Unknown)

Simply put, hoofed mammals have toes covered with a horny structure composed of keratin that helps them to run away from predators efficiently. Hooves, like the nails of primates, evolved from the keratinous claws of other mammals, such as cats and rodents. Like ballet dancers standing on point, these animals have their entire weight concentrated on their toes.

Hoofed mammals have traditionally been called ungulates (from the Latin *ungula,* "hoof"), a group that had a common origin sometime in the Late Cretaceous, 70–65 million years ago. Recent evidence that also places the nonhoofed ELEPHANTS, HYRAXES, and aardvarks in the Superorder Ungulata (as well as the aquatic CETACEANS and SIRENIANS derived from ungulates) makes it more convenient to group hoofed cryptids separately. Most are herbivorous.

The two major extant orders of hoofed mammals are:

(1) The Artiodactyla, the order of even-toed or cloven-hoofed animals that includes cattle, deer, antelopes, giraffes, pigs, hippos, and camels. First seen in the Early Eocene, 55 million years ago, they are characterized by their elongated third and fourth toes, which form the primary support for the limbs. The skulls of living artiodactyls are elaborately modified for defense, with canines, incisors, horns, and antlers.

(2) The Perissodactyla, the order of odd-toed animals, with the middle toe bearing the primary weight. These animals include horses, rhinos, tapirs, and the extinct chalicotheres and brontotheres. This group diversified in North America and Eurasia to become the most abundant herbivores between 55 and 25 million years ago.

Extinct orders of hoofed mammals are the

embrithopods of Oligocene Africa, which included the rhinolike, twin-horned *Arsinoitherium;* the notoungulates, South American ungulates that included the horse- or rhinolike toxodonts and the smaller typotheres; the litopterns, also endemic to South America, which incorporated the long-necked, camel-like macraucheniids; the uintatheres of North America and Asia, among them the huge *Uintatherium,* which had three pairs of bony swellings on its skull and powerful canine teeth; the carnivorous mesonychids such as *Andrewsarchus* that may have been ancestral to cetaceans; the astrapotheres, South American animals that resembled tapirs or rhinos; and the pyrotheres and xenungulates, little-known South American ungulates.

Of the thirty-eight mystery animals in this list, twenty seem related to pigs, hippos, camels, deer, antelopes, giraffes, or oxen; eleven can apparently be grouped with horses, rhinos, and tapirs; three may be surviving notoungulates; one could be a survival into historical times of a litoptern; and three are too problematic to classify.

Six are found in North America, six in South America, four in Europe, eight in Africa, twelve in Asia, and two in Australasia.

## Mystery Hoofed Mammals
### Artiodactyls
AUSTRALIAN CAMEL; CAITETU-MUNDÉ; CAMELOPS; CUINO; ESAKAR-PAKI; ETHIOPIAN DEER; IRISH DEER; MANGARSAHOC; MANGDEN; MONGOLIAN GOAT-ANTELOPE; MUSKOX OF NOYON UUL; PERSEPOLIS BEAST; PUKAU; QUANG KHEM; SCHELCH; SCHOMBURGK'S DEER; SIVATHERE OF KISH; SPOTTED BUSHBUCK; TSY-AOMBY-AOMBY; WHITE BROCKET DEER

### Perissodactyls
BADAK TANGGILING; BLACK MALAYAN TAPIR; BLOOD-SWEATING HORSE; BLUE HORSE; EMELA-NTOUKA; JUMAR; ONE-HORNED AFRICAN RHINOCEROS; QUAGGA; TIGELBOAT; VAN ROOSMALEN'S TAPIR; WEB-FOOTED HORSE

### Notoungulates
DOMENECH'S PSEUDO-GOAT; MIRAMAR TOXODONT; THUNDER HORSE

### Litopterns
FIVE-TOED LLAMA

### Unknown
DEVIL PIG; LASCAUX UNICORN; WOLF DEER

## Horn Head
A category of SEA MONSTER identified by Gary Mangiacopra.

*Physical description:* Long, round body. Length, 25–60 feet. Dark on top, underside lighter. Scales like a crocodile's. Flat, round head about 2 feet across. Horns. Two pairs of flippers. Sawlike projections on the back. Tail forked or tapering to a point.

*Behavior:* Seen with young. Spouts water.

*Distribution:* North Atlantic Ocean.

*Significant sightings:* On November 23, 1869, while 300 miles off the coast of New England, Captain Allen and the crew of the bark *Scottish Bride* watched a 25-foot-long animal with a large, flat head and thick scales like a crocodile's. A smaller animal, apparently a juvenile only a few feet long, accompanied the large one.

On June 26, 1904, passengers on the French Line steamer *La Lorraine* saw a huge animal that spouted, churned the water into a foam, and dived and resurfaced repeatedly for more than an hour about 560 miles off Brest, France. Its eyes were huge, it had horns about 20 inches in length, its head stood 12 feet out of the water, and one dorsal fin ran nearly the entire length of its back, which some estimated to be 150 feet long.

*Possible explanation:* Similar to Bernard Heuvelmans's LONGNECK.

*Sources:* "The Old 'Fishy' Story," *New York Herald,* November 30, 1869, p. 8; "Eyes as Big as Saucers," *New York Tribune,* July 2, 1904, p. 1; Gary S. Mangiacopra, "The Great Unknowns of the 19th Century," *Of Sea and Shore* 8, no. 3 (Fall 1977): 175–178.

## Horned Hare
Legendary RABBIT of West Asia, Europe, and the United States.

*Scientific name: Lepus cornutus.*

*Variant names:* Jackalope (in the United States, "jackrabbit" + "antelope"), Raurackl (Old German), Wolpertinger (in Bavaria).

*Physical description:* Rabbit with antlers or horns.

*Distribution:* Western and midwestern United States; southern Germany; West Asia.

*Significant sightings:* The horned Raurackl was generally known to Bavarian hunters of the sixteenth century and appeared in a contemporary print by Joris Hoefnagel.

German naturalist Peter Simon Pallas allegedly shot a Horned hare in Azerbaijan in the late eighteenth century.

Douglas, Wyoming, claims the dubious distinction of the first Jackalope taxidermist hoax, involving a model created in 1934 by Douglas Herrick.

*Present status:* Often the subject of obviously faked photographs and postcards, the Horned hare has a venerable history going back to third-century Persia. It was considered rare but real in the eighteenth century.

*Possible explanations:*

(1) Photographic hoaxes or taxidermist hoaxes in which deer antlers are attached to the head of a stuffed rabbit.

(2) Cranial tumors in the shape of horns or antlers, which is a disease of Cottontails (*Sylvilagus* spp.) and other rabbits. Called papillomatosis, the condition is caused by the Shope papillomavirus and is probably transmitted by the Rabbit tick (*Haemaphysalis leporis-palustris*) or mosquitos. The tumors are irregular in shape and can appear on the face, neck, and rump, as well as the top of the head.

*Sources:* Gaspar Schott, *Physica curiosa* (Würzburg, Germany: Johannis Andreae Endteri, 1667), frontispiece and p. 900; Walker D. Wyman, *Wisconsin Folklore* (River Falls: University of Wisconsin—Extension, Department of Arts Development, 1979), pp. 13–18; J. W. Kreider and G. L. Bartlett, "The Shope Papilloma-Carcinoma Complex of Rabbits," *Advances in Cancer Research* 35 (1981): 81–110; Daniel S. Simberloff, "A Funny Thing Happened on the Way to the Taxidermist: An Evolutionary Ecologist Ponders the Origins of America's 'Jackalope,'" *Natural History* 96 (August 1987): 50–54; Fritz Koreny, *Albrecht Dürer and the Animal and Plant Studies of the Renaissance* (Boston: Little, Brown, 1988), p. 138; "Folklore and Cryptozoology Subject of Joint Conference," *ISC Newsletter* 9, no. 3 (Autumn 1990): 4.

## Horned Jackal

Mystery DOG of the Indian subcontinent.

*Variant name:* Churail (in India).

*Physical description:* Jackal with a small, bony horn at the back of its head, usually hidden by hair. It grows to about half an inch long.

*Behavior:* Sri Lankan folklore suggests that only the leaders of a pack possess this horn, which is called a *narri-comboo*. The horn is revered as a powerful talisman.

*Distribution:* Sri Lanka; India.

*Significant sighting:* A skull with such a horn was housed at the Museum of the College of Surgeons in London during the nineteenth century.

*Possible explanation:* A genetic defect or physical injury might produce a hornlike growth in the Golden jackal (*Canis aureus*).

*Sources:* J. Emerson Tennent, *Sketches of the Natural History of Ceylon* (London: Longman, Green, Longman, and Roberts, 1861); Edward Balfour, *The Cyclopaedia of India and of Eastern and Southern Asia* (London: B. Quaritch, 1885); Norah Burke, *Eleven Leopards: A Journey through the Jungles of Ceylon* (London: Jarrolds, 1965).

## Horse's Head

FRESHWATER MONSTER of Québec, Canada.

*Variant name:* MISIGANEBIC.

*Physical description:* Length, 6–30 feet. Head is like a horse's.

*Behavior:* Swims swiftly. Travels on land between lakes. Tourists used to put cartons of cream in the water for the monster to drink.

*Distribution:* Baskatong Lake, Lac Bitobi, Lac Blue Sea, Lac-des-Cèdres, Lac Creux, Lac Désert, Gatineau River, Lac Pocknock, and Lac Trente-et-un-Milles, all in Québec.

*Significant sighting:* Around 1910, Olivier Garneau was fishing in Lac Blue Sea when he saw a 10-foot animal with a horse's head rise up out of the water.

*Source:* Michel Meurger and Claude Gagnon, *Lake Monster Traditions: A Cross-Cultural Analysis* (London: Fortean Tomes, 1988), pp. 104–110.

## Huáng Yao

Unknown WEASEL of East Asia.

*Etymology:* Mandarin Chinese (Sino-Tibetan) word; *huáng* means "yellow."

*Physical description:* Weasel-like body. Yellow above, black below. Head is like a cat's.

*Distribution:* China.

*Source:* Richard Muirhead, "Some Chinese Cryptids (Part Two)," *Cryptozoology Review* 4, no. 1 (Summer 2000): 19–20.

## Huia

A perching BIRD of the Wattlebird family (Callaeatidae) of New Zealand, supposed extinct since 1907.

*Etymology:* Maori (Austronesian) word, from its distinctive call.

*Scientific name: Heteralocha acutirostris,* given by John Gould in 1836.

*Physical description:* Black plumage with metallic green gloss. Length, 19 inches. Orange facial wattles. Males have medium-length, sturdy bills; females have long, curved bills. The difference in bills between the two sexes is unique among bird species. Both bills are ivory-colored. Large, black tail feathers with white tips.

*Behavior:* Call is soft and fluting. Formerly prized by the Maori for its tail feathers.

*Habitat:* Beech and podocarp forests.

*Distribution:* When Europeans arrived in New Zealand, Huias were found in the southern half of North Island, from East Cape to Wellington. Any survivors may have moved north to the Urewera State Forest or the Tarawera Range.

*Significant sightings:* Throughout the 1920s, some twenty-three unsubstantiated reports were logged. Signs of Huias were found during an official search in 1924, though no living birds were seen.

On October 12, 1961, Margaret Hutchinson spotted a Huia at Lake Waikareti in the Urewera State Forest, North Island, noting its distinctive tail.

In 1991, Danish zoologist Lars Thomas claimed to have seen a Huia in the Pureora Forest, North Island.

*Present status:* Presumed extinct since shortly after December 28, 1907, when W. W. Smith spotted two males and a female; this is considered the final official sighting. Scientists and ethicists meeting in New Zealand in July 1999 agreed to allow the cloning of a Huia, using preserved DNA samples.

*Sources:* William J. Phillipps, *The Book of the Huia* (Christchurch, New Zealand: Whitcombe and Tombs, 1963); Margaret Hutchinson, "I Thought I Saw a Huia Bird," *Birds* 3 (September-October 1970): 110–113; Karl Shuker, *Extraordinary Animals Worldwide* (London: Robert Hale, 1991), pp. 83–86; Lars Thomas, *Mysteriet om Havuhyrerne* (Copenhagen: Gyldendal Boghandel, 1992); "Cloning of Extinct Huia Bird Approved," Environmental News Network, July 20, 1999; Errol Fuller, *Extinct Birds* (Ithaca, N.Y.: Cornell University Press, 2001), pp. 367–372.

## Huilla

FRESHWATER MONSTER of the West Indies.

*Etymology: Huilla* is a common name for the Anaconda (*Eunectes murinus*) in South America.

*Physical description:* Serpentine. Length, 25–50 feet. Moss green. Scaly. Horselike head.

*Behavior:* Amphibious. Swims swiftly by flexing its body into arches. Migrates from one body of water to another. Emits a high-pitched whistle.

*Tracks:* Three-toed.

*Distribution:* Ortoire River, Trinidad.

*Sources:* Edward L. Joseph, *History of Trinidad* (London: A. K. Newman, 1838); John O. Brathwaite (letter), *Strange Magazine*, no. 18 (Summer 1997): 2.

## Hungarian Reedwolf

Unknown wild DOG of Central Europe.

*Scientific names: Canis lupus minor,* given by M. Mojsisovics in 1887; *Canis aureus hungaricus,* renamed by Gyula Éhik in 1938.

*Variant name:* Rohrwolf (German).

*Physical description:* Like a small wolf.

*Distribution:* Hungary; eastern Austria.

*Present status:* Apparently became extinct in the early twentieth century. Some museum specimens exist.

*Possible explanations:*

(1) A diminutive subspecies of Gray wolf (*Canis lupus*), first suggested by M. Mojsisovics and now the generally accepted identification.

(2) A large Golden jackal (*Canis aureus*), proposed by János Szunyoghy.

*Sources:* Eugen Nagy, "Der ausgerottete ungarische Rohrwolf (*Canis lupus*) war kein Schakal (*Canis aureus*)," *Säugetierkundliche Mitteilungen* 4 (1956): 165–167; János Szunyoghy, "Systematische Revision des ungarländischen Schakals, gleichzeitig eine Bemerkung über das Rohrwolf-Problem," *Annales Historico-Naturales, Musei Nationalis Hungarici,* new ser. 8 (1957): 425–433; Eduard-Paul Tratz, "Ein Betrag zum Kapitel 'Rohrwolf' *Canis lupus minor* Mojsisovics, 1887," *Säugetierkundliche Mitteilungen* 6 (1958): 160–162.

## HYENAS (Unknown)

Hyenas (Family Hyaenidae) are long-legged, long-necked scavengers and carnivores with large eyes and ears, blunt snouts, high shoulders and low hindquarters, shaggy coats, and short tails. They probably originated in the Late Oligocene, 25 million years ago, from relatives of the CIVETS in Africa or Eurasia, parallel to the development of DOGS in North America. They were a diverse group that ranged from Europe to Africa and Indonesia, and one species (*Pachycrocuta brevirostris*) grew as large as lions and ultimately specialized in the ability to crush, swallow, and digest large mammal bones and teeth.

The only hyaenid in North America was the extinct genus *Chasmaporthetes,* also known from Eurasia and Africa. It lived from the Pliocene to the Pleistocene, 2 million–10,000 years ago, and was not a bone-cracker; its teeth were more adapted to slicing flesh.

There are only four extant species, all found in Africa. The largest is the Spotted hyena (*Crocuta crocuta*), which grows to a body length of 5 feet 9 inches. The others are the Striped hyena (*Hyaena hyaena*), also found in Arabia and India; the Brown hyena (*H. brunnea*); and the more distantly related Aardwolf (*Proteles cristatus*), which feeds on about 250,000 termites in a single night.

Of the eight hyaenids in this section, three could be explained by the striped hyena, three could be surviving *Chasmaporthetes,* one could be a surviving *Pachycrocuta,* and one seems to be an unknown variety.

### Mystery Hyenas

BEAST OF GÉVAUDAN; BOOAA; CHUTI; CUITLAMIZTLI; NANDI BEAR; AL-SALAAWA; SET; SHUNKA WARAK'IN

### Hylophagos

WILDMAN of Central Africa.

*Etymology:* Greek, "wood eater."

*Behavior:* Arboreal. Eats wood and seeds.

*Possible explanations:*

(1) The Chimpanzee (*Pan troglodytes*), suggested by Vernon Reynolds, because it eats the leaves, bark, and stems of certain plants.

(2) A distorted rumor of apes or African wildmen, perhaps an etymological leap from "men of the woods" to "eaters of wood," suggested by Bernard Heuvelmans.

*Sources:* Diodorus Siculus, *Historical Library,* III. 24; Pomponius Mela, *De situ orbis,* III. 93; Vernon Reynolds, *The Apes* (New York: E. P. Dutton, 1967), pp. 31–32; Bernard Heuvelmans, *Les bêtes humaines d'Afrique* (Paris: Plon, 1980), pp. 139–140, 164.

## HYRAXES (Unknown)

Hyraxes (Order Hyracoidea) are rabbit-sized animals that have no visible tail and look like guinea pigs. The forefeet have four toes bearing blunt, hooflike nails; the hind feet have three toes, two with nails and the third with a curved claw. They were once grouped with the RODENTS and later with the ELEPHANTS to which they are closely related. Thomas Huxley was the first to put them into an independent order of their own. Current molecular studies show a close connection to both the proboscideans and the perissodactyls (HOOFED MAMMALS).

There are three existing genera of hyraxes, all living in Africa: Rock hyraxes (*Procavia*), Bush hyraxes (*Heterohyrax*), and Tree hyraxes (*Dendrohyrax*). All three flourish in forested and rocky environments that hoofed mammals have

difficulty with. These are only a small percentage of hyracoids that lived in the early Cenozoic, when they were the dominant herbivores in Africa. At that time, some were as large as rhinoceroses, including *Titanohyrax* and *Kvabebihyrax*. The earliest known hyraxes lived in Tunisia and Algeria in the Middle Eocene, 45 million years ago. The three cryptids in this section are ambiguously described but might reasonably be identified as hyraxes.

**Mystery Hyraxes**
ETHIOPIAN HYRAX; GIANT HYRAX OF SHAANXI; SANDEWAN

# I

## Ichthyophagos

WILDMAN of North Africa and India.

*Etymology:* Greek, "fish eater."

*Physical description:* Long fingernails. Shaggy, uncut hair.

*Behavior:* Eats dried fishes. Wears animal skins. Does not engage in trade.

*Distribution:* Sudan; Ethiopia; northwestern India.

*Possible explanation:* Any number of indigenous peoples in Africa and India have a high proportion of fish in their diets.

*Sources:* Herodotus, *The Histories,* trans. Aubrey de Sélincourt (London: Penguin, 1996), pp. 161–163 (III. 19–23); Diodorus Siculus, *Historical Library,* III. 53; Quintus Curtius Rufus, *The History of Alexander,* trans. John Yardley (New York: Penguin, 1984), p. 235 (IX. 10. 8–10); Arrian, *The Campaigns of Alexander,* ed. Aubrey de Sélincourt (New York: Penguin, 1971), pp. 334, 343 (VI. 24, 28).

## Iemisch

OTTER-like animal of South America.

*Etymology:* Tehuelche (Chon), "water tiger," also used for the Marine otter (*Lontra felina*). Probably not "little pebbles," as Florentino Ameghino claimed.

*Scientific names: Neomylodon listai,* given by Ameghino in 1898, based on the fossil *Mylodon* hide from the Cueva del Milodón and Ramón Lista's sighting; *Iemisch listai,* given by Santiago Roth in 1899 based on the femur of an extinct jaguar found in the cave.

*Variant names:* Chimchimen, Erefilú, Guarifilu, Hymché, Jemechim, Jemisch, Ñerrefilu, Nervelu, Ngúrüvilu, Ñiribilu, Nirribilu, Nürü-filu (Mapudungun/Araucanian), Yem'chen, Yemische, Zorro-víbora (Spanish, "fox-viper," also used by Araucanian speakers).

*Physical description:* Size of a puma. Covered in short, coarse hair. Bay or dark brown color. Short, round head. Circle of light hair around the eyes extending to the ear-hole. No external ears. Big canine teeth. Short, plantigrade feet. Three webbed toes on the forefeet, four webbed toes on the hind feet. Long, flat, otterlike, supple tail.

*Behavior:* Nocturnal. Aquatic. Digs a burrow. Seizes horses and drowns them. Said to drag humans into the water.

*Tracks:* Catlike.

*Distribution:* Lago Colhué Huapi, Río Senguer, and Estancia Valle Huemeles, in Chubut Province, Argentina; Santa Cruz Province, Argentina; Aisén del General Carlos Ibáñez del Campo Region, Chile. Formerly ranged north to the Río Negro Province, Argentina, and in the south to lakes on the eastern slopes of the Andes Mountains and the Straits of Magellan.

*Significant sighting:* Ramón Lista came across a large animal that looked like a giant Pangolin (*Manis* spp.) with hair instead of scales in Argentina's Santa Cruz Province in the 1870s. Bullets failed to penetrate the animal's skin.

*Possible explanations:*

(1) An undetermined species of giant ground sloth, according to Florentino Ameghino, who was determined to show that the bones and tough, red-haired skin found in a cave now known as the Cueva del Milodón (24 kilometers north of Puerto Natales in Chile) were from an animal the Indians knew as Iemisch. However, the amphibious, carnivorous, web-footed Iemisch doesn't seem to match a terrestrial, vegetarian, huge-clawed sloth. The *Mylodon*

remains have been reliably carbon-dated to 13,000–8,600 years ago, though some stratigraphic evidence indicates ground sloths survived as recently as 3000 B.C.

(2) An aquatic reptile with the head of a fox, suggested by Esteban Erize.

(3) An unknown species of large otter or a surviving population of the Giant otter (*Pteronura brasiliensis*), which is now largely restricted to the Amazon watershed and grows to a length greater than 5 feet, including the tail.

(4) A confusion between the Marine otter (*Lontra felina*) and the Jaguar (*Panthera onca*), which once existed in Patagonia and grows to a full length of 6 feet.

(5) Exaggerated accounts of the Southern river otter (*Lontra provocax*), widely distributed in southern Chile and Argentina until the early twentieth century but now endangered and officially found only in isolated pockets in the southwestern fjords area. However, this animal may persist as far north as the Río Colorado and La Pampa Province, Argentina. It has a long body, flat head, small ears, and a broad, whiskered muzzle. It grows to nearly 4 feet long, including the tail, and has strong claws on its webbed feet. Color is dark to very dark brown above, with a lighter cinnamon below.

*Sources:* Francisco P. Moreno, *Viaje á la Patagonia austral* (Buenos Aires: La Nacion, 1879); Florentino Ameghino, "An Existing Ground-Sloth in Patagonia," *Natural Science* 13 (1898): 324–326; Florentino Ameghino, "El mamífero misterioso de la Patagonia (*Neomylodon listai*)," *La Pirámide* (La Plata, Argentina) 1 (1899): 51–63, 83–84; "The Jemisch, or Great Ground Sloth," *English Mechanic* 72 (1900): 118–119; André Tournouër, "Sur le Neomylodon et l'animal mystérieux de la Patagonie," *Comptes Rendus Hebdomadaires des Séances de l'Académie des Sciences* 132 (1901): 96–97; Robert Lehmann-Nitsche, "La pretendida existencia actual del Grypotherium: Supersticiones araucanas referentes a la lutra y el tigre," *Revista del Museo de La Plata* 10 (1902): 269–279; H. Hesketh Prichard, *Through the Heart of Patagonia* (New York: Appleton, 1902); Bernard Heuvelmans, *On the Track of Unknown Animals* (New York: Hill and Wang, 1958), pp. 265–277; Alberto Vúletin, *Zoonimia andina* (Santiago del Estero, Argentina: Instituto de Linguistica, Folklore y Arqueologia, 1960); Bruce Chatwin, *In Patagonia* (New York: Summit Books, 1977), pp. 186–194; Roy P. Mackal, *Searching for Hidden Animals* (Garden City, N.Y.: Doubleday, 1980), pp. 161–168; Edgar Morisoli, "La presencia del Animal de Agua en la zona de Casa de Piedra: Caldenia, Diario La Arena, 13 de Febrero [1981]," in *Obra callada, 1974–1986* (Santa Rosa, Argentina: Editorial Pitanguá, 1994); Esteban Erize, *Mapuche* (Buenos Aires: Editorial Yepun, 1987); "Ground Sloth Survival Proposed Anew," *ISC Newsletter* 12, no. 1 (1993–1996): 1–5; Charles Jacoby home page, http://users.tinyonline.co.uk/jacoby/giantsloth.htm; Mariano Martín Fernández, "Nutrias en La Pampa," http://orbita.starmedia.com/~faunapampeana/ma/5nutriaslp.html.

## Igopogo

FRESHWATER MONSTER of Ontario, Canada.

*Etymology:* Named sometime in the 1950s in imitation of OGOPOGO. Possibly inspired by the Walt Kelly comic strip "Pogo," whose main character had a mock campaign for U.S. president in 1952 with the slogan, "I Go Pogo."

*Variant names:* Beaverton Bessie, Kempenfelt Kelly, Simcoe Kelly.

*Physical description:* Seal-like animal. Length, 12–70 feet. Charcoal-gray color. Dog- or horse-like face. Prominent eyes. Gaping mouth. Neck is like a stovepipe. Several dorsal fins. Fishlike tail.

*Behavior:* Basks in the sun.

*Distribution:* Kempenfelt Bay, Lake Simcoe, Ontario.

*Significant sightings:* Reports date back to the 1880s.

On July 22, 1963, the Rev. L. B. Williams and his family observed a charcoal-colored animal with dorsal fins.

A sonar sounding of a large animal was taken June 13, 1983, by William W. Skrypetz from the Government Dock and Marina.

A videotape of Igopogo was taken in 1991 by the captain of a powerboat whose vessel had broken down. A large, seal-like animal reared up out of the water twice, then submerged.

*Possible explanation:* An occasional seal that makes its way to the lake.

*Sources: New York Times,* July 22, 1881, p. 2; *Oakville (Ont.) Journal-Record,* July 27, 1963; *Toronto Sun,* March 13, 1978, and July 31, 1978; John Kirk, *In the Domain of Lake Monsters* (Toronto, Canada: Key Porter Books, 1998), pp. 28, 113; Igopogo, a Mystery Solved, http://www.ultranet.ca/bcscc/igopogo. htm.


## Ijiméré

LITTLE PEOPLE of West Africa and the West Indies.

*Etymology:* Yoruba (Benue-Congo) word.
*Behavior:* Dangerous.
*Distribution:* Nigeria; Togo; Benin; Trinidad.
*Source:* Bernard Heuvelmans, *Les bêtes humaines d'Afrique* (Paris: Plon, 1980), p. 496.


## Ikal

LITTLE PEOPLE of Mexico.

*Etymology:* Tzeltal (Mayan), "black spirit."
*Variant names:* ?Ihk'al, Kek (Kekchi/Mayan).
*Physical description:* Height, 3 feet. Skinny. Covered in long, curly hair. Black face, arms, and legs. Erect ears. Thin neck. Said to have cloven hooves.
*Behavior:* Active at dusk.
*Habitat:* Caves.
*Distribution:* Chiapas State, Mexico.
*Sources:* Brian Stross, "The ?Ihk'als," *Flying Saucer Review* 14, no. 3 (May-June 1968): 12; Brian Stross, "Demons and Monsters of Tzeltal Tales," *University of Missouri Museum Brief,* no. 24 (1978), pp. 2–6, 30–31; John E. Roth, *American Elves* (Jefferson, N.C.: McFarland, 1997), p. 101.


## Ikimizi

SPOTTED LION of Central Africa.

*Etymology:* Rwanda (Bantu) word.

*Physical description:* Looks like a cross between a lion and a leopard. Gray color with dark spots. Beard on the chin.
*Distribution:* Virunga Volcanos region of Rwanda.
*Source:* Wilhelm, Prince of Sweden, *Among Pygmies and Gorillas* (London: Gyldendal, 1923).


## Île du Levant Wildcat

Mystery CAT of Western Europe.

*Variant name:* Lynx de la paille (French, "haystack lynx").
*Scientific name: Felis silvestris levantina,* given in 1986 by Bernard Heuvelmans.
*Physical description:* Wildcat like the other Mediterranean species, though somewhat larger. Weight, 22 pounds or greater.
*Distribution:* Île du Levant and Îles d'Hyères, in southern France.
*Significant sighting:* Several times in 1958, Bernard Heuvelmans observed one attacking feral domestic cats.
*Possible explanation:* Large subspecies of African wildcat (*Felis silvestris libyca*), the probable ancestor of the domestic cat.
*Source:* Bernard Heuvelmans, "Annotated Checklist of Apparently Unknown Animals with Which Cryptozoology Is Concerned," *Cryptozoology* 5 (1986): 1–26.


## Iligan Dolphin

Mystery CETACEAN of the Philippines.

*Physical description:* Length, 7–9 feet. An oceanic dolphin with a brown back, pink underside, and yellow flanks.
*Distribution:* Iligan Bay, Philippines.
*Significant sighting:* Seen by W. F. J. Mörzer Bruyns in Iligan Bay, Mindinao Sea, Philippines, in schools of up to thirty individuals.
*Source:* W. F. J. Mörzer Bruyns, *Field Guide of Whales and Dolphins* (Amsterdam: Tor, 1971).


## Illie

FRESHWATER MONSTER of Alaska.

*Etymology:* After Lake Iliamna.

*Variant name:* Jig-ik-nak (Inuktitut/Eskimo-Aleut).

*Physical description:* Whalelike or seal-like. Length, 10–20 feet. Dull gray. Dorsal fin with a white stripe.

*Behavior:* Swallows boats, especially those with red hulls.

*Distribution:* Lake Iliamna, Alaska.

*Significant sightings:* Babe Alsworth and Bill Hammersley were flying over the lake in September 1942 when they noticed several dozen grayish animals with large, blunt heads. Their size was well over 10 feet.

In 1963, a biologist from the Alaska Department of Fish and Game was flying his plane over the lake and spotted an animal 25–30 feet long that was swimming below the surface. It did not come up for air during the ten minutes he watched it.

Air-taxi pilot Tim LaPorte was flying above Pedro Bay in 1977 when he and his two passengers saw a dark, 12- to 14-foot animal on the surface. It dived straight down and revealed a large vertical tail.

On July 27, 1987, Verna Kolyaha saw a 10-foot, black fish with a white stripe along its fin. The next day, Jerry Pippen observed a large seal spouting water.

*Possible explanation:* A large Green sturgeon (*Acipenser medirostris*), which grows to 7 feet long, or a White sturgeon (*A. transmontanus*), which grows to 20 feet. Sturgeons have not been recorded in Iliamna, but they are known in other Alaskan lakes and coastal waters as far north as Cook Inlet.

*Sources:* Gil Paust, "Alaska's Monster Mystery Fish," *Sports Afield,* January 1959, pp. 54–56, 65–67; "Alaska's Monster Mystery Fish," *American Legion Magazine* 80 (June 1966): 52; "The Iliamna Lake Monster," *Alaska Magazine* 54 (January 1988): 17; Loren Coleman, *Tom Slick and the Search for the Yeti* (Boston: Faber and Faber, 1989), pp. 125–127; Matt Bille, "What Lies beneath Lake Iliamna?" *Crypto Dracontology Special,* no. 1 (November 2001): 66–69.

## Imap Umassoursua

SEA MONSTER of the North Atlantic Ocean.

*Etymology:* Inuktitut (Eskimo-Aleut) word.

*Physical description:* Flat. As large as an island.

*Distribution:* Coast of Greenland.

*Source:* "Water Monsters: Greenland," *Fortean Times,* no. 46 (Spring 1986): 29.

## Ink Monkey

Small PRIMATE of East Asia.

*Etymology:* So named because it had been trained by Chinese scholars to grind and prepare ink, turn manuscript pages, and fetch brushes when needed.

*Variant name:* Pen monkey.

*Physical description:* Length, 4–5 inches. Soft, jet-black fur. Scarlet eyes.

*Behavior:* Intelligent enough to be trained as a scribal assistant. Sleeps in the scholar's desk drawer or brush pot. Drinks the india ink left over when the scholar is finished writing.

*Distribution:* China.

*Significant sightings:* Used by scholars from 2000 B.C. to at least the time of Zhu Xi (A.D. 1130–1200).

A news item in the Chinese *People's Daily* of April 22, 1996, announced the rediscovery of the Ink monkey in the Wuyi Shan Mountains, Fujian Province, China. Beyond noting that the animal was no larger than a mouse and weighed 7 ounces, no other details were released. However, the story may actually refer to an earlier announcement of the discovery in the Yuanqu basin, southern Shanxi Province, of a mandible of the fossil *Eosimias centennicus,* a tarsier-like primate weighing 3.5 ounces that lived in the Eocene, 40 million years ago.

*Possible explanations:*

(1) The Slow loris (*Nycticebus coucang*), suggested by Cyril Rosen, is found in southern China and Indonesia and grows up to 15 inches long.

(2) An unknown species of Tarsier (*Tarsius* spp.) indigenous to China, perhaps even a surviving *Eosimias centennicus.* First discovered in May 1995 by Chris Beard, this primate's chin was deep and robust like a monkey's, and its canine teeth projected

high above the others.

*Sources:* Walter Henry Medhurst, *A Glance at the Interior of China Obtained during a Journey through the Silk and Green Tea Districts Taken in 1845* (Shanghai, China: Mission Press, 1849); Evangeline D. Edwards, ed., *The Dragon Book* (London: William Hodge, 1938), p. 149; Jonathan Mirsky, "Ink Monkey of Ancient China Is Rediscovered," *Times* (London), April 23, 1996; Karl Shuker, "A Real Pen and Ink," *Fortean Times,* no. 90 (September 1996): 44; Chris Beard, Searching for Our Primate Ancestors in China, http://www.chineseprehistory.org/beard.htm.

## Inkanyamba

FRESHWATER MONSTER of South Africa.

*Etymology:* Xhosa (Bantu), "tornado."

*Variant names:* Howie.

*Physical description:* Serpentine. Length, up to 25 feet. Head is like a snake's or a horse's. Long neck. Mane of skin.

*Behavior:* Moves from one body of water to another in the summer. Often seen in misty conditions. Blamed for the loss of livestock and storm damage.

*Distribution:* The pool below Howick Falls, Midmar Dam in the Umgeni River, the Mkomazi River, and dams in the Dargle area, all in KwaZulu-Natal Province, South Africa.

*Significant sightings:* In 1962, a game ranger named Buthelezi saw a horse-headed animal lying on a sandbank in the Umgeni River.

Caretaker Johannes Hlongwane saw the Howick monster twice, both times in misty conditions, in 1974 and 1981.

In September 1995, restaurant owner Bob Teeney saw a large, serpentine animal from the viewing platform at Howick Falls. Teeney offered a reward to anyone who could produce a photo of the animal, which created much media interest.

*Possible explanations:*

(1) The Nile crocodile (*Crocodylus niloticus*) is found as far south as northern Natal, but it doesn't look particularly serpentine.

(2) The African longfin eel (*Anguilla mossambica*) grows up to 5 feet long, is olive to grayish-black, and has a long dorsal fin. It is found in eastern rivers of South Africa.

(3) The Giant mottled eel (*A. marmorata*) also has a long dorsal fin, grows up to 6 feet, and lives in rocky pools in freshwater rivers of South Africa.

(4) A hoax to promote tourism.

*Sources:* "Of Ducks and Plesiosaurs: Howick Falls' Monster," *Cryptozoology Review* 1, no. 2 (Autumn 1996): 9; Sian Hall, "Legend of the Falls," *Fortean Times,* no. 123 (June 1999): 42–44.

## INSECTIVORES (Unknown)

Insectivores are an odd lot of small mammals with primitive features and an uncertain lineage. Included in this grouping are hedgehogs, tenrecs, shrews, moles, golden moles, solenodons, and nesophontids. All share a simplified gut, a reduced area of contact between the pubic bones, and small teeth with pointed cusps. However, a common insectivore ancestor has not been identified in Cretaceous fossils.

Of the four cryptids in this group, two are from Madagascar, one is from South America, and the last is an entire family from the West Indies.

### Mystery Insectivores

FONTOYNONT'S TENREC; KAVAY; MACAS MAMMAL; NESOPHONTID INSECTIVORES

## INVERTEBRATES (Unknown)

Most unknown animals of interest to cryptozoology are vertebrates, or animals with backbones. Invertebrates encompass more than thirty phyla, ranging from sponges and jellyfish to worms, insects, spiders, mollusks, and starfish. Animals without backbones attract our attention when they are of unusual size or found in odd environments; this is especially true of the CEPHALOPODS (a class of mollusks containing octopuses and squids), which have been placed in their own section. The remaining thirteen invertebrate cryptids are in four phyla.

In the Phylum Hemichordata (acorn worms), the giant ACORN WORMS are the undiscovered adults of a known species of larvae, while the LOPHENTEROPNEUST belongs to an undescribed deep-sea group. In the Phylum Cnidaria (corals and jellyfish) are the GIANT JELLYFISH and the CUERO (which may also be an octopus). The Phylum Arthropoda (animals with jointed legs) includes the sea spiders (DEEP-SEA SPIDER), myriapods (GIANT CENTIPEDE), arachnids (GIANT SPIDER), insects (GOLDEN ANT and giant MADAGASCAN HAWK MOTH), and crustaceans (MAGGOT and SPECS). The Phylum Annelida (segmented worms) may be appropriate for the MONGOLIAN DEATH WORM (more likely a snake or lizard) and THE THING (a polychaete worm).

## Mystery Invertebrates

ACORN WORMS (GIANT); CUERO; DEEP-SEA SPIDER; GIANT CENTIPEDE; GIANT JELLYFISH; GIANT SPIDER; GOLDEN ANT; LOPHENTEROPNEUST; MADAGASCAN HAWK MOTH (GIANT); MAGGOT; MONGOLIAN DEATH WORM; SPECS; THE THING

## Ipupiara

MERBEING of South America.

*Etymology:* Guarani (Tupí), "water dweller."

*Variant names:* Hipupiara, Iara (for the female), Igpupiara, Oyara, Uiara.

*Physical description:* Covered in short brown hair. Females have long, beautiful head-hair. Deep, sunken eyes. Whiskers. Blowhole is at the back of the head. Tail fins.

*Behavior:* Moans. Kills victims by constriction. Eats only the eyes, nose, tips of toes and fingers, and genitals.

*Distribution:* Coast of Brazil.

*Significant sightings:* In 1554, Baltasar Ferreira encountered an Ipupiara moving along the beach near São Vicente, Brazil. He killed it with his sword, although it put up some resistance. It was completely covered with hair and had whiskers and tail fins.

Anatomist Pieter Pauw (1564–1617) dissected a merman brought to him by merchants of the Dutch East Indies Company, who had allegedly captured it off the coast of Brazil. The corpse had a human head and torso, but the lower extremity was a shapeless, tailless mass of flesh. One hand and some ribs wound up in Danish physician Thomas Bartholin's "cabinet of curiosities," an early museum of anatomical and zoological oddities. The hand's fingers were webbed, and the knucklebones were robust.

*Possible explanations:*

(1) The Boto dolphin (*Inia geoffrensis*) is found in the Amazon and Orinoco River basins. Its flipper might be mistaken for a human hand, according to Michel Meurger. In Amazonian myth, the Boto is said to be able to interbreed with humans, producing hybrids.

(2) The merman corpse may have been a genetically malformed human suffering from sirenomelia, in which the limbs are fused throughout their length and no separate feet are present.

*Sources:* Pero de Magalhães de Gandavo, *Historia da provincia sacta Cruz a que "vulgar mete" chamamos Brasil* (Lisbon: Antonio Gonsalves, 1576); Jean de Léry, *Histoire d'un voyage fait en la terre du Bresil* (La Rochelle, France: Antoine Chuppin, 1578); Fernão Cardim, *Tratados da terra e gente do Brasil* [1585] (Lisbon: Commissão Nacional para as Comemorações dos Descobrimentos Portugueses, 1997); Joannes de Laet, *Novus orbis, seu, Descriptionis Indiae Occidentalis, libri XVIII* (Leiden, the Netherlands: Elzevirios, 1633), p. 508; Thomas Bartholin, *Historiarum anatomicarum rariorum centuria I et II* (Copenhagen: Academicis Martzani, 1654), pp. 186–191; Luis da Camara-Cascudo, "Los mitos de las aguas del Brasil," *Annuario de la Sociedad Folklorica de Mexico* 5 (1945): 14–15; Michel Meurger and Claude Gagnon, *Lake Monster Traditions: A Cross-Cultural Analysis* (London: Fortean Tomes, 1988), pp. 199–205.

## Irish Deer

Mystery deerlike HOOFED MAMMAL of Ireland.

*Physical description:* Large, black deer.

*Distribution:* Ireland.

*Significant sighting:* A black deer was hunted by ancient Irish tribes who used its skin for clothing and its meat and milk for food. Centuries ago, a human body was found in gravel under 11 feet of peat and completely clothed in antique garments of hair, said to be that of the legendary deer.

*Possible explanations:*

(1) Surviving giant Irish deer (*Megaloceros giganteus*), which stood 4 feet 6 inches–5 feet at the shoulder and carried enormous antlers that attained a span of nearly 12 feet and weighed up to 100 pounds. Irish deer skulls found at Lough Gur, Limerick County, in 1846 seem to have been tampered with by humans; however, this could have occurred at any time during the recent past. Evidence exists elsewhere that this huge deer persisted beyond its extinction in Ireland 10,500 years ago, in the Late Pleistocene. Fossils on the Isle of Man have recently been dated at 9,200 years ago. *See also* SCHELCH.

(2) By contrast, some of the Irish skulls may actually belong to Moose (*Alces alces*) that formerly existed in Ireland.

(3) A large, black-coated variety of Red deer (*Cervus elaphas*) might have been hunted by the ancient Irish.

*Sources:* Letter from Countess of Moira, in Edward Lhuyd, *Archaeologia Britannica* (Oxford: Bateman, 1707); H. D. Richardson, *Facts Concerning the Natural History, &c. of the Gigantic Irish Deer (Vervus Giganteus Hibernicus)* (Dublin: J. M'Glashan, 1846), p. 25; Philip H. Gosse, *The Romance of Natural History, Second Series* (London: James Nisbet, 1861), pp. 46–52; Karl Shuker, *In Search of Prehistoric Survivors* (London: Blandford, 1995), pp. 167–169; Silvia Gonzalez, Andrew C. Kitchener, and Adrian M. Lister, "Survival of the Irish Elk into the Holocene," *Nature* 405 (2000): 753–754.

## Irish Wildcat

Undescribed small CAT of Ireland.

*Physical description:* Twice the size of a domestic cat. Dirty-gray color. Tapering tail.

*Distribution:* Western and northern Ireland.

*Significant sightings:* William Thompson examined a large cat weighing 10 pounds 9 ounces that had been shot in the early nineteenth century in Shane's Castle Park, near Randalstown, County Antrim, Northern Ireland. It had a pointed tail, not bushy at the tip.

In 1883, W. B. Tegetmeier exhibited a specimen at the Zoological Society of London that looked like a domestic cat but had a distinctive tail and feet. It had been obtained in County Donegal.

Subfossil remains of what were at first thought to be wildcats were found in Edenvale, Newhall, and Barntick Caves, County Clare, in 1904. These, however, turned out to be domestic cats from the Irish Bronze Age, 2000–500 B.C. In 1965, A. W. Stelfox reported that these were the earliest known cat remains in Ireland.

In the summer of 1968, Lionel Leslie and others saw a large cat from a distance of 100 yards near Lough Nahooin, County Galway.

*Present status:* Only a handful of sightings are on record.

*Possible explanations:*

(1) Feral Domestic cats (*Felis silvestris catus*) are the likeliest candidates, rather than a native Irish wildcat.

(2) Introduced African wildcat (*F. s. lybica*), the wild ancestor of domestic cats.

(3) Thompson thought the Shane's Castle cat was a European wildcat × Domestic cat hybrid, though there would have to be a native wildcat population for this to occur.

*Sources:* William Thompson, *The Natural History of Ireland*, vol. 4, *Mammalia* (London: H. G. Bohn, 1856); Robert Francis Scharff, "On the Former Occurrence of the African Wild Cat (*Felis ocreata*, Gmel.) in Ireland," *Proceedings of the Royal Irish Academy* 26, sect. B (1906): 1–12; William Hamilton Maxwell, *Wild Sports of the West* (New York: Frederick A. Stokes, 1915); A. W. Stelfox, "Notes on the Irish 'Wild Cat,'" *Irish Naturalists' Journal* 15 (1965): 57–60; F. W. Holiday, *The Goblin Universe* (St. Paul, Minn.: Llewellyn, 1986), pp. 130–131; Karl Shuker, *Mystery Cats of the World* (London: Robert Hale, 1989), pp. 84–89; Jonathan Downes, *The Smaller Mystery Carnivores of the Westcountry* (Exwick, England: CFZ Publications, 1996).

## Irizima

Dinosaur-like animal of Central Africa, similar to the MOKELE-MBEMBE.

*Etymology:* Unknown, "the thing that may not be spoken of."

*Physical description:* Larger than a hippopotamus. Black. Long neck. Said to have rhinolike horns.

*Behavior:* Produces 3-foot-high waves in the water with its breathing.

*Habitat:* Swamps.

*Distribution:* Lake Edward, in both the Democratic Republic of the Congo and Uganda.

*Possible explanation:* A waterspout, suggested by E. A. Temple-Perkins.

*Sources:* Fulahn [William Hichens], "On the Trail of the Brontosaurus: Encounters with Africa's Mystery Animals," *Chambers's Journal,* ser. 7, 17 (1927): 692–695; Roger Courtney, *Africa Calling* (London: Harrap, 1935), p. 200; Eric Arnold Temple-Perkins, *Kingdom of the Elephant* (London: Andrew Melrose, 1955), pp. 232–233.

## Irkuiem

Unknown BEAR of Siberia.

*Etymology:* Koryak or Chukot (Chukotko-Kamchatkan), "trousers pulled down."

*Variant names:* Irquiem, Kainyn-kutkho ("god-bear").

*Physical description:* General shape of a polar bear. Shoulder height, 4 feet 7 inches. Weight, more than 2,000 pounds. White coat. Narrow body. Small head. Long forelegs. A bulge of fat hangs down between the short hind legs.

*Behavior:* Does not run but is said to move by throwing down its front legs and heaving the hind legs forward to meet them. Said to cross the Chukchi Sea to Alaska via ice floes.

*Habitat:* Tundra.

*Distribution:* Olyutorskiy, Karaginskiy, and Tigil' areas of the Kamchatka Peninsula, Koryak Autonomous Province, Siberia.

*Significant sightings:* Soviet hunter Rodion Sivolobov collected eyewitness accounts of the Irkuiem in the 1970s and 1980s, and in the spring of 1987, he obtained a skin. A photo of this skin (or another one like it) was examined by Valerii Orlov, who thought it was that of a Brown bear (*Ursus arctos*). Reindeer hunters are said to have killed specimens in 1976, 1980, and 1982.

*Possible explanations:*

(1) The fossil Short-faced bear (*Arctodus simus*) was 9 feet 6 inches long and 5 feet 7 inches high at the shoulder. An active carnivore with a short, broad muzzle, *Arctodus* could run swiftly and was a fearsome predator. It lived 44,000–11,500 years ago and ranged from Alaska to Mexico. First suggested as a possibility by Soviet zoologist Nikolai K. Vereshchagin. However, the Irkuiem's peculiar locomotion does not match the swift movements of *Arctodus.*

(2) A stray Polar bear (*Ursus maritimus*) or an isolated population of these bears on Kamchatka, suggested by Valerii Orlov.

*Sources:* "Giant Bear Sought by Soviets," *ISC Newsletter* 6, no. 4 (Winter 1987): 6–7; Andrew D. Gable, "Bergman's Bear," December 19, 2000, http://www.cryptozoology.com/cryptids/godbear.php.

## Isiququmadevu

Dinosaur-like animal of Central Africa, similar to the MOKELE-MBEMBE.

*Etymology:* Lozi (Bantu) word.

*Variant names:* Ing'ondotuya, Lengolengole, Lingongole.

*Physical description:* Length, 20–40 feet. Larger than an elephant. Head is like a snake's. Long neck. Lizardlike legs.

*Behavior:* Amphibious.

*Tracks:* Makes wide furrows in the reeds and mud.

*Habitat:* Swamps, rivers.

*Distribution:* Zambezi River from the Barotse Floodplain to Victoria Falls, Zambia.

*Significant sightings:* Lewanika, king of Barotseland in western Zambia in the early twentieth century, went to the spot where a huge, aquatic reptile had been seen. He found a large space where reeds had been flattened and a channel as large as a trek wagon where it had crawled through the mud.

In the southern summer of 1925, a river-transport manager named V. Pare saw a 30- to 40-foot, snakelike animal with a slate-gray head resting on a rock in the flooded Zambezi River near Victoria Falls, Zambia. It disappeared into a deep cave.

E. C. Saunders watched two animals with long necks in the Zambezi River near Katombora, Zambia, in January 1960. He estimated they were 20–25 feet long and did not look like pythons.

*Possible explanations:*

(1) An outsize African clawless otter (*Aonyx capensis*) may be the source of Pare's 1925 sighting, according to Bernard Heuvelmans. It grows to 5 feet in length, including its tail.

(2) Two African rock pythons (*Python sebae*) in coitus probably explain the Saunders sighting.

(3) A surviving sauropod dinosaur.

*Sources:* David Livingstone, *Missionary Travels and Researches in South Africa* (London: John Murray, 1857), p. 517; John G. Millais, *Far Away up the Nile* (London: Longmans, Green, 1924), pp. 61–67; Vernon Brelsford, "Some Northern Rhodesian Monsters," *African Observer* 4, no. 6 (1936): 58–60; William Hichens, "African Mystery Beasts," *Discovery* 18 (1937): 369–373; "Le monstre des chutes Victoria," *Atlas,* June 1963, p. 108; Bernard Heuvelmans, *Les derniers dragons d'Afrique* (Paris: Plon, 1978), pp. 221–229, 387–388.

## Isnachi

Mystery PRIMATE of South America.

*Etymology:* Quechua (Quechuan), "strong man."

*Variant names:* Camuenare (Amuesha/Arawakan, "father of the monkeys"), Maemi (Machiguenga/Arawakan), Majero (Yine/Arawakan), Maquisapa maman (Spanish, "mother of the spider monkeys").

*Physical description:* Height, 4 feet, or about twice the size of a spider monkey. Covered in short, thick, black or dark-brown hair. Muscular. Black face. Snout is like a mandrill's. Long teeth. Barrel-chested. Thick arms. Hands have nails, not claws. Huge thighs. Thick tail, 6 inches long.

*Behavior:* Arboreal. Usually solitary but is said to travel in groups of up to twenty. Travels with spider monkey troops. Attacks by running on its hind legs. Feeds on wild fruits and the shoots of the Chonta palm (*Euterpe precatoria*), which it rips apart in a characteristic way. Makes platforms in trees for resting.

*Habitat:* Mountainous forest at altitudes of 1,600–5,000 feet.

*Distribution:* Peru, from Loreto Department in the north, through Yanachaga-Chemillén National Park, to the Cordillera Urubamba.

*Significant sighting:* Ecuadorean botanist Benigno Malo saw a large, black ape along the Ecuador-Peru border in 1985 and managed to take a photograph before it moved away. The location of the photo is currently unknown.

*Possible explanations:*

(1) The Spectacled bear (*Tremarctos ornatus*) is black and lives in the area. It occasionally climbs trees to reach fruit. However, its white eye rings would be hard to mistake.

(2) An unknown species of monkey.

*Sources:* Peter J. Hocking, "Large Peruvian Mammals Unknown to Zoology," *Cryptozoology* 11 (1992): 38–50; Peter J. Hocking, "Further Investigation into Unknown Peruvian Mammals," *Cryptozoology* 12 (1996): 50–57.

## Issie

FRESHWATER MONSTER of Japan.

*Physical description:* Length, 16–90 feet. Black, possibly striped. Two humps.

*Distribution:* Lake Ikeda, Kagoshima Prefecture, Kyushu, Japan.

*A model of ISSIE, the lake monster of Lake Ikeda, Japan. (Shin-ichiro Namiki/Fortean Picture Library)*

*Significant sightings:* On December 16, 1978, Toshiaki Matsubara saw a strange whirlpool in Lake Ikeda and took a series of photos of an animal with humps.

A nine-minute video of a long, dark object with two humps was taken January 4, 1991, by Hideaki Tomiyasu. The object submerged when a motorboat passed.

*Sources: Straits Times* (Singapore), October 2, 1978; Simon Welfare and John Fairley, *Arthur C. Clarke's Mysterious World* (London: Collins, 1980), pp. 107–108; Kenji Chōno, "Issie of Japan's Lake Ikeda," *Elsewhen* 2, no. 4 (1991): 9; "Long, Dark and Humpsome," *Fortean Times,* no. 61 (February-March 1992): 13; Kyoichi Tsuzuki, *Roadside Japan* (Tokyo: Aspect, 1997), pp. 256–257.

## Isturitz Scimitar Cat

Mystery big CAT of Western Europe.

*Physical description:* Short body. Spotted coat. Sheathed incisors. Robust lower jaw. Long limbs. Short tail.

*Distribution:* Southwestern France.

*Significant sighting:* In 1896, a 6.5-inch-long Upper Paleolithic stone statuette of a big cat with a short tail was discovered in a cave near Isturitz in the Pyrénées Mountains of France.

*Present status:* Extinct but of interest because it may have been contemporaneous with modern humans.

*Possible explanations:* It was long thought that the statuette showed a European cave lion (*Panthera leo spelaea*), which has also been depicted in cave paintings. Vratislav Mazak believes it more likely represents a Scimitar toothed cat (*Homotherium latidens*), a Pleistocene sabretooth, because of its tail and jaw. If so, this cat would have survived until around 25,000 years ago and not died out some 200,000 years ago as is currently thought.

*Sources:* Vratislav Mazak, in *Zeitschrift für Säugetierkunde* 35 (1970): 359–362; Michel Rousseau, in *Archéologia* 40 (May-June 1971): 81–82; Michel Rousseau, in *Mammalia* 35 (December 1971): 648–657; Karl Shuker, *Mystery Cats of the World* (London: Robert Hale, 1989), pp. 90–91.

## Itzcuintlipotzotli

Mystery DOG of Mexico.

*Etymology:* Nahuatl (Uto-Aztecan), "hunchback dog," from *itzcuintli* ("dog") + *potzotli* ("hunchback").

*Physical description:* Size of a small dog. Black, brown, and white spots. Small, wolflike head. Short neck. Lumpy muzzle. Small, hanging ears. Fatty hump extends the length of its back. Forelegs shorter than the hind legs.

*Distribution:* Michoacán State, Mexico.

*Significant sighting:* Frances Calderón de la Barca saw a dead specimen hanging from a hook near the door of an inn in the Guajimalco Valley.

*Sources:* Francesco Saverio Clavigero, *Historia antigua de México* [1780] (Mexico City: Editorial Porrúa, 1945); Frances Calderón de la Barca, *Life in Mexico, during a Residence of Two Years in That Country* (London: Chapman and Hall, 1843).

## Ivory-Billed Woodpecker

Large BIRD of the Woodpecker family (Picidae) in the southern United States and Cuba, presumed extinct.

*Scientific names: Campephilus principalis principalis* (in United States), given by Carl von Linné in 1758; *C. p. bairdii* (in Cuba), given by John Cassin in 1863.

*Physical description:* Length, 20 inches. Tall, scarlet crest (males); black crest (females). White bill. White stripes on either side of the neck. Large patches of white on the wings.

*Behavior:* Feeds on wood-boring beetle larvae that infest recently dead trees. Level flight.

*Habitat:* Tall bottomland, swamp forest.

*Distribution:* Historical range was from eastern Texas to North Carolina and north in the Mississippi Valley to Missouri, southern Illinois, and southern Indiana. Scattered sightings since 1966 have been claimed in Texas, Louisiana, Mississippi, Florida, and Cuba.

*Significant sightings:*

*In the United States*—John V. Dennis observed Ivory-bills in the Neches River valley, Texas, on December 10, 1966, and February 19, 1967.

Wildlife artist Frank Shields saw individual

The IVORY-BILLED WOODPECKER (Campephilus principalis), presumed extinct in the United States since the 1960s. (© 2002 ArtToday.com, Inc., an IMSI Company)

Ivory-bills near Interlachen, Florida, on April 4 and 15, 1969; on June 11, he found a distinctive, black-and-white feather that he identified as belonging to an Ivory-bill.

In May 1971, a pair of Ivory-bills was allegedly seen and one of them photographed by an amateur birder in the Atchafalaya River area, Louisiana. However, some have said the photo shows a mounted museum specimen.

In 1987, Jerome Jackson heard a bird respond to his Ivory-bill recordings north of Vicksburg, Mississippi, but he did not see it.

On April 1, 1999, zoology student David Kulivan saw a pair of Ivory-bills at close range in the Pearl River Wildlife Management Area, Louisiana. Members of an expedition to the area in the winter of 2002 heard and recorded the bird's distinctive rapping but made no sightings; however, Cornell University ornithologists confirmed in June 2002 that the sounds were made by distant gunshots.

*In Cuba*—In 1948, John V. Dennis and Davis Crompton discovered a population in the Cuchillas del Toa Range, and in 1956, George Lamb found six groups there. Since the 1959 Cuban Revolution, the status of the species is uncertain.

In 1985, Lester L. Short found indirect evidence of the Ivory-bill in the Cupeyal Reserve, and on April 16, 1986, he saw a male Ivory-bill in flight at a distance of only 18 feet.

Giraldo Alayón and Alberto Estrada found traces in Ojito de Agua in 1986. On the afternoon of March 16, 1987, the last positive record of the species was recorded in the Cuchillas del Toa Mountains by Alayón and Aimé Pasada when they saw a female woodpecker flying at a distance of about 600 yards.

Members of a 1988 National Geographic expedition, which included Ted Parker and Jerome Jackson, could not find the species, although one individual might have been glimpsed. Unsuccessful searches were conducted in 1991, 1992, and 1993, but in 1998 and 1999, new evidence indicating the bird's presence was discovered in the Sierra Maestra.

*Present status:* A major decline, associated with the cutting of lowland hardwood forests, began in the United States around 1885 and continued until the 1920s. Considered extinct in the United States by the 1960s and in Cuba by 1990.

*Possible explanation:* The Pileated woodpecker (*Dryocopus pileatus*) is slightly smaller and much more common in the United States. It has a dark bill and an undulating flight pattern.

*Sources:* John V. Dennis, "A Last Remnant of Ivory-Billed Woodpeckers in Cuba," *Auk* 65 (1948): 497–507; John V. Dennis, "Return of the Ivory-Bill," *Animals* 10 (March 1968): 492–497; "An Ivory-Billed Woodpecker," *Pursuit,* no. 7 (July 1969): 49; John V. Dennis, "The Ivory-Billed Woodpecker, *Campephilus principalis,*" *Avicultural Magazine* 85 (1979): 75–84; "Ivory-Billed Woodpecker Found Alive in Cuba," *ISC Newsletter* 5, no. 2 (Summer 1986): 3–5; Martjan Lammertink, "No More Hope for the Ivory-Billed Woodpecker," *Cotinga,* February 1995, at http://www.neotropicalbirdclub.org/feature/ivory.html;

Christopher Cokinos, *Hope Is the Thing with Feathers* (New York: Jeremy P. Tarcher, 2000), pp. 59–117; Orlando H. Garrido and Arturo Kirkconnell, *Field Guide to the Birds of Cuba* (Ithaca, N.Y.: Cornell University Press, 2000), p. 152; Karl Shuker, "Woodpecker Discovery?" *Fortean Times,* no. 139 (November 2000): 23; Errol Fuller, *Extinct Birds* (Ithaca, N.Y.: Cornell University, 2001), pp. 267–274; Scott Weidensaul, *The Ghost with Trembling Wings* (New York: North Point Press, 2002), pp. 45–64; Chester Moore Jr., "High Strangeness Report: Is the Ivory-Billed Woodpecker Extinct?" 2002, on line at http://www.anomalist.com/reports/woodpecker2.html; James Gorman, "Listening for the Call of a Vanished Bird," *New York Times,* March 5, 2002, p. F1; James Gorman, "Faint Hope for Survival of a Woodpecker Fades," *New York Times,* June 10, 2002, p. A14.

# J

## Jacko

Alleged small BIGFOOT captured in western Canada.

*Physical description:* Humanlike. Height, 4 feet 7 inches. Weight, 127 pounds. Covered in glossy hair 1 inch in length. Long, black head-hair. Hands and feet are hairless. Forearms are longer than a human's.

*Behavior:* Makes a half-bark, half-growl noise. Eats berries and milk.

*Distribution:* Yale, British Columbia.

*Significant sighting:* On June 30, 1884, train-men of the British Columbia Express line 20 miles north of Yale saw a wildman lying close to the tracks. They blew the whistle and applied the brake, and the creature jumped up and climbed a steep bluff. Conductor C. J. Craig and others gave chase and after five minutes trapped it on a ledge. Craig threw a rock at it and knocked it out, allowing them to haul it back down with a rope. After reaching Yale, the wildman, whom they started calling Jacko, was kept for a few days by George Tilbury. A report that Jacko had been sent to the jail at New Westminster turned out to be false.

*Present status:* Grover Krantz has suggested that Jacko was acquired by P. T. Barnum and exhibited as "Jo-Jo the Dog-Faced Boy" in his circus, beginning in 1884. However, it is fairly well established that Jo-Jo was a Russian man, Fedor (or Theodor) Jeftichew, born in 1868 and afflicted with hypertrichosis, which caused him to have long, silky facial hair.

*Possible explanation:* Probable hoax, based on later newspaper accounts.

*Sources:* "What Is It? A Strange Creature Captured above Yale," *Victoria (B.C.) Daily Colonist,* July 4, 1884; Rex, "The 'What Is It,'" *New Westminster (B.C.) Mainland Guardian,* July 9, 1884; "The Wild Man," *New Westminster British Columbian,* July 11, 1884; Ivan T. Sanderson, *Abominable Snowmen: Legend Come to Life* (Philadelphia: Chilton, 1961), pp. 23–29; John Green and Sabina W. Sanderson, "Alas, Poor Jacko," *Pursuit,* no. 29 (January 1975): 18–19; Grover S. Krantz, *Big Foot-Prints* (Boulder, Colo.: Johnson, 1992), pp. 202–204.

## Jago-Nini

Dinosaur-like animal of Central Africa, similar to the MOKELE-MBEMBE.

*Etymology:* Possibly Punu or Sira (Bantu), "giant diver."

*Behavior:* Amphibious. Feeds on West African manatees (*Trichechus senegalensis*). Said to attack and eat humans.

*Habitat:* Swamps, rivers.

*Distribution:* Gabon.

*Source:* Trader Horn, *Life and Works,* ed. Ethelreda Lewis (London: Jonathan Cape, 1927), vol. 1, pp. 272–273.

## Jaguareté

Mystery CAT of South America.

*Etymology:* From the Guaraní (Tupí) *yaguarete* ("great beast"). The common jaguar is also known by this name in Brazil.

*Variant name:* Cougar noire.

*Physical description:* Like a jaguar, except for black coloration on the head, back, sides, and tail. Near white underparts, lower jaw, and paws.

*Behavior:* Eats lizards, alligators, fishes, turtle eggs, and the buds and leaves of the Pricklypear cactus (*Opuntia* spp.).

*Habitat:* Favors the seashore.

*Distribution:* Brazil; Guyana.

*Significant sighting:* Thomas Pennant claims that two Jaguaretés were exhibited in London in the eighteenth century.

*Possible explanations:*

(1) Melanistic Jaguars (*Panthera onca*) are known, but the black color is also found on the underparts. The spots are always faintly visible at certain angles to the light.

(2) A black-and-tan or pseudomelanistic jaguar morph, as suggested by Karl Shuker, might explain the dual coloration. However, no modern instances have been reported.

*Sources:* Thomas Pennant, *History of Quadrupeds,* vol. 1 (London: B. White, 1781); Thomas Bewick, *A General History of Quadrupeds,* 5th ed. (London: E. Walker, 1807); Karl Shuker, *Mystery Cats of the World* (London: Robert Hale, 1989), pp. 191–192.


## Japanese Hairy Fish

Mystery FISH of Japan.

*Physical description:* Scaly, fishlike body. Length, 4–5 feet. Humanlike hair on the head.

*Behavior:* Aggressive. Emerges from the water to fight or play. Emits loud cries. Said to attack and kill humans by disemboweling them.

*Distribution:* Unspecified river in Japan.

*Possible explanation:* Distorted account of the Northern fur seal (*Callorhinus ursinus*), suggested by David Heppell. It is about the right size, has finlike flippers, barks like a dog, and can be very playful. The Kuril Islands serve as a breeding grounds from May to June; little is known of the animal's whereabouts outside the breeding season.

*Sources:* Frederic Shoberl, ed., *The World in Miniature: Japan* (London: R. Ackermann, 1823); Karl Shuker, "Hairy Reptiles and Furry Fish," *Strange Magazine,* no. 18 (Summer 1996): 26–27.


## Jenny Haniver

Fabricated MERBEING.

*Etymology:* Possibly after the cities where they were most often manufactured—Genoa (Jenny, in nautical slang) and Antwerp (Anvers, in

French)—or from the French *jeune fille d'Anvers* ("young girl of Antwerp").

*Variant names:* Diable de mer (French), Garadiavolo (in Puerto Rico), SEA MONK.

*Physical description:* Shriveled sea creature with a vaguely humanoid, fish-tailed appearance.

*Present status:* Still found in curio shops.

*Possible explanation:* Dried-out and varnished ray, skate, or guitarfish (Suborder Rajoidei) with its fins cut to resemble wings and arms, its neck constricted, its nostrils doctored to look like eyes, its claspers shaped as hind legs, and the tail twisted capriciously. Manufactured as a seaside curiosity since the sixteenth century.

*Sources:* Gilbert P. Whitley, "Jenny Hanivers," *Australian Museum Magazine* 3 (1928): 262–264; E. W. Gudger, "Jenny Hanivers, Dragons, and Basilisks in the Old Natural History Books and in Modern Times," *Scientific Monthly* 38 (1934): 511–523; Willy Ley, *The Lungfish, the Dodo, and the Unicorn* (New York: Viking, 1948), chap. 4; Richard Ellis, *Monsters of the Sea* (New York: Alfred A. Knopf, 1994), pp. 82–85; Rosamond Purcell, *Special Cases: Natural Anomalies and Historical Monsters* (San Francisco, Calif.: Chronicle Books, 1997), pp. 81–88.


## Jersey Devil

FLYING HUMANOID of the eastern United States.

*Variant name:* Leeds devil.

*Physical description:* Length, 3–11 feet. Alligator skin. Head like that of a horse, ram, or dog. Horns. Large eyes. Batlike wings, 2 feet wide. Small front legs with paws. Cloven hooves. Long tail.

*Behavior:* Loud nocturnal cry like a squawk, whistle, moo, or screech. Said to spew flames from its mouth and glow in the dark. Foul smell. Kills livestock and dogs.

*Tracks:* Length, 3 inches. Width, 2 inches. Like hooves or horseshoe prints.

*Habitat:* Pine woods.

*Distribution:* Southern New Jersey; eastern Pennsylvania; Delaware.

*Significant sightings:* Bristol, Pennsylvania, postmaster E. W. Minster, John McOwen, and police officer James Sackville all separately saw a

*Artist's conception of the* Jersey devil. *(Richard Svensson/Fortean Picture Library)*

winged, screaming creature on January 17, 1909. Minster described it as resembling a large crane with a head like a ram's, long thin wings, and short legs. On January 21, something apparently left hoofprints in the snow in Trenton, New Jersey, near the state arsenal building and in the yard of City Councilman E. P. Weeden. These events initiated a weeklong spate of sightings in New Jersey, Pennsylvania, and Delaware. Schools and businesses were shut down, posses roamed the pine woods, and trolley drivers in Trenton and New Brunswick armed themselves against an attack.

A minor set of sightings occurred in late November 1951 in the Gibbstown, New Jersey, area when a ten-year-old boy collapsed after seeing a horrible-looking creature outside the Dupont Clubhouse.

Park ranger John Irwin saw a 6-foot-tall biped with black fur step in front of his car as he drove along the Mullica River in the Wharton State Forest, New Jersey, in mid-December 1993. Its deerlike head had piercing red eyes.

In late 1995, Sue Dupre was driving near Pompton Lakes, New Jersey, when a hopping animal with an armadillo-like face raced across the highway.

*Present status:* Said to be the devilish offspring of a Mrs. Leeds (or Shrouds) of Leeds Point or Burlington, New Jersey, in about 1735, this creature only came into prominence in 1909, when a wave of sightings made headlines. Nowadays, any New Jersey cryptid, from an EASTERN PUMA to a HAIRY BIPED, is designated a Jersey devil by the media.

*Possible explanations:*

(1) An elaborate hoax in 1909 designed to lower real estate prices and create a buyer's market. Ivan T. Sanderson claimed to have discovered the fake feet used to make the footprints in the snow. Jersey devil pranks and hoaxes have been frequent over the subsequent years.

(2) A stray Sandhill crane (*Grus canadensis*) would be tall and weird-looking, but this bird no longer winters in the Pine Barrens.

The Great blue heron (*Ardea herodias*) is about the same size, however, and is a yearlong New Jersey wetlands resident.

(3) The nocturnal calls might originate from a Red fox (*Vulpes fulva*), an Eastern screech-owl (*Otus asio*), a Long-eared owl (*Asio otus*), or from ice breaking on rivers.

(4) A surviving pterosaur, a fossil FLYING REPTILE that supposedly died out at the end of the Cretaceous period, 65 million years ago.

(5) Some sightings might be attributable to a kangaroo, though no escapees were reported from local zoos.

*Sources:* Henry Charlton Beck, "The Jersey Devil and Other Legends of the Jersey Shore," *New York Folklore Quarterly* 3 (1947): 102–106; Jeremiah J. Sullivan and James F. McCloy, "The Jersey Devil's Finest Hour," *New York Folklore Quarterly* 30 (1974): 231–238; James F. McCloy and Ray Miller Jr., *The Jersey Devil* (Wallingford, Pa.: Middle Atlantic Press, 1976); William H. McMahon, *Pine Barrens Legends, Lore and Lies* (Wallingford, Pa.: Middle Atlantic Press, 1980), pp. 36–39; James Pontolillo, "An Interpretation of the Jersey Devil," *INFO Journal*, no. 57 (July 1989): 17–19; Loren Coleman, "Jersey Devil Walks Again," *Fortean Times*, no. 83 (October-November 1995): 49; James F. McCloy and Ray Miller Jr., *Phantom of the Pines: More Tales of the Jersey Devil* (Moorestown, N.J.: Middle Atlantic Press, 1998); Loren Coleman, *Mysterious America*, rev. ed. (New York: Paraview, 2001), pp. 232–244.

## Jez-Tyrmak

GIANT HOMINID of Central Asia.

*Etymology:* Mongolian (Altaic), "copper fingernails."

*Variant names:* Dzehez-tyrmak, Zes tyrmak.

*Physical description:* Covered in long, shaggy hair. Color is dark gray to black. Copper-colored fingernails.

*Distribution:* Tibetan Plateau, Tibet.

*Source:* Ivan T. Sanderson, *Abominable Snowmen: Legend Come to Life* (Philadelphia: Chilton, 1961), p. 321.

## Jhoor

Unknown LIZARD of the Indian subcontinent.

*Physical description:* Length, 20 feet.

*Behavior:* Amphibious. Lives symbiotically with the Saltwater crocodile (*Crocodylus porosus*).

*Distribution:* Gir National Park, Kathiawar Peninsula, Gujarat State, India; Sundarbans, Bangladesh.

*Source:* E. B. Fox, "The Mysterious 'Jhoor,'" *Journal of the Bombay Natural History Society* 27 (1920): 175–176.

## Jingara

LITTLE PEOPLE of Australia, sometimes confused with the YOWIE.

*Etymology:* After Jingera Mountain region, New South Wales.

*Variant names:* Janjurrie, Jingera, Jongari.

*Physical description:* Covered with 2-inch, black hair. Height, 3 feet. A larger variety may exist. Long, pointed muzzle.

*Behavior:* Bipedal but ambles.

*Tracks:* Bearlike.

*Distribution:* New South Wales; Queensland.

*Significant sightings:* In a study of Aboriginal dialects in Cooma, New South Wales, in 1904, Jingara was said to refer to a mountain haunted by a hairy man.

In December 1999, Allan H. Bucholz saw a bearlike creature as he was driving his tractor near Gayndah, Queensland. It had a long snout like a wallaby, but it lacked a long tail. Two days later, his sister, Shirley Humphries, saw a similar animal ambling across a sandbank by a river. Sam Hill, a local man who was part Aborigine, declared that it was a Jongari.

*Sources:* Graham Joyner, *The Hairy Man of South Eastern Australia* (Kingston, A.C.T., Australia: Graham Joyner, 1977), p. 4; Tony Healy and Paul Cropper, *Out of the Shadows: Mystery Animals of Australia* (Chippendale, N.S.W., Australia: Ironbark, 1994), p. 115; "Mysterious Hairy Monster Spotted in Oz," *BBC Newsround*, February 8, 2000; "Bear-Like Creature Sighted," *Fortean Times*, no. 136 (August 2000): 11; Malcolm Smith, "Update on the Jongari," *North American BioFortean*

*Review* 3, no. 2 (October 2001): 18–19, http://www.strangeark.com/nabr/NABR7.pdf.

## Jipijkmak

MERBEING of eastern Canada.

*Etymology:* Passamaquoddy (Algonquian), "horned serpent people."

*Variant names:* Jipijkma or Jipijkamiskw (Abnaki-Penobscot/Algonquian).

*Physical description:* Red-and-yellow horn.

*Behavior:* Lives in either salt- or freshwater. Can move through solid rock. Can pass as human.

*Distribution:* New Brunswick.

*Source:* Rod C. Mackay, *Discoveries and Recoveries of Eastern North America,* accessed in 2000, http://www.oldcelticbooks.com/Fundy/george5.html.

## Jogung

TRUE GIANT hominid of Australia.

*Etymology:* Unknown, although the Australian *jingy* or *chingah* were terms used in Western Australia during the nineteenth century for "devils" or "evil spirits."

*Variant names:* Barmi birgoo, Illankanpanka (in central Queensland), Jimbra (in Western Australia), Jingra, Jinka (in Western Australia), Kraitbull (in South Australia), Lo-an (in Yarra Flats, Victoria), Pankalanka (in Northern Territory), Tjangara (in South Australia), Wolumbin. In Victoria, Lowan (Lo-an) is used for the Mallee fowl (*Leipoa ocellata*), a large megapode with a loud, three-noted, booming call.

*Physical description:* Gorilla-like. Height, 7–10 feet. Covered in dark-brown or black hair. Large genitals. Females have large breasts.

*Behavior:* Bipedal. Makes gutteral sounds. Has a rotten smell. Carries a club or tree limb to kill people.

*Tracks:* Length, 24 inches. Splayed big toes.

*Distribution:* Great Sandy Desert, Western Australia; Arnhem Land, Northern Territory; Mount Kosciusko area, New South Wales; central Queensland; Murray region, South Australia; Yarra Flats, Victoria.

*Significant sightings:* In July 1861, explorers

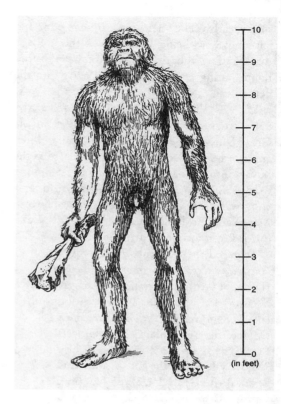

The JOGUNG, *a True Giant hominid of Australia. (Harry Trumbore/Loren Coleman)*

Dempster, Clarkson, and Harper heard from Aborigines at Lake Grace, Western Australia, about the Jimbra or Jingra, a fierce, monkeylike animal that killed solitary travelers.

Around 1960, Andy Hoad and Brett Taylor were prospecting in the Lake Ballard area, Western Australia, when they saw a group of huge, gorilla-like creatures emerge from a stand of scrub. One female was 7 feet tall and had long breasts and dark-brown hair, while a male stood 9 feet tall. Hoad and Taylor ran back to their hut, where they found a 10-foot gorilla in the process of tearing it down. They hid until the coast was clear, then jumped in their truck and drove away. Aborigines in Kalgoolrie-Boulder told them the gorillas were called Jimbra.

In June 1970, mountaineers Ron Bartlett and Frank Sinclair were breaking camp northwest of Mount Kosciusko, New South Wales, when they noticed huge, humanlike tracks in the snow nearby. They detected a strange odor and felt they were being watched. As they moved through the scrub, they saw a dark, hairy, 8-foot figure staring at them. It moved away into the dense bush.

In 1977, Vince and Trevor Collins were driving a truck north of Jimberingga Well in the Great Sandy Desert, Western Australia, when an enormous, black gorilla emerged from the bushes into the road, brandishing a tree limb.

In 1989, a 13-foot, hairy giant wielding a club was seen along Cooper Creek between Maree and Birdsville, South Australia, by two carloads of four-wheel-drive enthusiasts.

*Sources:* Ernest Favenc, *The History of Australian Exploration from 1788 to 1888* (London: Griffith, Farran, Okeden, and Welsh, 1888), pp. 188, 202; W. S. Ramson, ed., *The Australian National Dictionary* (Melbourne, Australia: Oxford University Press, 1988), pp. 335, 376; Rex Gilroy, "Mystery Lions in the Blue Mountains," *Nexus* 2, no. 8 (June-July 1992): 25–27, 64; Rex Gilroy, *Mysterious Australia* (Mapleton, Queensl., Australia: Nexus, 1995); Rex Gilroy, "Giants of the Dreamtime," *Australasian Ufologist* 3, no. 3 (1999), at http://www.internetezy.com.au/~mj129/Australasian_Ufologist2.html.

## Jumar

Supposed HOOFED MAMMAL of Western Europe.

*Etymology:* From the French *jument,* "mare."

*Physical description:* Mulelike. Head is like a bull's. Said to have small, knobby horns. Black eyes.

*Distribution:* Southern France; Italy.

*Significant sighting:* Neapolitan physicist Giambattista della Porta saw apparent mule × bull hybrids in Ferrara, Italy, in the sixteenth century.

*Possible explanations:*

(1) Konrad Gesner's suggestion that the Jumar is a Donkey (*Equus asinus*) × Bull (*Bos taurus*) hybrid is impossible, since the offspring would involve a mating between two separate ungulate orders, the artiodactyls and perissodactyls.

(2) The comte de Buffon supervised the dissection of two alleged Jumars but found them to be small donkeys.

*Sources:* Konrad Gesner, *Historiae animalium* (Zurich, Switzerland: Christ. Froschoverum, 1551–1587); Giambattista della Porta, *Natural Magick* [1558] (London: T. Young and S. Speed, 1658), II. 9; Georges, comte de Buffon, "De la dégénération des animaux" [1766], in *Œuvres complètes de Buffon* (Paris: Rapet, 1817– 1818).

## Jungli-Admi

Multipurpose name for the YETI, Hindu ascetics, or any group of people living in the mountains of Central Asia.

*Etymology:* From the Urdu (Indo-Aryan) *jangli* ("wild") + *admi* ("man").

*Behavior:* Said to use a bow and arrow.

*Distribution:* Nepal; Bhutan; Sikkim Province, northern India.

*Significant sighting:* In May 1940, C. Reginald Cooke and his wife, Margaret, were on the Sikkim-Nepal border at an altitude of 14,000 feet when they found and took photographs of tracks in the ground made by a heavy creature with an opposed toe. The Sherpa guides said they were made by Jungli-admi.

*Sources:* Donald Macintyre, *Hindu-Koh: Wanderings and Wild Sport on and beyond the Himalayas* (Edinburgh: William Blackwood, 1889), pp. 74–75; H. J. Elwes, "On the Possible Existence of a Large Ape, Unknown to Science, in Sikkim," *Proceedings of the Zoological Society of London,* 1915, p. 294; C. Reginald Cooke, "Yeti Country," *Mankind Quarterly* 15, no. 3 (1975): 178–192; C. Reginald Cooke, *Dust and Snow: Half a Lifetime in India* (Saffron Waldon, England: C. Reginald Cooke, 1988).

## Junjadee

LITTLE PEOPLE of Australia.

*Etymology:* Australian word.

*Variant names:* Bitarr (Kumbainggar/Australian), Brown Jack, Burgingin (Alawa/Australian, in Northern Territory), Dinderi (in Queensland), Junjuddi, Net-net (in Victoria),

Nimbunj, Nyol (in eastern Victoria), Waaki, Winambuu (Wiradhuri/Australian), Yuuri (in New South Wales).

*Physical description:* Height, 3–4 feet. Upright. Brown or red skin. Apelike limbs. Covered with dark brown or black hair.

*Behavior:* Runs on two legs. Incredibly strong. Calls are a series of three barks, "Arroo-ARROO-arroo," interspersed with a gurgling "gu-gu-gu-gu." Also said to cackle like a chicken. Guards certain locations. Has supernatural powers. Protects lost children.

*Tracks:* Variable. Three-toed or five-toed. In some cases, similar to a five-year-old child's; in others, each toe is about the size of a human big toe.

*Distribution:* Great Dividing Range of New South Wales and Queensland; eastern Victoria; Northern Territory north of the Roper River; the Western Australia coast between Shark Bay and Broome.

*Significant sightings:* Nathan Moilan's father had seen Junjadees in the woods several times while logging in the Great Dividing Range west of Tully, Queensland. One night, he and his brother were sleeping in a bush hut in the Kirrama Range when a little, hairy man entered and attacked them. They wrestled with it until it broke free and escaped by jumping out the window.

In 1956, a group of Aboriginal and Malay workmen were sitting by a campfire at Shark Bay, Western Australia, when a dark-skinned, little man about 4 feet tall approached them asking for food supplies. The Malay pearl fishermen refused to spend the night alone along the coast where these creatures live.

In September 1968, George Gray was sleeping in a bush hut near the sawmill settlement of Kookaburra in New South Wales when he woke up to find a little man trying to drag him to the door. It was 4 feet tall and covered with bristly, gray hair. They wrestled for several minutes, but the creature had loose skin and Gray couldn't get a firm grip on it.

Early in the morning of June 1, 1996, Gary Opit heard about ninety loud, barklike calls in the Koonyum Range of northeastern New South Wales.

*Possible explanations:*

(1) A mutant strain of stunted Aborigines, perhaps outcasts.

(2) A surviving group of pre-Aboriginal peoples.

(3) A dwarf variety or juvenile YOWIE.

(4) A supernatural entity.

*Sources:* Douglas Lockwood, *I, the Aboriginal* (Adelaide, South Australia: Rigby, 1962); Frank Povah, *You Kids Count Your Shadows: Hairymen and Other Aboriginal Folklore in New South Wales* (Wollar, N.S.W., Australia: Frank Povah, 1990); Karl Shuker, "Death Birds and Dragonets: In Search of Forgotten Monsters," *Fate* 46 (November 1993): 66–74; *Brisbane (Queensl.) Courier Mail,* January 24, 1994; Tony Healy and Paul Cropper, *Out of the Shadows: Mystery Animals of Australia* (Chippendale, N.S.W., Australia: Ironbark, 1994), pp. 117–118; Malcolm Smith, *Bunyips and Bigfoots: In Search of Australia's Mystery Animals* (Alexandria, N.S.W., Australia: Millennium Books, 1996), pp. 164–166; Gary Opit, "Understanding the Yowie Phenomena," May 1999, at http://www.yowiehunters.com/science/reports/understanding.htm.

# K

## Kadimakara

Legendary large reptile or MARSUPIAL of Australia.

*Etymology:* Dieri (Australian) word.

*Variant name:* Kadimurka.

*Physical description:* Reptilian monster. Horn on its forehead.

*Habitat:* Marshes and water holes.

*Distribution:* Lake Eyre, South Australia.

*Significant sighting:* Said by the Aborigines to have lived in the Dreamtime, the mythical period when the world was created.

*Possible explanations:*

(1) A memory based on the Saltwater crocodile (*Crocodylus porosus*) that formerly lived along the southern coast.

(2) A surviving *Diprotodon optatum,* a fossil marsupial (the largest known) that lived from 2.5 million years ago to as recently as 6,000 years ago. It was the size of a modern rhinoceros, about 10 feet long. Its heavy skull, nearly 3 feet long, featured massive jaws and a large lower incisor. Its bones have been found frequently near Lake Eyre and Cooper Creek, South Australia. Diprotodon fossils at Riversleigh, Queensland, bear marks suggestive of butchery by prehistoric tribes.

*Sources:* John Walter Gregory, *The Dead Heart of Australia: A Journey around Lake Eyre in the Summer of 1901–1902* (London: John Murray, 1906), pp. 3–4; Patricia Vickers-Rich and Gerard Van Tets, eds., *Kadimakara: Extinct Vertebrates of Australia* (Lilydale, Vic., Australia: Pioneer Design Studio, 1985), pp. 240–244.

## Kaha

Giant BIRD of Central Asia.

*Physical description:* Silver plumage.

*Behavior:* Its blood is said to contain a cure for blindness.

*Distribution:* Tajikistan.

*Present status:* Probably extinct.

*Sources:* Mirra Ginsburg, *The Kaha Bird: Tales from the Steppes of Central Asia* (New York: Crown, 1971); Joe Nigg, *A Guide to the Imaginary Birds of the World* (Cambridge, Mass.: Apple-Wood, 1984), pp. 77–79, 151.

## Kaigyet

CANNIBAL GIANT of western Canada.

*Etymology:* Carrier (Na-Dené) word.

*Physical description:* Very tall. Covered in long hair. Face is like a human's.

*Behavior:* Arrows do not hurt it.

*Tracks:* Leaves huge prints in the snow.

*Distribution:* Nechako River area, British Columbia.

*Source:* Diamond Jenness, "Myths of the Carrier Indians of British Columbia," *Journal of American Folklore* 47 (1934): 97, 220–222.

## Ka-Is-To-Wah-Ea

FRESHWATER MONSTER of New York.

*Etymology:* Seneca (Iroquoian) word.

*Physical description:* Serpentine. Two heads.

*Habitat:* Cave underneath a mountain.

*Distribution:* Western New York.

*Source:* Harriet M. Converse, "Myths and Legends of the New York Iroquois," *New York State Museum Bulletin,* no. 125 (1908): 113.

## Kakundakari

SMALL HOMINID of Central Africa.

*Etymology:* Konjo, Nyanga, and Kanu (Bantu) word.

*Scientific name: Congopithecus,* proposed by Charles Cordier in July 1960; amended to *Congopithecus cordieri* by Heini Hediger in October 1960.

*Variant names:* Amajungi (Komo/Bantu), KIKOMBA (possibly the male or adult of the species), Lisisingo (Poke/Bantu), Mbatcha (Tembo/Bantu), Niáka-ambúguza (Lega-Mwenga/Bantu).

*Physical description:* Height, 2–3 feet. Gray skin. Covered with thin hair except for the face. Long, black head-hair shaped in pageboy fashion. No large canine teeth. Short mane along the neck.

*Behavior:* Bipedal. Strong for its size. Travels alone or in pairs or threes. Does not climb trees. Does not swim but crosses streams by holding on to a floating log or by using a canoe. Horrible odor. Eats crabs, ginger-fruit, millipedes, snails, and birds. Carries a satchel made of leaves to hold gathered food. Sleeps in caves. Makes a bed of leaves. Dodges spears thrown at it. Possibly uses a machete. Gathers wood in the cave as if to make a fire but apparently cannot get it going.

*Tracks:* Four-toed. Length, under 5 inches.

*Distribution:* Kivu Region of the Democratic Republic of the Congo, north to the equator.

*Significant sightings:* In January 1957, a nearly dead Kakundakari was found by a hunter south of Kasese, near the Lugulu River. He brought it to a village, where it was kept caged until it escaped. It was seen by many blacks and dozens of whites.

Near Walikale, Charles Cordier found a 5-inch-long footprint close to a cavern said to be the home of a Kakundakari. The big toe was in the same proportion as that of a human print, but there was no trace of a fifth toe.

*Present status:* Always characterized as rare, it may be severely reduced in numbers or even extinct because of deforestation and warfare in the region.

*Possible explanations:*

(1) The PYGMIES of classical times, possibly the ancestors of the short-statured Mbuti of the Ituri Forest in the Democratic Republic of the Congo. However, the Pygmies seem more advanced physically and culturally than the Kakundakari.

(2) Confused folklore about the Chimpanzee (*Pan troglodytes*) or Bonobo (*Pan paniscus*).

(3) Surviving gracile australopith, suggested by Bernard Heuvelmans. Australopiths were a family of hominids known from the Pliocene to the Early Pleistocene, 4.4–1.4 million years ago. More than 2,000 individual fossils are known. Australopiths had apelike skulls, hominid teeth, pronounced cheekbones, and projecting jaws. The molars were heavy, with thick enamel. In Ethiopia and Tanzania, *Australopithecus afarensis* males probably stood about 4 feet 6 inches tall and weighed over 100 pounds; females stood about 3 feet 6 inches and weighed about 60 pounds. The arms were proportionately longer than those of humans but shorter than those of apes. The chest tapered sharply upward. Cranial capacity averaged 400–410 milliliters. They had an upright, bipedal gait but apparently also climbed trees. They were vegetarians, based on their molar size, and probably ate leaves, fruit, tubers, seeds, and insects.

The best-known specimen of *A. afarensis* is known as Lucy, found in 1974 at Olduvai Gorge, Tanzania, and considered by some anthropologists, notably Donald Johanson, to be ancestral to modern humans. Other evolutionary relationships have been proposed, but there is little consensus on this issue; it's likely that generic renamings will occur as relationships are further explored.

Bipedal footprints 3.8–3.6 million years old preserved in volcanic ash at Laetoli, Tanzania, and discovered by Mary Leakey in 1978 may belong to *afarensis*. The apelike and grasping big toe is in alignment with the others, and the print roughly resembles the sketch of a Kakundakari track drawn by Cordier nearly twenty years earlier.

*Sources:* Heini Hediger, "Auf der Spur eines neuen Menschenaffen," *Das Tier* 1 (October

1960): 49; Charles Cordier, "Deux anthropoïdes inconnus marchant debout, au Congo ex-Belge," *Genus* 29 (1963): 2–10; Charles Cordier, "Animaux inconnus au Congo," *Zoo* 38 (April 1973): 185–191; Bernard Heuvelmans, *Les bêtes humaines d'Afrique* (Paris: Plon, 1980), pp. 570–598.

## Kalanoro

LITTLE PEOPLE of Madagascar.

*Etymology:* Betsileo and Sakalava Malagasy (Austronesian) word.

*Variant names:* Biby olona, Kimos, Koko-lampo, Kotokely.

*Physical description:* Height, 2–3 feet. Covered with hair. Head-hair falls to the waist. Fingers with hooked nails. Three toes. Has many other mythical attributes similar to those of European FAIRIES and MERBEINGS. Characteristics vary from region to region.

*Behavior:* Amphibious. Sometimes malevolent. Voice like a woman's. Eats fishes and likes milk. Said to abduct children.

*Habitat:* Lakes, lagoons, caves, and forests.

*Distribution:* Lac Alaotra and Lac Kinkony; Ankazoabo District; Andoboara Cave; the central highlands, Madagascar.

*Possible explanations:*

(1) Surviving giant lemur, possibly *Archeolemur* or *Hadropithecus*. These lemurs had short limbs, hands, and feet and were powerfully built. They probably represented the same ecological niche for Madagascar as Africa's baboons. Weight was 30–55 pounds. However, they apparently were not amphibious.

(2) The Malagasy equivalent of a MERBEING.

*Sources:* Ch. Lamberton, "Les Hadropithèques," *Mémoires de l'Académie malgache* 20 (1937): 127–170; Raymond Decary, *La faune malgache, son rôle dans les croyances et les usages indigènes* (Paris: Payot, 1950), pp. 207–208; Bernard Heuvelmans, "Annotated Checklist of Apparently Unknown Animals with Which Cryptozoology Is Concerned," *Cryptozoology* 5 (1986): 1–26; Loren Coleman, *Mothman and Other Curious Encounters* (New York: Paraview, 2002), pp. 110–111.

## Kappa

MERBEING of Japan.

*Etymology:* Japanese, "river child."

*Variant names:* Kawachi, Kyuusenbou, Masunta, Mu jima, Ningyo.

*Physical description:* Said to be half human and half turtle or frog. Height, 3–4 feet. Weight, 20–50 pounds. Apelike face. Long hair. Dishlike hollow in the top of the head where water is kept. Scaly limbs. Webbed hands and feet.

*Behavior:* Active in summer. Changes color like a chameleon, according to surroundings. Can also transform into a human. Has superhuman strength. Favorite food is the cucumber. Cries pearls instead of tears. Often malicious. Tries to drown children and travelers. Likes sumo wrestling.

*Habitat:* Rivers, lakes, and ponds; seen less frequently in the ocean and mountains.

*Distribution:* Kyushu; the southern tip of Honshu; the Sarugaishi River, Honshu, Japan.

*Significant sightings:* A shrine in Kumamoto Prefecture is said to possess the mummified hand of a Kappa.

A Kappa mummy is on display in Imari, Saga Prefecture, at the Matsuura Brewery, where it was discovered inside a black box during some renovations in the 1950s.

*Sources:* Henri L. Joly, *Legend in Japanese Art* (New York: J. Lane, 1908); Donald Alexander Mackenzie, *Myths of China and Japan* (London: Gresham, 1923), pp. 350–351; Kunio Yanagita, *Tōno monogatari* (Tokyo: Bungei Shunjū

*Alleged mummy of a* KAPPA, *a Japanese merbeing. (From a postcard in the author's collection)*

Shinsha, 1948); Catrien Ross, *Supernatural and Mysterious Japan* (Tokyo: Yenbooks, 1996), pp. 31, 99; Kyoichi Tsuzuki, *Roadside Japan* (Tokyo: Aspect, 1997), pp. 272–274; Oniko, Kappa Quest 2000, http://www.sonic.net/ ~anomaly/oniko/epaug99.htm.

## Kapre

GIANT HOMINID of Southeast Asia.

*Etymology:* Bikol and Tagalog (Austronesian) word derived from the Spanish *kafre* ("Moor").

*Variant name:* Xuě-rén.

*Physical description:* Height, 8 feet or more. Covered with hair. Dark, rough skin. Large eyes; some accounts say there is only one. Big ears. Flat nose. Big mouth. Thick lips. Human-like face, hands, and feet.

*Behavior:* Nocturnal. Active during new moons or after rainfall. Upright gait. Has a pungent odor. Omnivorous but fond of mangoes, pineapples, tamarind fruit, coconuts, papayas, radishes, fishes, land crabs, and rats. Said to leave fresh fruit or fish in exchange for cooked rice left out for it. Jumps out of trees and scares people. Said to smoke cigars and carry off women.

*Tracks:* Twice the length and breadth of a human's, with broader toes.

*Habitat:* Caves.

*Distribution:* The area around Mount Banahao, Luzon, Philippines; Samar, Philippines.

*Significant sightings:* A one-eyed Kapre named Agyo was said to have fought against the first Spanish conquistadors.

Human skulls with a single eye socket were said to have been discovered in caves on the island of Bohol.

*Possible explanation:* A myth based on the fear of black slaves brought to the Philippines during Spanish colonial times.

*Sources:* Karl Shuker, "Keeping Up with the Kapre," *Fortean Times,* no. 122 (May 1999): 18–19; Bobbie Short, "The Kapre of the Philippines," October 2001, http://www.n2. net/prey/bigfoot/creatures/kapre.htm; "Keeping an Eye on Cyclops," *Fortean Times,* no 159 (July 2002): 13.

## Kaptar

WILDMAN of West Asia.

*Etymology:* Avar (North East Caucasian) word.

*Variant names:* Kara-pishik, Keetar, Kheeter, Meshe-adam (Azerbaijani/Turkic, "forest man"), Tukhli-adam, Veshshi-adam.

*Physical description:* Height, 5–7 feet. Covered in 1-inch-long hair. Color varies from reddish-brown in smaller individuals to dark brown, dark gray, black, and silvered. Head-hair reaches to shoulders. Only slightly hairy on the face. Wide shoulders. Slightly stoop-shouldered. Long arms. Palms and soles free of hair. Fingers thick and large. Females have breasts.

*Behavior:* Bipedal. Cry consists of several repeated high- and low-pitched sounds, somewhat plaintive. Prefers the cold; sweats in a warm environment. Harbors body lice unlike those found on humans. Bathes in rivers.

*Tracks:* Length, 9–10 inches. Large big toe, 3.5 inches long. Other four toes slightly splayed. Narrower at the heel.

*Distribution:* Dagestan Republic, Russia; Caucasus Mountains, Azerbaijan.

*Significant sightings:* In December 1941, Lt. Col. Vazghen Sergeyevich Karapetian of the Soviet Army Medical Corps inspected a captured wildman 18 miles from Buynaksk in Dagestan, Russia. It was brought to him because local authorities thought it might be a German spy in disguise. The creature was nearly 6 feet tall and covered with shaggy, brown fur like a bear's. Karapetian gave his opinion it was a wildman and no spy, and he never heard anything about the incident again.

In July 1957, V. K. Leontiev was inspecting game in an area near the head of the Jurmut River, Dagestan, when he came across huge footprints in the snow. That night, he heard something make a strange cry, and the next day, he saw and took a shot at a 7-foot-tall Kaptar. He pursued it for several minutes, but it outpaced him up a steep slope. He measured and sketched the many prints it left behind.

On August 20, 1959, veterinarian Ramazan Omarov encountered a male Kaptar on a mountain path in Dagestan.

On September 18, 1959, anthropologists Yuri I. Merezhinskiy and Marie-Jeanne Kofman

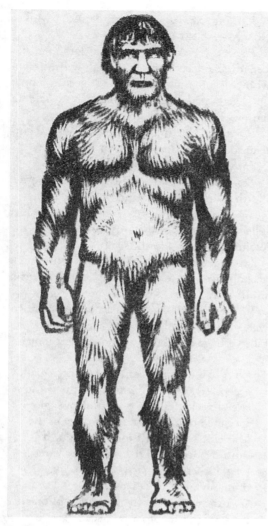

*A KAPTAR, or wildman, examined in 1941 by Lt. Col. Karapetian of the Russian Army Medical Corps in the Caucasus Mountains. (Fortean Picture Library)*

human mandible has been found at Azych, Azerbaijan, in association with Early Paleolithic hand axes. Neanderthal toolkits occur in several caves in the region, and a probable *Homo erectus* mandible 1.5 million–900,000 years old was found in 1991 at Dmanisi in Georgia.

*Sources:* Boris F. Porshnev and A. A. Shmakov, eds., *Informatsionnye materialy, Komissii po Izucheniyu Voprosa o "Snezhnom Cheloveke,"* 4 vols. (Moscow: Akademiia Nauk SSSR, 1958–1959); Ivan T. Sanderson, *Abominable Snowmen: Legend Come to Life* (Philadelphia: Chilton, 1961), pp. 289–296; *Tekhnika Molodezhi,* 1966, no. 8; Boris F. Porshnev, "Problema reliktovykh paleoantropov," *Sovetskaia Etnografiia* 2 (1969): 115–130; Bernard Heuvelmans and Boris F. Porshnev, *L'homme de Néanderthal est toujours vivant* (Paris: Plon, 1974), pp. 164–170; Dmitri Bayanov, *In the Footsteps of the Russian Snowman* (Moscow: Crypto-Logos, 1996), pp. 15–18, 24, 31–33.

## Kashehotapalo

CANNIBAL GIANT of the southern United States.

*Etymology:* Choctaw (Muskogean), *kasheho* ("woman") + *tapalo* ("call").

*Physical description:* Small head. Shriveled face. Hairy legs. Cloven hooves.

*Behavior:* Enjoys frightening hunters. Screams like a woman.

*Habitat:* Swamps.

*Distribution:* Louisiana.

*Source:* David I. Bushnell Jr., "The Choctaw of Bayou Lacomb," *Bulletin of the Bureau of American Ethnology* 48 (1909): 31.

## Kaurehe

Mystery otterlike mammal or reptile of Australasia, similar to the WAITOREKE.

*Etymology:* Maori (Austronesian), uncertain; may mean "very spiny," may be related to "swimming" or "ancestors," or may just be a generic "monster."

*Variant name:* Frequently misspelled as Kaureke.

*Physical description:* Otterlike. Whitish fur.

*Behavior:* Nocturnal. Seemingly less aquatic

staked out a spot near Balakon in northwestern Azerbaijan in the hopes of photographing an albino Kaptar said to frequent the area. They were accompanied by an old hunter named Hajji Magoma. Soon, they saw and heard the white-haired creature splashing in a stream and making laughing noises. Instead of taking its picture, however, Merezhinskiy shot at it with a revolver he had concealed. The Kaptar escaped, and Magoma never took visitors back to the area.

*Possible explanation:* Hominid fossils from this region are sparse but promising. An archaic

than the Waitoreke. Observed when it visits its watering places. Eats lizards. Lays eggs as large as a duck's.

*Distribution:* Arowhenua Bush, South Island, New Zealand.

*Possible explanations:* An unknown, egg-laying mammal (monotreme) or a spiny lizard of some kind.

*Sources:* Gideon A. Mantell, *Petrifactions and Their Teachings* (London: H. G. Bohn, 1851), p. 105; Richard Taylor, *Te Ika a Maui*, 2d ed. (London: W. Macintosh, 1870), p. 604; Roger Duff, *The Moa-Hunter Period of Maori Culture* (Wellington, New Zealand: Department of Internal Affairs, 1950), p. 289; J. S. Watson, "The New Zealand 'Otter,'" *Records of the Canterbury Museum* 7, no. 3 (1960): 175–183; James Herries Beattie, *Traditional Lifeways of the Southern Maori: The Otago University Museum Ethnological Project, 1920* (Dunedin, New Zealand: University of Otago Press, 1994), p. 354; H. W. Orsman, ed., *The Dictionary of New Zealand English* (Auckland, New Zealand: Oxford University Press, 1997), p. 398. *See also* WAITOREKE.

## Kavay

Unknown INSECTIVORE of Madagascar.

*Etymology:* Tsimihety Malagasy (Austronesian) word.

*Behavior:* Amphibious.

*Distribution:* Ankaizina Mountains, north-central Madagascar.

*Possible explanation:* A large variety of the nocturnal Aquatic tenrec (*Limnogale mergulus*), an insectivore that grows to not much more than 12 inches in length, including its tail. It has webbed feet and a flattened tail for swimming.

*Source:* Bernard Heuvelmans, *On the Track of Unknown Animals* (New York: Hill and Wang, 1958), p. 516.

## Kawekaweau

Unknown LIZARD of Australasia.

*Etymology:* Maori (Austronesian) word. Without the final "u" (Kawekawea), it refers to the Long-tailed cuckoo (*Eudynamis taitensis*).

*Variant names:* Kaweau, KUMI, Moko, NGARARA, TANIWHA.

*Physical description:* Like a large gecko. Length, 2–6 feet. Thick as a man's wrist. Brown with red, longitudinal stripes. Projecting upper lip. Large teeth. Serrated dorsal crest. Said to have six legs.

*Behavior:* Nocturnal. Arboreal. Makes guttural sounds. Eats roots and small birds.

*Habitat:* Forests, riverbanks, caves.

*Distribution:* Waoku Plateau; near Gisborne, North Island, New Zealand.

*Significant sightings:* A Urewera Maori chief killed one of these lizards in 1870 in the Waimana Valley, North Island.

Joe McClutchie claims to have seen a giant gecko twice—once in the late 1960s and again in the early 1980s—while driving at night.

In the late 1970s, Neil Farndale and others were driving at night when their car hit and killed a huge lizard. It was about 2 feet long.

*Possible explanations:*

(1) A living population of Delcourt's giant gecko (*Hoplodactylus delcourti*), the world's largest gecko, known only from a single specimen of unknown provenance in the Marseille Natural History Museum. It is a short-headed lizard that measures 2 feet long. Other *Hoplodactylus* geckos live primarily in New Zealand.

(2) Exaggeration of the size either of one of the sixty species of native gecko (*Hoplodactylus* or *Naultinus*), the largest of which is the 6-inch Giant gecko (*H. duvaucelii*), or of the 12-inch Otago skink (*Oligosoma otagense*).

(3) An out-of-place Tuatara (*Sphenodon punctatus*), a 2-foot-long, lizardlike reptile that is now confined to about twenty small islands off the northeast coast of New Zealand and in Cook Strait.

(4) A misidentified introduced animal, such as a Ferret (*Mustela furo*).

(5) A Maori mythological creature.

*Sources:* W. G. Mair, "Notes on Rurima Rocks," *Transactions of the New Zealand Institute* 5 (1873): 151–153; H. D. Skinner, "Crocodile and Lizard in New Zealand Myth and Material Culture," *Records of the Otago*

Museum, *Anthropology* 1 (1964): 1–43; Aaron M. Bauer and Anthony P. Russell, "*Hoplodactylus delcourti* (Reptilia: Gekkonidae) and the *Kawekaweau* of Maori Folklore," *Journal of Ethnobiology* 7 (1987): 83–91; Aaron M. Bauer and Anthony P. Russell, "Osteological Evidence for the Prior Occurrence of a Giant Gecko in Otago, New Zealand," *Cryptozoology* 7 (1988): 22–37; Aaron M. Bauer and Anthony P. Russell, "Recent Advances in the Search for the Living Giant Gecko of New Zealand," *Cryptozoology* 9 (1990): 66–73; Michel Raynal and Michel Dethier, "Lézards géants des Maoris: La verité derrière la légende," *Bulletin Mensuel de la Société Linnéene de Lyon* 59 (March 1990): 85–91; "Search for Giant Gecko Intensifies," *ISC Newsletter* 9, no. 4 (Winter 1990): 1–4.

## Kecleh-Kudleh

CANNIBAL GIANT of the southeastern United States.

*Etymology:* Cherokee (Iroquoian), "hairy man."

*Variant names:* Chickly cudly, Ke-cleah kudleah.

*Distribution:* Western North Carolina.

*Source:* Kyle Mizokami, Bigfoot-Like Figures in North American Folklore and Tradition, http://www.rain.org/campinternet/bigfoot/bigfoot-folklore.html.

## Kéédieki

WILDMAN of Siberia.

*Etymology:* Yakut (Turkic) word.

*Variant name:* Kheed'eki.

*Distribution:* Verkhoyansk Range, Sakha Republic, Siberian Russia.

*Source:* Bernard Heuvelmans and Boris F. Porshnev, *L'homme de Néanderthal est toujours vivant* (Paris: Plon, 1974), p. 143.

## Kellas Cat

All-black small CAT of Scotland.

*Etymology:* Named by Karl Shuker from specimens found near Kellas, Grampian, Scotland.

*Variant names:* Black beast of Moray, CAIT SITH, Wangie cat.

KELLAS CAT *known as Specimen K, shot in 1983 by Tomas Christie near Kellas, Scotland. (Andrew Barker/ Fortean Picture Library)*

*Physical description:* Slender and well muscled. Length, 2–3 feet, with a 12-inch tail. Weight, 5–15 pounds. Bristly, black fur, sprinkled with white primary guard hairs. Small, long head. Rounded ears. Short muzzle. Large nose. Large, prominent upper and lower canines. Paws are long and narrow. Claws are retractile. Tail is broad and thickly furred.

*Behavior:* Hunts in pairs during the daytime. Can swim well. Graceful, loping gait. Feeds on rabbits and birds.

*Habitat:* Woodlands.

*Distribution:* Highland and Grampian, Scotland.

*Significant sightings:* In June 1984, a black, male wildcat about 3 feet long was trapped in a fox snare on the grounds of the Revack Lodge near Grantown-on-Spey, Highland. The specimen was lost after it was taken to a taxidermist.

In October 1984, a second, smaller cat (known as Specimen K) came to light; it had been shot in January 1983 by Tomas Christie while crossing the River Lossie near Kellas, Grampian.

In April and October 1985, two other specimens were shot, near Avie and Kellas.

On February 28, 1988, a black wildcat was caught alive near Redcastle in northern Scotland. It measured 3 feet long and weighed 13 pounds.

*Possible explanations:*

(1) An unknown species of cat. This is unlikely, since the animal shares the same well-known habitat as the Scottish wildcat (*Felis silvestris grampia*) and would likely have been recognized long ago.

(2) A melanistic Scottish wildcat, especially possible for Specimen K, which is more slender than a typical wildcat, with longer limbs, head, body, and teeth.

(3) An isolated population of feral Domestic cats (*F. s. catus*). However, the skull, limb, and dental dimensions are too large for a domestic cat and closer to those of a wildcat.

(4) Introgressive domestic cat × Scottish wildcat hybrid, suggested by Karl Shuker. Continuous mating of hybrids with both ferals and wildcats could produce a breed with a distinctive Kellas appearance, whereas initial crossbreeds more closely resemble Scottish wildcats, though with longer tails. The increase in the Scottish wildcat population since World War I may actually have been jump-started by hybridization.

*Sources:* "The Black Beasts of Moray," *Fortean Times,* no. 45 (Winter 1985): 10–12; Karl Shuker, *Mystery Cats of the World* (London: Robert Hale, 1989), pp. 70–80; Karl Shuker, "The Kellas Cat: Reviewing an Enigma," *Cryptozoology* 9 (1990): 26–40; David Alderton, *Wild Cats of the World* (London: Blandford, 1998), pp. 96–98; Hybridisation and the Scottish Wildcat, http://www.scottishwildcats.co.uk/Scottish%20wildcat%20hybridisation.htm; Sarah Hartwell, "Domestic × Wild Hybrids in the Wild," 2001, http://messybeast.com/hybrids.htm; Scottish Big Cats, http://www.bigcats.org/abc/.

## Ke-Ló-Sumsh

CANNIBAL GIANT of the northwestern United States.

*Etymology:* Southern Puget Sound Salish (Salishan), "giant hunters of the mountains."

*Distribution:* Puget Sound, Washington.

*Source:* George Gibbs, "Tribes of Western Washington and Northwestern Oregon," *Contributions to North American Ethnology* 1 (1877): 308.

## Kelpie

WATER HORSE of Scotland.

*Etymology:* From the Gaelic *colpach* ("colt").

*Variant names:* Kelpy, Water kelpie.

*Physical description:* Like a young, black horse. Wild, staring eyes.

*Behavior:* Mischievous or malevolent. Howls and wails before a storm. Can change its shape into that of a rough, shaggy man. Lures women and youths into the water to drown them. Causes its home lake to swell and flood. Wears a magic bridle. Said to tear humans to pieces and devour them occasionally.

*Habitat:* Rivers and lakes, especially fast-moving streams.

*Distribution:* Scotland.

*Artist's conception of a KELPIE, a Scottish water horse. (Fortean Picture Library)*

*Sources:* William Grant Stewart, *Popular Superstitions and Festive Amusements of the Highlanders of Scotland* (Edinburgh: A. Constable, 1823); James M. Mackinlay, *Folklore of Scottish Lochs and Springs* (Glasgow: William Hodge, 1893), pp. 164–166, 171–187; Helen Drever, *The Lure of the Kelpie* (Edinburgh: Moray, 1937); Gwen Benwell and Arthur Waugh, *Sea Enchantress* (London: Hutchinson, 1961), pp. 174–176; Katharine M. Briggs, *A Dictionary of Fairies* (London: Allen Lane, 1976), p. 246.

## Kenaima

LITTLE PEOPLE of South America.

*Etymology:* Wapisianas (Arawakan) and Yecuana (Carib) word used for various demons and entities.

*Variant name:* Kanaima (Pemon/Carib).

*Behavior:* Nocturnal.

*Distribution:* Guyana; Venezuela.

*Sources:* Bernard Heuvelmans, "Annotated Checklist of Apparently Unknown Animals with Which Cryptozoology Is Concerned," *Cryptozoology* 5 (1986): 1–26; John E. Roth, *American Elves* (Jefferson, N.C.: McFarland, 1997), pp. 90, 94.

## Kènkob

SMALL HOMINID of West Africa.

*Variant name:* Bétsan.

*Physical description:* Height, 3–4 feet. Long beard.

*Behavior:* Excellent singer. Good marksman. Hunts apes, baboons, wild pig, antelopes, and elephants.

*Distribution:* Sierra Leone; the Fouta Djallon Mountains, Guinea.

*Significant sightings:* Gaspard Mollien reported a race of small people with good singing voices in the village of Faran in the interior of Guinea in 1818.

S. W. Kölle's informant, a chief in Sierra Leone named Yon, spoke of two kinds of short-statured groups in the interior, the Kènkob and the Bétsan.

*Possible explanation:* An undiscovered group of short-statured hunter-gatherers possibly related to the Mbenga Pygmies of Gabon and Cameroon.

*Sources:* Gaspard Mollien, *Voyage dans l'intérieur de l'Afrique aux sources du Sénégal et de la Gambie, fait en 1818* (Paris: Mme. Ve Courcier, 1820), vol. 2, p. 210; Sigismund Wilhelm Kölle, *Polyglotta Africana* (London: Church Missionary House, 1854), p. 12.

## Keshat

WILDMAN of West Asia.

*Etymology:* Adygey and Kabardian (Circassian), "mountain man."

*Physical description:* Height, 5 feet. Covered with brownish hair. Protruding face.

*Behavior:* Upright gait. Not aggressive but might attack a lone hunter. Agile. Occasionally raids crops in villages. Said to engage in bartering with humans and use a crude language.

*Tracks:* Humanlike but wider than a man's.

*Habitat:* Forests.

*Distribution:* Caucasus Mountains in the Adygey and Kabardin-Balkar Republics, Russia.

*Source:* John Colarusso, "Ethnographic Information on a Wild Man of the Caucasus," in Marjorie Halpin and Michael M. Ames, eds., *Manlike Monsters on Trial* (Vancouver, Canada: University of British Columbia Press, 1980), pp. 255–264.

## Ketos

SEA MONSTER of the Mediterranean Sea.

*Etymology:* Greek, "sea monster" or "whale."

*Variant name:* Cetus (Latin).

*Physical description:* Serpentine. Doglike head. Fishlike tail.

*Distribution:* Eastern Mediterranean.

*Significant sightings:* After a flood, a Ketos appeared on the coast near ancient Troy, Çanakkale Province, Turkey, and ravaged the countryside. King Laomedon sent his daughter Hesione as a sacrifice to appease the monster, but the hero Herakles arrived in time to rescue her and kill the beast. A Corinthian vase painting from the sixth century B.C. depicts the incident, including the Trojan monster, which looks like a huge skull with forward-projecting teeth and bony plates around the eye sockets.

In Greek mythology, Perseus rescued Andromeda, who was chained to a rock at Yafo, Israel, where she was beset by a Ketos. Marcus Aemilius Scaurus claimed to have found the bones of the monster in 58 B.C. and had them shipped to Rome, where they were reassembled for display. Pliny wrote that the backbone was 40 feet long and 18 inches thick.

*Possible explanations:*

(1) Adrienne Mayor and others have concluded the Corinthian vase artist must have used a fossil skull, perhaps that of an extinct giraffid such as *Samotherium,* as a model. However, the teeth look like they might have come from a reptile or whale skull, while the sclerotic eye ring is characteristic of birds and dinosaurs. Possibly, features from several fossils were combined.

(2) Scaurus might have found the skeleton of a beached Sperm whale (*Physeter catodon*). These whales are still seen regularly in the eastern Mediterranean.

*Sources:* Homer, *Odyssey,* V. 421, *Iliad,* XX. 147; Aristophanes, *Frogs,* 556, *The Thesmophoriazusae* 1033; Lycophron, *Alexandra,* 954; Diodorus Siculus, *Historical Library,* II. 54.3; Pliny, *Historia naturalis,* V. 69, IX. 4–11; Oppian, *Halieutica,* V. 113; Pausanias, *Guide to Greece,* I. 4.1, II. 10.2, II. 34.2, IV. 34.2, IV. 35.9, V. 17.11, V. 25.3, VIII. 2.7, IX. 26.5, X. 4.4, X. 12.1; Ælian, *De natura animalium,* XIII. 21, XV. 19; Katharine Shepard, *The Fish-Tailed Monster in Greek and Etruscan Art* (New York: Katharine Shepard, 1940); John Boardman, "Very Like a Whale: Classical Sea Monsters," in Ann E. Farkas, Prudence O. Harper, and Evelyn B. Harrison, eds., *Monsters and Demons in the Ancient and Medieval Worlds: Papers Presented in Honor of Edith Porada* (Mainz am Rhein, Germany: Philipp von Zabern, 1987); Adrienne Mayor, "Paleocryptozoology: A Call for Collaboration between Classicists and Cryptozoologists," *Cryptozoology* 8 (1989): 12–26; Adrienne Mayor, *The First Fossil Hunters: Paleontology in Greek and Roman Times* (Princeton, N.J.: Princeton University Press, 2000), pp. 138–139, 144–145, 157–163.

## Kheyak

GIANT HOMINID of far eastern Siberia.

*Physical description:* Height, 6–7 feet.

*Behavior:* Swift runner. Said to have hypnotic powers.

*Distribution:* Coastal mountains of the Khabarovsk Territory, Siberian Russia, on the Sea of Okhotsk.

*Source:* Dmitri Bayanov, *In the Footsteps of the Russian Snowman* (Moscow: Crypto-Logos, 1996), p. 230.

## Khodumodumo

Mystery animal of South Africa, similar to the NANDI BEAR.

*Etymology:* Unknown; said to mean "gaping-mouthed bush monster."

*Behavior:* Nocturnal. Breaks into livestock pens and steals sheep, goats, and calves. Attacks silently. Can climb or leap over 6-foot fences with an animal in its jaws.

*Tracks:* Round. Claw marks, 2 inches long.

*Distribution:* Northern Cape, Eastern Cape, and North-West Provinces of South Africa.

*Significant sighting:* Numerous attacks were made on kraals in the Graaff-Reinet area, Eastern Cape. A posse of more than 100 settlers failed to find it.

*Source:* William Hichens, "African Mystery Beasts," *Discovery* 18 (1937): 369–373.

## Khot-Sa-Pohl

CANNIBAL GIANT of the midwestern United States.

*Etymology:* Kiowa word.

*Physical description:* Covered with hair. Pointed head.

*Behavior:* Terrible odor. Eats human flesh. Said to be capable of speech.

*Habitat:* Mountains, plains, swamps, forests.

*Distribution:* West-central Oklahoma.

*Significant sighting:* Russell Bates and his brother saw a tall figure covered with dark hair when they were setting off fireworks on July 4, 1978, north of Anadarko, Oklahoma. The fireworks apparently annoyed the creature, and it stalked off into the woods.

*Source:* Russell Bates, "Legends of the Kiowa," *INFO Journal,* no. 52 (May 1987): 4—10.

## Khün Görüessü

WILDMAN of Central Asia.

*Etymology:* Mongolian (Altaic), "man-beast."

*Variant names:* Hün garees, Hün göröös, Hün har göröös, Khün har görüessü ("black man-beast"), Kümün görügesü, Zerleg khün.

*Distribution:* Mongolia; Xinjiang Uygur Autonomous Region, China.

*Significant sighting:* A wildman skin preserved in a Mongolian temple proved to be a bear's.

*Sources:* Nikolai M. Przheval'skii, *Mongolia, the Tangut Country, and the Solitudes of Northern Tibet* (London: S. Low, Marston, Searle and Rivington, 1876), pp. 249—250; Emanuel Vlček, "Old Literary Evidence for the Existence of the 'Snow Man' in Tibet and Mongolia," *Man* 59 (1959): 133—134; Bernard Heuvelmans and Boris F. Porshnev, *L'homme de Néanderthal est toujours vivant* (Paris: Plon, 1974), pp. 43, 52.

## Khya

Alternate name for the YETI of Central Asia.

*Etymology:* Newari (Sino-Tibetan), "famous"; also "joke."

*Physical description:* Trickster figure. Covered in long hair.

*Distribution:* Kathmandu Valley, Nepal.

*Source:* Kesar Lall, *Lore and Legend of the Yeti* (Kathmandu: Pilgrims Book House, 1988), pp. 5, 23.

## Kibambangwe

DARK LEOPARD of Central Africa.

*Etymology:* Bantu languages, "snatcher," a term also used for hyenas.

*Variant names:* Kikambangwe, Uruturangwe.

*Physical description:* Size of a leopard. Dark color. Blackish markings. Small ears. Retractile claws. Long tail.

*Behavior:* Mostly nocturnal. Cry is a few deep grunts followed by high-pitched shrieks. Comes down from the mountains to kill livestock. Ferocious if cornered. Said to enter huts and suffocate sleeping people by lying on them and putting its jaws around their faces.

*Habitat:* Lava caves in the mountains.

*Distribution:* Bufumbira County, southwestern Uganda; Virunga Volcanos region, Rwanda.

*Significant sighting:* In 1920, a dark mystery animal called a Kibambangwe killed livestock in Bufumbira County, Uganda, but it was never captured.

*Possible explanations:*

(1) A large, dark-colored Spotted hyena (*Crocuta crocuta*).

(2) A pseudomelanistic Leopard (*Panthera pardus* var. *melanotica*), with dark coloration.

(3) A nonexistent composite creature based on both hyenas and leopards.

*Sources:* Charles R. S. Pitman, *A Game Warden among His Charges* (London: Nisbet, 1931), pp. 304—308; Karl Shuker, *Mystery Cats of the World* (London: Robert Hale, 1989), pp. 135—136.

## Kidoky

Unknown PRIMATE of Madagascar.

*Etymology:* Malagasy (Austronesian) word.

*Physical description:* Height, 4—5 feet. Weight, about 55 pounds. Dark coat. White spots above and below the face. Face is round and very humanlike.

*Behavior:* Ground-dwelling, not arboreal. Solitary. Runs away by taking short leaps along the ground. Call is a long, single "whoo."

*Habitat:* Deciduous forest.

*Distribution:* Southwestern Madagascar.

*Significant sighting:* Jean Noelson Pascou saw a Kidoky in the forest in 1952.

*Possible explanation:* Surviving giant lemur, possibly *Archeolemur* or *Hadropithecus*. These lemurs had short limbs, hands, and feet and were powerfully built. They probably represented the same ecological niche in Madagascar as Africa's baboons. Weight was 30—55 pounds. *Hadropithecus* primarily ate grasses and seeds, while *Archeolemur* ate tough fruits and seeds.

*Source:* David A. Burney and Ramilisonina,

"The *Kilopilopitsofy, Kidoky,* and *Bokyboky:* Accounts of Strange Animals from Belo-sur-Mer, Madagascar, and the Megafaunal 'Extinction Window,'" *American Anthropologist* 100 (1998): 957—966.

## Kigezi Turaco

Unknown BIRD of East Africa.

*Physical description:* Slender, nonpasserine bird. Green, with very little red on the wings.

*Distribution:* Kabate (former Kigezi) District, Uganda.

*Significant sighting:* Seen briefly by John G. Williams and other ornithologists in Uganda.

*Possible explanation:* Fleeting glimpses of the southern race of Ruwenzori turaco (*Tauraco johnstoni*), which shows little of its red wings in short flights, was suggested by Jonathan Kingdon.

*Source:* John G. Williams, *A Field Guide to the Birds of East Africa* (London: Collins, 1980), p. 12.

## Kikiyaon

Unknown BIRD of West Africa.

*Etymology:* Bambara (Mande) word.

*Physical description:* Like a large owl. Covered in greenish-gray fur rather than feathers. Immense wings. A sharp spur juts from each of its two shoulder joints. Large talons. Short, tufted tail.

*Behavior:* Call is a deep grunt like the "uh-uh-uhu-hoom-hoom" of Pel's fishing owl (*Scotopelia peli*). Also makes a noise like a person being strangled.

*Habitat:* Dense forest.

*Distribution:* Senegal.

*Source:* Karl Shuker, "The Secret Animals of Senegambia," *Fate* 51 (November 1998): 46—51.

## Kikomba

WILDMAN of Central Africa.

*Etymology:* Konjo, Nyanga, and Kanu (Bantu) word.

*Scientific names: Paranthropus congensis,* pro-posed by Charles Cordier in 1963; *Kikomba leloupi,* suggested by Bernard Heuvelmans in 1980.

*Variant names:* Abamaánji, Apamándi (Komo/Bantu), KAKUNDAKARI (possibly the female or young individuals), Tshingombé (Tembo/Bantu), Zaluzúgu (Lega-Mwenga/Bantu).

*Physical description:* Height, 5 feet 2 inches. Light skin. Covered in black hair. Long, black head-hair. Broad shoulders. Pronounced sexual dimorphism, if the KAKUNDAKARI is indeed the female.

*Behavior:* Bipedal. Uses a walking stick. Holds its long hair away from its eyes while walking. Howls more terrifyingly than a gorilla. Sometimes screams or barks like a Water chevrotain (*Hyemoschus aquaticus*). Steals game from traps. Eats honey from beehives, roots, and ginger-fruit. Knocks down trees in search of insects. Said to attack humans either by hitting them with its fists or an old axe handle or by wrestling.

*Tracks:* Length, 8—12 inches. Second toe larger than the first and third.

*Distribution:* Kivu Region, Democratic Republic of the Congo; possibly in Kenya, if it corresponds to the cryptid designated by Jacqueline Roumeguère-Eberhardt as hominid X1.

*Significant sighting:* In January 1960, a local man encountered a Kikomba along a path near the Umaté gold mine in a mountainous area of Kivu, Democratic Republic of the Congo. Charles Cordier drove 45 miles to the spot, where he found a humanlike footprint 8 inches long. Another time, near Tulakwa, he found several tracks 12 inches long.

*Possible explanations:*

(1) A large, solitary male Chimpanzee (*Pan troglodytes*).

(2) Surviving robust australopith, suggested by Bernard Heuvelmans. Australopiths were a family of Pliocene fossil hominids that persisted into the Early Pleistocene, 4.4—1.4 million years ago. More than 2,000 individual fossils are known. Three species of "robust" hominids are known in the genus *Paranthropus: P. aethiopicus* (East Africa), *P. boisei* (East Africa), and *P.*

*robustus* (South Africa). Five other "gracile" species have been placed in the genus *Australopithecus.* The distinction between gracile and robust genera is now seen as unwarranted, since body size is largely speculative. Robustness originally referred to the heavy structure of the skulls. *Paranthropus* had apelike skulls, sagittal crests anchoring massive jaw muscles, small incisors and canine teeth, enormous cheek teeth, and molarized bicuspids. They were vegetarians, based on the molar size, and probably ate leaves, fruit, tubers, seeds, and insects. *P. robustus's* cranial capacity was 450—550 milliliters. The few postcranial bones that have been found indicate a wide range of body sizes, probably due to sexual dimorphism. *Paranthropus* may have had the ability to manipulate stone tools, which some think makes them responsible for the early Oldowan tool industry, dating from 2.6—2.5 million years ago.

(3) Surviving *Homo ergaster,* the first known hominid with an essentially human body form. A complete skeleton was discovered in West Turkana, Kenya, in 1984. It lived 1.8—1.5 million years ago in East Africa. Its close resemblance to the Asian *Homo erectus* has led some to equate the two. Adults may have been 5 feet 7 inches tall, with slender torsos, long limbs, and narrow hips and shoulders. The cranium was high and rounded, with distinct browridges. The chewing teeth were smaller than those of *Homo habilis.* Cranial capacity was 850 milliliters. A meat-eater, *ergaster* may have been the first hominid to fashion a hand axe, perhaps 1.5 million years ago (Acheulean culture).

*Sources:* Charles Cordier, "Deux anthropoïdes inconnus marchant debout, au Congo ex-Belge," *Genus* 29 (1963): 2—10; Charles Cordier, "Animaux inconnus au Congo," *Zoo* 38 (April 1973): 185—191; Bernard Heuvelmans, *Les bêtes humaines d'Afrique* (Paris: Plon, 1980), pp. 570—598; Jacqueline Roumeguère-Eberhardt, *Les hominidés non-identifiés des forêts d'Afrique: Dossier X* (Paris: Robert Laffont, 1990).

## Ki-Lin

The UNICORN of East Asia. One of the four sacred animals of Chinese mythology, symbolizing wisdom and justice.

*Etymology:* Chinese (Sino-Tibetan) word, composed of *ki* ("male") + *lin* ("female").

*Variant names:* Ch'i-lin, Ki-rin (Japanese), Qi-lin, Sin-yu (Japanese), Tso'po (in Tibet), Zhi.

*Physical description:* Deerlike, though covered with scales. Multicolored. Single horn with a fleshy tip. Has a flamelike mane. Sometimes portrayed as winged. Horselike hooves. Tail of an ox. The more goatlike Zhi also had a single horn.

*Behavior:* Solitary and elusive. Said to live for 1,000 years. Tame, gentle nature.

*Distribution:* China and Japan.

*Possible explanations:*

(1) The Indian rhinoceros (*Rhinoceros unicornis*) was well known to the Chinese and accurately described as a completely different animal.

(2) A surviving sivathere, a subfamily of ox-sized giraffids from Eurasia and Africa with hefty builds, relatively short legs and necks, and branching, skin-covered horns. They lived from 15 million years ago, in the Late Miocene, to the Late Pleistocene.

*Sources:* Charles Gould, *Mythical Monsters* (London: W. H. Allen, 1886), pp. 348—359; Odell Shepard, *The Lore of the Unicorn* (Boston: Houghton Mifflin, 1930), pp. 94—97, 210—211; Georges Margouliès, *Anthologie raisonnée de la litterature chinoise* (Paris: Payot, 1948); Jeannie Thomas Parker, The Mythic Chinese Unicorn Zhi, 2001, http://www.rom.on.ca/pub/unicorn/index.html.

## Kiltanya

GIANT HOMINID of far eastern Siberia.

*Etymology:* Lamut (Tungusic), "goggle-eye."

*Variant names:* Arynk, Arysa ("plainsman"), Dzhulin ("sharp head"), Girkychavyl'in ("swift runner"), Rekhem, Teryk ("dawn man").

*Physical description:* Big eyes. Narrow nose bridge.

*Behavior:* Scavenges fish and game left by hunters.

*The* KING CHEETAH *(Acinonyx jubatus var. rex), a variety of cheetah with a thick coat marked by stripes and blotches. (Fortean Picture Library)*

*Tracks:* Length, 18 inches. Humanlike toes. Narrow heel.

*Habitat:* Mountains.

*Distribution:* Eastern Siberia.

*Source:* Myra Shackley, *Still Living? Yeti, Sasquatch and the Neanderthal Enigma* (New York: Thames and Hudson, 1983), pp. 133–134.

## King Cheetah

Large striped CAT of South Africa, once thought to be a separate species of cheetah.

*Etymology:* After the regal splendor of its coat.

*Scientific names: Acinonyx rex,* proposed by Reginald Pocock in 1927; modified later to *A. jubatus* var. *rex.*

*Variant names:* Mazoe leopard, Nsui-fisi (from the Swahili/Bantu *chui-fisi,* "leopard-hyena"), Rhodesian cheetah.

*Physical description:* Like the common cheetah but with a thicker, silky coat. Marked with slightly raised black stripes on the spine and dark blotches on a cream-colored background. A more pronounced mane. Fully ringed tail.

King cheetah variants are found in the litters of normal cheetahs.

*Behavior:* Nocturnal, as opposed to the traditional cheetah preference for daytime hunting.

*Habitat:* Forests, whereas the cheetah prefers open country, from desert to dry savanna.

*Distribution:* Zimbabwe; Botswana; Mozambique; Northern Province, South Africa. There is also a report of a single skin recovered from Burkina Faso in West Africa.

*Significant sightings:* First brought to scientific attention in 1926 when A. C. Cooper noticed an unusual skin in Harare's Queen Victoria Memorial Library and Museum. Reginald Pocock identified it as a cheetah's but with a vastly different coat pattern. At least twenty-one other skins were obtained through 1974.

The first King cheetah born in captivity was born to normally marked parents in 1981 at the Seaview Game Park in Port Elizabeth, South Africa. The DeWildt Cheetah and Wildlife Centre in North-West Province, South Africa, obtained twelve live King cheetah specimens between 1981 and 1987, three of them cubs born from their breeding program.

*Possible explanations:*
(1) Now generally seen as a single-locus genetic morph of the common Cheetah (*Acinonyx jubatus*). Lena Bottriell considers it to be an instance of evolution in the making: the modified, striped coat provides better camouflage as the cheetah adapts to night hunting in dense forests. If the King cheetahs are separated reproductively from the rest of the cheetah population for an appropriate amount of time, they may actually become a distinct species.
(2) By contrast, King cheetahs may represent a genetic throwback to the time when Africa was colder and more forested.
*Sources:* Reginald I. Pocock, "Description of a New Species of Cheetah (*Acinonyx rex*)," *Proceedings of the Zoological Society of London,* 1927, pp. 245—251, 257; Daphne M. Hills and Reay H. N. Smithers, "The 'King Cheetah': A Historical Review," *Arnoldia Zimbabwe* 9, no. 1 (1980): 1—23; Lena Godsall Bottriell, *King Cheetah: The Story of the Quest* (Leiden, the Netherlands: E. J. Brill, 1987); Karl Shuker, *Mystery Cats of the World* (London: Robert Hale, 1989), pp. 118—122; G. W. Frame, "First Record of the King Cheetah in West Africa," *Cat News* 17 (1992): 2—3; David Alderton, *Wild Cats of the World* (London: Blandford, 1993), pp. 38—42.

## Kipumbubu

Unknown CROCODILIAN of East Africa.
*Physical description:* A very large crocodile.
*Behavior:* Climbs up on the rim of riverboats at night, seizes people in its jaws, and eats them. Said to eat about six people each year.
*Distribution:* Rufiji River, Tanzania.
*Possible explanation:* Nile crocodiles (*Crocodylus niloticus*) kill about 300 people every year in Africa, but jumping 3 feet up onto boat rims is not a standard feeding method.
*Sources:* Rufiji [Ronald Delabere Barker], *The Crowded Life of a Hermit* (Nairobi, Kenya: W. Boyd, 1942—1944); Bernard Heuvelmans, *Les derniers dragons d'Afrique* (Paris: Plon, 1978), pp. 181—185, 372.

## Kitanga

Unknown big CAT of East and West Africa.
*Etymology:* Embu (Bantu) word.
*Physical description:* Cheetah with short limbs and other lionlike characteristics.
*Habitat:* Highland forests.
*Distribution:* Near Embu, Kenya; also Senegal in West Africa.
*Sources:* Kenneth C. Gandar Dower, *The Spotted Lion* (Boston: Little, Brown, 1937); Karl Shuker, *Mystery Cats of the World* (London: Robert Hale, 1989), pp. 123—124.

## Kiwákwe

CANNIBAL GIANT of the northeastern United States.
*Etymology:* Abnaki-Penobscot (Algonquian) word.
*Physical description:* Covered with an impenetrable shell. Mop of grizzly-bear hair on head.
*Tracks:* Enormous.
*Distribution:* Maine.
*Source:* Frank G. Speck, "Penobscot Tales and Religious Beliefs," *Journal of American Folklore* 48 (1935): 81—82.

## Klato

FRESHWATER MONSTER of British Columbia, Canada.
*Variant names:* Klamahsosaurus, Klatomsaurus, Klematosaurus.
*Physical description:* Length, 25 feet. Three to six humps. Middle hump is 5 feet in diameter. Orange belly. Long tail. Two tail flukes, 6 feet long.
*Distribution:* Oyster River, Vancouver Island, British Columbia.
*Source:* Mary Moon, *Ogopogo* (Vancouver, Canada: J. J. Douglas, 1977), pp. 161—162.

## Koau

Unknown flightless BIRD of Oceania.
*Etymology:* Marquesan (Austronesian) word.
*Variant name:* Koao.
*Physical description:* Similar to a rail. Size of a chicken. Purplish-blue plumage. Yellow bill. Stumpy wings. Long, yellow legs.

*Behavior:* Runs rapidly. Burrows in the mud.

*Distribution:* Hiva Oa, Îles Marquises, French Polynesia.

*Significant sighting:* Thor Heyerdahl and a native named Terai were horseback riding on Hiva Oa in 1937 when they saw a seagull-sized, wingless bird run along the trail and disappear into some ferns.

*Present status:* Possibly recently extinct, the victim of hunters.

*Possible explanations:*

(1) Surviving flightless rail (*Porphyrio paepae*) related to the Takahe (*P. mantelli hochstetteri*) of New Zealand and known from subfossil bones discovered on Hiva Oa and Tahuata in 1988, proposed by Michel Raynal.

(2) The Spotless crake (*Porzana tabuensis*), a black rail 6—8 inches long, also called the *koao,* suggested by Jean-Jacques Barloy. It runs swiftly and lives in some valleys on Ua Pou and Fatu Hiva. However, *koao* may refer to both the crake and an unknown rail.

*Sources:* Francis Mazière, *Archipel du Tiki* (Paris: Robert Laffont, 1957), p. 261; Thor Heyerdahl, *Fatu-Hiva: Back to Nature* (London: Allen and Unwin, 1974), p. 225; Jean-Jacques Barloy, *Merveilles et mystères du monde animal* (Geneva, Switzerland: Famot, 1979), pp. 115—117; Michel Raynal, "'Koau,' l'oiseau insaisissable des Îles Marquises," *Bulletin de la Société d'Étude des Sciences Naturelles de Béziers,* new ser., 8 (1980—1981): 20—26; Michel Raynal and Michel Dethier, "Lézards géants des Maoris et oiseau énigmatique des Marquisiens: La vérité derrière la légende," *Bulletin Mensuel de la Société Linnéenne de Lyon* 59, no. 3 (1990): 85—91; Michel Raynal, "The Mysterious Bird of Hiva-Oa," *INFO Journal,* no. 73 (Summer 1995): 17—21, updated in http://perso.wanadoo.fr/cryptozoo/dossiers/hiva_eng.htm.

## Koddoelo

Mystery animal of East Africa, similar to the NANDI BEAR.

*Etymology:* Pokomo (Bantu) word.

*Physical description:* Looks like a large baboon. Length, 6 feet. Shoulder height, 3 feet 6 inches. Reddish or yellow fur. Long nose. Large canines. Thick mane. Thick forelegs. Long claws. Tail is 18 inches long, 4 inches wide.

*Behavior:* Nocturnal. Fierce. Walks on four legs, occasionally on two. Cannot climb trees. Raids sheep pens. Attacks humans on sight.

*Tracks:* Five-toed, with one deep claw mark.

*Distribution:* Lower and Middle Tana River, Kenya.

*Possible explanation:* Surviving Giant baboon (*Theropithecus oswaldi*), a fossil baboon that lived in Kenya 650,000 years ago. The male was roughly the size of a female gorilla and weighed 250 pounds.

*Sources:* C. W. Hobley, "Unidentified Beasts in East Africa," *Journal of the East Africa and Uganda Natural History Society,* no. 7 (1913): 85—86; Charles R. S. Pitman, *A Game Warden among His Charges* (London: Nisbet, 1931), pp. 287—302.

## Kolowisi

FRESHWATER MONSTER of New Mexico.

*Etymology:* Zuni word.

*Physical description:* Horned. Gaping jaws. Has feathers and fins.

*Behavior:* Can cause floods.

*Habitat:* Underground streams.

*Distribution:* Western New Mexico.

*Sources:* Ruth L. Bunzel, "An Introduction to Zuñi Ceremonialism," *Annual Report of the Bureau of American Ethnology* 47 (1930): 487, 515—516; Etienne B. Renaud, *Pictographs and Petroglyphs of the High Western Plains* (Denver, Colo.: University of Denver, Department of Anthropology, 1936).

## Kondlo

Unknown BIRD of South Africa.

*Etymology:* Zulu (Bantu) word.

*Variant name:* Inkondhlo.

*Physical description:* Glossy-black, turkeylike bird. Sexes are similar. Smooth crest. No comb or baldness. Red beak. Red legs. Red claws.

*Behavior:* Flies low. Seen in flocks of five to ten animals. Voiceless. Eaten by the Zulu people.

*Habitat:* Grassy, treeless hills.

*Distribution:* KwaZulu-Natal Province, South Africa.

*Significant sighting:* G. T. Court shot and ate some of these birds from 1912 to 1914 in the Entonjaneni Hills near Melmoth, South Africa. He saw a flock again in November 1960.

*Possible explanations:*

(1) The Southern ground hornbill (*Bucorvus leadbeateri*), which has distinctive red wattles and a black bill.

(2) The Bald ibis (*Geronticus calvus*), which is a glossy, green color and has a prominent, bald head.

(3) An unknown species of wildfowl (Order Galliformes).

*Sources:* G. T. Court, "Inkondhlo?" *African Wild Life* 16 (1962): 81; G. T. Court, "'Kondlo': A Wild Fowl," *African Wild Life* 16 (1962): 342; Karl Shuker, "Gallinaceous Mystery Birds," *World Pheasant Association News,* no. 32 (May 1991): 3—6.


## Kongamato

FLYING REPTILE of Central and South Africa.

*Etymology:* Kaonde (Bantu), "broken boats."

*Physical description:* Length, 2 feet 6 inches—4 feet 6 inches. Smooth skin. Black or red in color. Long beak with teeth. Batlike wings. Wingspan, 3—7 feet. Long, narrow tail.

*Behavior:* Said to capsize canoes by diving in the water. Said to attack and eat people occasionally. It is particularly fond of their little fingers, toes, earlobes, and noses.

*Habitat:* Caves near rivers and swamps.

*Distribution:* The Mwinilunga District, the Mutanda River, and the Bangweulu and Jiundu Swamps of northern Zambia; parts of Zimbabwe.

*Significant sightings:* In 1923, Frank Melland described the belief of the Kaonde people of Zambia that a huge flying reptile with bat wings lived in the Jiundu Swamp. When crossing rivers, some of them carried amulets that would protect them from a Kongamato. When he showed them pictures of pterodactyls in books, they identified them as looking like the Kongamato.

In 1925, G. Ward Price heard stories of a monstrous bird with a long beak that attacked people in the swamps of Zimbabwe. When a man who had been wounded by the animal was shown a picture of a pterodactyl, he screamed in terror.

Engineer J. P. F. Brown saw two flying reptiles in January 1956 near Mansa, Zambia. They had long, narrow tails and a wingspan of 3 feet—3 feet 6 inches. From beak to tail, they were about 4 feet 6 inches.

A man was brought into a hospital in Mansa in 1957, suffering from a chest wound. He claimed a huge bird in the Bangweulu Swamp had attacked him.

*Possible explanations:*

(1) The Shoebill (*Balaeniceps rex*) looks like a large, silver-gray stork, 4 feet long, with an 8-foot wingspan and a distinctive, 8-inch-long, hooked bill. In flight, it retracts its head and neck like a heron. It is a closer relative of the pelicans than true storks. Like all other living birds, the shoebill has no teeth.

(2) The Saddle-billed stork (*Ephippiorhynchus senegalensis*) is second only to an ostrich in standing height. It has a 9-foot wingspan and is 5 feet in length. It has black-and-white plumage and a black head and neck. The long, upturned bill is red with a black band in the middle and a brilliant yellow frontal shield. Its white breast has a bare, red "medal."

(3) The Southern ground hornbill (*Bucorvus leadbeateri*) is about 3 feet 6 inches long and dull black with white primary feathers. It has a heavy, downcurved, black bill and bright red skin around its eye and down its foreneck.

(4) Lord Derby's anomalure (*Anomalurus derbianus*), a small, gliding squirrel—like rodent, was proposed by museum director Reay Smithers. It is only 15 inches long.

(5) A surviving pterosaur, the flying reptiles of the Mesozoic era. Fossils of *Pterodactylus* (wingspan 1—8 feet, short tail), *Dsungaripterus* (wingspan 9—12 feet, short tail), and *Rhamphorynchus* (wingspan 1—6 feet, long tail) from the Jurassic have been

*The KONGAMATO, a huge flying reptile of Central Africa. (William M. Rebsamen)*

found at Tendaguru Hill, Tanzania. Only two pterosaur fossils from the Cretaceous have been discovered in Africa: a wing bone of an *Ornithocheirus* (wingspan 14—16 feet) in the Democratic Republic of the Congo and a neck vertebra from a species similar to the giant *Quetzalcoatlus* (wingspan 36—39 feet but no teeth). However, the fossil record in South America is much richer, and since the two continents were joined at the time, there is reason to suspect more specimens will turn up.

(6) Carl Wiman suggested that the Kongamato tradition originated with natives who assisted in the excavation of pterosaur bones at the Tendagaru fossil beds in Tanzania prior to World War I.

*Sources:* Frank H. Melland, *In Witchbound Africa* (London: Seeley, Service, 1923), pp. 236—242; Carl Wiman, "Ein Gerücht von einem lebenden Flugsaurier," *Natur und Museum* 58 (1928): 431—432; Vernon Brelsford, "Some Northern Rhodesian Monsters," *African Observer* 4, no. 6 (1936): 58—60; Charles R. S. Pitman, *A Game Warden Takes Stock* (London: J. Nisbet, 1942), pp. 202—203; Stany [Roger de Chateleux], *Loin des sentiers battus: Douze femmes* (Paris: La Table Ronde, 1953), vol. 4, pp. 217—232; "Pterodactyls Seen near Northern Rhodesian River," *Rhodesia Herald,* April 2, 1957; "Museum Director Says There Are No Flying Reptiles," *Rhodesia Herald,* April 5, 1957; G. Ward Price, *Extra-Special Correspondent* (London: George Harrap, 1957), p. 178; Zoé Spitz-Bombonnel, "Animaux perdus et non retrouvés," *Le Chasseur Français,* June 1959, p. 375; Maurice Burton and C. W. Benson, "The Whale-Headed Stork or Shoe-Bill: Legend and Fact," *Northern Rhodesia Journal* 4 (1961): 411—426; Tom Dobney, "Myths and Monsters," *Horizon* (Salisbury) 6 (September 1964): 24—26; Bernard Heuvelmans, *Les derniers dragons d'Afrique* (Paris: Plon, 1978), pp. 417—427, 445—456.

## Kooloo-Kamba

Mystery PRIMATE of Central Africa.

*Etymology:* Mbama (Bantu), either from *n'koula* ("chimpanzee") or from its call "kooloo" + *kamba* ("speak").

*Scientific name: Pan troglodytes koolokamba,* given by W. C. Osman-Hill in 1967.

*Variant names:* Choga, DEDIÉKA, Ebôt (Bulu/Bantu), Itsena, Koolakamba, Koulanguia (Kélé/Bantu, *koula* "chimpanzee" + *nguia* "gorilla"), Koulou-nguira, Kulu-kampa, Kulu-kanba, N'tchego, Sipandjee.

*Physical description:* Larger than a normal chimpanzee. Cranium is larger than a chimpanzee's, with some cresting. Ebony-black, prognathous face. Heavy browridge. Wide, flat, fleshy nose. Small ears. Powerful jaws. Upper and lower incisors meet squarely. Broad pelvic structure.

*Behavior:* Frequently walks bipedally. Call is "koola-kooloo koola-kooloo." Aggressive. Lives singly or in smaller groups than other chimpanzees.

*Habitat:* Primarily high-altitude forests, although stray individuals are apparently found elsewhere with normal chimpanzee groups.

*Distribution:* Gabon; Cameroon; Equatorial Guinea.

*Significant sightings:* In the 1850s, Paul Du Chaillu shot a male Kooloo-kamba in southwestern Gabon. It was smaller than an adult male gorilla but stockier than a female gorilla. It had a round head and face, a small nose, and large ears. The skull is housed in the British Museum of Natural History.

A 4-foot-tall female ape nicknamed "Mafuca" was taken to the Dresden Zoo in 1874 from the port of Loango in the Republic of the Congo. Several observers classified it as a young female gorilla, others were convinced it was a chimpanzee, and still others thought it could be a chimp-gorilla hybrid. Sir Arthur Keith in 1899 classed Mafuca with Du Chaillu's Kooloo-kamba. Some zoologists now think it likely that Mafuca was a bonobo, which can be stockier than some chimpanzees.

Louis de Lassaletta collected a Kooloo-kamba in the hilly Nsok region of Equatorial Guinea in 1954.

Individuals with Kooloo-kamba characteristics have been maintained in the Coulston Foundation's animal experimentation laboratory in Alamogordo, New Mexico, since the 1960s.

In 1993, Steve Holmes saw a "wildman" in the Gamba coastal area of Gabon. It was just under 5 feet tall and running with its arms held high above its head. Nearby villagers called it the Sipandjee and said it was aggressive.

*Possible explanations:*

(1) An unknown species or subspecies of Chimpanzee (*Pan troglodytes*), suggested by Du Chaillu and E. Franquet.

(2) A Lowland gorilla (*Gorilla gorilla*) with uncharacteristic individual variations. A supposed Kooloo-kamba was brought to the Basel Zoo in 1967, but it turned out to be a red-backed female gorilla.

(3) A misidentified Bonobo (*Pan paniscus*), which was not recognized as a separate chimpanzee species until 1933.

(4) A chimp × gorilla hybrid. Though these two apes are closely related, successful hybridization between them is unknown either in captivity or in the wild. Individuals with both chimp and gorilla characteristics merely reflect this close genetic relationship.

(5) A misidentified large male chimpanzee. Facial color in chimps darkens with age.

(6) A misidentified small female gorilla, the equivalent of a PYGMY GORILLA.

(7) An emergent variety or species with adaptations to a mountainous habitat, suggested by Karl Shuker.

*Sources:* E. Franquet, "Sur le Gabon et sur les diverses espèces de singes anthropomorphes d'origine africaine," *Archives du Muséum d'Histoire Naturelle* 10 (1858): 91—97; Paul B. Du Chaillu, *Explorations and Adventures in Equatorial Africa* (London: John Murray, 1861); Paul B. Du Chaillu, *Stories of the Gorilla Country* (New York: Harper, 1868); Arthur Keith, "On the Chimpanzees and Their Relationship to the Gorilla," *Proceedings of the Zoological Society of London,* 1899, pp. 296—312; Raingeard, "Note sur un anthropoïde africain: Le Koula-Nguia," *Mammalia* 2 (1938): 81—83; Ernst Schwarz, "A propos du Koula-Nguia," *Mammalia* 3 (1939): 52—58; Albert

Irwin Good, "Gorilla-Land," *Natural History* 56 (January 1947): 36—37, 44—46; W. C. Osman Hill, "The Nomenclature, Taxonomy and Distribution of Chimpanzees," in Geoffrey H. Bourne, ed., *The Chimpanzee* (Basel, Switzerland: S. Karger, 1969), vol. 1, pp. 22—46; Bernard Heuvelmans, *Les bêtes humaines d'Afrique* (Paris: Plon, 1980), pp. 301—304, 417—440; Don Cousins, "On the Koolakamba: A Legendary Ape," *Acta Zoologica et Pathologica Antverpiensia* 75 (1980): 79—93; Brian T. Shea, "Between the Gorilla and the Chimpanzee: A History of Debate Concerning the Existence of the Kooloo-kamba or Gorilla-Like Chimpanzee," *Journal of Ethnobiology* 4 (1984): 1—13; Elaine Jane Struthers, "Koolakamba," Primate Info Net, University of Wisconsin, Madison, July 17, 1996, http://www.primate.wisc.edu/pin/koola.html; Steve Holmes, "Incident in Gabon," *Fortean Times,* no. 113 (August 1998): 52; Don Cousins, "No More Monkey Business," *Fortean Times,* no. 136 (August 2000): 48.

## Koosh-Taa-Kaa

CANNIBAL GIANT of Alaska and northwestern Canada.

*Etymology:* Haida-Tlingit (Na-Dené), "land-otter man."

*Physical description:* Covered with long hair except for the face.

*Behavior:* Tries to steal the souls of people who are drowned or lost in the woods.

*Distribution:* British Columbia, Canada; Alaska.

*Sources:* John R. Swanton, "Tlingit Myths and Texts," *Bulletin of the Bureau of American Ethnology* 39 (1909): 86; Frederica de Laguna, "Under Mount Saint Elias: The History and Culture of the Yakutat Tlingit," *Smithsonian Contributions to Anthropology* 7 (1972): 744—749, 766; Grant R. Keddie, "On Creating Un-Humans," in Vladimir Markotic and Grover Krantz, eds., *The Sasquatch and Other Unknown Hominoids* (Calgary, Alta., Canada: Western Publishers, 1984), pp. 22—29.

## Kra-Dhan

Unknown PRIMATE of Southeast Asia.

*Etymology:* Bahnar (Austroasiatic) word.

*Variant names:* Bêc'-boc, Bekk-bok, Con lười ười (Vietnamese/Austroasiatic).

*Physical description:* Large, monkeylike animal.

*Behavior:* Bipedal. Vicious. Has a chameleon-like ability to change color. Call is an insane laugh. Attacks humans.

*Tracks:* Length, 18 inches. Width, 8 inches. Stride, 4 feet.

*Distribution:* Annam Highlands near Kon Tum and Pleiku, Vietnam.

*Significant sighting:* In 1943, a Kra-dhan killed a man near Kon Tum, Vietnam.

*Possible explanation:* A surviving mainland population of the Orangutan (*Pongo pygmaeus*), which is now limited to the islands of Borneo and Sumatra. Orangutan fossils from around 2 million years ago have been found in Laos, Vietnam, and southern China, as well as the islands of Sumatra, Java, and Borneo. These apes are more distantly related to humans than are Chimpanzees (*Pan troglodytes*) and Gorillas (*Gorilla gorilla*). The lineage is unclear, but the likeliest theory is that they derived from *Sivapithecus*, an extinct ape that lived in India and Pakistan in the Late Miocene, 12—8 million years ago.

*Sources:* Ivan T. Sanderson, *Abominable Snowmen: Legend Come to Life* (Philadelphia: Chilton, 1961), pp. 244—245; "Abominable Jungle-Men," *Pursuit,* no. 10 (April 1970): 36—37.

## Kraken

Giant CEPHALOPOD of the Atlantic and Pacific Oceans. Accepted by science after remains that washed up near Ålbæk, Denmark, in 1853 were examined, the Giant squid (*Architeuthis*) was described officially by Danish naturalist Japetus Steenstrup in 1857. The largest known specimens of giant squid have a total length, including their two long arms, of about 55 feet. The evidence for even larger animals is considered here as the legacy of the Kraken.

*Etymology:* The plural form of the Norwegian

*krake,* first mentioned by Francesco Negri in 1700. Possibly related to a word meaning "uprooted tree" because the squid's body and arms appear similar to the trunk and roots of a tree.

*Scientific name: Architeuthis dux,* given by Johannes Japetus Steenstrup in 1857.

*Variant names:* Aale tust (Norwegian, "tuft of eels"), Anker-trold ("anchor-troll"), Horv ("harrow"), Kolkrabbi, Krabbe ("crab"), Kraxen, Sciu-crak, Sæ-horven.

*Physical description:* In Norwegian mythology, the Kraken is a supergiant squid, with a body 1.5 miles in circumference. It appears like several small islands surrounded by seaweed. Dark brown with light speckles. High, broad forehead. Large eyes. Pointed snout (actually the tail). Its arms or tentacles are as big as medium-sized ships.

*Behavior:* Causes fishes to come closer to the surface when it rises; creates a huge eddy when it sinks. Said to attack ships and grasp their rigging with its arms.

*Distribution:* Atlantic and Pacific Oceans.

*Significant sightings:* In 1801, Pierre Denys de Montfort noted that in the chapel of St. Thomas at St.-Malo in Brittany, France, there was a votive picture showing a huge squid or octopus attacking a ship by winding its arms around the masts and rigging. The incident is said to have taken place off the coast of Angola. The ship's sailors made a vow to St. Thomas that they would make a pilgrimage if he would save them, then set to work with their axes and cutlasses to cut off the monster's tentacles. Later, they went directly to the chapel in St.-Malo, where a picture was hung illustrating their adventure.

Denys de Montfort also interviewed whalers at Dunkerque, Pas-de-Calais Department, France, who told him some squid stories. An American, Captain Reynolds, described a cut-off squid arm that was 45 feet long and 2 feet 6 inches in diameter. A retired Danish captain named Jean-Magnus Dens said he had encountered a huge squid, again off the coast of Angola, that had attacked and killed three men on board his ship. The crew sank five harpoons into the monster before it was finished. Dens estimated the animal's arms were more than 35 feet long.

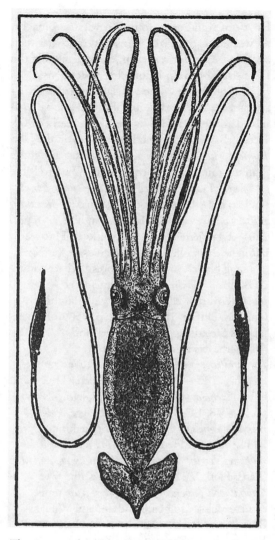

*The giant squid* (Architeuthis). *(© 2002 ArtToday.com, Inc., an IMSI Company)*

Frank Bullen's description of a moonlight battle between a sperm whale and a huge squid in the Strait of Malacca in the Indian Ocean in 1875 is, at best, exaggerated.

The size of toothed sucker marks on the skin of sperm whales has been offered as evidence of extremely large giant squids. However, marks greater than 1—2 inches in diameter are difficult to verify. The suckers of the 46-foot specimen that washed ashore at Bonavista Bay, Newfoundland, in 1872 measured 2.5 inches in diameter.

Bernard Heuvelmans also cites the great

length of squid arms found in whale stomachs, as well as a theorized constant ratio between sucker size and arm length, as evidence for the large size of certain incomplete specimens.

One of the few known sightings of a living specimen at the surface involved an animal estimated to be 100 feet in length. In early 1969, Dennis Braun and two other marines on the USS *Francis Marion* watched this monster for more than ten minutes off Vieques Island, Puerto Rico.

*Sources:* Francesco Negri, *Viaggio settentrionale* (Padua, Italy, 1700); Erik Pontoppidan, *Natural History of Norway* (London: A. Linde, 1755), pp. 210—218; Pierre Denys de Montfort, "Histoire naturelle des mollusques, animaux sans vertebres et a sang blanc," in Georges Louis Leclerc, comte de Buffon, *Histoire naturelle, générale et particulière,* new ed., edited by C. S. Sonnini (Paris: F. Dufart, 1801), vol. 2, p. 256; Johannes Japetus Steenstrup, "Oplysninger om Atlanterhavets colossale Blæksprutter," *Forhandlinger ved de Skandinaviske Naturforskeres* 8 (1857): 182—185; Frank T. Bullen, *The Cruise of the Cachalot* (New York: D. Appleton, 1899), pp. 77—78, 143—144; Kristian Brugge, *Folke-minneoptegnelser* (Oslo: Norsk Folkeminnelag, 1934); Bernard Heuvelmans, *Dans le sillage des monstres marins: Le kraken et le poulpe colossal* (Paris: Plon, 1958); Japetus Steenstrup, *The Cephalopod Papers of Japetus Steenstrup,* trans. Agnete Volsøe, Jørgen Knudsen, and William Rees (Copenhagen: Danish Science Press, 1962); Bernard Heuvelmans, *In the Wake of the Sea-Serpents* (New York: Hill and Wang, 1965), pp. 45—79; Tor Åge Bringsværd, *Phantoms and Fairies from Norwegian Folklore* (Oslo: Johan Grundt Tanum Forlag, 1970), pp. 67—71; Simon Welfare and John Fairley, *Arthur C. Clarke's Mysterious World* (London: Collins, 1980), pp. 71—74; Richard Ellis, *The Search for the Giant Squid* (New York: Lyons, 1998); Michel Meurger, "Francesco Negri, the Kraken, and the Sea Serpent," *Fortean Studies* 6 (1999): 238—244.

## Ksy-Gyik

WILDMAN of Central Asia.

*Scientific name: Primihomo asiaticus,* proposed by V. A. Khakhlov in 1914.

*Etymology:* From the Kyrgyz (Turkic) *kishi* ("man") + *giik* ("wild" or "powerful").

*Variant names:* Kiik-adam, Kiik-kish, Kish-kiik, Kishi-kiyik.

*Physical description:* Height, 5 feet. Covered with dark-brown or yellowish, shaggy hair. Sloping forehead. Arched browridges. Small nose with large nostrils. Ears are large, lobeless, and backward-pointing. Massive lower jaw. No chin. A hump on the back of the neck. Long arms. Stoop-shouldered. Female has breasts.

*Behavior:* Runs awkwardly, swinging its arms. Screeches. Eats raw meat, vegetables, and grain. Drinks water by lapping or by dipping its arm in water and lapping up the drips. Lives in rock shelters strewn with grass. Sleeps by squatting on its knees and elbows, resting its forehead on the ground, and placing its wrists over its head.

*Tracks:* Short and broad. Splayed toes. Large toe smaller than a human's.

*Distribution:* The Altai Mountains of Kazakhstan; the Kirgiz Steppe around Astana and Qaraghandy, Kazakhstan; the Chatkal and Alai Mountains of Kyrgyzstan; the Junggar Pendi depression in Xinjiang Uygur Autonomous Region, northern China.

*Significant sightings:* The young Russian zoologist V. A. Khakhlov spent much of 1911 and 1912 with Kazakh herders in the area around Zaysan Köli and the Tarbagatay Mountains, Kazakhstan, with an excursion into neighboring Xinjiang Uygur Autonomous Region, China. During this time, he collected valuable data from two Kazakhs who served as his guides and who had seen the creatures at various times.

In July 1948, a Kazakh herder named Mad'yer showed geologist A. P. Agafonov the preserved hand of a wildman that his grandfather had killed in the Chatkal Mountains of Kyrgyzstan, probably in the mid-nineteenth century. The creature had tried to carry off his young wife, so he killed it with his hunting knife and cut off the hand as a trophy. Agafonov said the hand was human and covered with long, sparse hairs except on the palm. Boris Porshnev attempted to locate the hand in 1963, but Mad'yer had died, and his heir pretended not to know of the relic.

In August 1948, geologist M. A. Stronin was

camped in a remote area of the Alai Mountains near the Inyl'chek River in Kyrgyzstan when his guides woke him in alarm. A Kiik-kish with thick, yellowish hair was trying to steal their horses, but it ran away on two legs down an extremely steep slope.

In the summer of 2001, a Kyrgyz frontier guard in the Alai Mountains discovered a set of human tracks 18 inches long and 12 inches wide in the clay bank of a river.

*Sources:* Vitaly A. Khakhlov, ["On the Question of Wild Men: Preliminary Note"], unpublished report in the Archives of the Akademiia Nauk, Historical-Philological Section, 1914, possibly still in the Akademiia archives, St. Petersburg or Moscow, Russia; Ivan T. Sanderson, *Abominable Snowmen: Legend Come to Life* (Philadelphia: Chilton, 1961), pp. 307—308, 313—318; Bernard Heuvelmans and Boris F. Porshnev, *L'homme de Néanderthal est toujours vivant* (Paris: Plon, 1974), pp. 49—64, 141, 150—161; "Bigfoot's Footprints Found in Kyrgyzia Republic," *Pravda,* August 29, 2001.

## Ktchi Pitchkayam
FRESHWATER MONSTER of eastern Canada.
*Etymology:* Micmac (Algonquian), "great snake."
*Variant names:* Chepitchkaam, Chepitkam, Ktchí at'husis, Ktchi kinepíkwa, Tcipitckaam.
*Distribution:* Nova Scotia and New Brunswick.
*Sources:* Silas Tertius Rand, *Legends of the Micmacs* (New York: Longmans, Green, 1894), p. 53; Albert S. Gatschet, "Water-Monsters of American Aborigines," *Journal of American Folklore* 12 (1899): 255—260.

## Kuddimudra
FRESHWATER MONSTER of Australia.
*Etymology:* Australian word.
*Variant name:* Coochie.
*Physical description:* Serpentine. Hair on the head.
*Behavior:* Agitates the water. Eats people.
*Distribution:* Water holes near the Diamantina River, South Australia.

*Source:* George Farwell, *Land of Mirage: The Story of Men, Cattle and Camels on the Birdsville Track* (London: Cassell, 1950).

## Kul
GIANT HOMINID of western Siberia.
*Etymology:* Khanty (Ob-Ugric) word, though seemingly related to the Tajik (Persian) GUL.
*Variant name:* Uten-ekhti-agen ("forest wanderer").
*Physical description:* Height, 6—7 feet. Body-hair. Black face. Glowing red eyes. Arms are longer than a man's.
*Behavior:* Excellent swimmer. Turns feet in when walking. Follows an annual migratory path.
*Distribution:* Northern Ob' River basin, Yamal-Nenets Autonomous Province, Siberia.
*Source:* Dmitri Bayanov, *In the Footsteps of the Russian Snowman* (Moscow: Crypto-Logos, 1996), pp. 126—129.

## Kumi
Giant LIZARD of Australasia.
*Etymology:* Maori (Austronesian) word.
*Physical description:* Length, 5—12 feet. Huge jaws with curved teeth. Six legs.
*Behavior:* Arboreal.
*Distribution:* North Island, New Zealand.
*Significant sighting:* In September 1898, a Maori bushman on W. D. Lysnar's ranch near Gisborne, North Island, was startled by the sight of a huge lizard, some 5 feet long, advancing toward him.
*Possible explanations:*
(1) An out-of-place Tuatara (*Sphenodon punctatus*), a 2-foot-long, lizardlike reptile that is now confined to about twenty small islands off the northeast coast of New Zealand and in Cook Strait.
(2) A living population of Delcourt's giant gecko (*Hoplodactylus delcourti*), a 2-foot-long gecko that once lived in New Zealand.
(3) Distorted folk memories of the Saltwater crocodile (*Crocodylus porosus*), found much farther to the west off the northern coast of Australia.

*Sources:* James Hector, "On the Kumi," *Transactions of the New Zealand Institute* 31 (1899): 717—718; H. W. Orsman, ed., *The Dictionary of New Zealand English* (Auckland, New Zealand: Oxford University Press, 1997), p. 430.

## Kung-Lu

TRUE GIANT of Southeast Asia.

*Etymology:* Unknown, said to mean "mouth man."

*Physical description:* Gorilla-like. Height, 20 feet.

*Behavior:* Said to eat humans.

*Tracks:* Leaves a trail of broken trees.

*Habitat:* Mountains.

*Distribution:* Myanmar, near the Thai border.

*Source:* Hassoldt Davis, *Land of the Eye* (New York: Henry Holt, 1940), p. 111.

## Kungstorn

Giant BIRD of Northern Europe.

*Etymology:* Swedish, "King's eagle."

*Variant name:* Svanhildørn ("Svanhild's eagle").

*Physical description:* Wingspan, 6 feet 6 inches.

*Behavior:* Can lift prey as large as a reindeer calf.

*Distribution:* Norway; Sweden.

*Significant sighting:* On June 5, 1932, three-year-old Svanhild Hansen was snatched by an eagle and carried for about 1.2 miles up to the mountain of Hagafjell on the island of Leka, Nord-Trøndelag County, Norway. The 42-pound girl was found essentially unharmed after a seven-hour search by 100 people. The incident was said to have been made into a film by Knut Vadseth and Skule Eriksen.

*Possible explanations:*

(1) Probably a *kungsörn,* or Golden eagle (*Aquila chrysaetos*), which has a wingspan of 6—7 feet.

(2) The bird that abducted Svanhild Hansen was said by some to be a White-tailed eagle (*Haliaeetus albicilla*), found along the Norwegian coast. It is 3 feet long and also has a wingspan of 6—8 feet.

*Sources:* Steinar Hunnestad, *Ørnerovet: Skildring med virkelighetsmotiv* (Bergen, Norway: Lunde, 1960); Karl Shuker, "Big Birds in Scandinavia," *Fortean Times,* no. 139 (November 2000): 23; Karl Shuker, "Scandinavian 'Big Birds' Update," *Fortean Times,* no. 141 (January 2001): 23.

## Kurrea

FRESHWATER MONSTER of Australia.

*Etymology:* Australian word.

*Physical description:* Serpentine.

*Behavior:* Can travel from one lake to another by digging its own water channel. Said to eat humans.

*Distribution:* Boobera Lagoon, south of Goodiwindi, New South Wales.

*Source:* Robert Hamilton Mathews, *Folklore of the Australian Aborigines* (Sydney, Australia: Hennessey, Harper, 1899).

## Kushii

FRESHWATER MONSTER of Japan.

*Etymology:* Japanese, after Lake Kussharo.

*Variant name:* Kussie.

*Physical description:* Eel-like. Dark color.

*Distribution:* Kussharo-ko, Hokkaido Territory, Japan.

*Significant sightings:* Toshio Komama took a distant photo of two animals in the lake on September 2, 1973.

On September 18, 1973, Yoshinori Kataoka saw a 50-foot animal with ridges along its back. It was moving very swiftly in the lake and creating a wake.

*Sources:* Ronald Yates, "Old Nessie Makes Room for Kussie," *Chicago Tribune,* June 17, 1976, p. 1; Simon Welfare and John Fairley, *Arthur C. Clarke's Mysterious World* (London: Collins, 1980), pp. 107—108; "Japan's Own Sea Serpent," *Newsweek,* August 11, 1997, p. 8.

## Kynoképhalos

SMALL HOMINID of ancient India or Africa.

*Etymology:* Greek, "dog-headed."

*Variant names:* Calinges, Calystrian, Choromanda, Cynocephalos, Dog-man, Kalystriai, Sunamukha.

*Physical description:* Height, 2—3 feet. Head is like a dog's. Snub-nosed. Big teeth. Long beard. Claws or long fingernails. Both men and women are said to have tails.

*Behavior:* Barking language. Eats raw meat. Drinks sheep's milk. Lives in caves in the mountains. Said to live to be 170—200 years old. Tends sheep and oxen. Uses bow and arrows and spears skillfully. Hunts hares with ravens, kites, crows, and vultures. Wears animal skins. Sails in boats on an oily lake.

*Distribution:* South and central India west to the Indus River; North Africa. (The names India and Ethiopia were widely used synonymously by ancient and medieval writers.)

*Possible explanations:*

(1) Various "hill tribes" of southern India, possibly the Kadar, Irular, Panniyan, or Kurumba peoples.

(2) The Hamadryas baboon (*Papio hamadryas*) has a doglike face and is found in Arabia, Ethiopia, and Sudan.

(3) The Hoolock gibbon (*Hylobates hoolock*) is the only ape found in India, standing nearly 3 feet when upright.

(4) A derogatory name for any disliked group of people.

*Sources:* Herodotus, *The Histories,* trans. Aubrey de Sélincourt (London: Penguin, 1996), p. 276 (IV. 191); Ctesias, *Indika,* in J. W. McCrindle, ed., *Ancient India* (Calcutta, India: Thacker, Spink, 1882), pp. 15—16, 21—25, 52—53, 63, 84—90; Pliny the Elder, *Historia naturalis,* in John F. Healy, ed., *Natural History: A Selection* (New York: Penguin Classics, 1991), pp. 78—79 (VII. 21—27); Ælian, *De natura animalium,* IV. 46, X. 25; David Gordon White, *Myths of the Dog-Man* (Chicago: University of Chicago Press, 1991), pp. 26—30, 47—70.

# L

## La La

CANNIBAL GIANT of western Canada.

*Etymology:* Heiltsuk (Wakashan) word.

*Distribution:* Central British Columbia.

*Source:* Susanne Storie and Jennifer Gould, *Bella Coola Stories* (Victoria, Canada: British Columbia Indian Advisory Committee Project, 1971), pp. 1, 44, 101.

## Lagarfljótsormurinn

FRESHWATER MONSTER of Iceland.

*Etymology:* Icelandic, "the serpent of Lagarfljót." (The Lagar Fljót River flows out of Lögurinn Lake.)

*Variant name:* SKRIMSL.

*Physical description:* Total length, 46 feet. Pale color. Head and neck, 6 feet long. Face has whiskers. One large hump. Body, 22 feet. Tail, 18 feet.

*Behavior:* Swims with undulations.

*Distribution:* Lögurinn Lake, Iceland.

*Significant sightings:* A huge, humped animal was seen in the lake in 1345. Other prominent sightings occurred in 1749, 1750, and 1819.

In 1998, a class of children and their teacher from Hallormsstadarskóli School witnessed a pale-colored streak undulating through the water for about twenty-five minutes close to the shore near the Geitagerdi farm. One of the students is said to have taken a photograph.

*Possible explanations:*

(1) Large bubbles of methane gas welling up from the bottom of the lake.

(2) Horses and cows swimming in the water, mistaken for the monster.

(3) Mats of leaves and other plant matter brought together by strong river currents.

*Sources:* Sabine Baring-Gould, *Iceland: Its Scenes and Sagas* (London: Smith, Elder, 1863), pp. 345—348; Jón Árnason, *Icelandic Legends* (London: Richard Bentley, 1864), pp. 106—108; Axel Olrik, *Ragnarok: Die Sagen vom Weltuntergang* (Berlin: W. de Gruyter, 1922); Simon Welfare and John Fairley, *Arthur C. Clarke's Mysterious World* (London: Collins, 1980), pp. 102—103; "The Lagarfljót Monster and Other Water Beasts," *Daily News from Iceland,* May 28, 1999; "Monster Alert," *Daily News from Iceland,* May 16, 2000.

## Lake Sentani Shark

Unknown FISH of Australasia.

*Distribution:* Lake Sentani, Irian Jaya, Indonesia.

*Significant sighting:* During World War II, George Agogino dropped a grenade into the lake in order to kill some fishes for his army unit to eat. A 12-foot shark was brought to the surface, and he was able to sketch it before it sank.

*Possible explanations:*

(1) The sharklike Largetooth sawfish (*Pristis microdon*), normally a marine or riverine species, is found in this lake. It can grow to a length of 19 feet. If the grenade had torn off its bladelike snout, Agogino might have mistaken it for a shark.

(2) The Bull shark (*Carcharhinus leucas*), which grows to 11 feet, has been reported from Lake Jamoer in Irian Jaya.

*Source:* Bernard Heuvelmans, "Annotated Checklist of Apparently Unknown Animals with Which Cryptozoology Is Concerned," *Cryptozoology* 5 (1986): 1, 12.

## Lake Titicaca Seal

Mystery SEAL or SIRENIAN of South America.

*Physical description:* Looks like a seal or manatee. Length, 12 feet.

*Distribution:* Lago Titicaca, especially around the Copacabana Peninsula and the Strait of Tiquina, Bolivia.

*Present status:* The only exclusively freshwater seals are the Baikal seal (*Phoca sibirica*), found in Lake Baikal, Buryatia Republic, Siberia, and the Caspian seal (*Phoca caspica*) of the Caspian Sea in West Asia. The Amazon manatee (*Trichechus inunguis*) of Brazil and Colombia is the only known sirenian completely confined to freshwater drainages, including lakes, rivers, and floodplains.

*Possible explanation:* Unknown seal or sirenian, too poorly described for a diagnosis.

*Source:* Adolph F. Bandelier, *The Islands of Titicaca and Koati* (New York: Hispanic Society of America, 1910).

## Lake Worth Monster

HAIRY BIPED of north Texas.

*Variant names:* Goatman, Hairy horror.

*Physical description:* A cross between a goat and a man. Height, 7 feet. Weight, 250—300 pounds. Covered with both scales and whitish-gray hair.

*Behavior:* Bipedal. An agile swimmer. Growls or makes a pitiful "yeeepe" or "yuuuu" cry. Has a foul odor. Said to attack cars. Kills sheep by breaking their necks.

*Tracks:* Length, 16 inches. Width, 8 inches at the toes.

*Distribution:* Lake Worth, east of Fort Worth, Texas.

*Significant sightings:* John Reichart, his wife, and two other couples were parked around Lake Worth on the north side of Fort Worth just after midnight on July 10, 1969, when a monster jumped out of the trees onto the Reicharts' car. Reichart drove away quickly after the creature tried to grab his wife. Police later found an 18-inch scratch in the side of the car. Sightings by more than 100 other people turned up throughout the summer and fall, until early November.

On November 7, 1969, Charles Buchanan was sleeping in the bed of his pickup truck on the shore of Lake Worth when he awoke to see a large humanoid looking down at him. After it pulled him from the truck while he was still in his sleeping bag, he shoved a bag of leftover chicken into its face. It took the bag in its mouth, jumped into the water, and swam toward Greer Island.

*Possible explanation:* Hoax and exaggeration have undoubtedly contributed to these reports. Sallie Ann Clarke admits that much of her book on the subject was written as fiction.

*Sources:* Jim Marrs, "Fishy Man-Goat Terrifies Couples Parked at Lake Worth," *Fort Worth (Tex.) Star-Telegram,* July 10, 1969, p. 2A; Jim Marrs, "Police, Residents Observe but Can't Identify 'Monster,'" *Fort Worth (Tex.) Star-Telegram,* July 11, 1969, p. 1A; Sallie Ann Clarke, *The Lake Worth Monster* (Fort Worth, Tex.: Sallie Ann Clarke, 1969); Mark Chorvinsky, "The Lake Worth Monster," *Fate* 45 (October 1992): 31—35; Peni R. Griffin (letter), *Strange Magazine,* no. 15 (Spring 1995): 3.

## Laocoön Serpent

SEA MONSTER of the Mediterranean Sea.

*Physical description:* Red crest or mane. Red (bloodshot) eyes. Flickering tongue. Venomous fangs. Immense coils.

*Behavior:* Kills by constriction.

*Distribution:* Aegean Sea.

*Significant sighting:* The Trojan priest Laocoön (responsible for the warning about "Greeks bearing gifts" in sending the Trojan horse) and his two sons were strangled by sea snakes while they were sacrificing at the altar of Poseidon on the seacoast. The snakes were said to have come from Bozca Ada Island, not far from ancient Troy along the coast of Çanakkale Province, Turkey.

*Possible explanation:* The harmless, crested Oarfish (*Regalecus glesne*), the most elongated bony fish in the world, is found in the Mediterranean. Though this is often a deep-sea fish, it is sometimes found dead or dying at the surface.

*Sources:* Vergil, *Æneid,* II. 199—231; Hermann Kleinknecht, "Laocoön," *Zeitschrift* für klassische Philologie 79 (1944): 66—111; Bernard Heuvelmans, *In the Wake of the Sea-Serpents* (New York: Hill and Wang, 1968), p. 84.

## Lascaux Unicorn

Mystery HOOFED MAMMAL of prehistoric Western Europe.

*Etymology: Unicorn* is a misnomer, since the animal clearly has two horns.

*Physical description:* Reddish. Large spots. Square head. Two long, straight horns. Large belly, possibly indicating a pregnancy. Short tail.

*Distribution:* Grotte de Lascaux, Dordogne Department, south-central France.

*Significant sighting:* A 17,000-year-old Upper Paleolithic painting of an unidentifiable animal exists in the cave of Lascaux, above the Vézère River valley near Montignac, France. The cave, with its many depictions of horses, bulls, and other animals, was discovered in September 1940.

*Present status:* The cave was closed to tourists in 1963, but a partial replica was opened nearby in 1983.

*Possible explanations:* Most commentators speculate that the 17,000-year-old painting shows an imaginary animal, though the other paintings in the cave are naturalistic depictions of real fauna. Others have suggested that it shows a hunter or shaman dressed in skins taken from different animals.

*Sources:* Peter Costello, *The Magic Zoo* (New York: St. Martin's, 1979), pp. 27—29; Mario Ruspoli, *The Cave of Lascaux: The Final Photographs* (New York: Harry Abrams, 1987).

## Lau

FRESHWATER MONSTER of East Africa.

*Etymology:* Nuer and Dinka (Nilo-Saharan) word.

*Variant names:* Jâk, Jâk-anywong ("punishing spirit," Dinka/Nilo-Saharan), Nyal (Shilluk/Nilo-Saharan), NYAMA.

*Physical description:* Serpentine but with legs. Size estimates vary widely: 40—100 feet long, as large as a donkey or a horse, or about 12 feet long and smaller than a python. Brown or dark yellow. Its snakelike head has a 3-inch-long crest like that of a crowned crane. Some say four bones united by a membrane appear around its mouth; others say it has barbels like a catfish.

*Behavior:* Call is a loud, booming cry, heard at night. Its stomach makes loud gurgles, especially in the rainy season. Lives in holes in riverbanks.

*Tracks:* Makes a furrow in swampy ground.

*Habitat:* Swamps.

*Distribution:* Bahr al 'Arab, Bahr al Ghazāl, Bahr al Zerāf, Bahr al Jabal, and other sources of the White Nile, from Malakāl south to Rajjāf and Lake No south to Shambe, Sudan.

*Significant sightings:* A 40-foot Lau was observed near Wāw, Sudan, in the late nineteenth century.

In 1914, the complete skeleton of a Lau was retrieved from the Bahr al Zerāf, and the bones were distributed among the Nuer people to wear as charms. A few years later, a 12-foot specimen was seen in the Bahr al Zerāf. Loud gurgles from a Lau were heard in the Bahr al 'Arab in 1918.

In 1937, William Hichens published a photo of a wooden effigy in the shape of a Lau's head. The effigy was apparently used in ritual dances and was carved by Mshengu she Gunda, who lived in the Iramba District of the Singida Region, Tanzania, and had hunted extensively in the Nile swamps.

*Possible explanations:*

(1) An unknown species of large Catfish (Family Siluridae) with long barbels, a dorsal fin that could be mistaken for a crest, and a long body. Some species of catfishes crawl out onto land at night. They have no vocal cords but can make a growling noise. Some have poisonous spines, and others produce electric shocks. The Electric catfish (*Malapterurus electricus*) of the Nile and tropical Africa is 5 feet long, but the Wels catfish (*Siluris glanis*) of Europe reaches nearly 10 feet long and can weigh more than 500 pounds.

(2) A large Marbled lungfish (*Protopterus aethiopicus*), a native of East African lakes and marginal swamps, including Lake No.

It can grow to more than 6 feet in length.

(3) A large, aquatic variety of the African rock python (*Python sebae*), which often attains a length of 30—33 feet.

(4) A composite animal, made up of the characteristics of several dangerous aquatic denizens.

(5) A generic name for any aquatic, elongated creature, possibly including the Nile bichir (*Polypterus bichir*), a 2-foot-long fish with nineteen to twenty-one dorsal spines that lives in lakes and rivers in Ethiopia and Chad; the Eel catfishes (*Channallabes apus* and *Gymnallabes typus*) of Central and West Africa; the North African catfish (*Clarias gariepinus*) that spends the dry season in burrows; and the Vundu (*Heterobranchus longifilis*), another air-breathing catfish of the Niger and Nile Rivers.

*Sources:* H. C. Jackson, "The Nuer of the Upper Nile Province," *Sudan Notes and Records* 6 (1923): 59, 187—189; John G. Millais, *Far Away up the Nile* (London: Longmans, Green, 1924), pp. 62—67; William Hichens, "African Mystery Beasts," *Discovery* 18 (1937): 369—373; Bernard Heuvelmans, *On the Track of Unknown Animals* (New York: Hill and Wang, 1958), pp. 447—449; Thomas Richard Hornby Owen, *Hunting Big Game with Gun and Camera* (London: Herbert Jenkins, 1960), pp. 92—95; Bernard Heuvelmans, *Les derniers dragons d'Afrique* (Paris: Plon, 1978), pp. 151—159, 363—370.

*Le Guat's giant, a large white bird reported on Mauritius in the 1690s by François Le Guat. From his* Voyage et avantures de François Leguat *(Amsterdam: Chez Jean Louis de Lorme, 1708). (From the original in the Special Collections of Northwestern University Library)*

## Le Guat's Giant

Unknown BIRD of Mauritius.

*Scientific name:* Leguatia gigantea, given by Hermann Schlegel in 1858.

*Variant name:* Giant water hen.

*Physical description:* Size of a goose. White plumage. Small red patch under the wings. Fat. Beak like a goose's but sharper. Long neck that extends to a height of 6 feet. Toes are long and widely separated.

*Behavior:* Ungainly. Often attacked by dogs. Good to eat.

*Habitat:* Marshes.

*Distribution:* Mascarene Islands in the Indian Ocean.

*Significant sighting:* Huguenot refugee François Le Guat saw large, white birds in the 1690s on Mauritius and Rodrigues. They have not been reported since.

*Possible explanations:*

(1) An unknown species of Rail (Family Rallidae), now extinct.

(2) A misidentified Gallinule (*Gallinula* spp.) or Flamingo (*Phoenicopterus ruber*), which is about 4 feet long and has a 5-foot

wingspan. Although an accurate observer in other matters, Le Guat may have failed to describe the bird properly.

*Sources:* François Le Guat, *Voyage et avantures de François Leguat* (Amsterdam: Chez Jean Louis de Lorme, 1708); Lionel Walter Rothschild, *Extinct Birds* (London: Hutchinson, 1907), pl. 31; Masauji Hachisuka, *The Dodo and Kindred Birds* (London: H. F. and G. Witherby, 1953); Errol Fuller, *Extinct Birds* (Ithaca, N.Y.: Cornell University Press, 2001), p. 386.

## Least Hominid

Term used by Mark A. Hall for the *Homo erectus* WILDMEN to distinguish them from his category of Neanderthaloid SHORTER HOMINIDS. He includes in this category the ALMAS and the BARMANU.

*Variant name:* Erectus hominid.
*Physical description:* Height, 5—6 feet.
*Tracks:* Length, 9.5 inches. Width, 4.5 inches.
*Distribution:* From the Caucasus Mountains to China.
*Possible explanation: Homo erectus* lived in Asia and Africa from the Early to Middle Pleistocene 2 million—125,000 years ago. It was squat, heavily muscled, and had a long, flat skull with browridges.

*Sources:* Mark A. Hall, *Living Fossils: The Survival of* Homo gardarensis, *Neandertal Man, and* Homo erectus (Minneapolis, Minn.: Mark A. Hall, 1999), pp. 63—68; Loren Coleman and Patrick Huyghe, *The Field Guide to Bigfoot, Yeti, and Other Mystery Primates Worldwide* (New York: Avon, 1999), pp. 26—28.

## Lechy

WILDMAN of European Russia and Central Asia.

*Variant name:* Leshi.
*Physical description:* Height, 4 feet 6 inches. Dark skin. Covered with hair.
*Distribution:* From the Vologda Region of European Russia east to the Buryat Republic, Russia.
*Significant sightings:* Seen once by psychologist K. K. Platonov east of Lake Baikal. Other sightings have taken place in the Saratov Region in 1989 and near Kargopol' in the Arkhangel'sk Region in 1992.

Physician V. V. Shatalov was attending to a patient one winter in a remote part of the Vologda Region, Russia, when he saw a naked, manlike creature in a courtyard. He observed that its legs were thin, but it had well-developed humeral muscles. It ran away in the snow and hid in an abandoned dwelling.

*Sources:* Konstantin K. Platonov, *Psikhologiia religii* (Moscow: Izd-vo Polit. Lit-ry, 1967); Paul Stonehill, "Russia's Unusual Bigfoot," *Fate* 48 (February 1995): 78; *Bigfoot Co-op,* no. 20, June 1999.

## Lenapízha

FRESHWATER MONSTER of the north-central United States.

*Etymology:* Peoria (Algonquian), "true tiger."
*Physical description:* Fiery dragon.
*Distribution:* Northwestern Illinois; northeastern Iowa; southwestern Wisconsin.
*Source:* Albert S. Gatschet, "Water-Monsters of American Aborigines," *Journal of American Folklore* 12 (1899): 255—260.

## Lenghee

CANNIBAL GIANT of Alaska.

*Etymology:* Tanaina (Na-Dené).
*Behavior:* Comes from the north occasionally and eats people.
*Distribution:* Cook Inlet, Alaska.
*Source:* Polaris, "Alaskan Mythology," *San Francisco Chronicle,* February 27, 1876, p. 1.

## Leviathan

SEA MONSTER of the Middle East; *see* SEMI-MYTHICAL BEASTS.

*Etymology:* Hebrew (Semitic), *livyatan,* from *livyah* ("twisted") + *tan* ("monster").
*Variant names:* Rahab, Tannin, Yam.
*Physical description:* Enormous size. Tightly joined scales. Glowing red eyes. Fiery breath. Large teeth. Strong neck. Has limbs or fins.
*Behavior:* Raises itself up on the water.

*Possible explanations:*
(1) A whale of some kind, especially the Sperm whale (*Physeter catodon*), which is found in the Mediterranean Sea and the Indian Ocean. Its fiery breath could be the whale spouting, while raising itself up on the water could be the animal's habit of breaching.

(2) The Nile crocodile (*Crocodylus niloticus*), suggested by Samuel Bochart, although this is a freshwater animal and does not live in the sea. The Egyptians occasionally hunted this crocodile with baited hooks.

(3) The Saltwater crocodile (*Crocodylus porosus*), but this reptile only lives in Southeast Asia and Australasia.

(4) A MULTIFINNED SEA MONSTER, advocated by Bernard Heuvelmans.

(5) Karl Shuker suggested a surviving mosasaur, a group of twenty genera that includes some of the largest marine reptiles ever, frequently exceeding 33 feet in length. They lived in the Late Cretaceous, 95—65 million years ago.

(6) A mythical composite of several large animals.

*Sources:* Bible, Old Testament (Job 3:8, 41:1—34; Pss. 74:14, 104:26; Isa. 27:1); Samuel Bochart, *Hierozoicon, sive, bipartitum opus De animalibus Sacrae Scripturae* (London: John Martin and Jacob Allestry, 1663); Bernard Heuvelmans, *In the Wake of the Sea-Serpents* (New York: Hill and Wang, 1968), pp. 80—83, 568; Karl Shuker, *In Search of Prehistoric Survivors* (London: Blandford, 1995), pp. 127—128.


# Lindorm

Wingless DRAGON or unknown LIZARD of Northern Europe.

*Etymology:* Swedish, *lind* ("flexible body") + *orm* ("serpent").

*Variant names:* Drage (Norway), Drake (Sweden), Lindwurm (Germany), Vassorm (Norway).

*Physical description:* Serpentine. Length, 10—28 feet. Body is heavy and as thick as a man's thigh. Black color. Head is like that of a pike fish. Older specimens have a crest or mane that is black to gray in color and parted in the middle. Large, saucer-shaped, red or yellow eyes that shine with reflected light. Short, protruding ears on the top of the head. Square nose. Forked tongue. Mouth is full of white teeth. Sacks of skin hang from the corners of the mouth. Bristles or whiskers are on the chin. Yellow belly. Tail is short and stubby.

*Behavior:* Primarily terrestrial but often seen in water. Swims with a horizontal movement, its head 2 feet above the water. Has a hypnotic or paralyzing gaze. Ill tempered and pugnacious. Snorts like a horse. Hisses when alarmed. Before attacking, it contracts its body and then rises up 4—6 feet on its tail and pounces. Spits a poisonous liquid. Its carcass has a repugnant stench.

*Tracks:* Makes well-worn trails that look as if a log had been dragged along the ground.

*Habitat:* Lakes, swamps, mountains, rocky areas.

*Distribution:* Kronoberg and Jönköping Counties, Sweden, including Asnen, Rottnen, Öjen, and Helgasjön Lakes; parts of Norway and Finland.

*Significant sightings:* In the fall of 1826, Daniel Nilsson, of Odensö, Kronoberg County, had a difficult and lengthy struggle with a Lindorm in the forest of Ulvehult.

Walking to their boathouse in August 1869, Magnus Bergström and Karin Svensdotter noticed a black snake in the grass. After poking it with a stick, Bergström realized it was a Lindorm when it opened its mouth 11 inches wide and showed its forked tongue; the creature hissed, rose upright, and rushed at him. After a long fight, Bergström killed it with a stick. Its mouth was full of fangs about the size of a man's little finger, and it had a mane of scales pointed like horsehair. The carcass began to stink almost immediately.

In November 1878, a Lindorm was killed in Husaby Forest, Kronoberg, by the farmer Johan Jonsson of Hakadal.

In 1883, Lindorms were seen near Hinneryd, Urshult, Kalvsvik, the estate of Skäggalösa Persgård, and Husaby Forest, all in Kronoberg County, Sweden.

Gunnar Olof Hyltén-Cavallius organized a

hunt for witnesses of the Lindorm from 1883 to 1885. He uncovered forty-eight individuals who had memories of seeing these animals from the 1820s to the 1880s.

*Possible explanations:*

(1) Said by Erik Pontoppidan to be a juvenile SEA MONSTER, which travels downstream to the sea like an eel when it matures.

(2) Hallucinations, folktales, or misidentifications.

*Sources:* Erik Pontoppidan, *The Natural History of Norway* (London: A. Linde, 1755), vol. 2, pp. 38—39, 195—208; Gunnar Olof Hyltén-Cavallius, *Om draken eller lindormen: Memoire till Kongliga Vetenskaps-akademien* (Växjö, Sweden, 1884—1885); Johan Theodor Storaker, *Naturrigerne i den norske Folketro* (Oslo: Norsk Folkeminnelag, 1926), pp. 243—249; Martin Bjørndal, *Segn og tru: Folkeminne fra Møre* (Oslo: Norsk Folkeminnelag, 1949), pp. 84—87; Reidar Thoralf Christiansen, *The Migratory Legends* (Helsinki: Suomalainen Tiedeakatemia, 1958), pp. 49—52; Aukusti V. Rantasalo, *Einige Zaubersteine und Zauberpflanzen im Volksaberglauben der Finnen* (Helsinki: Suomalainen Tiedeakatemia, 1959), pp. 26—31; Sven Rósen, "The Dragons of Sweden," *Fate* 35 (April 1982): 36—45; Michel Muerger, "In Jormungandra's Coils: A Cultural Archaeology of the Norse Sea Serpent," *Fortean Times,* no. 51 (Winter 1988—1989): 63—68; Karl Shuker, *Dragons: A Natural History* (New York: Simon and Schuster, 1995), pp. 40—43.

## Lipata

Unknown CROCODILIAN of Central Africa.

*Variant name:* Libata.

*Physical description:* Bulkier than a Nile crocodile. Length, 13—20 feet. Eyes are close together on the top of the head. Mouth is larger and throat is wider than a Nile crocodile's. Serrated scales along the tail.

*Behavior:* Most active at the beginning of the rainy season or the end of September. Seen on the water's surface in the morning and at dusk.

Comes on land only occasionally. Attacks and eats goats, pigs, cattle, crocodiles, and, from time to time, humans. Very shy of people. Women who fish in the river shout to scare the animal away.

*Distribution:* Chiumbe and Kasai Rivers, northeastern Angola.

*Significant sightings:* Around 1890, the inhabitants of the village of Tyipukungu, Angola, set a trap for a Lipata after it had taken three of their cattle. The animal took the bait and was killed.

On September 1, 1932, a man from Tyipukungu saw a Lipata sleeping on dry land around 9:00 A.M.

*Possible explanations:*

(1) Large or old and aggressive specimens of the Nile crocodile (*Crocodylus niloticus*). The official size record for this crocodile was set in 1953 at 19 feet 6 inches, though there are reports of larger specimens. Most rarely grow larger than 16 feet.

(2) An African slender-snouted crocodile (*Crocodylus cataphractus*) somewhat south of its normal range. This reptile grows to 13 feet long, although the average length is 8 feet. It has solitary habits and is most often found in open water in lakes and rivers.

(3) An unknown species of large Dwarf crocodile of the genus *Osteolaemus,* with a head that is shorter and rounder than that of the Nile crocodile, suggested by Bernard Heuvelmans.

*Sources:* Albert Monard, "Sur l'existence en Angola d'un grand reptile encore inconnu," *Bulletin de la Société Neuchâteloise de Sciences Naturelles* 57 (1932): 67—71; Bernard Heuvelmans, *On the Track of Unknown Animals* (New York: Hill and Wang, 1958), pp. 456—460, 470; Bernard Heuvelmans, *Les derniers dragons d'Afrique* (Paris: Plon, 1978), pp. 233—239, 372.

## Little People

Most cultures throughout the world have myths, legends, and folklore about small ENTITIES who stand anywhere from 4 feet 6 inches to only a few inches tall. In the vast majority of cases, these beings are regarded as primarily supernatural, although they may leave physical traces and do other things that humans do—eat food, wear clothes, use weapons, speak a language, and worship gods. Little people often represent the world lived in by children: they are imperfectly understood, inferior, and yet compelled to do the bidding of adults.

The literature on Little people is vast. Descriptions vary widely depending on the environment and local belief systems. Some cultures have difficulty distinguishing between the real and the mythic worlds, and the cryptozoologist trying to make sense of it all runs the risk of making the false assumption that these creatures have a basis in physical reality. Often, the legends are cited as evidence for SMALL HOMINIDS, which might include anything from an unknown race of human Pygmies to surviving australopiths or unclassified species of apes or monkeys. Perhaps some folktales are based on beings that went extinct thousands of years ago and have become distorted, amplified, or hopelessly entangled with other motifs.

In this category are found diminutive entities that could represent folk memories of genuine HOMINIDS or PRIMATES, as well as those that have been mentioned in the cryptozoological literature.

*Variant names:*

*Africa*– AZIZA, GNÉNA, IJIMÉRÉ, KALANORO, MMOATIA, TOKOLOSH.

*Asia*– DJINNI.

*Australasia*– JINGARA, JUNJADEE, MUMULOU, VUI, YAWT.

*Central and South America*– ALUX, CURUPIRA, DUENDE, GÜIJE, IKAL, KENAIMA, SHIRU, TRAUCO, WASHIPI.

*Europe*–ÆLF (Old English), Duergar (Scandinavia), Dwarf, Elf, Ellyllon (Wales), FAIRY, Gnome (Germany), Knocker (Cornwall, England), Kobold (Germany), Massariol (Italy), Nis (Scandinavia), Vila (Eastern Europe).

*North America*– AMAYPATHENYA, ATNAN, Ja-gen-oh (Iroquoian), MEMEGWESI, NINIMBE, NUNNEHI, PININI, PUKWUDGEE, SQUOLK-TY-MISH, YUNWI TSUNSDÍ.

*Oceania*– MENEHUNE, VÉLÉ.

*Sources:* Katharine M. Briggs, *A Dictionary of*

*Fairies* (London: Allen Lane, 1976); Nancy Arrowsmith and George Moorse, *A Field Guide to the Little People* (New York: Hill and Wang, 1977); Carol Rose, *Spirits, Fairies, Gnomes, and Goblins: An Encyclopedia of the Little People* (Santa Barbara, Calif.: ABC-CLIO, 1996); John E. Roth, *American Elves* (Jefferson, N.C.: McFarland, 1997).

## Lizard Man

Bipedal, reptilian ENTITY of North America.

*Variant names:* Jabberwok, Reptile man.

*Physical description:* Humanoid form. Height, 7 feet. Greenish, grayish, or brown color. Scales, sometimes in combination with or confused with hair. Froglike face. Glowing red eyes. Three-fingered hands.

*Behavior:* Amphibious, by some accounts. Pungent odor.

*Tracks:* Three-toed, clawed prints 14 inches long, 7 inches wide, and 1 inch deep. Stride, 40 inches.

*Habitat:* Swamps, rivers.

*Distribution:* Lake Thetis, British Columbia, Canada; Riverside, California; Dogtown, Indiana; Frederick County, Maryland; Newton, New Jersey; Cincinnati area and Mansfield, Ohio; Scape Ore Swamp, South Carolina.

*Significant sightings:* At 4:00 A.M. on May 25, 1955, Robert Hunnicutt saw three small figures, about 3 feet tall, kneeling by a road in Branch Hill, Ohio. They were grayish, with froglike faces, a bulge on the chest, and slender arms. In July 1955, a civil defense volunteer was driving across the Little Miami River in Loveland, Ohio, when he briefly saw four small figures on the riverbank beneath the bridge. These events have been classed with unidentified flying object (UFO) sightings, though no direct connection was established.

Again in Loveland, Ohio, police officers Ray Shockey and Mark Matthews saw a 4-foot-tall, leathery-skinned biped with a froglike face on Riverside Road on March 3, 1972. It jumped over a guardrail and down an embankment to the Little Miami River. A similar creature was seen on March 17.

On June 29, 1988, seventeen-year-old

Christopher Davis was changing a flat tire near the Scape Ore Swamp southwest of Bishopville, South Carolina, when he saw a green, 7-foot-tall, scaly creature with red eyes and three fingers. He jumped into his car and sped away, but the creature leaped on top of the vehicle. The report inspired many misidentifications and at least one hoax over the following weeks, but Davis stuck to his story and passed a polygraph test in September.

*Possible explanations:*

(1) Except for its scaly texture and color, Lizard man bears a certain resemblance to HAIRY BIPED entities and NORTH AMERICAN APES.

(2) Lizard man's alien look suggests to some an origin connected to the UFO phenomenon.

(3) Folklore or hoaxes based on the Gillman character from the 1954 movie *Creature from the Black Lagoon.*

(4) A surviving *Coelophysis,* a small, meat-eating dinosaur that lived in New Mexico in the Late Triassic, 210 million years ago, suggested by Erik Beckjord. However, this species did not resemble a humanoid (though it was bipedal), and it had an unmistakably long, slender, balancing tail.

(5) Chris Orrick has suggested that the 1972 case involved a misidentified Great gray owl (*Strix nebulosa*), though southern Ohio is outside its normal range.

*Sources:* Leonard H. Stringfield, *Situation Red: The UFO Siege!* (Garden City, N.Y.: Doubleday, 1977), pp. 87—92; "Loveland Frog Leaps Back," *Fortean Times,* no. 46 (Spring 1986): 19; "'Lizard Man' Facts," *Columbia (S.C.) State,* August 15, 1988; Mark Opsasnick and Mark Chorvinsky, "Lizard Man," *Strange Magazine,* no. 3 (1988): 32—33; Loren Coleman, "Other Lizard People Revisited," *Strange Magazine,* no. 3 (1988): 34, 36; Paul Sieveking, "Lizard Man," *Fortean Times,* no. 51 (Winter 1988—1989): 34—37; Loren Coleman, *Mothman and Other Curious Encounters* (New York: Paraview, 2002), pp. 88—100.

# LIZARDS (Unknown)

Lizards make up the Suborder Lacertilia of the large reptilian Order Squamata, which also includes SNAKES and Amphisbaenians (Worm lizards). In general, lizards are small- to medium-sized scaly reptiles with four clawed feet, elongated bodies, and tapering tails. Some are highly arboreal, others specialize in burrowing, and still others are occasionally bipedal. There are four lizard infraorders: Gekkota, Iguania, Scincomorpha, and Anguimorpha.

Infraorder Gekkota includes Geckos (Gekkonidae and Eublepharidae) and Australasian legless lizards (Pygopodidae). Geckos are known for their ability to climb up walls and across ceilings because of the microscopic suction cups on the bristles of their toe pads. They are widespread throughout tropical and subtropical regions of both the New and Old Worlds. Geckos can also vocalize, and their name derives from an Asian species with a cry that sounds like "geck-o." The earliest unequivocal gekkotan fossil is *Hoburogecko* from Mongolia in the Early Cretaceous, 105 million years ago. Most geckos are less than 6 inches long (not including the tail, which frequently breaks off).

Infraorder Iguania includes Iguanas (Iguanidae), Agamids (Agamidae), and Chameleons (Chameleonidae). In general, they have robust bodies, short necks, fleshy tongues, well-developed eyelids, distinct heads, and overlapping and noniridescent scales. Many species have well-developed ornamental crests, spines, frills, or colorful throat fans. Some, such as the Water dragon (*Physignathus*), are bipedal and run rapidly on only two legs. Others, such as the Flying lizards (*Draco*) of Asia, have ribs modified for arboreal gliding. The first unequivocal iguanian fossil is *Pristiguana* from South America in the Late Cretaceous, 80 million years ago.

Infraorder Scincomorpha includes Spectacled lizards (Gymnophthalmidae), Night lizards (Xantusiidae), Wall lizards (Lacertidae), Whiptails and Tegus (Teiidae), Spinytail lizards (Cordylidae), and True skinks (Scincidae). In general, these animals have slim bodies, with heads not clearly differentiated from the neck; if the scales overlap, they are iridescent. Except for the wall lizards, this group has a definite tendency toward limb reduction and development of a snakelike body. The Common lizard (*Lacerta vivipara*) of Europe exists above the Arctic Circle; no other lizard is found that far north. The largest animal in this group, the Prehensile-tailed skink (*Corucia zebrata*) of the Solomon Islands, is about 2 feet long and is the only true herbivorous skink. The earliest unambiguous scincomorph fossil is *Paramacellodus* from the United Kingdom in the Middle Jurassic, 170 million years ago.

Infraorder Anguimorpha includes Glass lizards and Alligator lizards (Anguidae), Legless lizards (Anniellidae), Rock lizards (Xenosauridae), Plated lizards (Gerrhosauridae), Blind lizards (Dibamidae), the venomous Heloderms (Helodermatidae), Monitors (Varanidae), and Earless monitors (Lanthanotidae). The group also includes a number of large, heavily armored extinct forms–notably, the aquatic mosasaurs. A diverse group, anguimorphs have two-part tongues and relatively solid teeth in common. The earliest known anguimorph is *Parviraptor* from the United Kingdom in the Middle Jurassic, 170 million years ago.

The largest living lizard is the Komodo dragon (*Varanus komodoensis*) of Indonesia, which averages 8 feet 6 inches in total length and weighs 175—200 pounds. The largest accurately measured Komodo was a male residing at the St. Louis Zoological Park that was 10 feet 2 inches long and weighed 365 pounds in 1937. Komodo dragons are excellent swimmers and can run at speeds up to 15 miles per hour. The Crocodile monitor (*Varanus salvadorii*) of Papua New Guinea regularly grows over 12 feet long, making it the longest lizard in the world. One specimen was measured at 15 feet 7 inches.

Lizards can be difficult to identify in the field without capturing a specimen, which makes it especially problematic to place lizardlike cryptids into their respective infraorders. Larger mystery lizards tend to be identified as monitors because of their size and general appearance. Of the twenty-five lizards in this list, eight (AFA, ARTRELLIA, AU ANGI ANGI, AUSTRALIAN GIANT MONITOR, BURU, DAS-ADDER, NGUMA-MONENE, and VENEZUELAN MONITOR) could be monitors.

Lizards have also been proposed as candidates for mystery CROCODILIANS, DINOSAURS, DRAGONS, FLYING REPTILES, FRESHWATER MONSTERS, and SNAKES.

## Mystery Lizards
### Africa
DAS-ADDER; GIANT ETHIOPIAN LIZARD; MUHURU; NGUMA-MONENE; OLDEANI MONSTER

### Asia
AFA; BIS-COBRA; BURU; JHOOR

### Australasia and Oceania
ARTRELLIA; AU ANGI-ANGI; AUSTRALIAN GIANT MONITOR; GIANT TONGAN SKINK; HAIRY LIZARD; KAWEKAWEAU; KUMI; NGARARA

### Europe
DARD; GENAPRUGWIRION; LINDORM; OSSUN LIZARD; TATZELWURM

### North America
GIANT NORTH AMERICAN LIZARD; GOWROW

### South America
VENEZUELAN MONITOR

## Lizzie
FRESHWATER MONSTER of Scotland.

*Physical description:* Length, 12—40 feet. Distinct humps. Light underside. Has fins or paddles.

*Distribution:* Loch Lochy, Highland.

*Significant sightings:* In 1937, a photo was taken of an object surfacing in the loch. Maurice Burton believes it shows a vegetable mat.

Eric Robinson saw an animal with a broad back and a fin on July 15, 1960.

A family named Sargent was driving near the Corriegour Lodge Hotel on September 30, 1975, when they saw a black, 20-foot hump causing waves.

Andy Brown and fifteen others at the Corriegour Lodge Hotel observed a 12-foot animal with a curved head and three humps moving from side to side on September 13, 1996.

*Sources:* Maurice Burton, *The Elusive Monster* (London: Rupert Hart-Davis, 1961);

Tim Dinsdale, *The Leviathans* (London: Routledge and Kegan Paul, 1966), p. 61; Graham J. McEwan, *Mystery Animals of Britain and Ireland* (London: Robert Hale, 1986), pp. 92—93; "Good Month for Monster Hunters," *Fortean Times,* no. 95 (February 1997): 18.

## Llamhigyn y Dwr

FRESHWATER MONSTER of Wales.

*Etymology:* Welsh, "water leaper."

*Physical description:* Like a large toad. Large wings instead of legs. Tail.

*Behavior:* Makes a hideous shriek. Eats sheep and other livestock.

*Habitat:* Rivers.

*Distribution:* Wales.

*Source:* John Rhys, *Celtic Folk-lore, Welsh and Manx* (London: Oxford University Press, 1901).

## Longneck

A category of SEA MONSTER identified by Bernard Heuvelmans. It looks and behaves like certain FRESHWATER MONSTERS in lakes and rivers around the world.

*Scientific name: Megalotaria longicollis,* given by Heuvelmans in 1965.

*Variant name:* Heuvelmans's seal.

*Physical description:* Shape varies from serpentine to thick and bunched up. Length, 15—65 feet. Dark brown on top, with black-and-gray or whitish mottling. Underside much lighter. Skin looks smooth when wet, but up close, it appears wrinkled and rough. Small, round head with two small horns. Small eyes. Tapered muzzle, sometimes described as like a seal's or dog's and at other times like a horse's or camel's. Long, slender, flexible neck. A collar behind the head is sometimes reported. There are one to three dorsal humps, with the middle being the largest. A slight ridge along the spine. Four webbed flippers. The hind pair sometimes resembles a bilobate tail.

*Behavior:* Amphibious. Most frequently seen between April and October in the north, year-round in the tropics, and from February to April in the south. Swims with vertical undulations. Exceptional speed, 15—35 knots. The illusion of more than three humps may be caused by turbulence waves generated by its speed. Submerges vertically, as if pulled under. Does not spout. On land, it moves by gathering its hind legs up toward the front, then leaping forward with the front legs in a manner similar to that of sea lions.

*Habitat:* Near coasts in cold temperate regions; midocean in warm temperate regions.

*Distribution:* Cosmopolitan, except for polar waters. Frequently seen around the British Isles, Newfoundland, Maine, and British Columbia in the Northern Hemisphere; in the Tasman Sea, off southeastern Australia, New Caledonia, New Zealand, and Tasmania in the Southern Hemisphere. It may also be the same animal reported in many lakes in cold temperate regions in both hemispheres.

*Significant sightings:* Some time before 1846, Captain Christmas of the Danish navy encountered a long-necked animal in the North Atlantic Ocean between Iceland and the Faroe Islands. Its neck moved like a swan's until it disappeared, head foremost, like a duck diving. The part above water seemed about 18 feet in length.

In September 1893, London physician Farquhar Matheson and his wife were boating in Loch Alsh, an arm of the sea between Skye and the Scottish mainland, when they saw a straight-necked animal moving toward them. Its neck was as tall as the mast of their yacht. It dived and reappeared every two to three minutes.

On December 4, 1893, Captain R. J. Cringle and the crew and passengers of the Natal Line steamer *Umfuli* watched a long-necked Sea monster for thirty minutes not far off the coast of Guerguerat, Western Sahara. It was 80 feet long and about as thick as a whale. The head and neck together were 7—15 feet long, and the body seemed to have three humps.

On August 5, 1919, J. Mackintosh Bell and others fishing for cod off the south coast of Hoy Island in the Orkney Islands of Scotland encountered a long-necked animal about 30 yards from their boat. When it swam alongside the boat at a depth of 10 feet, they were able to see its full outline, with four flippers and a total head-to-tail length of 18—20 feet. The head looked very much like a retriever dog's and was

6 inches long by 4 inches wide.

Big-game hunter Tromp Van Diggelen and other passengers on the *Dunbar Castle* saw an animal with a 12-foot neck in Table Bay off Cape Town, South Africa, in November 1930.

In 1945, Arthur Féré and others in a motorboat saw a strange animal sticking up above the water in a bay off Canala, New Caledonia, in the South Pacific. It had a big head on a black neck marked with yellow. When the boat approached to 200 yards, the animal dived, raising a plume of water.

Robert Duncan, a beachcomber on Bribie Island, Queensland, Australia, saw a whitish-gray monster 2 miles offshore twice in September 1962. It had a swan's neck, a whale's body, and a fish's tail and fins.

*Possible explanation:* An extremely elongated form of Sea lion or Fur seal (Family Otariidae) adapted for a purely marine existence, according to Heuvelmans. Cladistic studies now suggest that the true seals, sea lions, and walruses all are most closely related to the bears, emerging from that family 27—25 million years ago, in the Late Oligocene. So specialized a variety most probably represents a recent evolution. Robert Cornes speculates that it may not be related to any of the existing seal families.

*Sources:* Philip Henry Gosse, *The Romance of Natural History,* 4th ed. (London: Nisbet, 1861); Alfred T. Story, "The Sea-Serpent," *Strand Magazine* 10 (August 1895): 161—171; Rupert T. Gould, *The Case for the Sea-Serpent* (London: Philip Allan, 1930), pp. 188—194, 215—220; Tromp Van Diggelen, *Worthwhile Journey* (London: William Heinemann, 1955); "Un monstre marin identique à celui de Hook Island avait déjà été vu en 1945 dans la baie d'Ouengho à Canala," *France Australe,* June 26—27, 1965, pp. 1, 4; Bernard Heuvelmans, *In the Wake of the Sea-Serpents* (New York: Hill and Wang, 1968), pp. 506, 557—562, 565—566; Robert Cornes, "The Case for the Surreal Seal," *Crypto Dracontology Special,* no. 1 (November 2001): 39—45.

## Loo Poo Oi'yes
CANNIBAL GIANT of California.

*Etymology:* Miwok (Penutian) word.
*Distribution:* Northern California.
*Source:* Kyle Mizokami, Bigfoot-Like Figures in North American Folklore and Tradition, http://www.rain.org/campinternet/bigfoot/bigfoot-folklore.html.

## Lophenteropneust
Unknown INVERTEBRATE of the South Pacific Ocean.

*Etymology:* Greek, "ridged" enteropneust.
*Scientific name:* Lophenteropneusta, given by Henning Lemche.
*Physical description:* Length, 2—4 inches. Cylindrical, translucent body. A ring of tentacles surrounds the mouth. Terminal anus.
*Tracks:* Spirals and loops of fecal strings.
*Habitat:* Abyssal marine depths.
*Distribution:* Pacific Ocean.
*Significant sightings:* In 1962, the Scripps Institute of Oceanography's Proa Expedition in the research vessel *Spencer F. Baird* took some 4,000 photos of the sea bottom in five trenches in the western Pacific to a depth of 35,800 feet. Several photographs showed fecal coils made by a hemichordate of uncertain taxonomy. Similar animals were later photographed by the Disturbance and Recolonization Experiment in a Manganese Nodule Area of the Deep South Pacific (DISCOL) project in the Peru Trench.
*Possible explanation:* The cylindrical shape is like an Acorn worm (Class Enteropneusta), while the ring of tentacles is characteristic of the vase-shaped Class Pterobranchia.
*Sources:* Henning Lemche, et al., "Hadal Life as Analyzed from Photographs," *Videnskabelige Meddelelser fra Dansk Naturhistorisk Forening* 139 (1976): 263—336; O. S. Tendal, "What Became of Lemche's Lophenteropneust?" *Deep-Sea Newsletter* 27 (1998): 21—24; O. S. Tendal, "Lemche's Lophenteropneust Widely Known but Still an Enigma," *Deep-Sea Newsletter* 28 (1999): 8; DISCOL Megafauna Atlas, http://www.drbluhm.de/da_fig052.html.

## Lord of the Deep
Immense sharklike FISH of the South Pacific Ocean.

*A surviving Megalodon shark* (Carcharodon megalodon*) might explain stories of a huge Pacific shark called the* LORD OF THE DEEP *(William M. Rebsamen)*

*Etymology:* Possibly from a Melanesian (Austronesian) word.

*Physical description:* Gray above and white below or entirely pale white. Length, greater than 30 feet, perhaps up to 100 feet or more.

*Distribution:* South Pacific Ocean off eastern Australia.

*Significant sightings:* In 1918, fishermen in deep water near the Broughton Islands, New South Wales, Australia, watched as a gigantic, "ghostly white" shark made off with their crayfish pots. Their estimates of the animal's size ranged from 115 to 300 feet long.

Novelist and fisherman Zane Grey saw a yellow-and-green shark, 35—40 feet long, near Rangiroa Atoll in French Polynesia in 1927 or 1928. In 1933, he and his son saw a similar shark, perhaps as much as 10 feet longer, in the same area.

*Possible explanations:*

(1) The spotted Whale shark (*Rhincodon typus*) probably accounts for Zane Grey's observations.

(2) Outsize specimens of the Great white shark (*Carcharodon carcharias*), which averages 14—15 feet long. Individuals more than 20 feet long exist but are extremely rare. An unconfirmed 37-foot great white shark was found trapped in a herring weir at White Head Island near Grand Manan, New Brunswick, Canada, in June 1930.

(3) A surviving Megalodon shark (*Carcharodon megalodon*), a species that was thought to have become extinct about 1.5 million years ago, at the end of the Pliocene. (Fossil teeth dredged up by oceanographic expeditions have been described as "fresh-looking" and erroneously assumed to be fresh or subfossil.) This ancestor or relative of the great white shark is known to have been at least 40—50 feet long and weighed 55 tons. Its mouth was

large enough to swallow an entire cow, and its triangular teeth were 4—6 inches long.

*Sources:* Vadim D. Vladykov and R. A. McKenzie, "The Marine Fishes of Nova Scotia," *Proceedings of the Nova Scotian Institute of Science* 19 (1935): 17—113; David G. Stead, *Sharks and Rays of Australian Seas* (Sydney, Australia: Angus and Robertson, 1963), pp. 38—47; Gerald L. Wood, *The Guinness Book of Animal Facts and Feats,* 3d ed. (Enfield, England: Guinness Superlatives, 1982), pp. 129—135; Michael Goss, "Do Giant Prehistoric Sharks Survive?" *Fate* 40 (November 1987): 32—41; Karl Shuker, *In Search of Prehistoric Survivors* (London: Blandford, 1995), p. 123; Ben S. Roesch, "A Critical Evaluation of the Supposed Contemporary Existence of *Carcharodon megalodon,*" *Cryptozoology Review* 3, no. 2 (Autumn 1998): 14—24.


## Lukwata

FRESHWATER MONSTER of East Africa.

*Etymology:* Ganda (Bantu) word. The prefix *lu-* can mean "giant." It is remotely possible that the name originated from the exclamation "Look [at] water!" spoken in imperfect English.

*Variant names:* Lokwata, Luquata.

*Physical description:* Length, 20—30 feet. Size of a small porpoise. Dark color. Round or ovoid head. Neck is 4 feet long.

*Behavior:* Aggressive. Swims with head and neck out of the water. Moves with vertical undulations. Causes whirlpools. Loud, bellowing voice. Attempts to seize fishermen in boats or canoes. Said to fight with crocodiles. Pieces of its body are prized as charms by the local natives.

*Distribution:* Lake Victoria and tributary rivers in Uganda, Tanzania, and Kenya.

*Significant sightings:* W. Grant, provincial governor at Jinja, Uganda, once saw from a distance an animal swimming with its head out of the water in the Napoleon Gulf, Lake Victoria.

Sir Clement Hill observed a large, long-necked animal with a dark, roundish head off Homa Mountain, Lake Victoria, Kenya, around 1900. Hill insisted it was not a crocodile.

In the 1930s, E. G. Wayland, director of the Geological Survey of Uganda, was shown a fragment of bone that belonged to a Lukwata. Wayland stated he had heard the animal's bellowing roars.

In late 1959, T. E. Cox and his wife saw a large, black animal among some reeds near the shore of Mohoru Bay, Lake Victoria, Kenya. It was 20—30 feet long and had a thick body with two humps on its back, a thin neck, and a snakelike head. It swam with vertical undulations toward the center of the lake after noticing their presence.

*Possible explanations:*

(1) An unknown species of large Catfish (Family Siluridae), based on its barbels.

(2) An African rock python (*Python sebae*), suggested by Hector Duff.

(3) A freshwater LONGNECK similar to NESSIE, suggested by Bernard Heuvelmans and based on the Cox sighting.

*Sources:* Harry Johnston, *The Uganda Protectorate* (London: Hutchinson, 1902), vol. 1, pp. 79—80; C. W. Hobley, "On Some Unidentified Beasts," *Journal of the East Africa and Uganda Natural History Society,* no. 6 (1913): 48—52; Hector L. Duff, *African Small Chop* (London: Hodder and Stoughton, 1932), pp. 158—164; Stella and Edgar B. Worthington, *The Inland Waters of Africa* (London: Macmillan, 1933), pp. 126—127; William Hichens, "African Mystery Beasts," *Discovery* 18 (1937): 369—373; Henry Hesketh J. Bell, *Witches and Fishes* (London: Edward Arnold, 1948), pp. 156—159; Bernard Heuvelmans, *Les derniers dragons d'Afrique* (Paris: Plon, 1978), pp. 165—172, 176—177, 299, 306—307, 370—371.


## Lummis's Pichu-Cuate

Small, deadly SNAKE of southwestern North America.

*Etymology:* Pichucuate is a generic name given in the Southwest and Mexico to snakes believed to be venomous. It has been applied to the Cantil (*Agkistrodon bilineatus*) and Mexican lyre snake (*Trimorphodon tau*) of Mexico and the Narrow-headed garter snake (*Thamnophis rufipunctatus*) in Arizona.

*Physical description:* As thick as a pencil. Gray

above, rosy below. Head the size of a man's fin-gernail. Horns above the eyes. Tiny fangs. Extremely quick-acting, deadly venom.

*Behavior:* Buries itself in the sand to await prey.

*Habitat:* Desert.

*Distribution:* Mexico; Arizona; New Mexico.

*Significant sightings:* Charles Lummis met with this snake on three occasions, the first in June 1889 in Valencia County, New Mexico. The Pueblo Indians, for whom rattlesnakes are a familiar totem, avoid it entirely.

*Present status:* Possibly extinct.

*Possible explanations:*

(1) The Mexican horned pitviper (*Ophryacus undulatus*) has supraocular horns but is a semiarboreal snake found only in the mountains of southern Mexico. Its range may have been more extensive in the past.

(2) The Black-tailed montane pitviper (*Porthidium melanurum*) also has supraocular horns, but it has a distinctly black tail and lateral stripes and is also limited to Mexico.

*Sources:* Charles F. Lummis, *The King of the Broncos, and Other Stories of New Mexico* (New York: Charles Scribner's Sons, 1897); Chad Arment, "Notes on Lummis' Pichu-cuate," *North American BioFortean Review* 2, no. 3 (December 2000): 5—10, http://www.strangeark.com/nabr/NABR5.pdf.

## Lusca

Giant octopus-like CEPHALOPOD of the Caribbean Sea.

*Etymology:* Bahamas Creole word.

*Variant names:* Giant scuttle, Him of the Hairy Hands, Lucsa, Luska.

*Physical description:* Said to be half octopus and half shark or half eel and half squid. Width, 50 feet. Phosphorescent eyes. Tentacles with sucker tips.

*Behavior:* Moves with the speed of a shark. Surfaces at night when the moon is full. Possibly feeds on crabs and shrimp. Said to drag boats and people into the water with its tentacles.

*Habitat:* Blue holes, narrow pits that plunge as much as 200 feet straight down through rock and coral into deep water. Some are offshore, others are in interior lagoons.

*Distribution:* Andros and Grand Bahama Islands in the Bahamas; Caicos Islands; off the coast of Cuba.

*Possible explanations:*

(1) GIGANTIC OCTOPUS of the kind found washed ashore in Florida in 1896.

(2) Misidentifications of the Giant squid (*Architeuthis*), suggested by Bruce Wright.

(3) Tidal surges and vortices at the mouth of blue holes could be mistaken for the movement of giant tentacles.

*Sources:* J. S. George, "A Colossal Octopus," *American Naturalist* 6 (1872): 772; François Poli, *Sharks Are Caught at Night* (Chicago: Henry Regnery, 1959), pp. 102—103; Bruce S. Wright, "The Lusca of Andros," *Atlantic Advocate* 51 (June 1967): 32—39; Forrest G. Wood and Joseph G. Gennaro, "An Octopus Trilogy," *Natural History* 80 (March 1971): 15—24, 84—87; Gary S. Mangiacopra, "The Great Ones: A Fragmented History of the Giant and the Colossal Octopus," *Of Sea and Shore* 7, no. 2 (Summer 1976): 93—96; Robert Palmer, "In the Lair of the Lusca," *Natural History* 96 (January 1987): 42—47; Gary S. Mangiacopra, Michel Raynal, Dwight G. Smith, and David F. Avery, "*Octopus giganteus*: Still Alive and Hiding Where? Lusca and Scuttles of the Caribbean," *Of Sea and Shore* 18, no. 1 (Spring 1996): 5—12; Michel Raynal, "Des poulpes 'de dimension anormale'?" http://perso.wanadoo.fr/cryptozoo/floride/poulpe3. htm.

# M

## Maasie
FRESHWATER MONSTER of Belgium.

*Etymology:* After the Meuse River's Flemish name—Maas.

*Distribution:* Meuse River, Belgium.

*Significant sighting:* A 3-foot crocodile was seen in the river near Ombret-Rawsa, Belgium, on August 6, 1979.

*Possible explanation:* A discarded pet crocodile.

*Source: Saarbrücker Zeitung,* August 9, 1979.

## Macarena Bear
Unknown BEAR of South America.

*Physical description:* Bear with red fur.

*Distribution:* Serranía de la Macarena, Meta Department, Colombia.

*Possible explanation:* The Spectacled bear (*Tremarctos ornatus*) exhibits a wide variety of individual variations in general color and facial markings. Its coloration ranges from pure black to dark reddish-brown, and it is found throughout the Andes Mountains in western Colombia.

*Source:* Jim Halfpenny, "Tracking the Great Bear: Mystery Bears," *Bears and Other Top Predators Magazine,* Spring 1996, on line at http://www.cryptozoology.com/articles/mysterybears.php.

## Macas Mammal
Unknown MARSUPIAL or INSECTIVORE of South America.

*Physical description:* Molelike. Length, 14–16 inches. White fur, with three broad, brown bands across the back. Elongated, trunklike snout. No ventral pouch. Webbed feet.

*Behavior:* Amphibious.

*Habitat:* Rivers.

*Distribution:* Macas, Morona-Santiago Province, Ecuador.

*Significant sighting:* In July 1999, Angel Morant Forés ran across a stuffed specimen in Macas and took several photographs of it. Ecuadorian zoologist Didier Sanchez managed to purchase the specimen in November 1999.

*Present status:* Known from only one specimen.

*Possible explanations:*

(1) A Water opossum (*Chironectes minimus*) considerably altered by a taxidermist, suggested by Didier Sanchez.

(2) Undescribed species of long-nosed, aquatic marsupial lacking a pouch.

(3) Unknown species of amphibious insectivore, suggested by Karl Shuker.

*Sources:* Angel Morant Forés, "An Investigation into Some Unidentified Ecuadorian Mammals," October 1999, http://perso.wanadoo.fr/cryptozoo/expeditions/ecuador_eng.htm; Karl Shuker, "Have Trunk, Will Tantalize: A Mystifying Mammal from Macas," *Strange Magazine,* no. 21 (Fall 2000), on line at http://www.strangemag.com.

## MacFarlane's Bear
Unknown variety of BEAR of northern Canada.

*Scientific name: Vetularctos inopinatus,* proposed by C. Hart Merriam in 1918.

*Physical description:* Whitish buff to pale yellowish buff, darkening to pale reddish brown on the underside. Broad head. Ears set like a dog's. Square, long muzzle. Teeth are unlike the brown bear's, presenting a combination of long canines and well-developed cusps with broadly flattened surfaces; the cusps of the upper first and second molars are reduced, while the lower second molar lacks the posterior cusp and notch. Wide at the shoulders. Hair on the bottom of its paws. Hind claws are as big as the front claws.

*Distribution:* Canadian Arctic; Kodiak Island, Alaska.

*Significant sightings:* The only known specimen was killed near Rendezvous Lake, Northwest Territories (65°52′N, 127°01′W) by Inuit hunters on June 24, 1864. The skin and skull were obtained by Roderick MacFarlane and shipped to the Smithsonian Institution, where it was examined by C. Hart Merriam.

In the late nineteenth century, Caspar Whitney heard of a species of bear in the Canadian North that resembled a cross between a polar bear and a grizzly.

In 1943, Clara Helgason reminisced about an incident many years earlier when hunters on Kodiak Island, Alaska, shot a large, off-white bear with hair on the soles of its paws.

*Possible explanations:*

(1) A Polar bear (*Ursus maritimus*) × Brown bear (*Ursus arctos*) hybrid, which does occur sometimes in the wild.

(2) A brown bear with a whitish coat.

(3) A surviving Short-faced bear (*Arctodus simus*), an immense fossil bear and the largest North American carnivore of the Ice Age. Arctic specimens date from 44,000–20,000 years ago; a Wyoming skull dates from 11,500 years ago. C. Hart Merriam thought the teeth resembled *Arctodus* and its relative the Spectacled bear (*Tremarctos ornatus*) of South America more than *Ursus*.

*Sources:* Caspar Whitney, *On Snow-Shoes to the Barren Grounds* (New York: Harper, 1896); Charles Mair and Roderick MacFarlane, *Through the Mackenzie Basin* (London: Simpkin, Marshall, Hamilton, Kent, 1908), pp. 217–218; C. Hart Merriam, *Review of the Grizzly and Big Brown Bears of North America (Genus* Ursus*) with Description of a New Genus, Vetularctos* (Washington, D.C.: Government Printing Office, 1918), pp. 131–133; George G. Goodwin, "*Inopinatus:* The Unexpected," *Natural History* 55 (November 1946): 404–406; Jim Halfpenny, "Tracking the Great

Bear: Mystery Bears," *Bears and Other Top Predators Magazine,* Spring 1996, on line at http://www.cryptozoology.com/articles/mysterybears.php.

## Madagascan Hawk Moth (Giant)

Undiscovered insect (INVERTEBRATE) of Madagascar.

*Scientific name: Xanthopan* sp.

*Physical description:* Hawk moth with a 16-inch proboscis.

*Distribution:* Lake Itasy, Madagascar.

*Possible explanation:* The epiphytic Madagascan orchid *Angraecum longicalcar* has a rostrellum about 16 inches deep that leads to its nectar-producing organs. No known local moth has a proboscis that long. However, entomologist Gene Kritsky predicts that one must exist, since the plant manages to propagate itself.

*Present status:* The existence of a Madagascan hawk moth (*Xanthopan morgani praedicta*) with a 12-inch proboscis was predicted in 1862 by Charles Darwin, due to the physical requirement for reaching the nectar in the Comet orchid (*Angraecum sesquipedale*). The insect was finally discovered and described in 1903.

*Sources:* Gene Kritsky, "Darwin's Madagascan Hawk Moth Prediction," *American Entomologist* 37 (Winter 1991): 206–210; Natalie Angier, "It May Be Elusive, but Moth with 15-Inch Tongue Should Be Out There," *New York Times,* January 14, 1992.

## Madukarahat

CANNIBAL GIANT of the western United States.

*Etymology:* Karok (Hokan), "giant."

*Distribution:* Klamath River, California.

*Source:* Kyle Mizokami, Bigfoot-Like Figures in North American Folklore and Tradition, http://www.rain.org/campinternet/bigfoot/bigfoot-folklore.html.

## Maeroero

WILDMAN of New Zealand.

*Etymology:* Maori (Austronesian) word for a "wild man" or a "lost tribe."

*Variant names:* Macro (an occasional misspelling), Maero (North Island), Mairoero, Ngatimamaero (South Island), Ngatimamo.

*Physical description:* Smaller than a man. Covered with long, coarse hair. Bald forehead. Long, bony fingernails.

*Behavior:* Solitary. Climbs trees. Eats birds. Said to kidnap humans and stab them with its fingers.

*Habitat:* Mountains.

*Distribution:* Tararua Range, North Island, New Zealand; Fiordland National Park and the Tautuku Forest, South Island, New Zealand.

*Significant sightings:* Naked, humanlike tracks were found in 1974 near Dusky Sound and in 1993 near Lake Manapouri in the Fiordland National Park, South Island.

*Possible explanations:*

(1) The arboreal Silver-gray brushtail possum (*Trichosurus vulpecula*) was introduced to New Zealand in 1858, but the Maeroero tradition had been established by 1844. This possum is only 22 inches long at most, with a 15-inch tail.

(2) Surviving remnants of moa-hunting pre-Maori Polynesian or Melanesian aboriginals.

*Sources:* J. E. Gray, "Habits of the 'Kakapo' and 'Macro' of New Zealand," *Annals and Magazine of Natural History* 18 (1846): 427; Ferdinand von Hochstetter, *New Zealand: Its Physical Geography, Geology and Natural History* (Stuttgart, Germany: J. G. Cotta, 1867), p. 211; Alexander W. Reed, *Treasury of Maori Folklore* (Wellington, New Zealand: A. H. and A. W. Reed, 1963); Karl Shuker, "Death Birds and Dragonets: In Search of Forgotten Monsters," *Fate* 46 (November 1993): 66–74; Herries Beattie, *Traditional Lifeways of the Southern Maori: The Otago University Museum Ethnological Project, 1920,* ed. Atholl Anderson (Dunedin, New Zealand: University of Otago Press, 1994), p. 214; H. W. Orsman, ed., *The Dictionary of New Zealand English* (Auckland, New Zealand: Oxford University Press, 1997), pp. 458–459; Craig Heinselman, "Hairy Maeroero," *Crypto* 4, no. 1 (January 2001): 23–26.

## Magenta Whale

Unknown CETACEAN of the South Pacific and North Atlantic Oceans.

*Etymology:* After the scientific research ship *Magenta.*

*Scientific name: Amphiptera pacifica,* given by Enrico Hillyer Giglioli in 1870.

*Variant names:* MONGITORE'S MONSTROUS FISH, Rhinoceros whale.

*Physical description:* Length, 60 feet. Gray-green back. Lower parts are grayish-white. Muzzle is large and blunt. Lower jaw is slightly longer than the upper. Two dorsal fins, 6 feet apart.

*Distribution:* South Pacific and North Atlantic Oceans; possibly the Mediterranean Sea.

*Significant sightings:* A two-finned baleen whale was observed September 4, 1867, by Enrico Hillyer Giglioli on the ship *Magenta* in the South Pacific Ocean, about 1,000 miles off the coast of Chile. The distance between the animal's two fins was about 6.6 feet, and it was grayish-green above and grayish-white below. It had no ridges on the top of its head or on its throat.

A sea monster with two dorsal fins about 20 feet apart was seen by Alexander Taylor and the crew of the fishing boat *Lily* off the coast of Stonehaven, Aberdeenshire, Scotland, in October 1898. It spouted like a whale and was about 68 feet long.

On July 17, 1983, a sailboat was followed by a large animal with two dorsal fins, a trapezoidal head, and a white belly in the Mediterranean between Corsica and Cavalaire-sur-Mer, France.

*Sources:* Enrico Hillyer Giglioli, *Note intorno alla distribuzione della fauna vertebrata nell'oceano prese durante un viaggio intorno al globo, 1865–1868* (Florence, Italy: Giuseppe Civelli, 1870), pp. 75–76; Enrico Hillyer Giglioli, *I cetacei osservati durante il viaggio intorno al globo della R. pirocorvetta* Magenta, *1865–1868* (Naples, Italy: Stamperia della Regia Universita, 1874), pp. 59–72; "Scared by a Sea Serpent," *Daily Mail* (London), October 10, 1898, p. 3; Jacques Maigret, "Les cétacés sur les côtes ouest-africaines: Encore quelques énigmes!" *Notes Africaines* 189 (1986): 20–24; Michel Raynal, "Do Two-Finned Cetaceans Really Exist?" *INFO Journal,* no. 70 (January 1994): 7–13.

## Maggot

Mystery INVERTEBRATE of eastern Canada.

*Physical description:* Similar to a lobster. Length, 1 foot. Fishlike eyes. Pincers are 3 inches long. Three pairs of legs. No jointed, lobsterlike tail.

*Distribution:* Gander Lake and Swanger's Cove, Bay d'Espoir, Newfoundland.

*Significant sightings:* Seen at Gander Lake in the 1930s and at Swanger's Cove around 1952.

*Possible explanation:* Misidentified American lobster (*Homarus americanus*), found in Newfoundland waters.

*Source:* X, "A mari usque ad mare," *Fortean Times,* no. 46 (Spring 1986): 44–51.

## Mahamba

Giant CROCODILIAN of Central Africa.

*Etymology:* Lingala (Bantu) word.

*Physical description:* Length, 50 feet. Head like a Nile crocodile's but wider.

*Behavior:* Carnivorous. Said to eat humans. Lays eggs. Digs long, underground tunnels.

*Distribution:* Republic of the Congo; Maika marshes, northeast Democratic Republic of the Congo.

*Significant sighting:* In 1954, Guy de la Ruwière saw a crocodile that was at least 23 feet long in the Maika marshes. It lifted its large head out of the water several times, showing an abnormally long neck. It created a huge wave when it dived below the surface.

*Possible explanations:*

(1) Large or old and aggressive specimens of the Nile crocodile (*Crocodylus niloticus*). The official size record for this crocodile was set in 1953 at 19 feet 6 inches, though there are reports of larger specimens. Most rarely grow larger than 16 feet.

(2) A surviving *Deinosuchus,* which lived 80 million years ago in the Late Cretaceous of south Texas. Current estimates place its length at 33–50 feet and its weight at 2 tons, which means it was large enough to prey on sizable dinosaurs. Other crocodilians of comparable size included the caiman *Purussaurus* and the gavialoid *Rhamphosuchus,* both fossils from the

Miocene, 15 million years ago.

*Sources:* John Reinhardt Werner, *A Visit to Stanley's Rear-Guard at Major Barttelot's Camp on the Aruhwimi* (Edinburgh: W. Blackwood, 1889), pp. 108–109, 125; Roy P. Mackal, *A Living Dinosaur? In Search of Mokele-Mbembe* (Leiden, the Netherlands: E. J. Brill, 1987), pp. 273–282, 321–326.

## Maipolina

WATER TIGER of South America.

*Etymology:* Wayana (Carib) word.

*Variant names:* Popoké, Water mother.

*Physical description:* Length, 9 feet 9 inches. Width, 3 feet 3 inches. Short, fawn-colored fur. Whitish on chest. White stripe, 5 inches wide, along back. Large brown eyes like a tapir's. Drooping ears. Tusks like a walrus's. Clawed feet like an anteater's. Tufted tail like a cow's.

*Behavior:* Waits underwater for prey. Attacks humans and canoes.

*Habitat:* Caves and hollows in the riverbank.

*Distribution:* Maroni River, near Maripasoula, French Guiana.

*Significant sighting:* The body of a boy who drowned in the Maroni on October 21, 1962, was found partially eaten. The Maipolina was blamed.

*Possible explanation:* An aquatic Saber-toothed cat (*Smilodon* sp.) similar to the YAQUARU; possibly its female counterpart, suggested by Karl Shuker.

*Sources:* Richard Chapelle, *J'ai vécu l'enfer de Raymond Maufrais* (Paris: Flammarion, 1969); René Ricatte, *De l'Île du Diable aux Tumuc-Humac* (Paris: La Pensée Universelle, 1978); Karl Shuker, *Mystery Cats of the World* (London: Robert Hale, 1989), pp. 204–205.

## Makalala

Giant BIRD of East Africa.

*Etymology:* Uncertain; said to mean "noisy." Similar to *makalele* ("noise") in Lingala (Bantu).

*Scientific name: Megasagittarius clamosus,* given by Karl Shuker in 1995.

*Physical description:* Standing height, 7–8 feet. Head and beak like a bird of prey's. Horny plates or claws on the wing tips. Long legs.

*Behavior:* Capable of sustained, powerful flight. Makes a loud noise when it claps its wings together. Feeds on carrion.

*Distribution:* Tanzania.

*Significant sighting:* In Zanzibar, a Dr. Fischer saw a bird's rib that narrowed from 8 inches at one end to 1 inch at the other. August Friedrich graf von Marschall records that the "Wasequa" people use Makalala skulls as ceremonial helmets.

*Present status:* Not seen since the nineteenth century.

*Possible explanations:*

(1) Surviving phorusrhacid bird similar to the 6- to 9-foot *Titanus walleri,* though these lived in North and South America and were flightless. A *Titanus* toe bone found in Texas could be as recent as 15,000 years ago.

(2) The Secretary bird (*Sagittarius serpentarius*) inhabits much of Africa south of the Sahara and has adapted so well to a snake-eating life on the ground that it rarely flies. It is one of the few bird species to have claws on its wing tips. Karl Shuker suggests that a giant species of secretary bird might account for the Makalala.

*Sources:* August Friedrich, graf von Marschall, "Oiseau problematique," *Bulletin de la Société Philomatique,* 7th ser., 3 (1878): 176; Karl Shuker, *In Search of Prehistoric Survivors* (London: Blandford, 1995), pp. 72–73.

## Makara

Mythical Hindu SEA MONSTER.

*Etymology:* Sanskrit (Indo-Aryan), "neither one thing nor another" or "mythical beast."

*Physical description:* Ridden by the god Vishnu, the Makara is sometimes portrayed as a crocodile; as a dolphin, crab, or shark; or as half fish and half elephant.

*Distribution:* Borobudur Temple, Java, Indonesia.

*Significant sighting:* The Makara sculptures that serve as waterspouts on the ninth-century Buddhist temple of Borobudur are elephantine, with four unusual cheek teeth that more closely resemble those of a fossil gomphothere than the Asian elephant (*Elephas maximus*).

*Possible explanations:*

(1) The sculptures could have been based on fossil elephant skulls.

(2) A tetralophodon, an advanced gomphothere that lived in India and Java in the Pliocene, 5 million years ago, and survived into historical times.

(3) The myth of the Makara may be based in part on the Nile crocodile (*Crocodylus niloticus*) and the Hippopotamus (*Hippopotamus amphibius*).

*Sources:* Ermine C. Case, "The Mastodons of Baraboedaer," *Proceedings of the American Philosophical Society* 81 (1939): 569–572; Makara, 2001, http://www.khandro.net/mysterious_makaras1.htm.

## Malagasy Lion

Mystery CAT of Madagascar.

*Physical description:* Large, lionlike cat.

*Habitat:* Caves.

*Distribution:* Unexplored areas of Madagascar.

*Present status:* Since the island has no known native canids or felids, the discovery of a predatory carnivorous cat would be surprising.

*Possible explanation:* Surviving *Machairodus* saber-toothed cat, proposed by Paul Cazard.

*Source:* Paul Cazard, in *Le Chasseur Français,* October 1939, p. 664.

## Mala-Gilagé

SMALL HOMINID of Central Africa.

*Etymology:* Unknown origin; said to mean "tailed men."

*Physical description:* Short. Hairy. Reddish skin. Long head-hair. Short arms. Natural tail.

*Behavior:* Tends black camels the size of donkeys.

*Distribution:* Southern Chad.

*Significant sighting:* The comte d'Escayrac de Lauture heard of small, hairy, reddish men with tails south of the Chari River in Chad.

*Possible explanations:*

(1) An undiscovered group of short-statured

hunter-gatherers related to the Pygmy Mbenga people of Gabon and Cameroon. (2) Legends of men with tails might have arisen from the practice of wearing loincloths made of animal pelts with the tail still attached or tail-like ornaments made of leather. The Azande of Sudan were once thought tailed because they wound monkey skins around their waists.

*Source:* Stanislas, comte d'Escayrac de Lauture, *Mémoire sur le Sudan* (Paris: A. Bertrand, 1855–1856), pp. 50–53.


## Malagnira

Unknown PRIMATE of Madagascar.

*Etymology:* Malagasy (Austronesian) word, Sakalava du Menabe dialect.

*Physical description:* A lemur smaller than the Gray mouse lemur (*Microcebus murinus*).

*Behavior:* Said to have different habits than the mouse lemurs.

*Distribution:* Tsingy de Bemaraha Strict Nature Reserve, western Madagascar.

*Significant sighting:* Two informants described this animal in an early 1990s survey.

*Possible explanation:* The Pygmy mouse lemur (*Microcebus myoxinus*) is probably the world's smallest living primate and has a total length of less than 8 inches. First named in 1852, it became taxonomically confused with other mouse lemurs until 1994, when its species status was rehabilitated by Jutta Schmid and Peter Kappeler. Its presence in the Bemaraha is unconfirmed but possible.

*Source:* Nasolo Rakotoarison, T. Mutschler, and U. Thalmann, "Lemurs in Bemaraha (World Heritage Landscape, Western Madagascar)," *Oryx* 27 (January 1993): 35–40.


## Malpelo Monster

Mystery sharklike FISH of the eastern Pacific Ocean.

*Etymology:* After the island.

*Variant names:* Bongo, el Monstro.

*Physical description:* Length, 15 feet (female). Large eyes. Dorsal fin placed above the pectoral fins.

*Habitat:* Prefers colder water under the thermocline, below 160 feet.

*Distribution:* Pacific Ocean off Isla de Malpelo, an island 285 miles from the coast of Colombia.

*Significant sightings:* This shark has been seen and photographed on rare occasions by divers.

*Present status:* Colombian biologist Sandra Bessudo launched an investigation in March 2001 to determine its status.

*Possible explanations:*

(1) The Sand tiger shark (*Carcharias taurus*) is similar and found in all warm seas except perhaps the eastern Pacific. It grows to a length of about 10 feet 6 inches. However, its dorsal fin is placed farther back.

(2) The Smalltooth sand tiger (*Odontaspis ferox*) has been seen off southern California and Baja California. It has smaller eyes and grows to 12 feet long.

(3) An unknown species of Sand tiger shark (Family Odontaspididae), suggested by Sandra Bessudo.

*Source:* François Sarano, "The Malpelo Monster: A New Species of Shark?" August 23, 2001, http://www.photoceans.com/anglais/mag/index.cfm?id_act=243&id_rub=71.


## Mamantu

Legendary ELEPHANT-like animal of Siberia and East Asia.

*Etymology:* Word common to Yakut (Turkic), Khanty (Ob-Ugric), and Koryak (Chukotko-Kamchatkan) peoples, meaning "underground animal."

*Variant names:* Fén-shŭ (Chinese/Sino-Tibetan, "underground rat"), Jukhensinggheri (in Mongolia, "rat beneath the ice"), Kilu kpuk (Yupik/Eskimo-Aleut, "Kilu whale"), Shŭ-mu (Chinese/Sino-Tibetan, "rat mother"), Tai-shŭ, Tuilu (Itelmen/Chukotko-Kamchatkan), Xól-hut (Yukagir/Paleosiberian), Yen-shŭ (Chinese/Sino-Tibetan, "self-concealing rat").

*Physical description:* Large as an elephant. Grayish-red hair. Tiny eyes. Tusks. Short tail.

*Behavior:* Lives underground. Digs tunnels in the snow. Its wanderings are said to cause earthquakes.

*The Woolly mammoth* (Mammuthus primigenius). *(© 2002 ArtToday.com, Inc., an IMSI Company)*

*Tracks:* Oval. Width, 2 feet. Length, 18 inches. Spaced 12 feet apart.

*Habitat:* Pine and birch forests, tundra.

*Distribution:* Siberia; Mongolia; China.

*Significant sightings:* In 1581, the Cossack leader Yermak Timofeyevich reported meeting up with a hairy elephant east of the Ural Mountains.

A Russian hunter came across enormous tracks in a forested region near the Obskaya Gulf, Yamal-Nenets Autonomous Province, Siberia, in 1918. The prints were 24 inches long and 18 inches wide, with a stride of 12 feet. He found large droppings consisting of vegetable matter and noticed tree branches at a height of 9–10 feet that were apparently damaged by the animal's passing. After several days of tracking, he sighted two huge elephants with white, curved tusks and dark-chestnut hair that was longer on the flanks and shorter in front.

*Possible explanation:* A myth based on the subfossil remains of the Woolly mammoth (*Mammuthus primigenius*), an elephant that lived in Europe, Asia, and North America at the end of the last Ice Age. It was covered with thick, spiral locks of black or dark-brown guard hairs above shorter, silkier underwool. With a shoulder height of 9–12 feet, its weight has been estimated at 4–7 tons. Both males and females had tusks.

Rumors of mammoth survival seem primarily to be based upon subfossil specimens 40,000–10,000 years old found frozen in the permafrost, with muscles, skin, and hair intact. Carcasses found in defrosting peat by Siberian nomads may have been interpreted as contemporaneous fauna that lived underground. The observations of 1581 and 1918 are isolated and not strong evidence of the animal's persistence into historical times.

The most famous finds of frozen mammoths in the Siberian permafrost are: the Adams or Lena mammoth, discovered in 1799 in the Lena River delta, Sakha Republic (35,000 years old); the Berezovka mammoth, found in 1900 along the Berezovka River, Sakha Republic (40,000–30,000 years old); the Taymyr mammoth, recovered in 1949 (13,000 years old); the Dima mammoth, a complete carcass of a 6- to 12-month-old baby discovered in 1979 on the

Kirgilyakh River, Magadan Region (40,000–26,000 years old); a mummified baby mammoth, less than three months old, found in 1988; and the Jarkov mammoth, discovered in 1997 near the Bolchaya Balakhnya River, Taymyr Autonomous Province, and excavated nearly intact in 1999 by Bernard Buigues (20,000 years old). The Heilongjiang Province of China contains dozens of mammoth finds.

In the Crimea and the Caucasus, mammoths became extinct about 30,000–20,000 years ago; on the Russian plain, they were still present about 13,000 years ago. Based on radiocarbon dating, the latest mammoth remains found in Western Europe (northern France, Switzerland, and Great Britain) also date to 13,000–12,000 years ago.

Radiocarbon dating of teeth, tusks, and bones of dwarf mammoths found on Wrangel Island, Chukot Autonomous Province, between 1989 and 1991 proved that some mammoths survived into historical times, until about 2,000 B.C. With a shoulder height of only 6 feet and weighing only 4,400 pounds, these isolated animals constitute a distinct subspecies (*M. p. vrangeliensis*).

*Sources:* "Observations de physique et histoire naturelle de l'Empereur Kang-hi," in *Mémoires concernant l'histoire, les sciences, les arts, les moeurs, les usages, &c, des Chinois: Par les missionnaires de Pékin* (Paris: Nyon, 1776–1791), vol. 4, p. 481; Mikhail Adams, "Relation d'un voyage à la mer glaciale et découverte des restes d'un mammouth," *Journal du Nord* 32, suppl. (1807): 633–640, 621–628 (pages misnumbered); Edward Newman, "The Mammoth Still in the Land of the Living," *Zoologist,* ser. 2, 8 (1873): 3731–3733; "Chinese Accounts of the Mammoth," *American Naturalist* 24 (1890): 847–850; Waldemar Jochelson, "Some Notes on the Traditions of the Natives of North-Eastern Siberia about the Mammoth," *American Naturalist* 43 (1909): 48–50; I. P. Tolmachoff, "The Carcasses of the Mammoth and Rhinoceros Found in the Frozen Ground in Siberia," *Transactions of the American Philosophical Society,* new ser. 23, pt. 1 (1929): 1–74; Eugen W. Pfizenmayer, *Siberian Man and Mammoth* (London: Blackie and Sons, 1939); Marcel Marmet, "A la recherche des traces des derniers mammouths," *Science et Vie* 77 (January 1950): 10–12; Bernard Heuvelmans, *On the Track of Unknown Animals* (New York: Hill and Wang, 1958), pp. 330–353; Nikolai K. Vereshchagin and V. M. Mikhel'son, *Magadanskii mamontenok: Mammuthus primigenius (Blumenbach)* (Leningrad: Nauka, 1981); N. K. Vereshchagin and G. F. Baryshnikov, "Quaternary Mammalian Extinctions in Northern Eurasia," in Paul S. Martin and Richard G. Klein, eds., *Quaternary Extinctions: A Prehistoric Revolution* (Tucson: University of Arizona Press, 1984), pp. 483–516; Gary Haynes, *Mammoths, Mastodonts, and Elephants: Biology, Behavior, and the Fossil Record* (New York: Cambridge University Press, 1991); S. L. Vartanyan, Kh. A. Arslanov, T. V. Tertychnaya, and S. B. Chernov, "Radiocarbon Dating Evidence for Mammoths on Wrangel Island, Arctic Ocean, until 2000 B.C.," *Radiocarbon* 37 (1995): 1–6; *Raising the Mammoth* (video) (Discovery Channel, 2000); *Land of the Mammoth* (video) (Discovery Channel, 2001); Richard Stone, *Mammoth: The Resurrection of an Ice Age Giant* (Cambridge, Mass.: Perseus, 2001).

## Mamba Mutu

MERBEING of Central Africa.

*Etymology:* From the Swahili (Bantu) *mamba mtu* ("crocodile man").

*Variant name:* Mamba muntu.

*Physical description:* Half human, half fish.

*Behavior:* Sucks human blood and eats brains.

*Distribution:* Lake Tanganyika and Lukuga River, Democratic Republic of the Congo.

*Possible explanations:*

(1) Isolated population of the West African manatee (*Trichechus senegalensis*), though these animals are herbivorous.

(2) An unknown species of giant otter with a flat skull, suggested by zoologist Carlos Bonet.

*Sources:* Carlos Bonet, "Le *mamba mutu:* Un carnivore aquatique dans le lac Tanganyika?" *Cryptozoologia,* no. 10 (January 1995): 1–5; Karl Shuker, "Bloodsucking Mermaids and Vampire Fishes," *Strange Magazine,* no. 15 (Spring 1995): 32–33.

## Mami Water

MERBEING of West Africa.

*Variant names:* Mami wata, Tahbin, Tahbi-yin (for the female).

*Physical description:* Fair-skinned female water spirit.

*Behavior:* Represents eroticism.

*Habitat:* Rivers.

*Distribution:* Ghana; Niger River delta in Nigeria.

*Possible explanations:*

(1) The West African manatee (*Trichechus senegalensis*) is 7–12 feet long and found in coastal waters and rivers from Senegal to Angola. It feeds primarily at night, making chance observations mysterious.

(2) The oldest known Mami water wooden carvings date from about 1901 in riverine areas of southern Nigeria. There is artistic evidence that they derive from *Der Schlangenbandinger* (The Snake Charmer), a circa 1880–1887 chromolithograph of the exotic, long-haired, snake-charming wife of a Hamburg zookeeper. Copies of this popular print that were sold in West Africa in the mid- to late 1950s originated in Mumbai and England.

*Sources:* Edward Geoffrey Parrinder, *African Traditional Religion* (New York: Hutchinson's University Library, 1954); Jill Salmons, "Mamy Wata," *African Arts* 10, no. 3 (1977): 8–15, 87–88; Henry John Drewal, "Performing the Other: Mami Wata Worship in Africa," *TDR: The Drama Review* 32, no. 2 (1988): 160–185; David Hecht, "Mermaids and Other Things in Africa," *Arts Magazine* 65, no. 3 (1990): 80–86; "Ghanaian Scientists Unravel Mystery Mermaid's Being," Panafrican News Agency, April 5, 2001, http://allafrica.com/stories/200104050058.html.

## Mamlambo

FRESHWATER MONSTER of South Africa.

*Etymology:* From the Xhosa (Bantu) *umamlambo,* the name for a mythical river goddess or MERBEING who brings riches and whose true form is a snake.

*Physical description:* Length, 66 feet. Head is like a snake's or horse's. Eyes glow green at night. Crocodilian body. Short, stumpy legs.

*Behavior:* Said to kill people and suck out their blood and brains.

*Distribution:* Mzintlava River, near Mount Ayliff, Eastern Cape Province, South Africa.

*Significant sighting:* Eyewitnesses reported a "half-fish, half-horse" monster in the Mzintlava River that was held responsible for nine deaths in the first months of 1997.

*Possible explanations:*

(1) The alleged Mamlambo victims probably just drowned.

(2) The Electric catfish (*Malapterurus electricus*) attains a length of 5 feet and can deliver stunning or fatal shocks.

*Sources:* "Nature Conservation Called to Hunt East Cape 'Monster,'" *Johannesburg Star,* April 30, 1997; "Mamlambo on the Loose," *Cape Argus,* May 17, 1997; Ben S. Roesch, "Mamlambo: A 'Man-Eating' Reptile?" *Cryptozoology Review* 2, no. 1 (1997): 9–10; John Kirk, *In the Domain of Lake Monsters* (Toronto, Canada: Key Porter Books, 1998), pp. 259–260; Brian Siegel, "Water Spirits and Mermaids: The Copperbelt Case," paper presented at the Southeastern Regional Seminar in African Studies (SERSAS), April 14–15, 2000, Cullowhee, N.C., on line at http://www.ecu.edu/african/sersas/Siegel400.htm.

## Manaus Pterosaur

FLYING REPTILE of South America.

*Physical description:* Flat head. Long beak. Long neck. Ribbed wings. Wingspan, 12 feet.

*Behavior:* Flies in a V-formation.

*Distribution:* Manaus, Amazonas State, Brazil.

*Significant sighting:* Five winged animals flying in a V-formation were seen by J. Harrison near Manaus, Brazil, in 1947. Their wings resembled brown leather and seemed to lack feathers.

*Possible explanations:*

(1) A large stork native to Brazil, either the Jabiru (*Jabiru mycteria*), the Maguari (*Ciconia maguari*), or the Wood stork (*Mycteria americana*).

(2) Pelicans (Family Pelecanidae) often fly in V-formations.

*Source:* John Michell and Robert J. M. Rickard, *Living Wonders* (London: Thames and Hudson, 1982), p. 50.

## Man-Beast of Darién

Unknown PRIMATE of Central America.

*Physical description:* Height, 6 feet. Weight, about 300 pounds. Long, black hair. Apposed big toe.

*Behavior:* Bipedal. Threatening behavior. Chattering speech.

*Distribution:* Southern Panama.

*Significant sighting:* In 1920, an American prospector named Shea killed a large, apelike animal in the Serranía del Sapo, near Piñas Bay, Darién Province, Panama.

*Source:* Richard Oglesby Marsh, *White Indians of Darien* (New York: G. P. Putnam's Sons, 1934), pp. 19–21.

## Maned American Lion

Mystery CAT of North America.

*Variant names:* California lion, Lunkasoose (in Maine).

*Physical description:* Resembles a male African lion. Length, 5–8 feet. Shoulder height, 3 feet. Shaggy coat. Brown or tawny. Large head. Thick hair around the neck like a mane. Muscular shoulders. Long tail with a bushy tip. Some reports involve striped or partially striped animals.

*Behavior:* Tends to travel in pairs, sometimes with melanistic EASTERN PUMAS. Roars. Attacks, kills, and eats pigs, chickens, calves, colts, lambs, dogs, and cats.

*Tracks:* Length, 5 inches. Width, 4.75 inches. Prints are 40 inches apart.

*Distribution:* A partial list of places where Maned American lions have been reported follows:

*Arkansas*—Dierks, Dover.

*California*—El Toro, Fremont, Lake County.

*Florida*—Loxahatchee.

*Georgia*—Alapaha, Berrien County.

*Illinois*—Centralia, Decatur, Joliet, Peoria County, Piatt County, Rockford, Roscoe, Will County, Winnebago County.

*Indiana*—Abington, Elkhorn Falls, Warrick County.

*Iowa*—Muscatine, Wapello.

*Maine*—Penobscot County.

*Missouri*—Cross Timbers.

*Nebraska*—Ceresco, Surprise, Waterloo.

*New Brunswick, Canada*—Gagetown, McAdam.

*New Jersey*—North Brunswick.

*North Carolina*—Rutherfordton.

*Ohio*—Clinton County, Dodson Township, Geauga County, Groesbeck, Hillsboro, Lorain County, Mentor, Miami Township, Morning Sun, North Avondale, North Olmsted, Springboro.

*Oklahoma*—Craig County, Rogers County, Vinta.

*Ontario, Canada*—Kapuskasing.

*Pennsylvania*—Bald Eagle Mountain, Clinton County, Jackson, Lackawanna County, Newton Township, Nicholson, Pike County, Susquehanna County, Wyoming County.

*Texas*—Fort Worth.

*Washington*—Spokane, Tacoma.

*West Virginia*—Marlinton.

*Significant sightings:* In 1797, frontiersman Peter Pentz killed a big cat with a matted, yellow-brown mane in a cave on Bald Eagle Mountain, Pennsylvania. It had been killing local livestock for six years.

Hunter Archie McMath shot and killed a yellowish, lionlike animal in Lake County, California, in 1868. It had a total length of 11 feet and weighed more than 30 pounds. The front part was stockier than its hindquarters, and it had black stripes along its shoulders and back and down its fore parts. Its hair was darker and thicker around its neck.

A maned cat was seen in conjunction with a black panther around Elkhorn Falls, Indiana, from August 5 to 8, 1948. The pair apparently migrated east into Ohio by early September.

Around August 1, 1954, farmer Arnold Neujahr saw what looked like an African lion 2 miles west of Surprise, Nebraska. Similar incidents took place closer to town and near Rising City. Residents recalled other lion sightings around

Ceresco in November 1951 that sparked a lion hunt.

Near Kapuskasing, Ontario, in June 1960, Leo Paul Dallaire watched an animal resembling an African lion on his farm. It was light tan and had a mane and a 4-foot tail with a bushy tip.

On the evening of November 10, 1979, several residents of Fremont, California, reported that a large, male lion was on the loose in the Coyote Hills Regional Park. Police Officer William Fontes saw it in the Alameda County flood channel, and he estimated its weight as 300–400 pounds.

On July 30, 1986, Cindy Belmont and her brother saw a long-tailed, beige "tiger" near Jackson, Pennsylvania. It was accompanied by a shaggy animal that looked like a collie dog.

Several reports of a 7-foot-long, maned cat were phoned in to the Mentor, Ohio, police in June 1992. Although witnesses insisted it looked like a male lion, the police decided a large golden retriever dog was responsible.

On June 5, 1996, Belen Grabb was driving on Canyon Drive near Spokane, Washington, when a lion strolled off an adjacent golf course and into the road in front of her. She stopped her car 4 feet away from it. It was dark beige with a brown mane. The sighting sparked two days of intense searching for the animal.

*Possible explanations:*

(1) An African lion (*Panthera leo*) escaped from a zoo, circus, or exotic pet owner. Escapes do occur, and some may not be reported by private owners.

(2) The Domestic dog (*Canis familiaris*) is rarely a look-alike for a large, maned cat. However, Karl Shuker points out that both a chow chow and a Brittany spaniel have been put forward as candidates by authorities desperate for an explanation. A Newfoundland is a better match, although it looks more like a bear cub.

(3) A surviving American lion (*Panthera atrox*), a powerful Pleistocene predator with large canine teeth that died out about 9,000 years ago, has been suggested by Mark A. Hall and Loren Coleman. Fossils of this animal have been found from Alaska to Peru, but the richest source are the tar pits at Rancho La Brea, California. Males were about 25 percent larger than the African lion, with blunter faces and longer legs. The molar teeth indicate sexual dimorphism, leading Hall and Coleman to speculate that the female *P. atrox* might account for reports of a melanistic EASTERN PUMA. Such an extreme difference between the sexes is difficult to accept without further proof. (However, having two species of large, unknown felids occupying basically the same habitat in North America is at least equally unlikely.) Presumably, this species traveled in prides, but Maned American lions seem to travel in pairs.

*Sources:* Henry W. Shoemaker, *More Pennsylvania Mountain Stories* (Reading, Pa.: Bright, 1912); Henry W. Shoemaker, *Juniata Memories: Legends Collected in Central Pennsylvania* (Philadelphia: John Joseph McVey, 1916); *Toronto Telegram,* June 28, 1960; Loren Coleman, "Maned Mystery Cats," *Fortean Times,* no. 30 (Autumn 1979): 47–50; Loren Coleman, "Maned Mystery Cats," *Fortean Times,* no. 31 (Spring 1980): 24–27; Loren Coleman, "An Answer from the Pleistocene," *Fortean Times,* no. 32 (Summer 1980): 21–22; Karl Shuker, *Mystery Cats of the World* (London: Robert Hale, 1989), pp. 166–172; Mark A. Hall, "The American Lion (*Panthera atrox*)," *Wonders* 3, no. 1 (March 1994): 3–20; Loren Coleman, "Roaring at the Mane Event," *Fortean Times,* no. 92 (November 1996): 40; Loren Coleman, *Mysterious America,* rev. ed. (New York: Paraview, 2001), pp. 127–159.

## Maner

A category of SEA MONSTER identified by Gary Mangiacopra.

*Physical description:* Serpentine or eel-like. Length, 15–50 feet. Horselike or snakelike flat head, 3 feet long, tapering down to the muzzle. Enormous eyes. Slender neck, 10 feet long or more. A mane or beard has been reported. Round tail, either fanlike or tapering to a point.

*Behavior:* Swims rapidly by squirming. Churns up the water. Spouts. Curious and cautious; sometimes playful. Has been reported to circle a boat, jump completely out of the water, and land on its stomach.

*Distribution:* North Atlantic Ocean along the coast of the United States.

*Significant sightings:* On September 25, 1888, Captain Springs of the tug *Henry Buck* was towing a schooner in Winyah Bay, near Georgetown, South Carolina, when he spotted a 50-foot animal swimming on the surface with its head 3 feet in the air. The head was vermilion, and the neck was covered with a long mane. The captain's story was corroborated by others.

Pilot Alexander Banta watched a black creature larger than a whale as he was off City Island, New York, on August 10, 1902. It dived, came up under the boat, and struck it so that it nearly capsized. The monster had enormous eyes and a yellow mane.

*Present status:* Similar to Bernard Heuvelmans's MERHORSE.

*Possible explanation:* An unknown mammal, perhaps related to the Seals (Suborder Pinnipedia).

*Sources:* "The Sea Serpent," *St. Louis Globe-Democrat,* September 27, 1888, p. 6; "Sea Serpent Hits Hell Gate Pilot," *New York Herald,* August 11, 1902, p. 12; Gary S. Mangiacopra, "The Great Unknowns of the 19th Century," *Of Sea and Shore* 8, no. 3 (Fall 1977): 175–178.

## Manetúwi Msí-Pissí

FRESHWATER MONSTER of Ohio.

*Etymology:* Shawnee (Algonquian), "great miraculous tiger."

*Distribution:* Central Ohio.

*Source:* Albert S. Gatschet, "Water-Monsters of American Aborigines," *Journal of American Folklore* 12 (1899): 255–260.

## Mangarsahoc

Mystery HOOFED MAMMAL of Madagascar.

*Etymology:* Malagasy (Austronesian), "beast whose ears hide its chin."

*Variant names:* Mangarisoaka, Tokatongotra.

*Physical description:* Donkeylike animal. Long ears. Round hoof like a horse's.

*Behavior:* Brays like a donkey.

*Tracks:* Hoofed.

*Habitat:* Rainforest.

*Distribution:* Ankaizina Mountains; Bealanana and Manirenja Districts in the north-central highlands; north of Tôlagnaro in southeastern Madagascar.

*Possible explanations:*

(1) Surviving Malagasy pygmy hippopotamuses (*Hippopotamus lemerlei* and *H. madagascariensis*) that supposedly died off within the past 1,000 years or so. *See* TSY-AOMBY-AOMBY.

(2) Unknown wild donkey indigenous to Madagascar.

*Sources:* Etienne de Flacourt, *Histoire de la grande isle Madagascar* (Paris: G. de Luyne, 1658); Raymond Decary, *La faune malgache, son rôle dans les croyances et les usages indigènes* (Paris: Payot, 1950), pp. 203–208; Bernard Heuvelmans, *On the Track of Unknown Animals* (New York: Hill and Wang, 1958), pp. 515–516.

## Mangden

Unknown HOOFED MAMMAL of Southeast Asia.

*Etymology:* Vietnamese (Austroasiatic), "black deer."

*Physical description:* Antlers are distinct from those of other deer species.

*Distribution:* Pu Mat Reserve, Vietnam.

*Significant sighting:* In the 1990s, zoologist John MacKinnon found a pair of odd antlers in a box of specimens collected from Pu Mat as long ago as the late 1960s.

*Source:* Eugene Linden, "Ancient Creatures in a Lost World," *Time,* June 20, 1994, pp. 56–57.

## Manguruyú

Giant FISH of South America.

*Variant names:* Gums, Mysterious beast.

*Physical description:* Sluglike or snakelike. Wide as a horse. Doglike head. No teeth. Stumpy tail that conceals a poisonous, barbed spike.

*Behavior:* Said to pull dogs and humans underwater.

*Distribution:* Río Paraguay drainage; swamps of the Gran Chaco, Boquerón Department, Paraguay.

*Possible explanations:*

(1) Toothless, freshwater shark, according to Percy Fawcett. However, the only known shark to stay for extended periods in freshwater, the aggressive Bull shark (*Carcharhinus leucas*) of the Amazon and elsewhere, has triangular, serrated teeth.

(2) Giant Catfish (Suborder Siluroidea), said to grow to 18 feet in length and weigh half a ton. Older specimens of catfish sometimes are toothless.

(3) A large sturgeon, according to Mike Grayson, though none occur anywhere in the Southern Hemisphere.

*Sources:* Charles William Thurlow Craig, *Spinner's Delight* (London: Hutchinson, 1951); Percy Fawcett, *Exploration Fawcett* (London: Hutchinson, 1953); Charles William Thurlow Craig, *Black Jack's Spurs* (London: Hutchinson, 1954), pp. 159–168; Karl Shuker, "Close Encounters of the Cryptozoological Kind," *Fate* 53 (May 2000): 26–29.


## Manipogo

FRESHWATER MONSTER of Manitoba, Canada.

*Etymology:* Named by Tom Locke in 1960, in imitation of OGOPOGO.

*Variant name:* Manny.

*Physical description:* Serpentine. Length, 10–40 feet. Brownish-black upper body. At least one hump. Flat, diamond-shaped head.

*Behavior:* Bellows like a train whistle.

*Distribution:* Lake Manitoba, Manitoba. The animal's name is also used as a synonym for WINNIPOGO in other Manitoban lakes.

*Significant sightings:* Louis Betecher and Eddie Nipanik saw a serpentine animal in the lake in 1957.

On August 10, 1960, government land inspector Tom Locke and sixteen other witnesses saw three creatures swimming offshore near Manipogo Beach. They looked like huge, dark-brown snakes. Many other sightings were reported that summer. Zoologist James A.

McLeod led an expedition to Lake Manitoba later in the year and interviewed many residents.

Richard Vincent and John Konefall saw a "large black snake or eel" off Meadow Portage on August 12, 1962. Vincent took three photos, one of which shows an elongated, snakelike object with a hump. Unfortunately, some inconsistencies have undermined the credibility of this case.

In the summer of 1987, Allen McLean and his family were boating in Portage Bay when they saw a large, black object swimming toward them.

*Sources: Winnipeg Free Press,* August 5, 1961, and August 15, 1962; Chris Rutkowski, *Unnatural History: True Manitoba Mysteries* (Winnipeg, Canada: Chameleon, 1993), pp. 137–147.


## Man-Monkey

WILDMAN of central England.

*Physical description:* Black creature. Large, white eyes.

*Behavior:* Rides horses.

*Distribution:* Woodseaves, Staffordshire, England.

*Significant sighting:* On January 21, 1879, on the highroad over the Birmingham and Liverpool Canal between Ranton and Woodcote, 1 mile from the village of Woodseaves, a strange creature jumped on the back of a man's carthorse and rode it into a canter. The man's application of his whip had no effect on it.

*Possible explanation:* Ghost or other apparition.

*Source:* Charlotte Sophia Burne, *Shropshire Folk-Lore: A Sheaf of Gleanings* (London: Trübner, 1883), pp. 106–107.


## Mänsanzhí

Dwarf MERBEING of the eastern United States.

*Etymology:* Miami (Algonquian), "freshwater being."

*Variant name:* Mänsanzhí-kwa (for the female).

*Distribution:* Ohio.

*Source:* Albert S. Gatschet, "Water-Monsters

of American Aborigines," *Journal of American Folklore* 12 (1899): 255–260.

## Manticora

SEMIMYTHICAL BEAST of West Asia and the Indian subcontinent.

*Etymology:* Old Persian, "man-eater," from *martiya* ("man") + *khvar* ("to eat"). A corrupt reading of Aristotle turned the Greek variant *martikhora* to *manticora*.

*Variant names:* Manticore, Man-tiger, Martikhora (Greek).

*Physical description:* Size of a lion. Red color. Head is like a man's. Gray or blue eyes. Large ears. Three rows of teeth in each jaw. Stingers on a pointed, scorpion-like, 18-inch tail.

*Behavior:* Sting from tail is said to be fatal. Can shoot foot-long spines in its tail a distance of 100 feet. Hunted by locals mounted on elephants.

*Distribution:* India; Iran.

*Possible explanations:*

(1) According to Valentine Ball, the Tiger (*Panthera tigris*) has a small, clawlike dermal structure at the end of its tail. Its whiskers can also cause lacerations. The three rows of teeth might refer to the tiger's trilobate molars. Tigers were hunted in ancient times by local princes who rode elephants. Man-eating tigers are feared by villagers in India. A tiger wounded by porcupine quills may prevent it from taking its usual prey and force it to become a man-eater. Distorted accounts of the CASPIAN TIGER (*P. t. virgata*) could be another source of information.

(2) The Indian crested porcupine (*Hystrix indica*) may have confused early travelers because it wounds tigers and leopards with its quills. However, it does not shoot them from a distance.

(3) The Slender loris (*Loris tardigradus*) of southern India may have contributed to the tradition of a human face.

*Sources:* Ctesias, *Indika*, in J. W. McCrindle, ed., *Ancient India* (Calcutta, India: Thacker, Spink, 1882), pp. 11–12; Pausanias, *A Description of Greece,* trans. W. H. S. Jones (Cambridge, Mass.: Harvard University Press, 1918) (IX. 21.4); André Thevet, *Cosmographie universelle* (Paris: P. L'Huilier, 1575), vol. 1, p. 52; Jean de Thévenot, *The Travels of Monsieur Thévenot into the Levant* (London: H. Clark, 1687), pt. 3, chap. 4, p. 7; Valentine Ball, "The Identification of the Pygmies, the Martikhora, the Griffin, and the Dikarion of Ktesias," *The Academy* 23 (1884): 277; Peter Costello, *The Magic Zoo* (New York: St. Martin's, 1979), pp. 104–110.

## Mao-Rén

WILDMAN of East Asia.

*Etymology:* Mandarin Chinese (Sino-Tibetan), "hairy man."

*Variant names:* Dà-mao-rén ("big hairy man"), Mao-gong, Mao-jùen, Mo-zhyn (in Kyrgyzstan).

*Physical description:* Height, 6 feet. Covered in red hair. Long head-hair.

*Behavior:* Walks upright.

*Tracks:* Huge, humanlike footprints.

*Distribution:* Hubei and Sichuan Provinces, China; Nei Mongol Autonomous Region, China; Kyrgyzstan.

*Sources:* Odette Tchernine, *The Yeti* (London: Neville Spearman, 1970), p. 176; Zhou Guoxing, "The Status of Wildman Research in China," *Cryptozoology* 1 (1982): 13–23.

## Mapinguari

Mystery PRIMATE or SLOTH of South America.

*Variant names:* Capé-lobo ("wolf's cape"), Juma, Mão de pilão ("pestle hand"), Mapinguary, Ow-ow, PÉ DE GARRAFA.

*Physical description:* Height, about 5–6 feet when standing upright. Weighs about 500 pounds. Long, reddish fur or hair. Monkeylike face. Manelike hair along its back. Said to have another mouth in its belly. Its feet are said to turn backward.

*Behavior:* Nocturnal. Avoids water. Descends from the mountains in the autumn. Cry is either a deafening roar or like a human shout. Releases a foul-smelling stench when threatened. Kills cattle by pulling out their tongues. Eats bacaba palm hearts and berries. Twists palm trees to the ground to get the palm hearts. Travels with herds of White-lipped peccaries (*Tayassu pecari*). Said to be followed by an army of beetles. Cannot be wounded by weapons except around its navel.

*Tracks:* Either humanlike or like the bottom of a bottle stuck into the ground. Length, 11–21 inches. Stride, 3–4 feet. Feces similar to a horse's.

*Distribution:* The apelike variety is more often seen in Mato Grosso and Pará States, Brazil; the slothlike variety has been reported in Amazonas and Acre States, Brazil. Possible evidence also exists in Paraguay.

*Significant sightings:* An adventurer named Inocêncio was with ten friends on an expedition up the Rio Uatumã, Pará State, Brazil, in 1930 when he was separated from them and got lost. As he slept in a tree for the night, he heard loud cries coming from a thickset, black figure that stood upright like a man. He shot at it several times and apparently hit it, as there was a trail of blood below his tree.

In 1975, mine worker Mário Pereira de Souza claims he encountered a Mapinguari at a mining camp along the Rio Jamauchím south of Itaituba, Pará State, Brazil. He heard a scream and saw the creature coming toward him on its hind legs. It seemed unsteady and emitted a terrible stench.

In the 1980s and 1990s, David Oren conducted fifty interviews with Brazilian Indians, rubber planters, and miners who know about the animal. He interviewed seven hunters who claim to have shot specimens. One group of Kanamarí Indians living in the Rio Juruá Valley claimed to have raised two infant Mapinguaris on bananas and milk; after one or two years, the creatures' stench became unbearable, and they were released.

In the late 1990s, Dutch zoologist Marc van Roosmalen heard that people in one village along the Rio Purus, Amazonas State, Brazil, moved their homes across the river after Mapinguari tracks were found nearby.

*Possible explanations:*

(1) Unknown ape similar to DE LOYS'S APE or the DIDI.

(2) A surviving man-sized Patagonian cave-dwelling sloth of the genus *Mylodon*. All subfossil fur samples are red. *Mylodon* walked with its clawed feet curved toward the center of its body. Its dermal ossicles (except around the navel) might protect it from gunfire. The round tracks might be the impression of the heavy tail tip as the creature stands upright. David Oren suggests that the "second mouth" is a specialized, scent-secreting gland.

*Sources:* Paulo Saldanha Sobrinho, *Fatos, histórias e lendas do Guaporé,* as quoted at http://www.pakaas.com.br/lenda2.asp; Frank W. Lane, *Nature Parade* (London: Jarrolds, 1955), p. 241; Luís da Câmara Cascudo, *Dicionário do folclore Brasileiro* (Rio de Janeiro: Instituto Nacional do Livro, 1962), vol. 2, p. 456; David C. Oren, "Did Ground Sloths Survive to Recent Times in the Amazon Region?" *Goeldiana Zoologia,* no. 19 (August 20, 1993): 1–11; "The Mother of All Sloths," *Fortean Times,* no. 77 (October–November 1994): 17; Laurie Goering, "Amazon Primatologist Shakes Family Tree for New Monkeys," *Chicago Tribune,* July 11, 1999; Marguerite Holloway, "Beasts in the Mists," *Discover* 20 (September 1999): 57–65.

## Maribunda

Unknown PRIMATE of South America.

*Etymology:* Similar to *marimonda,* the name for the white-bellied spider monkey in Colombia, Ecuador, and Venezuela.

*Physical description:* Slim body. Height, 5 feet when standing upright. Prehensile tail.

*Behavior:* Can walk upright. Call sounds like a human's.

*Distribution:* Río Orinoco basin, Venezuela.

*Possible explanation:* White-bellied spider monkey (*Ateles belzebuth*), though these stand 3 feet 7 inches high at most.

*Source:* Robert, marquis de Wavrin, *Les bêtes sauvages de l'Amazonie et des autres régions de l'Amérique du Sud* (Paris: Payot, 1951).

## Maricoxi

WILDMAN of South America.

*Etymology:* Arikapú (Macro-Ge) word.

*Variant name:* Morocoxo (Rikbaktsa/Macro-Ge).

*Physical description:* Covered with hair. Ape-like. Sloping forehead. Heavy browridge. Long arms.

*Behavior:* Makes grunting noises. Bad odor. Uses bow-and-arrow weapons. Lives in villages. Uses a horn when hunting.

*Distribution:* Serra dos Parecis, Mato Grosso State, Brazil.

*Significant sighting:* On an expedition to the area in 1914, Percy H. Fawcett encountered two hairy people who threatened him with bows and arrows and then ran away. Later, he came across a village in a clearing where they lived and was again approached menacingly. Fawcett fired a pistol and managed to retreat.

*Sources:* Percy H. Fawcett, *Exploration Fawcett* (London: Hutchinson, 1953), pp. 200–202; Ivan T. Sanderson, "Hairy Primitives or Relic Submen in South America," *Genus* 18 (1962): 60–74; Fritz Tolksdorf and Christian Darby, "Great White Chief of the Cannibals," *Argosy,* July 1971, p. 42.


## Marine Saurian

A category of SEA MONSTER identified by Bernard Heuvelmans.

*Physical description:* Length, 50–60 feet. Smooth skin. Grayish- or reddish-brown. Scales form rings around the body. Elongated, crocodile-like head. Slight dorsal crest. Prominent eye sockets. Long mouth. Numerous, closely set teeth. Two pairs of flippers or legs. Webbed toes. Long tail.

*Behavior:* Favors both coastal and deep waters. Swims quickly with horizontal undulations.

*Distribution:* Tropical and subtropical waters of the Atlantic, Pacific, and Indian Oceans.

*Significant sightings:* Capt. George Hope of the HMS *Fly* observed a huge, alligator-like animal with a long neck swimming underwater in the Gulf of California in the late 1830s.

A controversial sighting occurred January 13, 1852, in the South Pacific Ocean about 700 miles northeast of the Îles Marquises. Capt. Charles (or Jason) Seabury, of the New Bedford whaler *Monongahela,* claimed to have har-pooned and killed a reptilian monster 103 feet long with a 10-foot-long, alligator-like head. Seabury had the animal's head cut off and managed to preserve some bones, an eye, and the heart. However, after sending back a report on the encounter via another ship (variously identified as the brig *Gipsy* or the *Rebecca Sims*), the *Monongahela* was apparently lost at sea.

On July 30, 1877, Capt. W. H. Nelson and officers of the *Sacramento* sighted a sea monster in the mid–North Atlantic. It had a flat head raised several feet above the surface, was yellowish or reddish-brown in color, and appeared to be 40–60 feet long.

A crocodile-shaped, 60-foot-long sea monster was allegedly thrown into the air by an underwater explosion after the German submarine *U-28* torpedoed the British steamer *Iberian* off County Cork, Ireland, on July 30, 1915. Some discrepancies in the account given by the U-boat commander eighteen years later cast some doubt on the incident.

*Possible explanations:*

(1) The long beak of the Garpike (*Belone belone*) is slightly reptilian, but this fish has a maximum length of only 3 feet.

(2) A surviving thalattosuchian, a group of long-snouted crocodilians known mostly from European marine sediments of the Early Jurassic to the Early Cretaceous, 190–100 million years ago. Their elongated jaws made it easy for them to catch and eat fish, probably by ambush rather than pursuit. It is now thought, however, that they swam with vertical undulations like whales. There were two types: the teleosaurs, which had webbed feet, dermal armor, and a tapered tail, and the metriorhynchids, which had flippers, no armor, and possibly a tail fin. They appeared to favor open water.

(3) A surviving mosasaur, a group of twenty genera related to monitor lizards that included some of the largest marine reptiles ever, frequently exceeding 33 feet in length. They lived in the Late Cretaceous, 95–65 million years ago, and had large, conical teeth, each set in a deep socket. There is reasonable evidence to indicate that

mosasaurs swam upstream to breed in freshwater rivers and lakes, thus preferring coastal waters.

(4) A surviving pliosaur, a group of short-necked plesiosaurs with large heads, elongated jaws with massive teeth, two sets of flippers, and pointed tails. In some larger species, such as *Kronosaurus queenslandicus* (over 40 feet), the skull was 10 feet long. The animals lived 200–65 million years ago, from the Early Jurassic to the end of the Cretaceous; they swam underwater aerodynamically like penguins and were probably pursuit predators.

*Sources:* Edward Newman, "Enormous Undescribed Animal, Apparently Allied to the Enaliosauri, Seen in the Gulf of California," *Zoologist* 7 (1849): 2356; "Reported Capture of the Sea-Serpent," *Zoologist* 10 (1852): 3426–3429; *Australasian Sketcher* (Melbourne), November 24, 1877; Freiherr von Forstner, "Das schottische Seeungeheuer schon von *U 28* gesichtet," *Deutsche Allgemeine Zeitung,* December 19, 1933; C. O. Clark, "The Monongahela and the Sea Serpent," *Fate* 11 (December 1958): 31–33; Bernard Heuvelmans, *In the Wake of the Sea-Serpents* (New York: Hill and Wang, 1968), pp. 564, 566; Ulrich Magin, "Forstner Sea Serpent Sighting: A Possible Hoax?" *Strange Magazine,* no. 2 (1988): 4.

## Marked Hominid

A subcategory of GIANT HOMINID, HAIRY BIPED, or WILDMAN.

*Etymology:* Term used by Loren Coleman to characterize hairy creatures with distinct markings.

*Physical description:* Short, dark body-hair. More human-looking and slightly smaller than the GIANT HOMINIDS. Piebald or two-toned coloration. Sometimes has a light-colored mane, a near albino appearance, or a white patch in the middle of dark hair. Head-hair slightly longer. Facial hair from the eyes down. Large eyes. Rounded face. No neck. Well-developed leg muscles.

*Tracks:* Length, 10–15 inches. Width, 3–5 inches. Five splayed toes.

*Distribution:* North America; Siberia; Northern Europe.

*Source:* Loren Coleman and Patrick Huyghe, *The Field Guide to Bigfoot, Yeti, and Other Mystery Primates Worldwide* (New York: Avon, 1999), pp. 20–23.

## Marsabit Swift

Unknown BIRD of East Africa.

*Physical description:* Large, all-black, swiftlike nonpasserine bird.

*Distribution:* Marsabit National Reserve, Kenya.

*Possible explanation:* The Scarce swift (*Schoutedenapus myoptilus*) is uniformly gray or brownish and occasionally wanders as far north as the Marsabit Reserve.

*Source:* John G. Williams, *A Field Guide to the Birds of East Africa* (London: Collins, 1980), p. 12.

## MARSUPIALS (Unknown)

The young of most Marsupials (Infraclass Metatheria) are born extremely undeveloped compared to other mammals; after birth, they fasten themselves more or less permanently to teats in the mother's brood pouch. The epipubic bones that support this pouch, along with a specific dentition of four molars and five incisors in the upper jaw, are diagnostic traits of this diverse group. Some of the better-known marsupials are the American opossums, the Australian kangaroos and wallabies, the THYLACINE, TASMANIAN DEVIL, bandicoots, koalas, wombats, and possums.

The earliest marsupial fossils date from the Early Cretaceous, 110 million years ago, in North America. From there, they quickly spread to South America in one direction and Europe and Africa in another (where they eventually died out). The South American groups differentiated and radiated into Antarctica, where they diversified further, spreading to Australia by the Early Oligocene, 30 million years ago, when it became isolated from the other southern landmasses. All of the 282 living species are endemic to Australia and New Guinea or South America, with the sole exception of the Virginia opossum (*Didelphis virginiana*) of North America, a relative newcomer in the Pleistocene.

Marsupials achieved their greatest diversity in Australia, where they faced no competition from eutherian (nonmarsupial) mammals. The largest was the extinct, hippopotamus-sized *Diprotodon optatum,* which may have lingered long enough for the earliest Australians to prey on them 18,000–6,000 years ago. Other oddities were the horse-sized, huge-clawed *Palorchestes,* another Pleistocene survivor; *Thylacoleo,* the marsupial equivalent of a lion, which had saber-tooth incisors, huge molars, and clawed thumbs; a giant wombat, *Phascolonus;* and the sthenurine kangaroos, which may have looked like giant rabbits.

Of the eleven cryptids in this section, only one is South American (the MACAS MAMMAL); the PHANTOM KANGAROO is an out-of-place visitor reported with some frequency in North America and Europe; the others are from Australia or New Guinea and may represent survivals of supposedly extinct species or folk memories of them. A marsupial origin for the AUSTRALIAN BIG CAT has also been suggested, but this animal looks too much like a real cat to argue otherwise with any confidence.

## Mystery Marsupials

BUNYIP; DEVIL PIG; GIANT KANGAROO; GIANT RABBIT; GYEDARRA; KADIMAKARA; MACAS MAMMAL; PHANTOM KANGAROO; QUEENSLAND TIGER; TASMANIAN DEVIL (MAINLAND); THYLACINE

## Matah Kagmi

A Native American name for BIGFOOT in California.

*Etymology:* Klamath-Modoc (Penutian) word. The word bears a curious resemblance to METOH-KANGMI, a Tibetan name for the YETI.

*Physical description:* Height, 8–10 feet. Covered in coarse hair. Brown eyes.

*Behavior:* Call is a drawn-out "agooumm." Musky odor. Knows how to treat snakebites. Trades with the Indians upon occasion.

*Distribution:* Mount Shasta, California.

*Significant sighting:* Tawani Wakawa's grandfather was helped by three Matah Kagmi when he was bitten by a rattlesnake near Mount Shasta, California, around 1900.

*Source:* Tawani Wakawa, "Tawani Wakawa Tells of the Sasquatch," *Many Smokes* 3 (Fall 1968): 8–10.

## Mathews Range Starling

Mystery BIRD of East Africa.

*Physical description:* Grayish plumage. Long tail. Undertail coverts are red or chestnut.

*Distribution:* Mathews Range, Kenya.

*Possible explanation:* Female Red-winged starling (*Onychognathus morio*), which has a gray head and rufous primary feathers, suggested by Jonathan Kingdon. It is found at high elevations in the area.

*Source:* John G. Williams, *A Field Guide to the Birds of East Africa* (London: Collins, 1980), p. 12.

## Matlox

CANNIBAL GIANT of western Canada.

*Etymology:* Nootka (Wakashan) word.

*Variant name:* Matlose.

*Physical description:* Covered with stiff, black hair. Large head. Sharp fangs. Long arms. Claws on fingers and toes.

*Behavior:* Emits terrifying shouts.

*Distribution:* Nootka Sound, British Columbia.

*Source:* José Mariano Maziño, *Noticias de Nutka: An Account of Nootka Sound in 1792* (Seattle: University of Washington Press, 1970), p. 25.

## Matuyú

WILDMAN of South America.

*Physical description:* Feet are said to be turned the wrong way around.

*Distribution:* Brazil.

*Sources:* Simão de Vasconcellos, *Noticias curiosas, y necessarias das cousas do Brasil* (Lisbon: I. da Costa, 1668); Luís da Câmara Cascudo, *Dicionário do folclore Brasileiro* (Rio de Janeiro: Instituto Nacional do Livro, 1962), vol. 2, pp. 472–473.

## Mau

SMALL HOMINID of East Africa.

*Etymology:* After the Mau Escarpment, Kenya.

*Physical description:* Height, about 4 feet. Covered in reddish or black hair. White skin. Long head-hair.

*Behavior:* Upright gait. Said to steal cattle. Uses caves for shelter. Can use stones as weapons.

*Habitat:* Highlands and mountains.

*Distribution:* The Mau Escarpment, Mount Longonot, and the highlands east of Embu in Kenya.

*Significant sighting:* In the 1920s, S. V. Cook heard stories of little red men in the Kwa Ngombe Hills, Kenya.

*Possible explanations:*

(1) An undiscovered population of forest Pygmies.

(2) Surviving australopith (*see* KAKUNDAKARI), possibly corresponding to one of the Kenyan cryptids, designated as hominid X4, described by Jacqueline Roumeguère-Eberhardt in 1990.

*Sources:* S. V. Cook, "The Leprechauns of Kwa Ngombe," *Journal of the East Africa and Uganda Natural History Society,* no. 20 (November 1924): 24; Roger Courtney, *A Greenhorn in Africa* (London: H. Jenkins, 1940), pp. 37–49; Jacqueline Roumeguère-Eberhardt, *Les hominidés non-identifiés des forêts d'Afrique: Dossier X* (Paris: Robert Laffont, 1990).

## Mawas

Unknown PRIMATE of Southeast Asia.

*Etymology:* Malay (Austronesian), "orangutan."

*Variant names:* HANTU SAKAI, Orang mawas.

*Physical description:* Apelike. Height, 5–6 feet. Long, shaggy hair. Both black and brown colors reported, possibly indicating sexual dimorphism.

*Distribution:* Malaysia.

*Significant sighting:* Liong Chong Shen watched two shaggy-haired, apelike creatures in his durian orchard near Kampung Chennah, Negeri Sembilan State, Malaysia, in late December 1999. They were 5–6 feet tall and covered in black and brown hair, respectively.

*Possible explanations:*

(1) A mainland population of the Orangutan (*Pongo pygmaeus*).

(2) An extant population of *Homo erectus,* fossils of which have been found in Java, Indonesia.

*Sources:* Walter William Skeat and Charles Otto Blagden, *Pagan Races of the Malay Peninsula* (London: Macmillan, 1906), vol. 2, p. 283; Tony Healy and Paul Cropper, *Out of the Shadows: Mystery Animals of Australia* (Chippendale, N.S.W., Australia: Ironbark, 1994), pp. 152, 156; Hah Foong Lian, "Village Abuzz over Sighting of 'Mawas,'" *Star* (Kuala Lumpur), January 2, 2000.

## Mbielu-Mbielu-Mbielu

Unknown DINOSAUR-like animal of Central Africa.

*Etymology:* Possibly Lingala (Bantu).

*Physical description:* Has large "planks" growing out of its back, which is covered with green, algal growth.

*Behavior:* Almost always seen in water with its back protruding. Active in the late afternoon.

*Distribution:* Likouala River, Republic of the Congo.

*Significant sightings:* A Mbielu-mbielu-mbielu was seen near Epéna at a place called Ikekesse. Green vegetable growth was visible on its back as it came out of the water.

Odette Gesonget, a woman of Bouanila, Republic of the Congo, selected a picture of a *Stegosaurus* from books provided by Roy Mackal in 1980 as an animal that her parents had told her about.

*Possible explanation:* A surviving stegosaur, a group of dinosaurs best known from the Late Jurassic, 150 million years ago, suggested by Roy Mackal. An African fossil genus, the 20-foot *Kentrosaurus,* has been found in abundance in the Tendaguru beds of Tanzania and had six pairs of erect plates along the neck and upper back, followed by three pairs of flat spines and five pairs of large spines on the tail.

*Source:* Roy P. Mackal, *A Living Dinosaur? In Search of Mokele-Mbembe* (Leiden, the Netherlands: E. J. Brill, 1987), pp. 84, 139, 250–254.

## Mecheny

GIANT HOMINID of western Siberia.

*Etymology:* Mansi (Ob-Ugric) nickname for an individual, meaning "marked" and referring to its white forearm.

*Variant name:* Kompolen.

*Physical description:* Height, 6–7 feet. Covered in 2-inch, red-brown hair. Round head. Prominent browridges. Shorter hair covers the face, including the ears and nose. Glowing red eyes. No neck. Shoulders are wide and muscular. Powerful, barrellike chest. Left forearm is white. Large, scoop-shaped hands. Reddish skin on palms. Hair is longer in the groin. Enormous feet are covered with hair.

*Behavior:* Seen most frequently in August. Apparently follows a seasonal migratory path. Call is "khe!" like clearing the throat. Knocks on windows to announce its presence. Dislikes dogs.

*Distribution:* Tyumen' Region, western Siberia, Russia.

*Significant sightings:* Researcher Maya Bykova twice observed Mecheny outside a remote hunting cabin. She had traveled to the area after meeting a young Mansi man whose family had regularly encountered a large wildman near their hunting lodge since the 1940s. In the first encounter on August 16, 1987, Mecheny knocked on the cabin window at early dawn. Bykova and the family rushed outside and saw a figure more than 6 feet tall leaning against a dead tree. After about a minute of looking back at Bykova, Mecheny departed when the dog came out barking. In mid-October 1987, the dog disappeared and was found the next morning, its skull crushed and body ripped apart. Bykova's second encounter took place outside the cabin on the night of August 22, 1988, when she and the Mansi hunter watched Mecheny for seventy-five minutes in the woods, where it was apparently looking for frogs or mice. The next morning, they found trampled grass and narrow trails where the creature had walked.

*Source:* Dmitri Bayanov, *In the Footsteps of the Russian Snowman* (Moscow: Crypto-Logos, 1996), pp. 131–150.

## Mediterranean Giant Snake

Large SNAKE of Southern Europe.

*Variant name:* COLOVIA.

*Physical description:* Length, 6–33 feet. Green.

*Distribution:* Southern Spain; southern France; northern and central Italy; Greece; Serbia.

*Significant sightings:* On July 22, 1969, a 7-foot, green snake caused a traffic accident when it crossed a road near Chinchilla de Monte Aragón, Albicete Province, Spain.

A 6-foot snake with a huge head was seen several times on a farm in Orihuela, Alicante Province, Spain, in June 1970.

A monstrous serpent with a mane and a head like a baby's was seen in July 1973 near Aceuche, Cáceres Province, Spain.

Snakes up to 33 feet long have been seen on Ovčar Mountain near Čačak, Serbia. Near Ivanjica in the summer of 2000, a bus had to stop because a 33-foot snake was crossing the road.

*Possible explanation:* Stray specimens of the poisonous Montpellier snake (*Malpolon monspessulanus*), a gray, brownish, or olive-colored colubrid snake that can attain a length of 9 feet. It lives along the coasts of Spain, southern France, and Liguria in Italy; in North Africa from Morocco to Tunisia; and in Cyprus, Greece, and the Balkans. However, it may be expanding its range.

*Sources:* Ulrich Magin, "European Dragons: The Tatzelwurm," *Pursuit*, no. 73 (1986): 16–22; Bernard Heuvelmans, "Annotated Checklist of Apparently Unknown Animals with Which Cryptozoology Is Concerned," *Cryptozoology* 5 (1986): 1–26; Paolo Cortesi, "The Big Serpent," *INFO Journal*, no. 71 (Autumn 1994): 49–50; Marcus Scibanicus, "Strange Creatures from Slavic Folklore," *North American BioFortean Review* 3, no. 2 (October 2001): 56–63, http://www.strangeark.com/nabr/NABR7.pdf.

## Memegwesi

LITTLE PEOPLE of northern North America.

*Etymology:* Ojibwa (Algonquian), "hairy-faced dwarf." Plural, *Memegwesiwag.*

*Variant names:* Maymaygwayshi, Mee'meg-wee'ssi, Mekumwasuck (Passamaquoddy/Algonquian), Memegwecio (Cree/Algonquian), Memegwicio, Memekwesiw, Nagumwasuck (Passamaquoddy/Algonquian).

*Physical description:* Monkeylike, old-looking, and ugly. Height, 3–4 feet. Completely hairy, including the face. Big head. Flat nose or no nose at all. Long beard. Short arms. Bowed legs.

*Behavior:* High, insectlike voice with a nasal twang. Swims underwater. Raises arms out of the water when surfacing. Eats fishes and wild rice. Plays pranks on humans on stormy nights. Smokes tobacco. Makes stone projectile points, skin drums, and baskets. Wears chickadee skins. Carves rock art.

*Habitat:* Mountains, grottos, rocks, riverbanks, and caverns.

*Distribution:* Lake Superior area of central Ontario, Canada; northern Minnesota; northern Wisconsin; northern Michigan; Maine. Also in northern Manitoba and Saskatchewan, Canada.

*Significant sighting:* Memegwesiwag are often shown in pictographs as stick figures with lines running from their heads. One on the Semple River near Oxford House in northeast Manitoba marks the spot where the dwarfs cured a woman.

*Sources:* Johann G. Kohl, *Kitchi-Gami: Wanderings round Lake Superior* (London: Chapman and Hall, 1860), pp. 358–366; Frank G. Speck, "Myths and Folk-Lore of the Timiskaming Algonquin and Timagami Ojibwa," *Anthropological Series, Memoirs of the Geological Survey of Canada* 71, no. 9 (1915): 82; Regina Flannery-Herzfeld, "A Study of the Distribution and Development of the Memegwicio Concept in Algonquian Folklore," master's thesis, Department of Anthropology, Catholic University of America, 1931; Sr. Bernard Coleman, "The Religion of the Ojibwa of Northern Minnesota," *Primitive Man* 10 (1937): 33–57; Selwyn Dewdney and Kenneth E. Kidd, *Indian Rock Paintings of the Great Lakes* (Toronto, Canada: University of Toronto Press, 1967), pp. 12–24; Katharine M. Briggs, *A Dictionary of Fairies* (London: Allen Lane, 1976), pp. 268–270; John E. Roth, *American Elves* (Jefferson, N.C.: McFarland, 1997), pp. 9–10, 38–40, 114, 137–141.

# Memphré

FRESHWATER MONSTER of Québec, Canada.

*Etymology:* After Lake Memphrémagog. Coined in 1987 by Jacques Boisvert, an insurance broker of Magog, Québec.

*Physical description:* Serpentine. Length, 25–50 feet. Dark color. Smooth skin. Cow- or horselike head. Horns. Oval, red eyes, set 14 inches apart. Long neck. Has one to four humps.

*Behavior:* Swims in vertical undulations. Sometimes ventures on land.

*Distribution:* Lake Memphrémagog, Québec and Vermont.

*Significant sightings:* The first report of a monster in the lake appears to be in an 1816 journal entry of pioneer Ralph Merry, who mentioned having met several persons who saw a great serpent there.

In 1854, Henry Wadleigh saw an animal as large as a log that held its head 2 feet above the water.

In the summer of 1976, a black, seallike animal with a long neck was seen near Fitchbay.

On August 12, 1983, Barbara Malloy snapped a photo of a huge, black hump as she was boating on the lake with her family.

Mayor Denis Lacasse, of Magog, Québec, was skeptical until he saw Memphré on June 19, 1996, near Cummins Bay.

Patricia de Broin Fournier shot a video of an elongated, 16-foot animal creating a wave on August 12, 1997, near Les Trois Soeurs Island.

On June 4, 2000, Bruno, Johanne, and Serge Nadeau saw a 75-foot animal with a head like a horse's from their boat in Sergeant's Bay.

*Sources:* John Ross Dix, *A Hand Book for Lake Memphremagog* (Boston: Evans, 1860), p. 48; George C. Merrill, *Uriah Jewett and the Sea Serpent of Lake Memphremagog* (Newport, Vt.: George C. Merrill, 1917); William Bryant Bullock, *Beautiful Waters* (Newport, Vt.: Memphremagog Press, 1926); "Memphré Christened, Given Dual Citizenship," *ISC Newsletter* 6, no. 2 (Summer 1987): 7–8; Michel Meurger and Claude Gagnon, *Lake Monster Traditions: A Cross-Cultural Analysis* (London: Fortean Tomes, 1988), pp. 84–93, 270–272; John Kirk, *In Search of Lake*

Monsters (Toronto, Canada: Key Porter Books, 1998), pp. 136–143; Jacques Boisvert, "The Sea Serpent of Lake Memphrémagog," *Crypto Dracontology Special*, no. 1 (November 2001): 70–74; International Dracontology Society, Memphré, http://www.interlinx.qc.ca/memphre/ang.html.

## Mene Mamma

Fish-tailed MERBEING of South America.

*Etymology:* Quechuan or possibly Creole, "mother of waters."

*Variant names:* Femme poisson (in Martinique), Mae do rio, Mayuj-mamma, Orehu (Arawakan).

*Physical description:* Half woman, half fish.

*Behavior:* Sometimes drags canoes underwater.

*Distribution:* Guyana; the Caribbean; Brazil; Argentina.

*Significant sighting:* In 1793, Gov. A. I. van Imbyse van Battenburg of Berbice (now Guyana) told the British doctor Colin Chisholm of the half-women, half-fish seen in the rivers of his country. The creatures were generally observed in a sitting posture in the water; when disturbed, they swam away, creating a disturbance with their tails.

*Possible explanation:* Van Battenberg's animals are almost certainly the West Indian manatee (*Trichechus manatus*), often seen at the mouths of Guyanese rivers.

*Sources:* Colin Chisholm, *An Essay on the Malignant Pestilential Fever, Introduced into the West Indian Islands from Boullam, on the Coast of Guinea* (London: Mawman, 1801); Everard F. Im Thurn, *Among the Indians of Guiana* (London: Kegan, Paul, Trench, 1883); Gertrude Shaw, *West Indian Fairy Tales* (London: Francis Griffiths, 1914); Elsie Clews Parsons, *Folk-Lore of the Antilles, French and English* (New York: American Folk-Lore Society, 1933); Tobías Rosemberg, *El alma de la montaña* (Buenos Aires: Editorial Raigal, 1953), pp. 44–48.

## Menehune

LITTLE PEOPLE of Oceania.

*Etymology:* Hawaiian (Austronesian), "to get together to work and complete a task." The older Polynesian term *Manahuna* was used by Society Islanders and others to denote a specific class of people in their hierarchical system. The Manahuna were the lowest class, or common people.

*Variant names:* Manahuna, Nawao, People of Mu.

*Physical description:* Height, 2–3 feet. Nawao and Mu people are said to be taller. Hairy. Stout and muscular. Red or dark skin. Protruding forehead. Big eyes. Long eyebrows. Short, thick nose. Sharp ears. Small mouth. Broad shoulders. Round belly.

*Behavior:* Nocturnal. Has a deep voice. Normal language is telepathic, expressed with whispers or growls. Said to be able to learn English. Eats bananas, fish, shrimp, milk, squash, berries, sugarcane, and sweet potatoes. Lives in caves, lava tubes, hollow logs, or banana-leaf huts. Usually well dressed. Works at night to build fishponds, stoneworks, irrigation ditches, houses, and monuments. Carves petroglyphs. Enjoys playing games, music, dancing, singing, diving, and sports. Afraid of owls and dogs. Learned how to cook from humans.

*Habitat:* Mountain forests.

*Distribution:* Hawaiian Islands, especially on Kauai.

*Significant sightings:* In the late eighteenth century, a census of the island of Kauai by King Kaumualii counted sixty-five Menehune in the Wainiha Valley.

About forty-five elementary-school children and their school superintendent saw a group of Menehune jumping up and down among some trees on the Waimea Parish property in the 1940s. When they sensed they were being watched, the Menehune apparently disappeared into a secret tunnel near the parish house.

*Possible explanation:* A second wave of colonizing Polynesians around A.D. 1100–1300 found that those in the first wave, who had arrived in A.D. 500–800, were already established in Hawaii. These earlier individuals may have been treated as a common, or Manahuna, social class. Over time, as the two waves of colonists intermingled, the name may have become a reference to a mythical race of Little people somehow connected to ancient times.

*Sources:* Thomas G. Thrum, *Hawaiian Folk Tales* (Chicago: A. C. McClurg, 1917), pp. 19–30, 107–117, 133–138; J. H. Kaiwi, "Story of the Race of Menehunes of Kauai," *Thrum's Annual,* 1921, pp. 114–118; Padraic Colum, *At the Gateways of the Day* (New Haven, Conn.: Yale University Press, 1924), pp. 149–164; Martha Beckwith, *Hawaiian Mythology* (New Haven, Conn.: Yale University Press, 1940); Betty Allen, "Didja Ever See a Menehune?" *Honolulu Advertiser,* July 27, 1941; Katharine Luomala, "The Menehune of Polynesia and Other Mythical Little People of Oceania," *Bernice P. Bishop Museum Bulletin,* no. 203 (1951): 3–51; James T. Fitzpatrick, "The Leprechauns of the Pacific," *Asian Adventure,* August 1967, pp. 26–29; C. Alexander Stames, *Hawaiian Folklore Tales* (Hicksville, N.Y.: Exposition, 1975); Mary Kawena Pukui and Caroline Curtis, *Tales of the Menehune* (Honolulu, Hawaii: Kamehameha Schools, 1985); Frederick B. Wichman, *Kauai Tales* (Honolulu, Hawaii: Bamboo Ridge, 1985); Loren Coleman, "The Menehune: Little People of the Pacific," *Fate* 42 (July 1989): 78–89; John E. Roth, *American Elves* (Jefferson, N.C.: McFarland, 1997), pp. 132–137.

# MERBEINGS

Mermaids, mermen, merbeings, or merfolk encompass a wide variety of aquatic creatures that either are at least partially human in appearance or take human form at certain times. Some are ultimately based on the fish-tailed gods, goddesses, and entities represented in the mythologies of many cultures, including Mesopotamia, Japan, China, India, and Greece. Some of the earliest god-myths, such as the Babylonian EA and the Phoenician DERKETO, percolated through oral history via cultural dispersion and intermingled with European legends about the sea and streams; they took on regional lore about seals and other aquatic animals and ultimately contributed to tales of the SELKIES and RUSÁLKAS of European tradition. North American and African varieties were either imported from Europe and Asia or developed locally.

It makes sense for any maritime or riverine culture to have an amphibious deity for all sorts of reasons—to assure good weather, safe trips, and successful fishing or to serve as a scapegoat for bad luck—whether there is a zoological model for a fish-tailed human or not. Nonetheless, as the most human-looking of sea creatures, SEALS and SIRENIANS have undoubtedly influenced merfolk-lore for centuries. Both mammals can appear to rest vertically in the water in the posture of a classic MERMAID. Female manatees and dugongs have mammary glands located in the axilla, or armpit, with which they nurse their calves as they graze or move about, though this activity is relatively brief. The seal's fishy shape, smooth coat, expressive eyes, vocalizations, and fondness for basking are all elements that could foster a MERMAID myth.

However, scattered throughout history are seemingly reliable descriptions of unknown aquatic animals with a peculiar resemblance to human beings. These accounts (primarily found in the MERMAID entry) indicate an unknown variety of seal with light fur, oddly flexible front flippers, and a long mane. It is this seeming naturalism that has kept Merbeings from falling into the red-eyed, apparitional class of high-strangeness ENTITIES that keep one foot (or flipper) in a paraphysical realm. If these observations represent a real species that became extinct in recent times, then a physical archetype for half-man/half-fish myths worldwide could some day emerge from the fossil record.

Of the twenty-nine Merbeings listed here, eleven have European origins, five are Asian or Middle Eastern, two are African, six are North American, three are South American, one is Australasian, and one (JENNY HANIVER) is a generic name for Merbeing fakes. (Loren Coleman and Patrick Huyghe are more inclusive in their Merbeing category, adding in an assorted lot of web-footed ENTITIES such as CHUPACABRAS and some HAIRY BIPEDS.)

## Mystery Merbeings

APSARĀS; BLUE MEN OF THE MINCH; DERKETO; EA; HAVMAND; IPUPIARA; JENNY HANIVER; JIP-IJKMAK; KAPPA; MAMBA MUTU; MAMI WATER; MÄNSANZHÍ; MENE MAMMA; MERMAID; NEREID; NIX; NYKKJEN; RI; RUSÁLKA; SABA-

WAELNU; SELKIE; SHOMPALLHUE; SILENUS; SIREN; STELLER'S SEA APE; TCHIMOSE; TRITON; UNÁGEMES; VODYANY

## Merhorse

A category of SEA MONSTER identified by Bernard Heuvelmans.

*Scientific name:* Halshippus olaimagni, given by Heuvelmans in 1965.

*Variant names:* Hippokampos, MANER.

*Physical description:* Elongated, with smooth, shiny skin. Length, 15–100 feet, though rarely exceeding 60 feet. Dark-brown or steel-gray to black in northern regions; mahogany in warmer regions. Skin is smooth and shiny, possibly with short fur. Wide, flat, diamond-shaped head, described as similar to that of a horse, camel, snake, or hog. Head, 3 feet long. Wide mouth, perhaps edged with light-colored lips. Has whiskery bristles like a mustache. Enormous, forward-pointing, black eyes. Slender neck, 10 feet long or more. Often, a long, flowing, reddish mane hangs down its neck. Jagged crest on the back. Pair of frontal flippers. Possibly a hind pair of flippers that form a false tail; alternatively, a fanlike tail.

*Behavior:* Swims with pronounced vertical undulations. Rapid speed. Hisses. Feeds on fishes and possibly giant squid.

*Habitat:* Semiabyssal depths of 50–100 fathoms in the daytime, coming to the surface at night. Frequents coastal areas in temperate regions and moves further out on the continental shelf in warmer zones.

*Distribution:* Nearly cosmopolitan, except for polar seas and the Indian Ocean. At various times, it has been seen regularly off New England and Nova Scotia, the British Isles, Norway (especially Møre og Romsdal and Trøndelag Counties), British Columbia and southeastern Alaska, Portugal and the Canary Islands, southern California, La Plata in Argentina, the coast of South Africa, and in the Coral Sea.

*Significant sightings:* A description of this type of animal was first published in 1554 by the Scandinavian archbishop Olaus Magnus, who wrote that it was frequently seen in the fjords around Bergen, Norway. He mentioned the visible mane, large eyes, and elevated head and neck as prominent features.

In the spring of 1835, Captain Shibbles of the brig *Mangehan* reported an animal with large eyes and a long, maned neck 10 miles off Provincetown, Massachusetts.

In the summer of 1846, James Wilson and James Boehner were in a schooner near the western shore of St. Margaret's Bay, Nova Scotia, Canada, when they saw a 70-foot animal with a barrel-sized head and a mane. George Dauphiney spotted a similar animal near Hackett's Cove about the same time.

Officers and passengers of the British mail-packet *Athenian* observed a 100-foot, dark-brown sea serpent between the Canary and Cape Verde Islands in the North Atlantic on May 6, 1863. Its head and tail were out of the water, and it had something like a mane or seaweed on its head.

A "sea-giraffe" was observed by the crew of the steamer *Corinthian* east of Newfoundland, Canada, on August 30, 1913. It first appeared as a large head with finlike ears and huge blue eyes, followed by a 20-foot neck. It appeared attached to a large, seal-like body with smooth fur colored light brownish-yellow with darker spots.

Sports fisherman Ralph Bandini saw a maned animal about a mile west of Mosquito Harbor on San Clemente Island, California, in September 1920. Its neck was 5–6 feet thick, and the eyes were 12 inches in diameter.

Around 1938, some 100 yards off the coast of Skeffling, East Riding of Yorkshire, England, Joan Borgeest watched a huge, green creature with a flat head, protruding eyes, and a long mouth that opened and closed. When she called out to other people in the area, it dived and did not reappear.

George W. Saggers watched a head and neck with huge black eyes off Ucluelet, Vancouver Island, British Columbia, Canada, in November 1947. Its dark-brown mane looked like a bundle of warts.

*Possible explanation:* An elongated Seal (Suborder Pinnipedia) adapted for a semiabyssal marine existence.

*Sources:* Olaus Magnus, *A Compendious*

*History of the Goths, Swedes and Vandals* [1554] (London: J. Streater, 1658), pp. 225, 227, 231; "A Sea Serpent," *American Journal of Science* 28 (1835): 372–373; "The Great Sea-Serpent," *Zoologist* 21 (1863): 8727; John Ambrose, "Some Account of the Petrel—the Sea-Serpent—and the Albicore as Observed at St. Margaret's Bay," *Transactions of the Nova Scotia Institute of Natural Science* 1 (1864): 37–40; "Sea Monster's Bonny Blue Eyes," *Daily Sketch* (London), September 25, 1913, p. 6; Ralph Bandini, "I Saw a Sea Monster," *Esquire* 2 (June 1934): 90–92; George W. Saggers, "Sea Serpent off Vancouver," *Fate* 1 (Summer 1948): 124–125; Bernard Heuvelmans, *In the Wake of the Sea-Serpents* (New York: Hill and Wang, 1968), pp. 459, 552–557, 566.

## Mermaid

Fish-tailed MERBEING of the Atlantic Ocean.

*Etymology:* Middle English, *mere* ("sea") + *maide* ("maid").

*Variant names:* Merfolk (plural), Merman (for the male). Also Ben-varrey (Manx), Boctogaí (Irish), Ceasg (Gaelic), Dinny-mara (Manx), Doinney varrey (Manx, for the male), HAVMAND, Homen marinho (Portuguese), Liban, Maighdean mhara (Irish), Merrow (Irish), Merrymaid (in Cornwall, England), Mhaidan mhare (Scots Gaelic), Morgan (Breton, "sea woman"), Morverch (Breton, "sea daughter"), Muirgheilt (Irish), Murdhucha'n (Irish), Muruch (Irish).

*Physical description:* Head and torso of a woman or man. Tapering fish body and tail instead of legs.

*Behavior:* Fond of combing its hair and basking on the rocks. Said to be able to assume human form and go ashore to markets and fairs. Lures mariners to destruction on rocks. Said to gather the souls of the drowned. Some families claim to be descended from Mermaids.

*Habitat:* Underwater cities.

*Distribution:* Atlantic Ocean, especially off the coast of Scotland.

*Significant sightings:* The Irish annals mention several Mermaid encounters: in A.D. 558, when the legendary Liban was caught in a net in Ulster; in 887, when a 195-foot, white Mermaid (more likely a whale) was washed up on the coast; and in 1118, when two Mermaids were caught near Waterford.

A Merman was caught by fishermen at Orford, Suffolk, England, in 1197. It was bearded and hairy (but bald) and otherwise like a human. Sir Bartholomew de Glanville kept it at the castle for about two months before it managed to escape into the sea.

The medieval FAIRY Mélusine of Lusignan, in the Poitou-Charentes Region of France, turned into a Mermaid (in some traditions, a serpent) every Saturday, a fact she was able to conceal from her husband, Raymond of Poitou, for many years, until he spoiled everything one weekend by spying on her in the bath.

In 1403, when the dikes near Edam, the Netherlands, broke in a storm, some young girls found a Mermaid floundering in shallow water. They got it into their boat, took it home, and gave it clothes, but they were unable to make it speak. It was said to have lived for fifteen years afterward.

On his second voyage in search of a northeast passage, two of Henry Hudson's crew, Thomas Hilles and Robert Raynar, saw a Mermaid off Novaya Zemlya, Russia, at about 75° north latitude on June 15, 1608. It was as big as a human and had a woman's back and breasts, white skin, and long, black hair. When it dived, they could see its speckled, porpoiselike tail.

Capt. Richard Whitbourne was one of several witnesses to a Mermaid in the harbor at St. Johns, Newfoundland, in 1610. Its shoulders and back were square, white, and smooth, while its lower part was like a "broad hooked arrow."

In about 1667, Thomas Glover saw an unidentified seal or Merman with a fishlike tail on the Rappahannock River, Virginia.

Around 1698, a Merman with the bearded face of an old man was seen off the Orkney Islands, Scotland.

Sometime before 1791, Henry Reynolds ran across what looked to be a youth of about sixteen sitting in the sea near Castlemartin, Dyfed, Wales. As he came closer, he realized it had a huge, eel-like tail that moved constantly in a circular pattern. Its arms and hands seemed thick

*Stylized depiction of a MERMAID. (© 2002 ArtToday.com, Inc., an IMSI Company)*

and short. Brownish, ribbonlike streamers came out of its forehead and flowed over its back. Reynolds watched it for about an hour as it swam near a rock only 35 feet away.

Schoolmaster William Munro of Thurso, Highland, Scotland, was walking along the coast at Sandside Bay in 1798 when he came across what seemed to be a naked human female sitting on a partially submerged rock and combing its long, light-brown hair. He watched for about three minutes before it dropped into the sea and disappeared. Had it not been for the dangerous place where it was sitting and other reports in which people had seen a Mermaid for an hour or more, he might have thought it was human.

On October 18, 1811, farmer John M'Isaac saw a classic Mermaid sitting on a rock on the Kintyre Peninsula near Cambeltown, Argyll and Bute, Scotland. Its human-shaped upper half was white, while the lower half was a reddish-gray tail covered with hair. It seemed to be combing its head-hair with its arms, which seemed to be short in proportion to its body. M'Isaac watched the animal for two hours, after which it tumbled clumsily into the water and remained some minutes stroking and washing its chest. He then saw its face clearly, which he described as human, with hollow eyes and a short neck. Other witnesses came forward, and they all signed depositions testifying to the truth of the incident.

On August 15, 1814, two fishermen saw a black Merman with a flat nose, curly hair, and long arms swimming upright in the water off Portgordon, Moray, Scotland. It was accompanied by a female, who had breasts and long, straight hair.

In May 1817, somewhere in the North Atlantic Ocean at latitude 44°6', the crew of the ship *Leonidas* observed a strange animal with a fish's tail and humanlike upper parts, swimming erect about 2 feet out of the water for about six hours. It had a whitish face, short arms, and black hair. It remained looking at them for fifteen minutes at a time, then dived underneath and appeared on the other side of the ship. It was about 5 feet long from head to tail.

Around 1830, some people cutting seaweed near Griminis on Benbecula in the Outer Hebrides of Scotland discovered a small Mermaid splashing about in the sea. They tried chasing it, but it swam farther away until a boy killed it by throwing a rock at it. The creature washed up on the shore a few days later. Its skin was white, it had long, dark hair, and it looked like a three- or four-year-old child with abnormally developed breasts and a salmonlike tail, though without scales. The villagers made a coffin for it and buried it nearby.

In July 1833, three fishermen swore before a justice of the peace that they had caught a Mermaid some 30 miles off the coast of Yell in the Shetland Islands, Scotland. It was about 3 feet long, had breasts like a woman, arms about 9 inches long, and small hands with webbed fingers. There were fins on each shoulder. Its head was pointed, and it had blue eyes, two nostrils between which was a thick facial bristle, a wide mouth, and no ears or chin. The skin was white on the front and light gray on the back. Its navel was 9 inches below the breasts. The tail had two lobes and resembled a halibut's. It moaned pitifully, so they threw it back after three hours.

P. T. Barnum exhibited a 3-foot, faked "Feejee mermaid" both in his American Museum in New York and on his "Greatest Show on Earth" circus tour from 1842 until 1864, when the museum burned down. A similar monkey-fish was donated to Harvard's Peabody Museum in 1897.

In 1900, Alexander Gunn saw a Mermaid near Sandwood, Highland, Scotland, as he went to rescue a sheep lodged in a gully. It was human-sized, with curly, reddish-yellow hair, greenish-blue eyes, and arched eyebrows. Its back was arched, and it looked frightened and angry.

Around 1921, an animal with a fishlike tail and a woman's head and breasts was seen by a fisherman at Dassen Island off Western Cape Province, South Africa.

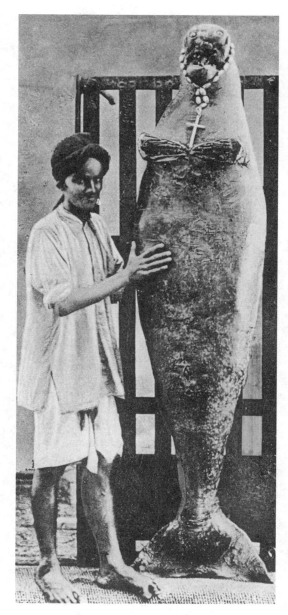

Fake MERMAID at Aden, apparently constructed from a dugong's body. From an old postcard. (Fortean Picture Library)

Sometime before 1936, a Scandinavian hunter encountered a Mermaid with green hair, beaming eyes, and a sad voice in the Strait of Magellan near Punta Arenas, Chile.

Between 1960 and 1962, a Mermaid resembling a normal woman was seen frequently off Kilconly Point, County Kerry, Ireland.

*Possible explanations:*

(1) Manatees and Dugongs (Suborder Sirenia) have been held responsible for Mermaid stories for hundreds of years, but it is difficult to see how these bulky, small-headed, flippered mammals could be mistaken for slender, long-haired, distinctly human females, even at a distance by lonely sailors. The explanation has always seemed too glib and ironic. Sirenians live in the warm waters of the Caribbean, the Amazon, West Africa, Indian Ocean, and the South Pacific, while Mermaids of the European type are most often reported in the North Atlantic. However, manatees and dugongs often sit vertically in the water to hold their young, which suckle the pectoral mammary glands located at the base of each flipper. Seaweed could conceivably masquerade as hair. Their fishlike tails are reminiscent of the SIREN and are the primary reason for this order's scientific name.

(2) Seals (Suborder Pinnipedia) are much more likely contenders, both physically and behaviorally. The seal's head is round, the flippers flexible, the body sleek, and the vocalizations expressive. Seals also like to bask on rocks. Many cultures have myths of seal-folk—humans descended from or changed into seals (*see* SELKIE).

(3) Some Mermaid sightings could be based on occasional visits to the British Isles by sealskin-clad, kayaking, nomadic Saami peoples from northern Norway. This might well explain lore about the Mermaid's upright appearance in the water, remarkably human appearance, and liaisons with the locals.

(4) An unknown species of seal with strikingly humanlike characteristics.

(5) A surviving primitive ape that at some point took to the water, perhaps an evolved *Oreopithecus,* suggested by Mark A. Hall. It is true that *Oreopithecus* lived in the Late Miocene, 8–7 million years ago, in a swampy forest habitat and perhaps subsisted on aquatic plants. Presumably, such an evolved ape would have developed webbed hands and feet rather than a fish tail, so it couldn't account for seal- or manatee-shaped animals.

(6) In the nineteenth century, many of the fake mermaids exhibited in traveling shows in Europe and the United States were said to be manufactured by Japanese taxidermists or Javan fishermen, who skillfully grafted monkey torsos onto the bodies and tails of large salmon or other fishes, augmented by papier-mâché.

(7) An expression of the myth of the fish-tailed gods and goddesses of antiquity.

*Sources:* Four Masters, *Annals of the Kingdom of Ireland,* ed. John O'Donovan (Dublin: Hodges and Smith, 1851), vol. 1, pp. 201–203, 541; Henry Hudson, "A Second Voyage or Employment of Master Henry Hudson," in Samuel Purchas, ed., *Purchas His Pilgrimes* [1625] (Glasgow: James MacLehose, 1905–1907), vol. 13, p. 318; Richard Whitbourne, "Captaine Richard Whitbourne's Voyages to New-found-land," in Samuel Purchas, ed., *Purchas His Pilgrimes* [1625] (Glasgow: James MacLehose, 1905–1907), vol. 19, pp. 439–440; John Swan, *Speculum mundi: or, A Glasse Representing the Face of the World* (Cambridge: Printers to the Universitie, 1635); Thomas Glover, "An Account of Virginia," *Philosophical Transactions* 11 (1676): 623, 625–676; John Brand, *A Brief Description of Orkney, Zetland, Pightland-Firth & Caithness* (Edinburgh: G. Mosman, 1701); George Waldron, *Description of the Isle of Man* [1731] (Douglas, Isle of Man: Manx Society, 1864); Benoît de Maillet, *Talliamed* (London: T. Osborne, 1750); Mary Morgan, *A Tour to Milford Haven in the Year 1791* (London: John Stockdale, 1795), pp. 302–306; *Times* (London), September 8, 1809; Asa Swift, "Mermaid," *American Journal of Science,* ser. 1, 2 (1820): 178–179; Robert Hamilton, "Amphibious Carnivora," in William Jardine, ed., *The Naturalist's Library* (Edinburgh: W. H. Lizars, 1845), vol. 25, pp. 280–283; Andrew Steinmetz, *Japan and Her People* (London: Routledge, Warnes, and Routledge, 1859), pp. 193–194; Philip

Henry Gosse, *The Romance of Natural History, Second Series* (London: James Nisbet, 1862); Henry Lee, *Sea Fables Explained* (London: William Clowes and Sons, 1883); Fletcher S. Bassett, *Legends and Superstitions of the Sea and Sailors* (Chicago: Belford, Clarke, 1885), pp. 148–201, 445, 451; Sabine Baring-Gould, *Curious Myths of the Middle Ages* (London: Longmans, Green, 1892), pp. 471–523; Alexander Carmichael, *Carmina Gadelica* (Edinburgh: T. and A. Constable, 1900); John Livingston Lowes, *The Road to Xanadu* (Boston: Houghton Mifflin, 1927); J. M. McPherson, *Primitive Beliefs in the North-East of Scotland* (London: Longmans, Green, 1929), pp. 72–73; W. Walter Gill, *A Manx Scrapbook* (London: Arrowsmith, 1929), p. 241; Cherry Kearton, *The Island of Penguins* (New York: R. M. McBride, 1931); Hakon Mielche, *Journey to the World's End* (New York: Doubleday, Doran, 1941); R. MacDonald Robertson, *Wade the River, Drift the Loch* (Edinburgh: Oliver and Boyd, 1948); R. MacDonald Robertson, *Selected Highland Folktales* (Isle of Colonsay, Scotland: House of Lochar, 1961), pp. 148–170; Gwen Benwell and Arthur Waugh, *Sea Enchantress* (London: Hutchinson, 1961); *Dublin Evening Press,* August 4, 1962, p. 5a; Tim Dinsdale, *The Leviathans* (London: Routledge and Kegan Paul, 1966), pp. 138–146; Ernest W. Marwick, *The Folklore of Orkney and Shetland* (London: B. T. Batsford, 1975), pp. 24–25; A. H. Saxon, *P. T. Barnum: The Legend and the Man* (New York: Columbia University Press, 1989); Richard Ellis, *Monsters of the Sea* (New York: Alfred A. Knopf, 1994), pp. 75–112; Jim Higgins, *Irish Mermaids: Sirens, Temptresses and Their Symbolism in Art, Architecture and Folklore* (Galway, Ireland: Crow's Rock Press, 1995); John M. MacAulay, *Seal Folk and Ocean Paddlers* (Cambridge: White Horse, 1998); Meri Lao, *Sirens: Symbols of Seduction* (Rochester, Vt.: Park Street Press, 1999); Jan Bondeson, *The Feejee Mermaid and Other Essays in Natural and Unnatural History* (Ithaca, N.Y.: Cornell University Press, 1999); Marc Potts, *The Mythology of the Mermaid and Her Kin* (Chieveley, England: Capall Bann, 2000).

## Mesingw

CANNIBAL GIANT of the eastern United States.

*Etymology:* Unami (Algonquian), "mask being."

*Variant names:* Mee sing, Misinghalikun, Wsinkhoalican.

*Physical description:* Covered with hair. Face is part red, part black.

*Behavior:* Cry is "ho ho ho." Rides a deer. Guards forest animals.

*Distribution:* New Jersey; eastern Pennsylvania.

*Sources:* Mark Raymond Harrington, *Religion and Ceremonies of the Lenape* (New York: Museum of the American Indian, Heye Foundation, 1921); Mark Raymond Harrington, *Dickon among the Lenape* (Philadelphia: John C. Winston, 1938).

## Messie

FRESHWATER MONSTER of South Carolina.

*Variant name:* Loch Murray monster.

*Physical description:* A large fish or humped animal. Length, 30 feet.

*Distribution:* Lake Murray, South Carolina.

*Significant sightings:* In 1973, Buddy and Shirley Browning and a friend were fishing when a large fish attacked their boat. It did not look like a sturgeon to them.

In late April 2000, Mary S. Shealey saw a large lump resembling an overturned boat near the Lake Murray Dam. She estimated it was 30 feet long and 10 feet high.

*Possible explanation:* The Atlantic sturgeon (*Acipenser oxyrhynchus*) grows to 14 feet in length and ascends coastal rivers to spawn. It is found in some South Carolina rivers, including the Santee and Edisto, but none have been confirmed in Lake Murray.

*Sources:* John Kirk, *In the Domain of Lake Monsters* (Toronto, Canada: Key Porter Books, 1998), pp. 172–173; *Lexington (S.C.) News,* May 14, 2000; Loch Murray Monster, http://www.geocities.com/CapitolHill/1171/irmo019.html.

## Metoh-Kangmi

Alternate name for the YETI of Central Asia.

*Etymology: Kangmi* (a contracted form of *gongs mi*) is Sherpa (Sino-Tibetan) for "snow or glacier man." *Metoh* is an unknown or corrupt word, though Ram Kumar Panday says it means "unwashed." Swami Pranavananda claims *metoh* means "abominable" and is a variant of MI-TEH, which he defines as "man-bear" because the creature can walk bipedally; he says the term refers to the red or isabelline variety of the Brown bear (*Ursos arctos*). In modern Tibetan, *mi sdug* means "disgusting." *See* ABOMINABLE SNOWMAN.

*Variant name:* Kangmi.

*Distribution:* Tibet; Nepal.

*Sources:* V. d'Auvergne, "My Experiences in Tibet," *Bihar and Orissa Research Society Journal* 26, no. 2 (1940): 101–119; Ralph Izzard, *The Abominable Snowman Adventure* (London: Hodder and Stoughton, 1955), pp. 28–29; Swami Pranavananda, "The Abominable Snowman," *Journal of the Bombay Natural History Society* 54 (1957): 358–364; Ram Kumar Panday, *Yeti Accounts: Snowman's Mystery and Fantasy* (Kathmandu: Ratna Pustak Bhandar, 1994).

## Mi-Chen-Po

Variant name for the YETI of Central Asia.

*Etymology:* Tibetan (Sino-Tibetan), "big man."

*Variant names:* Mi-bompo, Mi-shom-po ("strong man").

*Physical description:* Height, 7 feet. Dark brown or reddish hair.

*Distribution:* Tibet.

*Source:* René de Nebesky-Wojkowitz, *Oracles and Demons of Tibet* (The Hague, the Netherlands: Mouton, 1956), p. 344.

## Miga

Unknown SIRENIAN of Central Africa.

*Variant name:* Guidiara (in Guinea, "water lion").

*Physical description:* A large fish or octopus. Head is like a gorgon's. Tentacles.

*Behavior:* Sucks blood and eats the brains of infants. Hides among rocks in the river and attacks passing canoes.

*Distribution:* Mbomou River, Central African Republic; Uele and Dungu Rivers, Democratic Republic of the Congo; Niger River, Guinea.

*Possible explanations:*

(1) A FRESHWATER OCTOPUS, though all known cephalopod species are exclusively marine, and none are sanguinivorous.

(2) Evidence for an extended range of the West African manatee (*Trichechus senegalensis*), proposed by Bernard Heuvelmans. Adults are generally 9–10 feet long. This animal is found in rivers, estuaries, swamps, and lagoons from the Senegal River in the north to the Cuanza River, Angola, in the south, and it occurs as far as 1,200 miles from the sea along the Niger River. Its presence in certain tributaries of the Congo has been suspected but never confirmed. Its reputation as a brain-eater is probably fear-based, since all known sirenians are herbivorous.

(3) A giant Catfish (Family Siluridae), suggested by Marc Micha.

*Sources:* "Congo belge," *Le Temps* (Paris), August 22, 1900, p. 2; Charles Alexandre d'Ollone, *Mission Hostains-d'Ollone, 1898–1900: De la Côte d'Ivoire au Soudan et à la Guinée* (Paris: Hachette, 1901), p. 241; Raymond Colrat de Montrozier, *Deux ans chez les anthropophages ey les sultans de centre africain* (Paris: Plon-Nourrit, 1902), p. 147; Bernard Heuvelmans, *Les derniers dragons d'Afrique* (Paris: Plon, 1978), pp. 272–274, 358–363.

## Migo

Probable CROCODILIAN of Australasia.

*Variant names:* Masalai ("spirit"), Massali, Migaua, Mussali, Rui.

*Physical description:* Crocodile-like body. Length, 30–35 feet. Gray skin. Horselike head and neck. Sharp fangs. Ridged back. Turtlelike legs.

*Behavior:* Swift swimmer.

*Distribution:* Lake Dakataua, on New Britain in the Bismarck Archipelago, Papua New Guinea.

*Significant sightings:* During World War II, Wilfred T. Neill noticed crocodilians at the edges of some inland lakes of New Britain he was flying over.

In January and February 1994, a Japanese television crew accompanied by Roy Mackal took about five minutes of video footage apparently showing a Migo in Lake Dakataua. One segment featured three different sections of a long animal moving through the water.

*Possible explanations:*

(1) A surviving mosasaur, a group of large, marine lizards related to modern monitors, suggested by oceanographer Shohei Shirai. Mosasaurs died out by the end of the Cretaceous period, 65 million years ago.

(2) Roy Mackal at first thought the Migo might be an evolved basilosaurid, a member of a family of early whales that lived 42–33 million years ago, in the Middle to Late Eocene.

(3) Mackal now thinks the video shows three Saltwater crocodiles (*Crocodylus porosus*) in a mating ritual.

*Sources:* Wilfred T. Neill, "The Possibility of an Undescribed Crocodile on New Britain," *Herpetologica* 12 (1956): 174–176; *Brisbane Courier-Mail,* February 4, 1972; Karl Shuker, "Close Encounters of the Serpentine Kind: Monitoring the Migo," *Strange Magazine,* no. 15 (Spring 1995): 31–32; Karl Shuker, "New Britain's Lake Monster," *Fortean Times,* no. 82 (August-September 1995): 38–39; Darren Naish, "Analysing Video Footage Purporting to Show the 'Migo': A Lake Monster from Lake Dakataua, New Britain," *Cryptozoology Review* 1, no. 2 (Autumn 1996): 18–21; Karl Shuker, "Making the Most of the Migo," *Fortean Times,* no. 106 (January 1998): 15.

## Mi-Gö

Alternate name for the YETI of Central Asia.

*Etymology:* Tibetan (Sino-Tibetan) word. Probably means "wild man," though it also has a sense of "robber" or "ruffian." In modern usage, *mi go ba* means to "not notice"; *mi gos* means "untainted"; and *mi g.yo* is an "unwavering state of meditation."

*Variant names:* Me-gu, Megur, Miegye, Migeye, Mighu (Bhutanese pronunciation), Migiö, Migu, Migyur (Bhutan), Mirgod, Mirka, Ui-go.

*Physical description:* Height, 6–7 feet. Covered in dark brown hair. Powerful build. Oval head running to a point at the top. Apelike face sparsely covered with hair.

*Behavior:* Nocturnal. Bipedal in open country. Walks with an unsteady gait. Moves on all fours in the forest or swings in the trees. Visits snowfields in search of saline moss on exposed rocks. Makes a peculiar, whistling call. Has a bad odor. Fears fire. Usually runs away when it encounters humans and only attacks if it is wounded.

*Tracks:* Imprints are 6–7 inches long and 4 inches wide at the broadest, though this seems small for the animal's height. Five distinct toes. Clear instep. Pointed heel.

*Habitat:* Forested mountains.

*Distribution:* Eastern Himalaya Mountains of Tibet; Nepal; Bhutan; Sikkim State, India.

*Significant sighting:* A female relative of the third king of Bhutan, Jigme Dorji Wangchuck (1952–1972), saw a Mighu in the mountains.

*Sources:* René de Nebesky-Wojkowitz, *Where the Gods Are Mountains* (London: Weidenfeld and Nicholson, 1956), pp. 151–161; Swami Pranavananda, "The Abominable Snowman," *Journal of the Bombay Natural History Society* 54 (1957): 358–364; Boris F. Porshnev and A. A. Shmakov, eds., *Informatsionnye materialy, Komissii po Izucheniyu Voprosa o "Snezhnom Cheloveke,"* 4 vols. (Moscow: Akademiia Nauk SSSR, 1958–1959); Rory Nugent, *The Search for the Pink-Headed Duck* (Boston: Houghton Mifflin, 1991), pp. 70–82.

## Mihirung Paringmal

Giant flightless BIRD of Australia.

*Etymology:* Tjapwurong (Australian), "giant emu."

*Physical description:* Like the Emu (*Dromaius novaehollandiae*) but larger.

*Distribution:* Western Victoria; New South Wales; Queensland.

*Significant sighting:* Known from rock paintings in Cape York, Queensland, and rock carvings at Mootwingee, New South Wales. Aboriginal legends tell of giant birds that were still alive when volcanos were erupting in the Western District of Victoria, most recently in approximately 3000–2000 B.C.

*Possible explanations:*

(1) Aboriginal memories of the recently extinct *Genyornis newtoni,* a Late Pleistocene species of dromornithid bird related to ducks and geese that stood about 6 feet 7 inches tall and survived at least as recently as 26,000 years ago. Known from complete skeletons near Lake Callabonna in South Australia and isolated bones elsewhere.

(2) Aboriginal memories of *Dromornis stirtoni,* one of the largest flightless birds that ever lived, known from fossils in the Northern Territory dating from the Late Miocene to the Pleistocene, 15 million–30,000 years ago. Also a dromornithid, it weighed more than 1,300 pounds, stood 9 feet tall, and had a huge beak and jaw.

*Sources:* James Dawson, *Australian Aborigines: The Languages and Customs of Several Tribes of Aborigines in the Western District of Victoria, Australia* (Melbourne, Australia: G. Robertson, 1881), pp. 92–93; Patricia Vickers-Rich and Gerard Van Tets, eds., *Kadimakara: Extinct Vertebrates of Australia* (Lilydale, Vic., Australia: Pioneer Design Studio, 1985), pp. 17, 188–194.

## Miitiipi

CANNIBAL GIANT of the southwestern United States.

*Etymology:* Kawaiisu (Uto-Aztecan), "bad luck" or "disaster."

*Distribution:* Mojave Desert, California.

*Source:* Kyle Mizokami, Bigfoot-Like Figures in North American Folklore and Tradition, http://www.rain.org/campinternet/bigfoot/bigfoot-folklore.html.

## Milne

Unknown BEAR of South America.

*Etymology:* Ashéninca (Arawakan) word.

*Physical description:* Large bear. All-black color. Powerful claws.

*Behavior:* Eats ants.

*Distribution:* Gran Pajonal Range, Ucayli Department, Peru.

*Significant sighting:* Leonard Clark came across a large, black bear tearing apart an ants' nest in a decaying tree in northern Peru. It dived into the river ahead of his raft. He was able to shoot and kill it, but piranhas got to the body before he could retrieve it.

*Possible explanation:* The Spectacled bear (*Tremarctos ornatus*), South America's only bear, is smaller and has distinctive white markings on its face.

*Source:* Leonard Clark, *The Rivers Ran East* (New York: Funk and Wagnalls, 1953).

## Mindi

Giant SNAKE of Australia.

*Etymology:* From the Wemba (Australian) *mirndayi.*

*Variant names:* Mallee snake, Mindai, Mindyè, Myndie.

*Physical description:* Length, 18–30 feet. Hairy scales. Large head. Tongue with three points. Black mane.

*Behavior:* Hangs on tree branches. Lies in wait at water holes. Feeds on emus. Gives off a disgusting smell and leaves smallpox in its wake. Able to kill with its glance. Oviparous. Not poisonous.

*Habitat:* Eucalyptus scrub.

*Distribution:* Lower Murray River, South Australia; the Grampians, Victoria.

*Possible explanations:*

(1) Distorted memory of the Diamond python (*Morelia spilota spilota*), found only on the coast of New South Wales. It grows to 15 feet long and has olive-black scales with cream or yellow spots.

(2) A surviving Pleistocene madtsoiid snake, the Giant Australian python (*Wonambi naracoortensis*), known from fossil deposits in South Australia. It ranged from 10 to 20 feet in length.

*Sources:* J. C. Byrne, *Twelve Years' Wanderings in the British Colonies* (London:

Richard Bentley, 1848), vol. 2, p. 274; *Sydney Empire,* February 17, 1851, p. 3; A Resident [John Hunter Kerr], *Glimpses of Life in Victoria* (Edinburgh: Edmonston and Douglas, 1872), p. 251; Georgiana McCrae, *Georgiana's Journal: Melbourne, a Hundred Years Ago* (Sydney, Australia: Angus and Robertson, 1934), p. 129; Gilbert Whitley, "Mystery Animals of Australia," *Australian Museum Magazine* 7 (1940): 132–139; Charles Barrett, *The Bunyip and Other Mythical Monsters and Legends* (Melbourne, Australia: Reed and Harris, 1946), pp. 35–45.

## Minhocão

Mystery AMPHIBIAN of Central and South America.

*Etymology:* Portuguese, "giant earthworm."

*Variant names:* Miñocao, Sierpe ("snake," in Nicaragua).

*Physical description:* Serpentine. Length, up to 150 feet. Width, 15 feet. Black. Covered in thick, bony armored skin or scales. Two horns on its head. Piglike snout.

*Behavior:* Amphibious and subterranean. Knocks over trees, collapses roads, and creates new river channels with its burrowing in the ground. Most active after rainy weather. Overturns boats. Attacks and eats horses and cattle when fording rivers or lakes.

*Tracks:* Leaves a trail of deep, grooved furrows 3–10 feet wide.

*Distribution:* Mato Grosso do Sul, Goiás, Santa Catarina, Paraná, and Bahia States, Brazil; near San Rafael de Norte, Nicaragua; Río Madidi, Bolivia; Arapey area, Salto Department, Uruguay.

*Significant sightings:* In 1849, Lebino José dos Santos was traveling near Termas del Arapey, Uruguay, when he heard of a dead Minhocão that had caught itself in a narrow cleft of rock. Its skin was as thick as pine-tree bark, and it had scales like an armadillo.

About 1864, Antonio José Branco found the road close to his home near Curitibanos, Santa Catarina State, Brazil, undermined with 6-foot-wide trenches about 3,000 feet in extent.

In the late 1860s, a Minhocão appeared on the banks of the Rio das Caveiras near Laje, Bahia State, Brazil, and was seen by Francisco de Amaral Varella and Friedrich Kelling. It left a wide trench in swampy ground.

*Present status:* Not reported since the nineteenth century.

*Possible explanations:*

(1) A giant, scaly lungfish related to *Lepidosiren,* according to Auguste de Saint Hilaire. The pectoral fins of the fish might be confused with horns. The South American lungfish (*L. paradoxa*) grows up to 4 feet long and remains buried in the mud of riverbeds, hibernating until the rainy season when its reproductive cycle begins. It prefers stagnant water where there is little current.

(2) A surviving glyptodont, according to Emil Budde, who supposes it was a burrowing animal like the aardvark. This heavily armored genus of giant armadillos grew to 10 feet long and died out in North and South America around 10,000 years ago. However, the glyptodont's fused body armor would have interfered with tunneling and negated any need to flee from predators, especially since its macelike tail could have delivered a deadly blow even to a giant ground sloth. There is no fossil evidence that glyptodonts burrowed.

(3) A surviving pampathere, a fossil American armadillo that also disappeared about 10,000 years ago; it had more flexible armor than the glyptodont. The adult *Holmesina septentrionalis* was 6 feet long and weighed more than 500 pounds. Similarly, there is no evidence that this animal produced underground tunnels.

(4) A surviving archaic basilosaurid whale that lives in freshwater lakes and swamps, according to Bernard Heuvelmans. However, it is now known that this animal lacked armored skin.

(5) An unknown giant species of caecilian, a wormlike amphibian that burrows underground, suggested by Karl Shuker. The chemosensory tentacles on its head resemble horns. Some species have scales embedded in their skin. The largest

underground species, *Caecilia thompsoni,* lives in Colombia, is nearly 5 feet long, and feeds on earthworms; the aquatic genus *Typhlonectes* lives in South American rivers and lakes and feeds on fishes and invertebrates.

(6) A GIANT ANACONDA.

(7) Earthquake damage could be attributed to the Minhocão's tunneling.

*Sources:* Auguste de Saint Hilaire, "On the Minhocão of the Goyanes," *American Journal of Science,* ser. 2, 4 (1847): 130–131; Fritz Müller, "Der Minhocão," *Der Zoologische Garten* 18 (1877): 298–302; "A New Underground Monster," *Nature* 17 (1878): 325–326; "Underground Monsters," *Nature* 18 (1878): 389; Emil A. Budde, *Naturwissenschaftliche Plaudereien* (Berlin: G. Reimer, 1898); Florenzio de Basaldúa, *Pasado—presente—porvenir del territorio nacional de misiones* (La Plata, Argentina, 1901), p. 80; Robert, marquis de Wavrin, *Les bêtes sauvages de l'Amazonie et des autres régions de l'Amérique du Sud* (Paris: Payot, 1951); Bernard Heuvelmans, *On the Track of Unknown Animals* (New York: Hill and Wang, 1958), pp. 298–304; Karl Shuker, *In Search of Prehistoric Survivors* (London: Blandford, 1995), pp. 145–148.

## Mi-Ni-Wa-Tu

FRESHWATER MONSTER of South Dakota.

*Etymology:* Lakota (Siouan), "river monster."

*Physical description:* Body is like a bison's. Red hair. One horn in forehead. One eye. Sawtooth back. Grooved, flat tail.

*Behavior:* Swift swimmer. Creates a huge wake.

*Distribution:* Missouri River, North and South Dakota.

*Source:* J. Owen Dorsey, "Teton Folk-Lore Notes," *Journal of American Folklore* 2 (1889): 133–139.

## Minnesota Iceman

WILDMAN exhibited in a U.S. carnival. Said to have come from Southeast Asia but possibly a hoax.

*Scientific name: Homo pongoides,* given by Bernard Heuvelmans in 1969.

*Variant name:* Bozo.

*Physical description:* Adult male hominid. Height, 5 feet 10 inches. Covered in long brown hair (3–4 inches long) except on the face. The longer individual hairs seem to have an agouti pattern of light bands. Pinkish skin. Large eye sockets. Eyeballs are apparently missing. Pugged nose with nostrils pointing upward. Wide mouth with no eversion of the lips. Folds and wrinkle lines around the mouth are apparently humanlike. Short neck. A "cape" of long, black hair flows around the neck. Wide shoulders. Rounded ribcage. Trunk is ovoid in shape rather than hourglass. Relatively long arms. Wide wrist. Prominent pad on the heel of the palm. Hands are spatulate and disproportionately large, 11 inches long and more than 7 inches wide. Knuckles are poorly defined. Slender thumb is fully opposed. Fingernails are square, flat, and yellow but missing on the thumb. Fingers and toes are larger and more robust than a man's. Penis is slender, tapering, and pale yellow. Scrotum is wrinkled and brownish. Knees are only sparsely haired. Feet are 8–10 inches wide, measured across the toes. Profuse hair on the feet. Big toe is apposed, not opposed as in apes. All toes are nearly the same size, with bulbous terminal pads.

*Significant sighting:* After being alerted by university student Terry Cullen that a carnival was exhibiting a hairy man encased in ice, Ivan Sanderson and Bernard Heuvelmans visited the exhibit from December 16 to 18, 1968, at the home of its owner, Frank Hansen, near Rollingstone, Minnesota. For three days, they examined the "iceman" and took detailed notes, sketches, and photographs. Most significantly, they noted that the creature had seemingly been shot in its right eye, with the bullet displacing the left eye; a pool of blood apparently surrounded the back of its head. One forearm seemed to have an open fracture caused by a wound. Heuvelmans and Sanderson both detected an unmistakable odor of rotting flesh exuding from one corner of the exhibit. Gaseous exudations apparently formed bursts of semiopaque, crystalline ice within the transparent block. Sanderson claimed to have tracked down hair samples taken from the iceman when it was imported, "which," he wrote, "turned up in a university in the south."

Hansen claimed that the iceman was the property of an anonymous millionaire, who later withdrew the real specimen from public view and substituted a model, made from latex and hair, that went on the carnival route for several years afterward. His account of where the original came from varies: Russian seal hunters (or Japanese whalers) found it floating in Russia's Sea of Okhotsk, already entombed in ice; Hansen shot it himself, near Whiteface in northern Minnesota; or he purchased it from an exporter in Hong Kong.

*Possible explanations:*

(1) A synthetic fake, manufactured by Hansen or a Hollywood special-effects expert he hired. In 1981, it was claimed that the late Howard Ball, who made models for Disneyland, had created the original with his son Kenneth, based on an artist's conception of Cro-Magnon man.
In 1973, I worked for two individuals who were formerly associated with Hansen in the carnival exhibition business; they told me (with a certain amount of calculated reluctance) that Hansen had a model made "on his living-room floor." However, this information is still hearsay evidence, and although I had, at the time, established a limited amount of confidence as a "carney," I probably would not have been told the truth if Hansen had actually imported a Wildman from Southeast Asia. My sources, after all, were in the business of exhibiting a Brown agouti (*Dasyprocta variegata*) as a "giant rat from the sewers of Paris."
(2) Heuvelmans believed Hansen, who was a captain in the U.S. Air Force during the Vietnam War, had shot or at least acquired a NGUOI RUNG in Vietnam in the mid-1960s and smuggled it into the United States in a military body bag. The creature cannot be a normal human of any race or even a composite produced by assembling several species. However, Heuvelmans's identification of the animal as Neanderthal-like (*Homo neanderthalensis*) misses the mark, since there is no characteristic browridge or sloping forehead; the robust arms and legs do match, though. Heuvelmans apparently obtained photographs showing the face before and after it was thawed out for a short time.
(3) Mark Hall considers the iceman to be closer in form to *Homo erectus,* which he thinks is represented by the KSY-GYIK, ALMAS, or BAR-MANU of Central Asia.

*Sources:* Bernard Heuvelmans, "Note préliminaire sur un spécimen conservé dans la glace d'une forme encore inconnue d'hominidé vivant: *Homo pongoides* (sp. seu subsp. nov.)," *Bulletin de l'Institut Royal des Sciences Naturelles de Belgique* 45 (February 1969): 1–24; Ivan T. Sanderson, "The Missing Link," *Argosy,* May 1969, pp. 23–31, on line at http://www.n2.net/prey/bigfoot/articles/argosy2.htm; Ivan T. Sanderson, "Preliminary Description of the External Morphology of What Appeared to Be the Fresh Corpse of a Hitherto Unknown Form of Living Hominid," *Genus* 25 (1969): 249–78 (reprinted in *Pursuit,* no. 30 [April 1975]: 41–47, and no. 31 [July 1975]: 62–66); "Bozo, the Iceman," *Pursuit* 3 (April 1970): 45–46, and (October 1970): 89; Frank Hansen, "I Killed the Ape-Man Creature of Whiteface," *Saga,* July 1970, pp. 8–11, 55–60; John Napier, *Bigfoot* (New York: E. P. Dutton, 1972), pp. 92–107; Bernard Heuvelmans and Boris F. Porshnev, *L'homme de Néanderthal est toujours vivant* (Paris: Plon, 1974), pp. 209–467; C. Eugene Emery, "Sasquatch-Sickle: The Monster, the Model, and the Myth," *Skeptical Enquirer* 6 (Winter 1981–1982): 2–4; Russell Ciochon, John Olsen, and Jamie James, *Other Origins: The Search for the Giant Ape in Human Prehistory* (New York: Bantam, 1990), pp. 230–233; Dao Van Tien, "Wildman in Vietnam," *Tap Chi' Lâm Nghiêp,* 1990, no. 6, pp. 39–40, and no. 7, p. 12, at http://coombs.anu.edu.au/~vern/wildman/tien.txt; Ian Simmons, "The Abominable Showman," *Fortean Times,* no. 83 (October-November 1995): 34–37; Mark Chorvinsky, "The Burbank Bigfoot," *Strange Magazine,* no. 17 (Summer 1996): 9; Mike Quast, *The Sasquatch in Minnesota,* 2d ed. (Moorhead, Minn.: Mike Quast, 1996); Mark

A. Hall, *Living Fossils: The Survival of* Homo gardarensis, *Neandertal Man, and* Homo erectus (Minneapolis, Minn.: Mark A. Hall, 1999), pp. 69–86; Burt Gilyard, "The Hairy Truth," *Minneapolis City Pages,* October 4, 2000, at http://www.citypages.com/databank/21/1035/article9026.asp; Loren Coleman, *Mysterious America,* rev. ed. (New York: Paraview, 2001), pp. 221–230.

## Miramar Toxodont

Mystery HOOFED MAMMAL of South America.

*Significant sighting:* From 1912 to 1914, paleontologist Carlos Ameghino uncovered stone tools in Late Pliocene strata (2 million years old) along a cliff near Miramar, Buenos Aires Province, Argentina. Among them was a stone arrow or spear point embedded in the femur of a toxodont, a member of a family of large, horse- or rhinolike hoofed mammals that persisted in South America until about 10,000 years ago.

*Possible explanations:*

(1) Unless humans were in South America nearly 2 million years before the currently accepted date, the artifacts (and presumably the femur) must have been displaced from later strata. A large, grazing toxodont surviving into the Holocene would be a likely food source for early hunters.

(2) The arrow was shot into the femur hundreds of thousands of years after the animal died.

*Sources:* Carlos Ameghino, "El femur de Miramar," *Anales del Museo Nacional de Historia Natural de Buenos Aires* 26 (1915): 433–450; Antonio Romero, "El *Homo pampaeus,*" *Anales de la Sociedad Científica Argentina* 85 (1918): 5–48; Michael A. Cremo and Richard L. Thompson, *Forbidden Archeology* (San Diego, Calif.: Bhaktivedanta Institute, 1993), pp. 313–334.

## Mirrii

BLACK DOG of Australia.

*Etymology:* Wiradhuri (Australian) word. Plural, *Mirriuula.*

*Variant name:* Water dog.

*Physical description:* Size of a calf or pony. Black coat. Red eyes. Pointed ears.

*Behavior:* Grows visibly bigger as it is watched. Associated with water. Follows people home at night.

*Distribution:* Western New South Wales.

*Source:* Frank Povah, *You Kids Count Your Shadows: Hairymen and Other Aboriginal Folklore in New South Wales* (Wollar, N.S.W., Australia: Frank Povah, 1990).

## Mirygdy

GIANT HOMINID of far eastern Siberia.

*Etymology:* Lamut (Tungusic), "broad-shoulders."

*Physical description:* Height, 7 feet. Small head. No visible neck. Wide shoulders. Robust arms and legs.

*Behavior:* Active in summer. Scavenges game killed by hunters. Uses its hands to tear off and eat raw meat.

*Distribution:* Chukotskiy Peninsula, eastern Siberia, Russia.

*Significant sighting:* Victor Chebotarev and two other hunters saw a large, hairy, humanlike creature near the Amguema River in the Chukotskiy Region in August 1970.

*Sources:* Alexandra Bourtseva, "Zolotoi sled na Chukotke," *Tekhnika Molodezhi,* 1978, no. 6, pp. 52–53; Dmitri Bayanov, *In the Footsteps of the Russian Snowman* (Moscow: Crypto-Logos, 1996), pp. 231–232.

## Misaabe

CANNIBAL GIANT of eastern Canada.

*Etymology:* Kitcisakik or Ojibwa (Algonquian) word. Plural, *Misaabeg.*

*Physical description:* Tall. Long hair.

*Distribution:* Northern Minnesota; Grand Lake Victoria, Québec.

*Source:* D. S. Davidson, "Folktales from Grand Lake Victoria, Quebec," *Journal of American Folklore* 14 (1928): 275–277.

## Mishipizhiw

FRESHWATER MONSTER of Ontario, Canada.

*Etymology:* Ojibwa (Algonquian), "great lynx."

*Variant names:* Mishibizhii, Mitchipissy.

*Physical description:* Serpentine. Horned. Saw-toothed back. Sometimes characterized as feline.

*Distribution:* Lake Superior, Lake Nipigon, and other lakes of Ontario.

*Possible explanations:*

(1) Migrating Harbor seals (*Phoca vitulina*) or Ringed seals (*P. hispida*) that may have made their way upriver from James Bay.

(2) Folk knowledge of the Walrus (*Odobenus rosmarus*) from Hudson Bay.

*Sources:* Nicolas Perrot, *Mémoire sur les moeurs, coustumes, et relligion des Sauvages de l'Amerique septentrionale* [1705] (Paris: A. Franck, 1864); Selwyn Dewdney and Kenneth E. Kidd, *Indian Rock Paintings of the Great Lakes* (Toronto, Canada: University of Toronto Press, 1967), pp. 33, 123–128; Norval Morrisseau, *Legends of My People, the Great Ojibway* (Toronto, Canada: McGraw-Hill Ryerson, 1977).

## Misiganebic

FRESHWATER MONSTER of northern North America.

*Etymology:* Algonquian, "great serpent."

*Variant names:* HORSE'S HEAD, Misikinubick (Menomini/Algonquian).

*Distribution:* Ontario, Canada; Lac Blue Sea and Lac St.-Jean, Québec, Canada; Wisconsin.

*Sources:* A. Skinner, "Menomini Associations and Ceremonials," *Anthropological Papers of the American Museum of Natural History* 13 (1913); Michel Meurger and Claude Gagnon, *Lake Monster Traditions: A Cross-Cultural Analysis* (London: Fortean Tomes, 1988), pp. 109, 243.

## Mi-Teh

Alternate name for the YETI of Central Asia.

*Etymology:* Sino-Tibetan word. Meaning and origin not established, though one derivation is *mi* ("man") + *teh* ("animal"). Another is that *teh* is the same as *dred* ("bear"). In modern Tibetan, *te* is a particle attached to a verb and means

"when," "after," "thus," "although," or forms a gerund ("-ing"). *Mi ma yin* or *Mi min* are Tibetan ghosts or nonhumans.

*Variant names:* Meh-teh, Metay, Mih-teh, Mi-tre.

*Physical description:* Height of an adolescent boy, 5 feet, but heavily built. Covered with shaggy, reddish-brown, black, or red hair, with longer head-hair. Wide mouth. Prognathous jaw. Thick neck. Conical head. Males have a long mane. Females have pendulous breasts. Short, broad feet, said to be turned back to front.

*Behavior:* Bipedal. Eats pikas and small rodents, young birds, snails, and plants. Has a loud, wailing, yelping call. Also chatters and whistles. Uses sticks occasionally. Shy unless provoked.

*Habitat:* Elevations of 15,000–18,000 feet.

*Distribution:* Himalaya Mountains, Nepal.

*Sources:* Swami Pranavananda, *Kailas-Manasarovar* (Calcutta, India: S. P. League, 1949), p. 69; Ralph Izzard, *The Abominable Snowman Adventure* (London: Hodder and Stoughton, 1955), pp. 100–101; Bernard Heuvelmans, *On the Track of Unknown Animals* (New York: Hill and Wang, 1958), pp. 164–165; Ivan T. Sanderson, *Abominable Snowmen: Legend Come to Life* (Philadelphia: Chilton, 1961), pp. 267–268; Edmund Hillary and Desmond Doig, *High in the Thin Cold Air* (Garden City, N.Y.: Doubleday, 1962), pp. 31, 118.

## Mitla

Mystery DOG or CAT of South America.

*Etymology:* Unknown. In Nahuatl (Uto-Aztecan), *mitla* is the "abode of the dead," hence the name of the Mitla ruins near Oaxaca, Mexico.

*Physical description:* Doglike cat the size of an English foxhound. Black.

*Distribution:* Río Madidi area, Bolivia.

*Possible explanations:*

(1) The dark-brown to black Bush dog (*Speothos venaticus*), suggested by Roy Mackal, is found in this part of Bolivia. It is rare and little seen over much of its range, which extends from Panama to Brazil.

(2) The blackish-gray Short-eared dog (*Atelocynus microtis*), suggested by Karl Shuker, is slightly larger (2–3 feet long) and is also found east of the Andes in Bolivia. It moves with a feline grace.

*Sources:* Percy H. Fawcett, *Exploration Fawcett* (London: Hutchinson, 1953); Roy P. Mackal, *Searching for Hidden Animals* (Garden City, N.Y.: Doubleday, 1980), p. 177; Jeremy Mallinson, *Travels in Search of Endangered Species* (Newton Abbott, England: David and Charles, 1987); Karl Shuker, *Mystery Cats of the World* (London: Robert Hale, 1989), pp. 195–196; Karl Shuker, "South American Mystery Cats," *Cat World,* January 1996, pp. 36–37.

## Mjossie

FRESHWATER MONSTER of Norway.

*Variant name:* Mjoes orm.

*Physical description:* Length, 75 feet. Horse-like head. Black mane. Large eyes.

*Behavior:* Most active in the summer months. Eats livestock.

*Distribution:* Mjøsa, Hedmark County, Norway.

*Significant sightings:* In 1522, the monster was stranded on a rock, where it was killed by a daring youth who shot a volley of arrows into its eye. It washed up on land, and villagers burned it. The skeleton lay undisturbed on the beach for years until some German merchants obtained permission to haul away the bones.

A young couple discovered a Mjoes orm partially beached near their farm. It was brown-black in color, appeared 1.5 feet thick, and had a horselike head with no visible ears. About 30 feet of the body was out of the water. When the man approached, it rose up and quickly moved back into the water.

*Sources:* Jacob Ziegler, *Quae intus continentur* (Strasbourg, France: P. Opilionem, 1532); Olaus Magnus, *Historia Ioannis Magni Gothi Sedis apostolicae legati Svetiae et Gotiae primatis ac archiepiscopi upsalensis* (Rome: I. M. de Viottis, 1554); *The Hamar Chronicle* [1617–1624] (Hamar, Norway: Association and Friends of the Hedmark Museum, 1993);

Elizabeth Skjelsvik, "Norwegian Lake and Sea Monsters," *Norveg* 7 (1960): 29–48; Reidar Christiansen, ed., *Folktales of Norway* (Chicago: University of Chicago Press, 1964), pp. 70–71; "On the Beach," *Fortean Times,* no. 159 (July 2002): 17; Erik Knatterud, The Mjoes Orm, http://www. mjoesormen.no/themjoesorm.htm.

## Mlularuka

Supposed DOG-like flying jackal of East Africa.

*Physical description:* Batlike wings.

*Behavior:* Active at dusk. Flies or glides. Utters loud cries while flying. Raids mango and pomegranate orchards.

*Distribution:* Tanzania.

*Probable explanation:* Misidentified Lord Derby's scaly-tailed squirrel (*Anomalurus derbianus*), a 2 foot 6 inch gliding rodent found throughout Central Africa, though less consistently in Tanzania. This animal favors the bark of some trees, removing one narrow strip each night immediately adjacent to the previous night's strip.

*Source:* William Hichens, "African Mystery Beasts," *Discovery* 18 (1937): 369–373.

## Mmoatia

LITTLE PEOPLE of West Africa.

*Etymology:* Akan (Kwa), "little animal."

*Variant name:* Mmotia.

*Physical description:* Height, only 12 inches. Black, red, or white in color. Feet are turned the wrong way around.

*Behavior:* Whistles or hisses.

*Distribution:* Ghana.

*Sources:* Allan Wolsey Cardinall, *The Natives of the Northern Territories of the Gold Coast* (New York: E. P. Dutton, 1920), p. 27; Robert Sutherland Rattray, *Religion and Art in Ashanti* (Oxford: Clarendon, 1927), pp. 25–26.

## Mngwa

Unknown big CAT of East Africa.

*Etymology:* From the Swahili (Bantu) *mungwa* ("strange one").

*Variant name:* Nunda ("fierce animal," "cruel man," or "something heavy").

The MNGWA, a striped big cat of East Africa. (William M. Rebsamen)

*Physical description:* Size of a donkey. Gray stripes like a tabby cat. Small ears. Thick tail.

*Behavior:* Nocturnal. Has been heard to purr. Known to have raided villages in order to kill adults and carry off children.

*Tracks:* Leopardlike prints as big as a large lion's.

*Distribution:* The Tanzania coast near Lindi and Mchinga.

*Significant sightings:* In 1922, William Hichens was magistrate of Lindi, Tanzania, when several constables were killed or mangled by a huge cat with gray fur. Another outbreak of maulings took place at Mchinga in the 1930s.

*Possible explanations:*

(1) A surviving species of one of several large African fossil cats from the Pleistocene.

(2) An unknown, giant subspecies of the African golden cat (*Felis aurata*), which has a wide variety of coloration, from golden to dark gray, and is reputed to be highly aggressive when cornered. It occasionally raids villages for poultry. It is not known from Tanzania, though its range extends into Kenya and Uganda.

*Sources:* Edward Steere, *Swahili Tales, as Told by Natives of Zanzibar* (London: Bell and Daldy, 1870); Fulahn [William Hichens], "On the Trail of the Brontosaurus: Encounters with Africa's Mystery Animals," *Chambers's Journal,* ser. 7, 17 (1927): 692–695; Charles R. S. Pitman, *A Game Warden among His Charges* (London: Nisbet, 1931), p. 309; William Hichens, "African Mystery Beasts," *Discovery* 18 (1937): 369–373; Frank W. Lane, *Nature Parade* (London: Jarrolds, 1955), pp. 253–256, 266–268; Bernard Heuvelmans, *On the Track of Unknown Animals* (New York: Hill and Wang, 1958), pp. 415–420; Karl Shuker, *Mystery Cats of the World* (London: Robert Hale, 1989), pp. 137–141.

## Mochel Mochel

Alternate name for the BUNYIP of Australia.

*Physical description:* Size and shape of a sheep-dog. Brown fur. Head and whiskers like an otter's.

*Distribution:* Around the Darling Downs, Queensland.

*Source:* Thomas Hall, *A Short History of the Downs Blacks, Known as "The Blucher Tribe"* (Warwick, Queensl., Australia, n.d.).

## Moddey Dhoe

BLACK DOG of the Isle of Man.

*Etymology:* Manx, "black dog."

*Variant names:* Mauthe doog, Moddey dhoo.

*Physical description:* Size of a calf. Curled, shaggy hair. Large, fiery eyes.

*Distribution:* Isle of Man, England.

*Significant sightings:* In the seventeenth century, when Peel Castle, Isle of Man, was garrisoned, a shaggy, black dog came silently into the guardhouse and made itself at home.

In 1927, near Ramsey, at the Milntown corner, Isle of Man, a friend of Walter Gill saw a black dog with red eyes. He and the dog looked at each other until the dog moved aside and allowed him to pass. In 1931, at the same spot, a doctor passed a dog nearly the size of a calf, with bright, staring eyes.

*Sources:* W. Walter Gill, *A Manx Scrapbook* (London: Arrowsmith, 1929); W. Walter Gill, *A Second Manx Scrapbook* (London: Arrowsmith, 1932); James MacKillop, *Oxford Dictionary of Celtic Mythology* (New York: Oxford University Press, 1998), pp. 325–326, 332.

## Moha-Moha

Unknown FISH of Australia.

*Etymology:* Said to be an Australian word meaning "saucy fellow" or "dangerous turtle."

*Scientific name: Chelosauria lovelli,* given by William Saville-Kent in 1893.

*Variant name:* Moka moka.

*Physical description:* Length, 28–30 feet. Greenish-white head and neck, with large white spots. Band of white around the black eyes. Visible teeth. Round jaws, 18 inches long. Long neck. Dome-shaped, slate-gray central body, 8 feet long and 5 feet high. Said to have alligator-like legs and fingers. Dark-brown dorsal fin. Silvery, scaly, fish-shaped tail, 12 feet long.

*Behavior:* Said to be able to stand on its hind feet and attack Aboriginal camps. Edible.

*Distribution:* Sandy Cape, Fraser Island, Queensland, Australia.

*Significant sighting:* S. Lovell was walking along the beach at Sandy Cape, Queensland, in June 1890 when she saw the head and neck of a huge animal resting partly on the shore. She stood observing it for thirty minutes along with two schoolgirls; then it twisted its tail fluke, submerged, and vanished out to sea. Lovell said the animal had been seen the previous week by an Aboriginal boy.

*Present status:* Not seen since 1890.

*Possible explanations:*

(1) The Pitted-shelled turtle (*Carettochelys insculpta*) is a freshwater turtle found in the river systems of the Northern Territory. It has a distinctive, pale streak behind each eye and a longish neck, and it grows to about 2 feet 5 inches long.

(2) The Eastern snake-necked turtle (*Chelodina longicollis*) occurs throughout eastern Queensland in swamps, rivers, and billabongs. It has a relatively long neck, but its total length is only 10 inches.

(3) A surviving placoderm fish, suggested by David Alderton. *Pterichthyodes* of the Middle Devonian, 375 million years ago, had an armored head and trunk shield that resembled a turtle shell, but the fish was only 6 inches long. Also, it did not have a long neck. Larger placoderms such as *Dunkleosteus* reached up to 18 feet in length but were shaped more like sharks or bony fishes.

*Sources: Land and Water,* January 3, 1891, and April 25, 1891; William Saville-Kent, *The Great Barrier Reef of Australia: Its Products and Potentialities* (London: W. H. Allen, 1893), pp. 322–327; Rupert T. Gould, *The Case for the Sea-Serpent* (London: Philip Allen, 1930), pp. 173–183; Bernard Heuvelmans, *In the Wake of the Sea-Serpents* (New York: Hill and Wang, 1968), pp. 295–302.

## Mohán

WILDMAN of South America.

*Variant names:* Muan, Tigre mono.

*Physical description:* Covered with black hair. Thick head-hair. Coarse facial features. Muscular or stocky.

*Behavior:* Seeks lakes or rivers as a refuge when disturbed. Can hold its breath underwater for a considerable time. Enjoys drinking alcoholic beverages and smoking tobacco. Said to assault young women.

*Habitat:* Caverns.

*Distribution:* Huila and Tolima Departments, Colombia.

*Sources:* Juan de Castellanos, *Elegías de varones ilustres de Indias* [1588] (Bogotá: G. Rivas Moreno, 1997); Iván Salazar Duque, *Mitos y mensajes* (Medellín, Colombia: Grafoprint, 1990); "The Mohan," *Cryptozoology Review* 1, no. 2 (Autumn 1996): 8; Fabio Silva Vallejo, *Mitas y leyendas colombianas* (Santafé de Bogotá, Colombia: Panamericana Editorial, 2000).

## Mohin-Goué

SMALL HOMINID of West Africa.

*Etymology:* Guéré (Kru), "like a chimpanzee."

*Physical description:* Small, black, and hairy.

*Behavior:* Eats fruit. Has an incomprehensible language. Said to abduct human females into the forest.

*Distribution:* Forest between Duékoué and Buyo, Côte d'Ivoire.

*Significant sighting:* In 1932, a French hunter named Boisard was crossing the Cavally River from Liberia into Côte d'Ivoire when he saw a small, black, humanlike creature walking on all fours about 130 feet away. It was wearing a red loincloth and had apparently been mashing fruit with stones when he came upon it.

*Sources:* M. Jacquier, "Note sur l'existence probable de Négrilles dans les forêts vierges de l'ouest de la Côte d'Ivoire," *Bulletin de la Comité d'Études Historiques et Scientifiques d'A.O.F.* 18 (1935): 57–62; Prof. Roubaud, "L'existence probable de Négrilles dans les forêts de la Côte d'Ivoire," *Bulletin de la Comité d'Études Historiques et Scientifiques*

*d'A.O.F.* 18 (1935): 540; Charles Lavallée, "Encore les pygmées," *Notes Africaines,* no. 4 (October 1939): 46–47.

## Mokele-Mbembe

Unknown DINOSAUR-like animal of Central Africa.

*Etymology:* Lingala (Bantu), "water monster" or "one who stops the flow of rivers."

*Variant names:* AMALI, BADIGUI, IRIZIMA, ISIQUQUMADEVU, JAGO-NINI, Le'kela-bembe (Baka/Ubangi), Mbokälemuembe (in Cameroon), Mbulu-em'bembe or M'kuoo-m'bemboo (Denya/Bantu), M'(O)KÉ-N'BÉ, Nwe (Ewondo/Bantu), N'YAMALA.

*Physical description:* Size of an elephant or larger. Length, up to 35 feet. Shoulder height, 5–7 feet. Smooth, reddish-brown or brownish-gray skin. The male has a single long horn or tusk. Serpentine head. Flexible neck, 6–12 feet long and as thick as a man's thigh. Feet are like an elephant's. Long, muscular tail.

*Behavior:* Amphibious. Moves singly or in pairs. Active early in the morning or late in the afternoon. Said to live in the forest but feed in the lake. Makes a deep-throated, trumpeting growl. Vegetarian diet. Prefers the applelike fruit of lianas (*Landolphia mannii* and *L. owariensis*) with white blossoms, known locally as Malombo. Digs caves in the riverbank. Aggressively defends its territory. Kills hippopotamuses, elephants, and crocodiles. Said to overturn canoes and destroy the occupants by lashing its tail. Its flesh is said to be poisonous.

*Tracks:* Hippopotamus-like but bigger than an elephant's, or 12 inches in diameter. Three clawed toes. Also makes a furrow like that made by a large snake or a wagon wheel.

*Habitat:* Caves in river banks.

*Distribution:* Bai River, Likouala aux Herbes River, Likouala Swamp, Lake Makele, Sangha River, Lake Tebeki, Lake Télé, and Lower Ubangi River, Republic of the Congo; Ikelemba River, Democratic Republic of the Congo; Boumba, Cross, Loponji, Mbamé, Ngoko, Ntem, and Sanaga Rivers, Cameroon.

*Significant sightings:* In the mid-eighteenth century, French missionaries in the area of

*Pygmy hunters are said to have speared and killed a* MOKELE-MBEMBE *at Lake Télé around 1959. (William M. Rebsamen/Fortean Picture Library)*

Gabon or the western Republic of the Congo reported finding clawed tracks about 3 feet in circumference and 7–8 feet apart.

Capt. Freiherr von Stein zu Lausnitz collected information on the Mokele-mbembe in the Republic of the Congo for the German government during the Likuala-Kongo Expedition of 1913. Natives told him it had smooth skin, was the size of an elephant, had a long and flexible neck, and had a long tusk or horn. He was shown a path made by the animal to get at its preferred food, a white liana blossom.

Ivan T. Sanderson and Gerald Russell heard a loud roar and saw a huge animal swim out from a submerged cave in Mamfe Pool on the Cross River, Cameroon, in 1932 or 1933. All they could see was a dark head larger than a hippo's, which created a wave when it submerged. Several months earlier, they had come across large, hippolike tracks near the river.

About 1935, Firman Mosomele saw a Mokele-mbembe in the Likouala aux Herbes River near Epéna, Republic of the Congo. It had a reddish-brown, snakelike head, and its neck was 6–8 feet long.

Around 1959, a Mokele-mbembe was killed by Pygmies at Lake Télé, Republic of the Congo, by putting up a barrier in a waterway that the animal used to enter the lake; the cornered animal was then speared to death. They cut it up and ate the meat, but everyone is said to have died shortly afterward.

In the 1960s, Nicolas Mondongo was hunting for monkeys along the Likouala aux Herbes River between Bandéko and Mokengui when a huge animal reared out of the water about 40 feet away. Its head and neck together were 6 feet in length, and it had four sturdy legs and a long tail. Mondongo watched it for three minutes before it submerged.

In February 1980, Roy Mackal and James Powell went on a reconnaissance expedition that reached Epéna on the Likouala aux Herbes River, Republic of the Congo, and they collected firsthand reports of the Mokele-mbembe.

The Herman Regusters Expedition to Lake Télé, Republic of the Congo, from October 9 to December 9, 1981, made several observations of disturbances in the water caused by a large animal. A long neck was seen for five minutes during one encounter and for a few seconds on another occasion. On November 4, Regusters heard and recorded an animal making a loud growl.

Roy Mackal, Richard Greenwell, and Justin Wilkinson conducted an expedition to the Likouala Region, Republic of the Congo, from October 27 to December 3, 1981. They encountered an odd wake made by a large animal in the Likouala River between Itanga and Mahounda and examined the trail made by an unknown animal upstream from Djeké months earlier and discovered by Emmanuel Moungoumela.

A Congolese expedition led by zoologist Marcellin Agnagna surveyed the Likouala Swamp and Lake Télé area from April 3 to May 17, 1983. For twenty minutes on May 1, Agnagna and others saw a 15-foot animal with a wide back and long neck swimming in the lake; though the animal was observed through the telephoto lens of a movie camera, the film was on an incorrect setting and proved worthless. The expedition also found recent footprints near Djeké.

The British Operation Congo, led by William Gibbons from January to June 1986, returned from Lake Télé with little evidence, though it confirmed the existence of turtles, pythons, and crocodiles in the lake.

A Japanese film crew led by Tatsuo Watanabe shot a controversial video in September 1992 showing fifteen seconds of what they thought was a Mokele-mbembe crossing Lake Télé.

A village security officer at Moloundou, Cameroon, saw a Le'kela-bembe in the Boumba River in February 2000. The animal stopped swimming downstream when it saw a ferry and moved away upstream.

*Possible explanations:*

(1) Sauropod dinosaurs, herbivorous quadrupeds that ranged in total body length from 20 to 145 feet, had small heads, long necks, long tails, and massive limbs. They had five toes on all four limbs, with at most a single clawed toe on each forefoot and perhaps three on the hind feet. There were two types of sauropods, distinguished primarily from characteristics of the teeth: large animals with thick, spoon-shaped teeth, such as *Brachiosaurus,* and smaller animals with longer snouts and thin, peg-shaped teeth, such as *Diplodocus.* The earliest sauropod fossil is *Vulcanodon,* a 33-foot animal from Zimbabwe and dating from the Early Jurassic, 200 million years ago; other early species have been found in Germany and China. Sub-Saharan African sauropods include *Barosaurus, Brachiosaurus,* and *Dicraeosaurus* from Tanzania and *Janenschia* and *Malawisaurus* from Malawi. Presumably, the last sauropods died off at the end of the Cretaceous, 65 million years ago.

(2) *Ouranosaurus,* a 24-foot, bipedal iguanodontid dinosaur, was excavated in the Sahara Desert in Niger in 1966. Its distinctive dorsal spines are 2 feet high and may have supported a sail-like membrane. This explanation was proposed by Herman Regusters, who misidentified the fossil as a sauropod and alleged that one vertebra was radiocarbon-dated as only a few thousand years old. In fact, the remains date from the early Cretaceous, some 110 million years ago.

(3) An unknown species of giant Monitor (Varanidae) or Iguana (Iguanidae) lizard. Both groups include semiaquatic species, and some iguanas are herbivorous.

(4) Large African softshell turtle (*Trionyx triunguis*), called NDENDEKI by locals living in the Lake Télé area and said to grow up to 15 feet in diameter. Marcellin Agnagna's 1983 sighting may have involved this turtle.

(5) An African elephant (*Loxodonta africana*) swimming with its trunk raised.

(6) The Nile crocodile (*Crocodylus niloticus*), which can grow to over 20 feet long.

(7) During the rainy season,

Hippopotamuses (*Hippopotamus amphibius*) are said to hibernate in caves along the riverbanks. If disturbed, one of them might surprise and confuse the unwary traveler. This might explain Ivan Sanderson's sighting in Mamfe Pool, Cameroon.

(8) The West African manatee (*Trichechus senegalensis*) grows to about 12 feet in length and might be mistaken for a larger animal if encountered suddenly. It may be found in certain rivers of the Republic of the Congo.

*Sources:* Abbé Proyart, *Histoire de Loango, Kakongo, et autres royaumes d'Afrique* (Paris: C. P. Berton, N. Crapart, 1776), pp. 38–39; Wilhelm Bölsche, *Drachen: Sage und Naturwissenschaft* (Stuttgart, Germany: Franckh'sche Verlagshandlung, 1929), pp. 49–54; Leo von Boxberger, "Ein unentdecktes Grosstierart in Innerafrika," *Die Umschau,* 42 Jahr, Heft 49 (1938): 1133; Ivan T. Sanderson, "There Could Be Dinosaurs," *Saturday Evening Post* 220 (January 3, 1948): 17, 53–56; Bernard Heuvelmans, *On the Track of Unknown Animals* (New York: Hill and Wang, 1958), pp. 461–467, 475–478; Bernard Heuvelmans, *Les derniers dragons d'Afrique* (Paris: Plon, 1978), pp. 248–261, 269–270, 299–301; Herman A. Regusters, "Mokele-Mbembe: An Investigation into Rumors Concerning a Strange Animal in the Republic of the Congo, 1981," *Munger Africana Library Notes,* no. 64 (1981): 1–27; Charles W. Weber, James W. Berry, and J. Richard Greenwell, "Mokele-Mbembe: Proximate Analysis of Its Supposed Food Source," *Cryptozoology* 1 (1982): 49–53; Roy P. Mackal, J. Richard Greenwell, and M. Justin Wilkinson, "The Search for Evidence of Mokele-Mbembe in the People's Republic of the Congo," *Cryptozoology* 1 (1982): 62–72; Marcellin Agnagna, "Results of the First Congolese Mokele-Mbembe Expedition," *Cryptozoology* 2 (1983): 103–112; Herman A. Regusters and Kia L. Vandusen, "An Interim Report on the Search for Mokele Mbembe," *Pursuit,* no. 72 (1985): 174–180; "Mokele-Mbembe: New Searches, New Claims," *ISC Newsletter* 5, no. 3 (Autumn 1986): 1–7; Roy P. Mackal, *A Living Dinosaur? In Search of Mokele-Mbembe* (Leiden, the Netherlands: E. J. Brill,

1987); Rory Nugent, *Drums along the Congo* (Boston: Houghton Mifflin, 1993); Redmond O'Hanlon, *Congo Journey* (London: Hamish Hamilton, 1996); Mike Dash, "Dinosaur Caught on Film?" *Fortean Times,* no. 86 (May 1996): 32–35; Adam Davies, "I Thought I Saw a Sauropod," *Fortean Times,* no. 145 (May 2001): 30–32; Karl Shuker, "Mokele-Mbembe Goes West!" *Fortean Times,* no. 146 (June 2001): 20; David Woetzel, Behemoth or Bust: An Expedition into Cameroon Investigating Reports of a Sauropod Dinosaur, August 2001, at http://www.genesispark.org/genpark/expedition/report.htm; William Gibbons, "Cameroon Field Investigation Report," unpublished report, 2001.

## M'(o)ké-n'bé

Dinosaur-like animal of West Africa, similar to the MOKELE-MBEMBE.

*Etymology:* Waci-Gbe (Kwa) word. Perhaps a contraction of *Mokele-mbembe.*

*Physical description:* Size of an elephant. Gray. Small head. Long neck. Long tail.

*Behavior:* Aquatic.

*Distribution:* Swampy western tributaries of the Ouémé or Mekrou Rivers, Benin.

*Significant sighting:* Animal collector W. T. Roth heard stories about this animal in 1959 after his Waci guides refused to cross a swamp where it lived.

*Sources:* "He Have Head for Trunk," *Pursuit,* no. 9 (January 1970): 16–17; Bernard Heuvelmans, *Les derniers dragons d'Afrique* (Paris: Plon, 1978), pp. 280–282.

## Momo

HAIRY BIPED of Missouri.

*Etymology:* "Missouri monster" (Mo. + monster abbreviated), given by newspaper reporters.

*Physical description:* Height, 6–12 feet. Black body-hair. Facial features obscured by hair. No neck.

*Behavior:* Bipedal. Aggressive toward dogs. Growls, gurgling sound. Pungent, fetid odor.

*Tracks:* Three-toed or five-toed. Hind feet, 10–12 inches long and 2–5 inches wide. Handprint, 5 inches long and curved.

*Left: Witness Edgar Harrison's sketch of* MOMO. *Right: Drawing of creature seen by Doris and Terry Harrison, July 11, 1972. (Loren Coleman)*

*Distribution:* Louisiana, Missouri.

*Significant sightings:* Terry and Wally Harrison, ages eight and five, were playing in the woods near their property in Louisiana, Missouri, on the afternoon of July 11, 1972. Their sister Doris, fifteen, heard them scream and looked out to see a manlike creature, about 6–7 feet tall, with long, black hair carrying a dead dog under its arm and standing next to a tree. Their father, Edgar Harrison, found a footprint and a handprint on nearby Marzolf Hill on July 19.

Bill Suddarth and his wife found some narrower, three-toed tracks in their garden after hearing a high-pitched howl on August 3, 1972.

*Possible explanations:*

(1) The footprints were pronounced a hoax by the director of the Oklahoma City Zoo.

(2) Unidentified flying object (UFO) reports in the area at the same time encouraged some of the early investigators to speculate on an alien origin.

(3) Some reports may have been American black bears (*Ursus americanus*).

*Sources:* Richard Crowe, "Missouri Monster," *Fate* 25 (December 1972): 58–66; "'Momo' & Others," *INFO Journal*, no. 9 (Fall 1972): 49–51; Jerome Clark and Loren Coleman, "Anthropoids, Monsters and UFOs," *Flying Saucer Review* 19 (January-February 1973): 18–24.

## Mongitore's Monstrous Fish

Mystery FISH of the Mediterranean Sea.

*Scientific name: Oxypterus mongitori,* given by Constantin Samuel Rafinesque in 1814.

*Physical description:* Length, 45 feet 6 inches. Circumference, 23 feet 7 inches. Blowhole. Two dorsal fins. Tail, 10 feet 2 inches long.

*Distribution:* Mediterranean Sea, off Sicily.

*Significant sighting:* In September 1741, a monstrous fish was stranded near Licata, Sicily, Italy. It had a blowhole and powerful teeth.

*Possible explanations:*

(1) A Basking shark (*Cetorhinus maximus*), since no known whales have two dorsal fins. These sharks are known to grow to more than 40 feet.

(2) A Mediterranean occurrence of the unknown MAGENTA WHALE, which has two dorsal fins.

*Sources:* Antonino Mongitore, *Della Sicilia*

*ricercata nelle cose piu memorabili* (Palermo, Italy: Francesco Valenza, 1742–1743), vol. 2, pp. 98–99; Constantin Samuel Rafinesque, *Précis des découvertes et travaux somiologiques de C. S. Rafinesque-Schmaltz entre 1800 et 1814* (Palermo, Italy: Royale Typographie Militaire, 1814), p. 13; Michel Raynal, "Do Two-Finned Cetaceans Really Exist?" *INFO Journal,* no. 70 (January 1994): 7–13.

## Mongolian Death Worm

Huge SNAKE-like animal or INVERTEBRATE of Central Asia.

*Variant names:* Allergorhai-horhai, Allghoi khorkhoi, Olgoi khorkhoi (Mongolian/Altaic, "intestine worm"), Shar khorkhoi.

*Physical description:* Length, 2–5 feet. Dark red. A yellow variety (Shar khorkhoi) is also said to exist, though it is rarer. Thick. No differentiated head, tail, or feet.

*Behavior:* Comes to the surface in June and July. Squirts a bubbly, acidic, lethal poison from one end of its body. Said to have killed a geologist when it touched an iron rod he was holding, possibly by electrical shock.

*Habitat:* Areas of the desert where a parasitic herb, Goyo (*Cynomorium songaricum*), is found in the roots of saxaul bushes.

*Distribution:* Region around Ihbulag and Dalandzadgad, southern Gobi Desert, Mongolia; in the Nemegt Uul, Mongolia. Marie-Jeanne Kofman heard rumors of a similar animal in the Kalmyk Republic, Russia.

*Possible explanations:*

(1) Unknown species of giant Worm lizard (Amphisbaenidae), part of a family of limbless lizards with no external eyes and ears. These creatures live in underground burrows and move in a serpentine fashion. In the 1990s, there were several discoveries of fossil amphisbaenids in Central Asia, including Mongolia.

(2) A highly specialized giant form of Earthworm (Subclass Oligochaeta).

(3) Unknown species of venomous Sand boa (*Eryx* spp.).

(4) Unknown species of venomous elapid snake similar to the Southern death adder

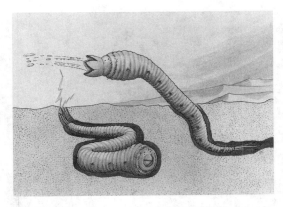

*The* MONGOLIAN DEATH WORM *of the Gobi Desert, said to squirt poison and electrocute its victims. (Phillippa Foster/Fortean Picture Library)*

(*Acanthopis antarcticus*) of Australia, which grows to over 3 feet; suggested by John L. Cloudsley-Thompson.

*Sources:* Roy Chapman Andrews, *The New Conquest of Central Asia* (New York: American Museum of Natural History, 1932), vol. 1, p. 62; Ivan Antonovich Efremov, "Olgoi-Khorkhoi," in *Stories* (Moscow: Foreign Languages Publishing House, 1954); Ivan Antonovich Efremov, *Doroga vetrov* (Moscow, 1958); Dongodiin TSevegmid, *Altayn tsaadakh govd* (Ulaanbaatar, Mongolia, 1987), pp. 5–7; Ivan Mackerle, "In Search of the Killer Worm," *Fate* 49 (June 1996): 22–27; Karl Shuker, "Meet Mongolia's Death Worm: The Shock of the New," *Fortean Studies* 4 (1997): 190–218; Angel Morant and Carlos Bonet, "Olgoi-khhorkhhoï: El gusano-intestino de Mongolia," *Biologica,* no. 17 (February 1998): 66–67; Michel Raynal, "Olgoï-Khorkhoï: Le 'Ver-intestin' Mongol," August 10, 1999, Institut Virtuel de Cryptozoologie, http://perso.wanadoo.fr/cryptozoo/olgoi.htm; Ivan Mackerle, *Mongolské Záhady* (Prague: Ivo Zelezny, 2001).

## Mongolian Goat-Antelope

Mystery HOOFED MAMMAL of Central Asia.

*Physical description:* Size of a goat. Brown. Small, spreading horns. High back.

*Behavior:* Runs faster than a sheep or goat but

slower than an antelope. Grazes in groups of four or five.

*Habitat:* Rocky ravines.

*Distribution:* Dundgovi Province, Mongolia.

*Significant sighting:* Discovered by local people, who keep them secret from outsiders.

*Source:* "Unknown Animals," *Mongolia Online,* March 20, 2001.

## Monkey Man

Monkeylike ENTITY of the Indian subcontinent.

*Variant name:* Bear man.

*Physical description:* Variously reported as a monkey, a man with a monkey face, a man with a mask and helmet, or even an alien or robot. Height, 4 feet 6 inches–6 feet. Glowing red eyes. Said to have flashing green and red lights on its chest. Metallic claws.

*Behavior:* Can jump 20 feet into the air from a crouching position. Bites people while they are sleeping. Sometimes dressed in white, black, or silver. Said to speak the Bhojpuri language.

*Distribution:* Around New Delhi, Noida, and Ghaziabad, Uttar Pradesh State, India; Nalbari, Assam State, India; Ahmadabad, Gujarat State, India.

*Significant sightings:* Reports of a belligerent monkey or a masked man began in early April 2001 in the suburbs of Ghaziabad, India. People went to the police with what appeared to be deep scratches or bites. By May 16, the panic had spread to Noida and New Delhi; there had been hundreds of reports and many injuries, three panic deaths had occurred, several people were mistaken for the entity and beaten, and more than 3,000 policemen on motorcycles, armed with rifles, were patrolling east Delhi to calm down residents. Typical reports were vague and varied: On May 10, a masked man struck the stomach of Saroj Sharma in Chhaprola; the next night, residents saw a shadowy figure that jumped like a monkey and had glowing red eyes. On May 13, an intruder dressed in white bandages attacked the wife and sister of M. P. Singh in Noida.

Bearlike or wolflike humanoids were reported in late May 2001 in the Nalbari District, Assam State. The Assam Science Society interviewed sixteen witnesses, who admitted they were half asleep when they felt something with sharp nails trying to grab them.

Another Monkey man was reported in the Khanpur suburb of Ahmadabad in early February 2002. The creature was dressed in black, had curly hair, wore a mask, and hopped from roof to roof.

*Possible explanations:*

(1) The police at first suspected many of the attacks were made by a man wearing a monkey mask.

(2) The final report by the New Delhi police, issued on June 19, 2001, concluded that fear and panic were behind all the sightings, ruled out a conspiracy by pranksters, and said there was no evidence for any bizarre creature that could have caused the attacks. Forensic specialists noted that most of the injuries were superficial and self-inflicted accidentally during panic attacks.

*Sources:* Parmindar Singh, "Masked Man or Monkey, It's a Menace," *Times of India,* May 7, 2001; "Monkey Man Cocks a Snook at Delhi Police," *Times of India,* May 17, 2001; "Police File Final Report on 'Monkeyman,'" *Times of India,* June 20, 2001; "Man, Myth or Monkey?" *Fortean Times,* no. 148 (August 2001): 8–9; "Monkey Madness," *Fortean Times,* no. 149 (September 2001): 7; "'Monkeyman' Creates Scare in Khanpur," *Indian Express* (Mumbai), February 20, 2002.

## Mono Grande

Unknown PRIMATE of South America.

*Etymology:* Spanish, "big monkey."

*Variant names:* Mojan (Arawakan), Mono rey ("king monkey").

*Physical description:* Tailless, apelike creature. Height, 5–6 feet.

*Behavior:* Arboreal. Runs with an odd, leaping gait. Call is an eerie howling sound. Throws stones at huts at night. Uses branches as weapons. Said to interbreed with the Indians.

*Distribution:* Serranía de Parijá of Colombia and Venezuela; eastern Venezuela; Río Paute, eastern Ecuador; Río Madidi area, Bolivia; possibly in Peru during colonial times.

*Significant sightings:* In 1968, archaeologist Pino Turolla glimpsed two apelike creatures in the Venezuelan jungle.

In 1997, British travel writer Simon Chapman searched for the Mono rey of northern Bolivia but found no compelling evidence. He heard rumors that a pelt had been purchased by a foreigner for DNA analysis and that a living animal had been exhibited at the zoo in Santa Cruz, Bolivia.

*Sources:* Pedro de Cieza de León, *The Travels of Pedro de Cieza de Leon,* A.D. *1532–50* [1553], Hakluyt Society Works, vol. 33, p. 339 (New York: Burt Franklin, 1864); Bernard Heuvelmans, *On the Track of Unknown Animals* (New York: Hill and Wang, 1958), pp. 307–308, 328; Pino Turolla, *Beyond the Andes* (New York: Harper and Row, 1980), pp. 123–124, 132–136, 253–254, 287–289, 293; Michael T. Shoemaker, "The Mystery of the Mono Grande," *Strange Magazine,* no. 7 (April 1991): 2–5, 56–60; Simon Chapman, *The Monster of the Madidi: Searching for the Giant Ape of the Bolivian Jungle* (London: Aurum, 2001).

## Montana Nessie

FRESHWATER MONSTER of Montana.

*Etymology:* After Scotland's NESSIE.

*Variant name:* Flattie.

*Physical description:* Apparently, two types of animal are involved. *Large animal*—Eel-shaped. Length, 20–40 feet. Brown to bluish-black. Head like a snake's. Distinctive, steel-black eyes. Dorsal fin sometimes reported. *Small animal*—Like a large fish. Length, 6–10 feet.

*Behavior:* Swims on the surface with vertical undulations. Often seen cavorting in the water, which could be either playfulness or feeding behavior.

*Distribution:* Flathead Lake, Montana.

*Significant sightings:* In 1889, Capt. James C. Kerr, piloting the steamboat *U.S. Grant,* noticed a whalelike object. Thinking it to be an approaching boat, he kept an eye on it but soon realized the object was an animal 20 feet long. His passengers became frightened, and one gun-toting man pulled out his rifle and shot at it. He missed, and the creature submerged.

On May 27, 1937, L. J. Eakins saw what looked like a large dog swimming down the Flathead River. It held its head out of the water and looked at him once as he shouted at it.

H. W. "Buck" Black and his family were returning to Polson by boat on July 10, 1949, when they saw a large fish about 150 feet away. Six feet of its back was showing, but the head was submerged. Black is convinced it was a 10- to 12-foot sturgeon.

Howard Gilbert and his wife were driving along the eastern shore on June 12, 1955, when they saw two large fishes, one 8 to 10 feet long, cavorting in the water. After two minutes, they submerged, one disappearing and the other swimming underwater toward the shore.

The E. E. Funke family saw a shiny, black animal with a dorsal fin swimming toward their boat dock on Indian Bay on August 19, 1965. It created a boat-sized wake.

U.S. Army Maj. George Cote and his son Neal saw a black object, as long as a telephone pole and twice as thick, surfacing and diving in Yellow Bay on May 25, 1985. It had a head like a snake's and a tail like an eel's. They saw a similar animal on July 1, 1987, near Lakeside.

Rich Gaffney and his family watched a NESSIE-like animal surface about 50 yards away amid a school of fish in Skeeko Bay on July 29, 1993. It had shiny humps and was 15–20 feet long.

On June 20, 1998, three brothers were wakeboarding on the lake when they saw three dark-green humps in the water ahead. The animal the humps belonged to was about 30 feet long, with rough scales. It dropped below the surface and swam past their boat, leaving a small wake.

*Present status:* Local investigator Paul Fugleberg has identified ninety different sightings from 1889 to 1999. The Montana Department of Fish, Wildlife, and Parks had collected seventy-eight sightings by 1998; 32 percent fit the description of a large sturgeon, while the other 68 percent describe a snakelike animal with humps.

*Possible explanations:*

(1) The White sturgeon (*Acipenser transmontanus*), though not officially known from this lake, reaches lengths of 20 feet

and is the largest freshwater fish in North America. It is dark gray or pale olive above and white to pale gray on the underside. The report that a sturgeon 7 feet 6 inches long was caught here on May 28, 1955, by C. Leslie Griffith is not accepted by all investigators; the fish can still be seen in the Polson-Flathead Historical Museum, however. The smaller, fishlike animal could be a sturgeon.

(2) The Northern elephant seal (*Mirounga angustirostris*) matches the size, but unfortunately, this is a marine mammal.

(3) A miniature submarine has been suggested.

*Sources:* Tim Church, "Flathead Lake Monster," *Pursuit,* no. 32 (October 1975): 89–92; Tim Church, "Flathead Lake Monster Update," *Pursuit,* no. 35 (Summer 1976): 62; Gary S. Mangiacopra, "The Two Monsters of Flathead Lake, Montana," *Of Sea and Shore* 12, no. 2 (1981): 93–96, 114; Paul Fugleberg, *Montana Nessie of Flathead Lake* (Polson, Mont.: Treasure State Publishing, 1992); Paul Fugleberg, *Flathead Lake Nessie Log: A Listing of Reported Sightings, 1889–1999* (Polson, Mont.: Paul Fugleberg, 2000); Paul Fugleberg, "Montana *Nessie?* The Creature from Flathead Lake, Montana," *Crypto Dracontology Special,* no. 1 (November 2001): 20–25; Monsterwatch Project, http://www.monsterwatch.itgo.com.

## Moolgewanke

Alternate name for the BUNYIP of Australia.

*Etymology:* Australian word.

*Physical description:* Half man and half fish. Has a matted crop of reeds instead of hair.

*Behavior:* Call is a booming sound.

*Distribution:* Lake Alexandrina, South Australia.

*Possible explanations:*

(1) The booming calls might be the cry of the Brown bittern (*Botaurus poiciloptilus*).

(2) Episodic explosions like guns firing have been heard in at least three seismically active parts of Australia: near Kooroocheang, Victoria; in Western Australia; and along the Darling River in New South Wales.

They have been attributed to micro-earthquakes.

*Sources:* George Taplin, *The Narrinyeri* (Adelaide, S.A., Australia: J. T. Shawyer, 1874); William R. Corliss, *Earthquakes, Tides, Unidentified Sounds and Related Phenomena* (Glen Arm, Md.: Sourcebook Project, 1983), pp. 151–158.

## Morag

FRESHWATER MONSTER of Scotland.

*Etymology:* From the Gaelic *a' Mhorag,* a feminine diminutive derived from the name of the loch.

*Variant names:* Maggie, Mhorag.

*Physical description:* Most common description is of a large, rounded object like an overturned boat. Length, 20–30 feet. Dark gray or brown. Rough skin. Flat, snakelike head. Slit eyes. Wide mouth. Neck, 5–6 feet long. From three to five humps, 15–20 feet in length. Four legs with three toes.

*Behavior:* Swims swiftly.

*Distribution:* Loch Morar, Highland, Scotland.

*Significant sightings:* James Macdonald told the story of his encounter with a creature called Mhorag in January 1887, but it appeared to be more like a MERMAID or KELPIE rather than a NESSIE-like animal.

John Gillies, Noel O'Donnell, and a boatload of tourists saw a 20-foot monster with five distinct humps off the southern shore of the loch on July 30, 1948.

In September 1958, George R. Cooper sketched a portrait of the animal, which had the appearance of a drifting log.

John MacVarish, barman at the Morar Hotel, was in his boat opposite Bracora on August 27, 1968, when he saw a long neck moving slowly down the loch. It stretched 5–6 feet out of the water and had a flat, snakelike head.

On the evening of August 16, 1969, Morag struck the stern of a motorboat piloted by Duncan McDonell and William Simpson, knocking a teakettle off the stove. McDonell hit the animal with an oar, while Simpson shot at it with a rifle, apparently frightening it off. It was 25–30 feet long with three low humps and had a brown, snakelike head about 12 inches across the top.

On March 3, 1981, Sydney Wignall, Bryan Woodward, and John Evans were in an inflatable boat west of Brinacory Island when they saw two black humps traveling at the same speed as their boat. They were visible for about 20 seconds.

*Sources:* Seton Gordon, *Afoot in Wild Places* (London: Cassell, 1937); Constance Whyte, *More than a Legend* (London: Hamish Hamilton, 1957), pp. 129–131; R. Macdonald Robertson, *Selected Highland Folktales* (Isle of Colonsay, Scotland: House of Lochar, 1961), p. 117; Elizabeth Montgomery Campbell, *Report, Loch Morar Survey 1970* (London: Loch Morar Survey, 1970); Loch Morar Survey, *Report of the 1971 Loch Morar Survey* (London: Loch Morar Survey, 1971); Elizabeth Montgomery Campbell and David Solomon, *The Search for Morag* (New York: Walker, 1973); "Unidentifieds," *Fortean Times,* no. 22 (Summer 1977): 18–26; GUST Zoology, accessed in 2001, http://www.bahnhof.se/ ~wizard/cryptoworld/index209.html.

## Morgawr

SEA MONSTER of Cornwall.

*Etymology:* Cornish, "sea monster." Probably coined by Noel Wain of the *Falmouth Packet* in 1976.

*Variant name:* Durgan Dragon.

*Physical description:* Length, 17–20 feet. Gray skin. Long neck. Several large humps. Long tail.

*Behavior:* Said (by Tony Shiels) to be attracted to naked women, especially young Wiccans.

*Distribution:* Cornwall, England, especially Falmouth Bay.

*Significant sightings:* On August 3, 1906, officers Spicer and Cuming of the American transatlantic liner *St. Andrew* saw a sea monster with a head 18 feet long and huge jaws and teeth while rounding Land's End, Cornwall.

The anonymous "Mary F." submitted two photographs of an apparent sea monster allegedly taken in February 1976. Published in the *Falmouth Packet* on March 5, they show a three-humped object with a long, downward-curving neck. Subsequent research showed that the photos were probably hoaxed by Irish busker and trickster Tony "Doc" Shiels.

On July 10, 1985, Sheila Bird and her brother Eric watched a 20-foot, gray animal off Porthscatho in Falmouth Bay for several minutes before it suddenly sunk vertically and vanished.

Morgawr was spotted in early September 1995 off Rosemullion Head by Gertrude Stevens. She saw a small head on a long neck and a broad, flat tail.

In August 1999, John Holmes took some video footage of an animal with a snakelike head and neck in Gerran's Bay, Cornwall.

Derek and Irene Brown saw a series of humps and a periscope-like head and neck off Falmouth on May 16, 2000.

*Possible explanations:*

(1) Largely a series of hoaxes and dubious stories created or encouraged by Tony Shiels beginning in the late 1970s, with some possibly genuine events and traditions mixed in.

(2) Basking sharks (*Cetorhinus maximus*) visit the coast frequently from May to July. The second-largest shark in the world, it reportedly grows to 40–50 feet and has distinctive gill slits that nearly encircle the head.

(3) Various types of whales and porpoises, unexpectedly seen.

*Sources:* "Serpent with 18-Foot Head Seen," *New York Herald,* August 11, 1906, p. 7; "The Durgan Dragon," *Fortean Times,* no. 15 (April 1976): 13, 16–17; Anthony Mawnan-Peller [pseud.], *Morgawr: The Monster of Falmouth Bay* (Falmouth, England: Morgawr Productions, 1976); "Unidentifieds," *Fortean Times,* no. 16 (June 1976): 17–19, no. 19 (December 1976): 12–17, no. 22 (Summer 1977): 18–26; Janet and Colin Bord, *Alien Animals* (Harrisburg, Pa.: Stackpole, 1981), pp. 26–32; Mark Chorvinsky, "The 'Mary F.' Morgawr Photographs Investigation," *Strange Magazine,* no. 8 (Fall 1991): 8–11, 46–49, and, in the same issue, "Interview: Michael McCormick," 15, 58–59; Ulrich Magin, "Morgawr Unborn: The Genesis of the

Falmouth Bay Serpent," *Strange Magazine,* no. 8 (Fall 1991): 18, 56–57; "Morgawr Is Back," *Fortean Times,* no. 84 (December 1995-January 1996): 8; Jonathan Downes, *The Owlman and Others* (Corby, England: Domra Publications, 1998); Paul Harrison, *Sea Serpents and Lake Monsters of the British Isles* (London: Robert Hale, 2001), pp. 94–104; "'Nessie Spotted' in Cornwall," *BBC News,* June 27, 2002.

## Mosqueto

FRESHWATER MONSTER of New York.

*Etymology:* Oneida (Iroquoian) word.

*Distribution:* Lake Onondaga, New York.

*Significant sighting:* The Oneidas had a legend that a great animal came from the lake and killed many Indians.

*Possible explanation:* Seals apparently visit this lake from time to time. A 6-foot, 100-pound seal was shot in the lake by George F. Kennedy on April 28, 1882.

*Sources:* David Cusick, *Sketches of Ancient History of the Six Nations* (Lewiston, N.Y.: David Cusick, 1827); *New York Times,* May 2, 1882, p. 2.

The MOTHMAN of Point Pleasant, West Virginia. (Richard Svensson)

## Mothman

FLYING HUMANOID or BIG BIRD of Ohio and West Virginia.

*Etymology:* Coined by a journalist on November 16, 1966, supposedly after the Killer Moth character in the *Batman* comics series.

*Variant name:* Birdman.

*Physical description:* Height, 5–7 feet. Gray or brown. No head. Luminous, bright-red eyes, apparently set in the shoulders. No arms. Broader than a man. Wingspan of 10 feet. Wings are folded against the back when not in use. Humanlike legs.

*Behavior:* Walks with a shuffle. Very swift flight (up to 100 miles per hour). Wings make a squeaking noise but do not flap when in flight.

*Distribution:* The Ohio River valley centering on Point Pleasant, West Virginia, and Gallipolis, Ohio. Also Salem and St. Albans, West Virginia; Lowell, Ohio; Maysville, Kentucky.

*Significant sightings:* In the first half of the twentieth century, a large bird with a wingspan of 12 feet and dark-red feathers was seen occasionally around the Ohio and Kanawha Rivers in West Virginia.

Roger and Linda Scarberry and Steve and Mary Mallette were driving near the TNT area, a World War II–era munitions dump north of Point Pleasant, West Virginia, on November 15, 1966, when they encountered a 7-foot, humanoid figure with glowing red eyes. Scarberry floored his Chevy, but the creature took wing and pursued them, at speeds up to 100 miles an hour, all the way to the city limits. It flew without flapping its wings.

Two Point Pleasant firemen, Paul Yoder and Benjamin Enochs, were in the TNT area on November 18, 1966, when they encountered a giant bird with red eyes.

On November 26, 1966, Marvin Shock and Ewing Tilton watched four large birds for several hours in daylight as they flew around and perched in trees near Lowell, Ohio. They were 4–5 feet long with wingspans of 10 feet. Their breasts were charcoal gray, their backs were dark brown with light flecks, their heads were reddish, and their bills were straight and about 6 inches long.

On November 28, 1966, Richard West of Charleston, West Virginia, saw what looked like Batman on the roof next door. It was about 6 feet tall with wings 6–8 feet wide and big red eyes.

*Present status:* The Mothman Prophecies was released in 2002 as a feature film starring Richard Gere as the investigator.

*Possible explanations:*

(1) The Sandhill crane (*Grus canadensis*) is about the size of a great blue heron but is gray, mottled with rust stains. It has a wingspan up to 6 feet 5 inches. Its migration route is not through West Virginia, but strays sometimes are reported.

(2) A Snowy owl (*Nyctea scandiaca*) was shot by a farmer in Gallipolis Ferry, West Virginia, in December 1966. This nearly pure-white bird is 2 feet long, with a 4-foot wingspread. Normally found no farther south than Michigan, individual owls are occasionally seen as far south as Louisiana, though usually in a feeble condition.

(3) A Barn owl (*Tyto alba*), suggested by Joe Nickell, can have a wingspan of 3 feet 6 inches and sometimes appears deceptively large.

(4) The Turkey vulture (*Cathartes aura*) is common throughout Ohio and West Virginia in the summer. A red-headed carrion-feeder with a wingspan of nearly 6 feet, it is more than 2 feet long and has a distinctive, bare, red head.

(5) A GIANT OWL has sometimes been reported in the area.

(6) An extraterrestrial entity, because unidentified flying objects (UFOs) and other odd phenomena were frequently reported in the vicinity at the time.

*Sources:* John A. Keel, "North America 1966: Development of a Great Wave," *Flying Saucer Review* 13 (March-April 1967): 3, 6–7; Helen M. White, "Do Birds Come This Big?" *Fate* 20 (August 1967): 74–77; John A. Keel, "West Virginia's Enigmatic 'Bird,'" *Flying Saucer Review* 14 (July-August 1968): 7–14; John A. Keel, *Strange Creatures from Time and Space* (Greenwich, Conn.: Fawcett, 1970), pp. 11, 195, 203–237; John A. Keel, *The Mothman Prophecies* (New York: Saturday Review, 1975); James Gay Jones, *Haunted Valley, and More Folk Tales* (Parsons, W. Va.: McClain, 1979), pp. 31–32; John A. Keel, "UFOs, Mothman, and Me," *High Times,* no. 57 (May 1980): 42–45, 72–75; Robert A. Goerman, "Mothmania," *Fate* 54 (June 2001): 8–12; Loren Coleman, "Why Mothman Belongs in Cryptozoology," Fortean Times Online, 2001, at http://www.forteantimes.com/exclusive/lcreplies.shtml; Loren Coleman, *Mothman and Other Curious Encounters* (New York: Paraview, 2002), pp. 38–64; Joe Nickell, "'Mothman' Solved!" *Skeptical Inquirer* 26 (March-April 2002): 20–21; Donnie Sergent Jr. and Jeff Wamsley, *Mothman: The Facts behind the Legend* (Point Pleasant, W. Va.: Mothman Lives Publishing, 2002); Bob Rickard, Rick Moran, Doug Skinner, Colin Bennett, Jerome Clark, and Loren Coleman, "The Mothman Special," *Fortean Times,* no. 156 (April 2002): 26–53.

## Mourou-Ngou

WATER LION of Central Africa.

*Etymology:* Banda (Ubangi), "water leopard." The name is also used for the Giant otter shrew (*Potomogale velox*), an aquatic insectivore nearly 2 feet long.

*Variant names:* Muru-ngu, Nze-ti-gou (Sango/Creole, "water leopard"), Ze-ti-ngu.

*Physical description:* Shaped like a leopard. Length, about 8–12 feet. Brownish, striped, or dappled with blue and white spots. Small head. Glowing eyes. Large fangs. Tail like a leopard's, though hairier.

*Behavior:* Amphibious. Nocturnal. Roaring cry like a strong wind. Hunts in pairs. Kills hippopotamuses and elephants. Inflicts long, deep wounds on its prey, even hippos. Said to capsize canoes and seize humans.

*Tracks:* Larger than a lion's. Claw marks. Described as "containing a circle in the middle."

*Habitat:* Caves in riverbanks.

*Distribution:* Bamingui, Bangoran, Gribingui, Iomba, Kotto, Koukourou, Mbari, and Ouaka Rivers in the Central African Republic; Chari River, Chad.

*Significant sightings:* In August 1910 or 1911, a boat containing French-African riflemen was overturned in the Bamingui River by a Mourou-ngou, which seized one of the men in its mouth and dragged him underwater. The animal looked like a leopard but with stripes. The records at the outpost at Ndélé confirm that a rifleman had been lost.

In 1920 and 1970, hippos were found slashed by an unknown animal along the Chari River, Chad.

On May 26, 1930, French civil servant Lucien Blancou shot a hippo on the River Mbari. During the night, a roaring animal that was not a Nile crocodile bit into the carcass.

In 1936, Lucien Blancou was told at Kaga Bandoro, Central African Republic, that a Mourou-ngou had carried off men from the village of Dogolomandji, about 20 miles to the southeast on the Gribingui River. Unlike a crocodile, it left no trace of its victims.

In the 1950s, a Water lion was caught in a fishnet on the Bangoran River. The villagers killed it and retained the cranium, which may still be kept by the village headman.

A fisherman was nearly knocked into the Bamingui River when a large animal swam past him in February 1985.

*Possible explanations:*
(1) An unknown species of large crocodile of the genus *Osteolaemus,* with a head that is shorter and rounder than the Nile crocodile (*Crocodylus niloticus*), suggested by Bernard Heuvelmans in 1955.
(2) A surviving saber-toothed cat, adapted for an aquatic lifestyle, suggested by Heuvelmans in 1978.

*Sources:* Bernard Heuvelmans, *On the Track of Unknown Animals* (New York: Hill and Wang, 1958), pp. 463–467, 468–470; Bernard Heuvelmans, *Les derniers dragons d'Afrique* (Paris: Plon, 1978), pp. 262–265, 356, 377–378, 382, 395; Karl Shuker, "Operation Mourou N'gou," *Strange Magazine,* no. 15 (Spring 1995): 33; Christian Le Noël, "Le Tigre des montagnes: Des félins à dents en sabre au coeur de l'Afrique?" Institut Virtuel de Cryptozoologie, http://perso.wanadoo.fr/cryptozoo/dossiers/tigrmont.htm.

## Muhlambela

Mystery SNAKE of South Africa.

*Physical description:* Length, nearly 12 feet. Crest.

*Behavior:* Emits a deerlike bleat. Strikes at people's heads by hanging from a tree.

*Distribution:* South Africa.

*Possible explanation:* A Black mamba (*Dendroaspis polylepis*) with molted skin on its head that forms a crestlike ornament. This deadly snake grows to 10 feet in length.

*Source:* Harry C. Wolhuter, *Memories of a Game-Ranger* (Johannesburg: Wild Life Protection Society of South Africa, 1948).

## Muhuru

Unknown LIZARD of East Africa.

*Etymology:* Unknown. Muhuru is the name of a bay in Lake Victoria as well as a dialect of Suba (Bantu), spoken by about 15,000 people in Kenya.

*Physical description:* Length, 9–12 feet. Dark gray. Sail-like structure on its back.

*Distribution:* Eastern Rift Valley, Kenya.

*Significant sighting:* Cal Bombay and his wife were driving through the Rift Valley in 1963 when their path was blocked by a gray, sail-backed reptile lying across the road.

*Possible explanation:* A surviving pelycosaur reptile, such as *Edaphosaurus* or *Dimetrodon,* which died out in the Late Permian, 275 million years ago. They both had tall neural spines between the neck and pelvis that supported a sail-like membrane; the membrane presumably provided a surface for body-temperature control and may also have been useful in defense or sexual display behavior.

*Source:* Karl Shuker, "From Dodos to Dimetrodons," *Strange Magazine,* no. 19 (Spring 1998): 22–23.

*Mystery MUHURU lizard seen by Cal Bombay in Kenya in 1963. (William M. Rebsamen)*

## Mulahu

GIANT HOMINID of Central Africa.

*Etymology:* Possibly Banda (Ubangi) word.

*Physical description:* Height, 7–8 feet. Weight, 800 pounds. Covered with dark hair. Black on the back. Long white hairs on the face.

*Behavior:* Bipedal. Aggressive nature.

*Distribution:* Ituri Forest, Democratic Republic of the Congo.

*Significant sighting:* The Italian explorer Attilio Gatti heard of the Mulahu in the Congo in the 1930s and decided to look for it after it had attacked a man. According to one account, he discovered huge tracks and found black hairs in the hollow of a tree; according to another (perhaps embellished) account, he actually glimpsed the animal and obtained three long, white hairs from the son of the man who found them clenched in the teeth and under the fingernails of an Australian photographer killed by the Mulahu sometime during World War I.

*Sources:* George Witten, "He's After the Real King Kong: A Mulahu?" *Family Circle,* September 16, 1938, pp. 10–11, 18; Ellen Gatti, *Exploring We Would Go* (London: Robert Hale, 1950), pp. 282–283; Attilio Gatti, *Sangoma* (London: Frederick Muller, 1962), pp. 26–29, 34–48, 70–73, 97–99; Bernard Heuvelmans, *Les bêtes humaines d'Afrique* (Paris: Plon, 1980), pp. 563–569.

## Mulilo

Unknown AMPHIBIAN of Central Africa.

*Physical description:* Black and sluglike. Length, 6 feet. Width, 12 inches.

*Behavior:* Appears only when a rainbow is seen. Its poisonous breath is fatal.

*Distribution:* Democratic Republic of the Congo; Zambia.

*Possible explanation:* An unknown Caecilian (Order Gymnophiona), a group of wormlike tropical amphibians that burrow through soil and mud. The largest underground species, *Caecilia thompsoni*, lives in Colombia, is nearly 5 feet long, and eats insects, larvae, and earthworms. Several species are known from sub-Saharan Africa.

*Source:* W. L. Speight, "Mystery Monsters in Africa," *Empire Review* 71 (1940): 223–228.

## Multicoiled Sea Monster

A category of SEA MONSTER identified by Gary Mangiacopra.

*Variant name:* Many-coiled sea monster.

*Physical description:* Serpentine. Large curves looping in and out of the water. Length, 20–100 feet. Small diameter for its length. Dark above, whitish below. Square-shaped, seal-like head, 2 feet in diameter. Some reports place a horn on the head. Large, bright eyes. Two types of dorsal fin reported: a continuous fin starting 8 feet in back of the head and single fins on each loop of the body, about 3–4 feet apart.

*Behavior:* Swims rapidly by arching its body out of the water like a snake. Makes a noise like steam escaping.

*Distribution:* North Atlantic Ocean.

*Significant sightings:* On September 22, 1895, Willard P. Shaw and his family and neighbors saw a huge snake off their front porch at Spring Lake, New Jersey. It appeared to be 75–100 feet long, with its head sticking 6 feet out of the water. The head was flat, with an alligator-like snout. It moved about 40 miles an hour by up-and-down writhing movements.

Professor R. H. Mohr of Boston and his son were sailing at sunrise on August 10, 1896, off Nahant, Massachusetts, when they saw a huge, seal-like head with a foot-long horn projecting from its forehead. Soon, they saw a series of loops, each capped by a single fin, circling their boat. The skin looked like that of a porpoise.

The crew of the ship *Livingston* sighted a sea serpent 50 miles north of Frontera, Tabasco State, Mexico, on June 21, 1908. It was allegedly 200 feet long, with a head 6 feet long by 3 feet wide. As it swam away with its tail erect, observers heard a rattling noise like a Gatling gun.

*Present status:* Similar to Bernard Heuvelmans's MULTIHUMPED SEA MONSTER.

*Possible explanation:* An archaic whale, possibly an elongated basilosaurid.

*Sources:* "Was It a Sea Serpent?" *New Haven (Conn.) Evening Register,* September 24, 1895, p. 1; "Faced Sea Serpents," *New York Herald,* August 16, 1896, p. 4; "200-foot Sea Serpent," *New York Times,* July 1, 1908, p. 1; Gary S. Mangiacopra, "The Great Unknowns of the 19th Century," *Of Sea and Shore* 8, no. 3 (Fall 1977): 175–178.

## Multifinned Sea Monster

A category of SEA MONSTER identified by Bernard Heuvelmans.

*Scientific name: Cetioscolopendra aeliani,* given by Heuvelmans in 1965.

*Variant names:* Ælian's sea centipede, Cetacean centipede, CON RÍT, Many-finned sea monster, TOMPONDRANO.

*Physical description:* Elongated body with peculiar lateral projections that look like forward-pointing fins, only four to twelve of which are usually seen above water. Length, 30–100 feet, with a probable average of 60–70 feet. Skin is smooth like tanned leather. Brown with dirty-yellow speckles or greenish gray. Covered with large scales or bony plates that form segmented armor. Round head, said to look like a walrus, seal, or calf. Small but prominent eyes, placed high on the head. Wide mouth like a turtle's. Visible nostrils, surrounded by hairs. Short, slender neck. An apparent saw-toothed crest along the spine, probably formed from the body armor. Pectoral flippers sometimes reported. Flat tail, possibly trilobate and only slightly spread horizontally.

*Behavior:* Seen throughout the year. Swims in vertical undulations. Turns by rolling to one side, making its lateral fins visible. Can reach a speed of 10 knots. Spouting or breath is almost always seen.

*Distribution:* Tropical and subtropical waters worldwide.

*Significant sightings:* This animal was apparently first described by the Roman rhetorician Ælian of the third century A.D. as a "sea centipede."

On August 28, 1852, Captain Steele and the Ninth Lancers regiment, on the British ship *Barham* in the Mozambique Channel, watched a green animal with the head and neck of an enormous snake. Its head was 16–20 feet out of the water, and it had a huge, saw-shaped crest down its back. It spouted water a long distance away from its head.

On July 8, 1856, about 50 miles south of the Cape of Good Hope, South Africa, Capt. A. R. N. Tremearne and the crew of the *Princess* saw a large fish with a walruslike head and twelve forward-pointing dorsal fins. Someone fired a rifle at the animal, hitting it in the head.

Lieutenant Lagrésille and the crew of the gunboat *Avalanche* observed two 65-foot, undulating animals in Halong Bay, Vietnam, in July 1897. They dived after the crew fired on them. The *Avalanche* chased two similar animals for ninety minutes on February 15, 1898, but they outpaced the boat. A similar incident occurred on February 26, when the officers and crew of the *Bayard* were also on board. The animals had seal-like heads and three large bodily coils with saw-tooth crests.

A monster with an "immense number of fins" was seen in the Mediterranean by the crew of the HMS *Narcissus* off Cape Falcon, Algeria, on May 21, 1899. It was more than 150 feet long and swam by means of an "immense" number of fins on both sides of its body. It spouted water like a whale from several points.

In July 1920, off the Florida coast between Miami and Fort Lauderdale, the captain and crew of the merchant ship *Craigsmere* watched a large sea animal with several porpoiselike dorsal fins.

In 1935, Lt. W. C. Hogan of the U.S. Coast Guard vessel *Electra* saw a 40- to 50-foot animal with six fins on its back off the coast of Norfolk, Virginia. Each fin was 2 feet high and 2 feet 6 inches wide at the base. The crew fired at it unsuccessfully.

*Possible explanations:*

(1) An archaic whale, possibly a basilosaurid, with heavy, armored scales, suggested by Bernard Heuvelmans. However, scales found with these fossil whales are now known to come from other animals.

(2) A giant crustacean of an unknown type, suggested by Karl Shuker, especially for the CON RÍT.

*Sources:* Ælian, *De natura animalium*, XIII. 23; Guillaume Rondelet, *Libri de piscibus marinus* (Lyon, France: Matthiam Bonhomme, 1554); "The Sea Serpent," *Times* (London), November 17, 1852, p. 6; Edmund J. Wheeler, "The 'Sea-Serpent' Again," *Illustrated London News* 29 (1856): 347–348; "Sea Serpent at It Again," *Daily Mail* (London), May 31, 1899; Bernard Heuvelmans, *In the Wake of the Sea-Serpents* (New York: Hill and Wang, 1968), pp. 463–464, 550–552, 567–568; Paul H. LeBlond, "A Previously Unreported 'Sea Serpent' Sighting in the South Atlantic," *Cryptozoology* 2 (1983): 82–84.

## Multihumped Sea Monster

A category of SEA MONSTER identified by Bernard Heuvelmans.

*Scientific names:* Misnamed *Scoliophis atlanticus* in 1817, based on confusion with a deformed black snake; *Plurigibbosus novaeangliae*, given by Heuvelmans in 1965.

*Variant names:* American sea serpent, CADDY, CASSIE, DORSAL FINNER, MULTICOILED SEA MONSTER, Many-humped sea monster.

*Physical description:* Elongated body, with many regularly placed humps that form a conspicuous ridge along the spine. Length, 60–115 feet. Diameter, 9–15 feet. Dull green to dark brown or black on top. Throat and underside pure white. Skin is usually smooth, though sometimes reported as rough. Scales are occasionally mentioned. Ovoid head, flat on top. Large eyes, 6 inches in diameter. Broad snout like an ox's. Slender neck, with one or two white stripes on the side. Throat and underside are white. Small, triangular fin sometimes seen on the shoulder. Single pair of frontal flippers. Sometimes seen with a straight fin or fan on its back. Bilobate tail, one lobe of which sometimes appears at the surface.

*Behavior:* Usually appears in the summer in New England and in the spring farther north. Swims with vertical undulations that resemble a caterpillar's motion. Splashes and lashes the

*Possibly the first depiction of a* MULTIHUMPED SEA MONSTER *in the Americas. From Johann Ludwig Gottfried,* Newe Welt und americanische Historien *(Frankfurt, 1655). (Fortean Picture Library)*

water. Reaches speeds of 22–40 knots. Spouting not reported.

*Distribution:* East coast of North America from New York to Newfoundland, with a preference prior to the twentieth century for Massachusetts Bay and the Gulf of Maine. Also seen south of Iceland and west of Scotland; elsewhere in the Gulf Stream; off British Columbia, Canada (CADDY); and elsewhere on the Pacific coast. This animal is very similar to the serpentine, coiled FRESHWATER MONSTERS seen in certain Canadian lakes, especially OGOPOGO and MANIPOGO.

*Significant sightings:* Capt. Elkanah Finney observed a classic American sea serpent for about five minutes in Warren Cove near Ply-

mouth, Massachusetts, around June 20, 1815. It looked like a string of thirty to forty buoys, with a head about 6–8 feet long. The total length was 100–120 feet. Finney got up early the next morning in hopes of seeing it again and watched it for two hours as it dived repeatedly. This time, it appeared to be only 20–30 feet long.

On August 10, 1817, Lydia Wonson watched a 70-foot sea serpent through a spyglass for nearly thirty minutes from her home near Rocky Neck, Gloucester Harbor, Massachusetts. It drew itself up in coils that looked like the buoys of a fishing net. The same day, Amos Story saw the animal off nearby Ten Pound Island. It carried its turtle-shaped head about a foot above the surface, but he could see only about 12 feet of the body. Ship's carpenter Matthew Gaffney shot at the same sea creature on August 14, 1817, when it was only 30 feet away in Gloucester Harbor. It sank down and went directly under his boat, surfacing 100 yards away. By August 28, hundreds of residents and tourists had seen one or even two multi-humped animals, and whale fishermen unsuccessfully tried to harpoon them.

On September 27, 1817, at Loblolly Cove in Rockport, Massachusetts, Goreham Norwood and a Mr. Colbey, coming to the aid of Colbey's screaming young son, killed a deformed Northern black racer (*Coluber constrictor constrictor*) about 3 feet 6 inches long. The Linnean Society of New England hastily concluded that this was a juvenile sea serpent and gave it the scientific name *Scoliophis atlanticus* (Atlantic humped snake).

James Prince, his family, Samuel Cabot, and a large crowd of other people watched a 60-foot sea serpent with thirteen to fifteen humps not far from the beach at Nahant, Massachusetts, for more than two hours on the morning of August 14, 1819.

On June 17, 1826, Capt. Henry Holdredge and passengers of the *Silas Richards* saw a 60-foot animal swimming 50 yards away from the ship off Georges Bank, about 134 miles east of Nantucket, Massachusetts.

On May 15, 1833, Capt. W. Sullivan and four other members of a Canadian rifle brigade saw a sea serpent in Mahone Bay, Nova Scotia, as they were sailing to the fishing grounds. They described it as a "common snake" about 80 feet long, with its head elevated and leaving a regular wake. Its head was about 6 feet long and was nearly black, irregularly streaked with white.

In August 1845, two observers saw a 100-foot monster in the Northumberland Strait 200 feet off Merigomish, Nova Scotia. It appeared to nearly be run aground, as it got away with difficulty only after thirty minutes. It occasionally raised its seal-like head out of the water. It had several protuberances, which one witness thought were humps and the other thought were vertical coils. Its skin was black, with a rough appearance.

The Rev. Arthur Lawrence and others on the yacht *Princess* saw a strange creature on July 30, 1875, between Swampscott and Egg Rock, Massachusetts. Only its head and neck were visible, but it was distinctly dark above and white below. It raised its head several times 6–8 feet out of the water. It had a dorsal fin at the back of its neck and what appeared to be fins or flippers at the front of its throat. They followed it for two hours, taking potshots at it.

B. M. Baylis saw an animal with four or five rounded humps off the coast at Hilston, Yorkshire, England, in early August 1945.

On October 16, 1966, George and May Ashton were strolling near the beach at Chapel St. Leonards, Lincolnshire, England, when they spotted an animal with six or seven pointed humps skimming through the water less than 100 yards offshore.

On October 31, 1983, five construction workers saw a dark, 100-foot animal swimming toward the cliffs at Stinson Beach, California. Mark Ratto watched it through binoculars and said it appeared to be followed by about 100 birds and two dozen sea lions. He saw a head and neck that "came up to look around" and three coils or body humps.

*Present status:* Much rarer in the North Atlantic since the beginning of the twentieth century.

*Possible explanation:* An archaic basilosaurid whale or another unknown family of early cetaceans, suggested by Heuvelmans.

*Sources:* Linnean Society of New England, *Report of a Committee of the Linnean Society of New England Relative to a Large Marine Animal Supposed to Be a Serpent, Seen near Cape Ann, Massachusetts, in August 1817* (Boston: Cummings and Hilliard, 1817); Charles Alexander Lesueur, "Sur le serpent nommé *Scoliophis*," *Journal de Physique, de Chimie, et d'Histoire Naturelle* 86 (June 1818): 466–469; Jacob Bigelow, "Documents and Remarks Respecting the Sea Serpent," *American Journal of Science* 2 (1820): 147–164; "Sea Serpent," *American Journal of Science* 11 (1826): 196; W. Sullivan, A. Maclachlan, G. P. Malcolm, B. O'Neal Lyster, and Henry Ince, "The Sea-Serpent," *Zoologist* 5 (1847): 1714–1715; Charles Lyell, *A Second Visit to the*

United States of North America (London: John Murray, 1849), vol. 1, pp. 132–140; John George Wood, "The Trail of the Sea-Serpent," *Atlantic Monthly* 53 (1884): 799–814; "Saw Monster in Sea—Claim," *Skegness Standard,* October 19, 1966; B. M. Baylis, "Those Sea Monsters," *Skegness Standard,* October 26, 1966; Bernard Heuvelmans, *In the Wake of the Sea-Serpents* (New York: Hill and Wang, 1968), pp. 548–550, 566–568; "'Sea Serpents' Seen off California Coast," *ISC Newsletter* 2, no. 4 (Winter 1983): 9–10; June Pusbach O'Neill, *The Great New England Sea Serpent* (Camden, Maine: Down East Books, 1999); Paul Harrison, "Loch Ness: The Tip of the Iceberg," *Crypto Dracontology Special,* no. 1 (November 2001): 49–54.

## Mumulou

LITTLE PEOPLE of Australasia.

*Etymology:* Ghari (Austronesian) word.

*Variant names:* Kakamora (on San Cristobal), Moka, Mola, Mumu (on Maramasike).

*Physical description:* Height, 3–4 feet. Fair to dark skin. Long, straight head-hair down to the knees. Tiny teeth. Long beard. Long nails.

*Behavior:* Moves by jumping on its toes. Cry is a half wail, half bark. Disgusting odor. Said to be a cannibal. Doesn't know fire or tools.

*Habitat:* Caves.

*Distribution:* Laudari Mountains of Guadalcanal; Maramasike and San Cristobal in the Solomon Islands.

*Significant sighting:* Charles Elliot Fox was traveling with a party of natives on San Cristobal when one of them saw a Kakamora in a river they were about to ford. When they came to the river, Fox found a raw, half-eaten fish and a few small, wet footprints on a dry stone in the river.

*Sources:* Charles Elliot Fox, *The Threshold of the Pacific* (New York: A. A. Knopf, 1924); Stanley G. C. Knibbs, *The Savage Solomons as They Were and Are* (London: Seeley, Service, 1929), pp. 48–50, 259; "Is There an Undiscovered People in Central Guadalcanal?" *Pacific Islands Monthly* 20 (February 1950): 96; A. H. Wilson, "Guadalcanal's Undiscovered People Just Another Tall Tale," *Pacific Islands*

*Monthly* 20 (March 1950): 7; John E. Roth, *American Elves* (Jefferson, N.C.: McFarland, 1997), pp. 132–136.

## Murung River Bear

Unknown BEAR-like animal of Southeast Asia.

*Physical description:* Like a large bear.

*Behavior:* Gathers in large groups along the river once a year to feed on river berries. Can swim. Will attack boats and claw humans to death.

*Habitat:* Jungle.

*Distribution:* Kedang Murung River, eastern Borneo, Indonesia.

*Source:* Leonard Clark, *A Wanderer till I Die* (New York: Funk and Wagnalls, 1937), p. 189.

## Muskox of Noyon Uul

Unknown HOOFED MAMMAL of ancient Central Asia.

*Physical description:* Long, dense hair. Broad head. Horns like a muskox's. Bull-like muzzle.

*Habitat:* Mountainous terrain.

*Distribution:* Mongolia; north-central Siberia.

*Significant sightings:* Carvings depicting an animal like a muskox were discovered in 1924 on two silver plaques excavated from Xiongnu (Hun) burial tombs dating from the first century B.C. in the Noyon Uul Mountains, Mongolia.

Muskox skulls dating from 1800–900 B.C. were found in 1948 on the Taymyr Peninsula, Siberia; they appear to have drill holes. Another subfossil skull was found in 1984 in the same vicinity.

*Possible explanations:*

(1) The Muskox (*Ovibos moschatus*), a large bovid with broad, flat horns and long, silky hair, is thought to have died out in Eurasia at the end of the Pleistocene, though it has survived in North America. Its apparent persistence nearly 8,000 years later has not been demonstrated conclusively.

(2) Other explanations for the carvings have included an Argali wild sheep (*Ovis ammon*), a Yak (*Bos grunniens*), and a Takin (*Budorcas taxicolor*).

*Sources:* Nikolai Spassov, "The Musk Ox in Eurasia: Extinct at the Pleistocene-Holocene Boundary or Survivor to Historical Times?" *Cryptozoology* 10 (1991): 4–15; Peter C. Lent, "More on Muskoxen," *Cryptozoology* 12 (1996): 91–94.